Principles of Seismology

AGUSTÍN UDÍAS

Professor of Geophysics, Universidad Complutense de Madrid

CAMBRIDGE
UNIVERSITY PRESS

PUBLISHED BY THE PRESS SYNDICATE OF THE UNIVERSITY OF CAMBRIDGE
The Pitt Building, Trumpington Street, Cambridge, United Kingdom

CAMBRIDGE UNIVERSITY PRESS
The Edinburgh Building, Cambridge CB2 2RU, UK www.cup.cam.ac.uk
40 West 20th Street, New York, NY 10011-4211, USA www.cup.org
10 Stamford Road, Oakleigh, Melbourne 3166, Australia
Ruiz de Alarcón 13, 28014 Madrid, Spain

First published 1999

Printed in the United Kingdom at the University Press, Cambridge

Typeface Times 10/12pt   *System* 3B2 [wv]

A catalogue record for this book is available from the British Library

Library of Congress Cataloguing in Publication data

Udías Vallina, Agustín.
    Principles of seismology/Agustín Udías.
        p.   cm.
    Includes bibliographical references and index.
    ISBN 0 521 62434 7. – ISBN 0 521 62478 9 (pbk.)
    1. Seismology.   I. Title.
  QE534.2.U35   1999
  551.22–dc21   98-32174 CIP

ISBN 0 521 62434 7 hardback
ISBN 0 521 62478 9 paperback

This book is dedicated to C. Kisslinger, O. Nuttli, and W. Stauder, my professors of seismology at Saint Louis University.

Contents

Preface

This textbook has been developed from 25 years of experience teaching seismology at the universities of Madrid and Barcelona. The text is at an introductory level for students in the last years of European licentiate or American upper-division undergraduate courses and at similar levels in other countries. As a first book, no previous knowledge of seismology, as such, is assumed of the student. The book's emphasis is on fundamental concepts and basic developments and for this reason a selection of topics has been made. It has been noticed that sometimes even graduate students lack a true grasp of the very fundamental ideas underlying some aspects of seismology. The most fundamental concepts are developed in detail. Simple cases such as one-dimensional problems and those in liquid media are used as introductory topics. Mathematical developments are worked out in complete detail for the most fundamental problems. Sometimes more difficult subjects are introduced, but not fully developed. In these cases references to more advanced books are given.

The book presupposes a certain amount of knowledge of mathematics and physics. Knowledge of mathematics at the level of calculus and ordinary and partial differential equations as well as a certain facility for vector and tensor analysis are assumed. Cartesian, spherical, and cylindrical coordinates and some functions such as Legendre and Bessel functions are used. Tensor index notation is used preferentially throughout the book. Fundamental ideas about certain mathematical subjects are given briefly in Appendixes 1–4. Basic knowledge of the mechanics of a continuous medium and of the theory of elasticity is also presupposed. The reader is reminded about the basic equations of elasticity in chapter 2, but they are not all fully explained. The student is referred to textbooks on elasticity that are cited in the bibliography.

Throughout the book there is an emphasis on the fundamental theoretical aspects of seismology and observations are treated briefly. Thus, some readers will miss discussion of recent results; I refer them to the excellent recent book by Lay and Wallace (1995). Also advanced developments of the theory of wave propagation and generation are not treated; see Pilant (1979), Aki and Richards (1980), and Ben Menahem and Singh (1981). I hope that my book is a good introduction to these excellent advanced books. It is difficult to decide where to stop in the subjects treated in a textbook that is designed as an introduction. I have selected to develop only, but with all mathematical detail, the very basic problems. In this sense, this book is different than those that already exist. The style and approaches are also sometimes different, and reflect those of the author

The first chapters are dedicated to the fundamentals of elasticity theory, solutions of the wave equation, normal modes, and ray theory. The following chapters are dedicated to the propagation of body and surface waves, and free oscillations. Four chapters are devoted to the study of the source. One chapter gives an introduction to anelasticity and anisotropy and the two final ones introduce the reader to seismicity, seismotectonics, and seismic risk, and to seismologic instrumentation. Appendixes 1–4 cover some mathematical tools, Appendixes 5–7 give some helpful information, and Appendix 8 is a collection of problems and exercises divided into seven topics. These exercises are related to the theoretical developments in the book. The bibliography includes books on seismology and related topics. Other references cited in the text are given separately. Some books are listed as references, so one must use both lists.

I wish to thank in the first place all my students, to whom I am indebted for their questions and suggestions that have helped me to write this book and their patience during my lectures. I must thank also a long list of Spanish seismologists, many of them former students, who will be difficult to name without omitting some of them, especially E. Buforn and D. Muñoz (Universidad Complutense de Madrid), E. Suriñach and A. Correig (Universidad de Barcelona), and A. Lopez Arroyo, G. Payo, and J. Mezcua (Instituto Geográfico Nacional). Revision of some chapters was aided and valuable suggestions were given by B. A. Bolt (University of California, Berkeley), R. Madariaga (Ecole Normale Supérieure, Paris), A. Cisternas (Institut de Physique du Globe, Strasbourg), and H. Kanamori (California Institute of Technology). I am especially indebted to S. Das (Oxford University) who encouraged me to write the English version and put me in contact with Cambridge University Press and S. Holt who revised the manuscript. Naturally, I am aware that I am leaving out many names that I should have listed and I hope that they all feel included in my thanks.

1 SEISMOLOGY, THE SCIENCE OF EARTHQUAKES

1.1 The historical development

The term seismology is derived from two Greek words, *seismos*, shaking, and *logos*, science or treatise. Earthquakes were called *seismos tes ges* in Greek, literally shaking of the Earth; the Latin term is *terrae motus*, and from the equivalents of these two terms come the words used in occidental languages. Seismology means, then, the science of the shaking of the Earth or the science of earthquakes. The term seismology and similar ones in other occidental languages (séismologie, sismología, sismologia, Seismologie, etc.) started to be used around the middle of the nineteenth century. Information about the main historical developments of seismology can be found in each chapter; a very short overview is given in the following paragraphs.

In antiquity, the first rational explanations of earthquakes, beyond mythical stories, are from Greek natural philosophers. Aristotle (in the fourth century BC) discussed the nature and origin of earthquakes in the second book of his treatise on meteors (*Meteorologicorum libri IV*). The term meteors was used by the ancient Greeks for a variety of phenomena believed to take place somewhere above the Earth's surface, such as rain, wind, thunder, lightning, comets, and also earthquakes and volcanic eruptions. The term meteorology derives from this word, but in modern use it refers only to atmospheric phenomena. Aristotle, following other Greek authors, such as Anaxagoras, Empedocles, and Democritus, proposed that the cause of earthquakes consists in the shaking of the Earth due to dry heated vapors underground or winds trapped in its interior and trying to leave toward the exterior. This explanation was part of his general theory for all meteors caused by various exhalations of gas or vapor (*anathymiaseis*) that extend from inside the Earth to the Lunar sphere. This theory was spread more widely by the encyclopedic Roman authors Seneca and Plinius. It was commented upon by medieval philosophers such as Albert the Great and Thomas of Aquinus, and, with small changes, was accepted in the West until the seventeenth century. For example, in 1678 A. Kircher related earthquakes and volcanoes to a system of fire conduits (*pyrophylacii*) inside the Earth. In the eighteenth century, M. Lister and N. Lesmery proposed that earthquakes are caused by explosions of flammable material concentrated in some interior regions. This explanation was accepted by Newton and Buffon.

The great Lisbon earthquake of 1 November 1755, which caused widespread destruction in that city and produced a large tsunami, may be considered the starting point of modern seismology. In 1760 J. Mitchell was the first to relate the shaking due to earthquakes to the propagation of elastic waves inside the Earth. This idea was further developed by, among others, T. Young, R. Mallet, and J. Milne. Descriptions of damage

1

due to earthquakes and the compilation of lists of their occurrence can be traced back to very early dates. Sometimes these lists include other natural disasters such as floods, famines, and plagues. Among the first catalogs of earthquakes published in Europe are those of J. Zahn in 1696 and J. J. Moreira de Mendonça in 1758. Modern catalogs started to be published around 1850 by A. Perrey and R. Mallet (Chapter 15).

Mallet's study of the earthquake of Naples of 1857 constitutes one of the first basic works of modern seismology. Mallet developed the theory of the seismic focus from which elastic waves are propagated in all directions and connected the occurrence of earthquakes with changes in the Earth's crust that are often attended by dislocations and fractures, abandoning the explosive theory. C. Lyell and E. Suess related earthquakes to volcanic and tectonic motions, and, at the beginning of this century, F. Montessus de Ballore and A. Sieberg assigned the cause of earthquakes to orogenic processes and contributed to many aspects of observational seismology (Chapter 15). R. D. Oldham, K. Zöppritz, and E. Wiechert published the first studies of the propagation of seismic waves, based on early work on the theory of elasticity (Chapter 2). The first models of the interior of the Earth based on seismologic observations were proposed between 1900 and 1940 by, among others, R. D. Oldham, B. Gutenberg, H. Jeffreys, K. Bullen, and J. B. Macelwane (Chapter 9). The first instruments used to observe the shaking of the ground produced by earthquakes were based on the oscillations of a pendulum and started to be used around 1830. By the end of the century, the first seismographic continuous recordings had been produced. In 1889, E. von Rebeur Paschwitz recorded in Potsdam an earthquake that took place in Tokyo; this was the first seismogram recorded at a large distance. Among the first names related to the development of seismologic instrumentation are those of J. Milne and F. Omori with the inclined pendulum, E. Wiechert with the inverted pendulum, B. B. Galitzin with the electromagnetic seismograph, and H. Benioff with that of variable magnetic reluctance (Chapter 21).

Since 1945, seismology has experienced a very rapid development. Details of this development and names associated with it can be found in the introductions to each chapter and elsewhere in this book. In this rapid evolution, two important subjects are the propagation of elastic waves in the Earth and the mechanism of the generation of earthquakes. Both include theoretical and observational aspects. In the study of the propagation of seismic waves, the Earth is approximated by models that have progressed from the early very simple models of homogeneous elastic media or media divided into layers to those with three-dimensional heterogeneity including anelasticity and anisotropy (Chapters 9 and 12–14). Models of the source of earthquakes developed from simple models of point foci to those including the complex process of fracture of crustal rocks (Chapters 18 and 19). Observations of seismic waves have improved and multiplied with the development of seismologic instrumentation from early mechanical seismographs with analog recording to the present systems with a broad-band response, electronic amplification, and digital recording (Chapter 21). These developments have contributed to an increase in knowledge about the complex processes which cause earthquakes and the properties and composition of the materials of the Earth's interior. Other aspects of seismology concerning the occurrence of earthquakes, its relation to tectonic processes, and the evaluation of seismic risk have also significantly expanded (Chapter 20).

1.2   Seismology, a multidisciplinary science

Recent trends in seismology tend to overemphasize those aspects related to the generation and propagation of seismic waves. With this emphasis, Aki and Richards (1980) define seismology as a science based on data called seismograms, which are the records of mechanical vibrations. Lay and Wallace (1995), following this point of view, define seismology as the study of the generation, propagation, and recording of elastic waves in the Earth (and other celestial bodies) and of the sources that produce them, and conclude that recordings of ground motion as a function of time, or seismograms, provide the basic data for seismologists. Lomnitz (1994) considers this approach rather narrow, because seismograms provide us with much less information about earthquakes than is needed. Moreover, this definition downplays many other aspects present in the complex phenomena of earthquakes.

In a more traditional approach, seismology is defined in a broader sense, as the science of the study of earthquakes. The analysis of seismic waves forms a very important part of seismology, but not its totality. Bolt (1978) considers the task of seismologists the study of all aspects of earthquakes, including their causes, occurrence, and properties. For Bullen (1947), it is evident that the study of earthquakes belongs to many fields of knowledge such as physics, chemistry, geology, engineering, and even philosophy. For this reason, Macelwane and Sohon (1936), Madariaga and Perrier (1991), and many other authors consider seismology a multidisciplinary science.

There is no doubt that the study of seismic waves recorded by seismographs is fundamental to seismology, for example, to the study of the mechanism causing earthquakes and the constitution of the Earth's interior. However, this does not imply that wave analysis is the only source of information about earthquakes. The seismicity of a region, for example, can not be understood correctly if solely instrumentally recorded earthquakes are considered. Owing to the long return periods for large earthquakes, the study of historical earthquakes is essential. The need to go even farther back into the past has promoted the study of other types of information from archeoseismicity and paleoseismicity. The characteristics of large earthquakes can not be fully understood without geologic field observations after their occurrence. Comparison of geodesic measurements before and after earthquakes is another important source of knowledge. All these types of data are important in helping one to interpret the nature of earthquakes and their consequences, and must be integrated with the information obtained from the analysis of seismic waves.

Two parts of seismology with a marked multidisciplinary character are the evaluation of seismic risk and work toward the prediction of earthquakes. In the first case, the interaction of seismologists with geologists and engineers is necessary in order to correctly assess earthquake hazards, expected ground motion, soil conditions, seismic zonation, and the responses of structures and buildings. In the second, many of the suggested precursory phenomena (electromagnetic signals, changes in resistivity, emissions of radon gas, and changes in geodesic measurements) are not directly related to seismic waves. Progress in the problem of earthquake prediction can not be achieved without a great multidisciplinary effort involving scientists working in many fields, such as seismologists, engineers, geologists, and physicists. Finally, we must not forget that earthquakes are natural disasters that affect human lives. Depending on the correct assessment of the seismic risk and the adequacy of the design and construction of buildings, the damage from earthquakes, especially loss of human lives, may vary greatly. The response of the population to the occurrence of an earthquake must also be taken into account, with all its serious psychological, social, and economic consequences. Seismologists can not be indifferent to all these problems.

1.3   Divisions of seismology

Seismology can be divided into three disciplines: seismology in the strict sense, seismic engineering, and seismic exploration. Seismology treats the occurrence of earthquakes and their related phenomena and is primarily based on the application of the principles of the mechanics of a continuous medium and of the theory of elasticity to them. As has already been mentioned, its two main subjects are the generation of earthquakes and the vibrations and propagation of seismic waves inside the Earth. From observations of these vibrations together with other types of data, we derive our knowledge about the nature of earthquakes, the structure of the Earth's interior, and its dynamic characteristics. The part of seismology that deals with seismologic instrumentation, called seismometry, studies the physical theory of the various types of instruments used to measure seismic motion.

Seismic or earthquake engineering is an applied science that treats how the motion produced by earthquakes affects buildings and other man-made structures. Starting from the characterization of ground displacement, velocity, and acceleration, seismic engineering proceeds to consider their effects on structures and seeks to design them to resist such motions. If earthquake-resistant structures are not to be unnecessarily expensive, a reliable evaluation of the expected ground motion at a particular site is necessary. For this task, an assessment of the seismic risk for a particular zone is needed. This assessment includes the consideration of many factors, such as the occurrence of earthquakes near a site, their source mechanism, seismic-wave attenuation and soil conditions, and the vulnerability of structures. The complete evaluation of seismic risk implies the statistical analysis of all these factors and requires the collaboration of seismologists, engineers, and geologists.

In seismic exploration, seismologic methods are applied to the search for mineral resources, especially oil deposits. These methods are based on the reflection and refraction of artificially generated seismic waves in geologic structures associated with the presence of such deposits. The methods that have proved to be the most effective are those based on vertical reflection of waves. Closely spaced distributions of wave generators and detectors together with complex processing of the digital data allow one to obtain detailed images of the upper part of the Earth's crust. The increasing demand for energy resources makes this work more and more important.

1.4   Theory and observation

Just like in all experimental sciences, theoretical and observational aspects of seismology must be considered. The first are based on the principles of the mechanics of continuous media with the assumption that the Earth is an imperfectly elastic body in which vibrations are produced by earthquakes. The study of the generation of these vibrations constitutes the theory of the source mechanism. In this theory, models of the processes occurring at the focus of earthquakes are proposed. They range from the more simple ones of instantaneous point sources to the more complex. The aim is to approximate the process of fracture along geologic faults.

Vibrations in the Earth can be treated using two approaches: wave propagation and normal modes theory. The first approach considers waves propagating inside the Earth or on its surface. The second considers the eigenvibrations or oscillations of the Earth as a whole. This approach is necessary when wave lengths are near the dimensions of the Earth. In the simplest models, the Earth behaves like a homogeneous isotropic perfectly elastic medium. For some problems the flat-Earth approximation may be sufficient, whereas others require the treatment of its sphericity. Heterogeneity in the Earth can be treated using layered models with different elastic properties for each layer or models in which these properties vary with the spatial coordinates. The assumption of a spherical radially heterogeneous medium is useful in providing a close approximation to the real Earth. Ray theory is used as a high-frequency approximation to wave propagation in heterogeneous media. Surface waves in layered media describe wave dispersion with the separation of phase and group velocities. The lack of perfect elasticity is accounted for by introducing the attenuation of vibrations and waves and by considering viscoelastic models. Isotropic models are adopted as a first approximation but further analysis needs to consider anisotropic conditions. By proceeding through these successive modifications in models of the Earth, its imperfect elasticity, heterogeneity, and anisotropy can be adequately considered.

An important part of seismologic observations consists in the recording of the ground's motion by instruments installed on its surface. Nowadays classical analog seismograms on photographic paper have largely been replaced by digital data kept on magnetic tapes or compact disks, which can be obtained directly from world data banks through the Internet. Previous to their interpretation through the use of digital computers, seismologic observations usually need careful complex numerical processing. As has already been mentioned, important seismologic data are also provided by other sources, for example, historical records of damage from pre-instrumental earthquakes, field observations of structural damage and ground deformation after earthquakes, geodesic measurements related to the occurrence of earthquakes, *in situ* stress measurements and geologic and tectonic implications. Progress in the methods of observation of all kinds of seismologic data has allowed one to apply models of increasing complexity to the problems of the generation of earthquakes and the structure of the Earth's interior.

The relation between observations and theories or models can be approached through direct and inverse problems. The direct problem refers to the determination of ground displacements from theoretical models of the generation and propagation of seismic waves. In the direct problem, theoretical models are assumed *a priori* and from them synthetic displacements are determined, which are compared with observations. If they agree, we consider the model well-suited to observations. However, in many instances, there is no assurance of its uniqueness and many other models may equally well satisfy the same observations. The inverse problem consists in the estimation of the parameters of a theoretical model from observations. This is often a more complicated problem than the direct one. Observations are always incomplete and contain errors, so that a solution of the inverse problem may exist only in a probabilistic sense. In general, inverse problems become more intractable as the number of parameters of the model increases. The mathematics of inverse problems requires, generally, the solution of nonlinear integral equations. Linearization of the problem is a standard procedure that leads, very often, to large unstable systems of equations. Difficulties in the solution of inverse problems lead to their substitution by repeated solutions of direct problems until sufficient agreement between observations and synthetic data predicted by the assumed models is reached.

1.5   International cooperation

The main objectives of seismology require the cooperation of, and exchange of observations among, scientists from different parts of the world. This collaboration was accomplished from early times through private initiatives. The global character of large earthquakes soon required the establishment of institutional cooperation at national and international levels. The first organizations were national ones such as the Seismological Society of Japan, created after the earthquake of 1880 with J. Milne as first secretary. In 1890, the Committee for the Investigation of Earthquakes was founded, also in Japan, of which F. Omori was president from 1897 to 1923. In Italy, the Italian Seismological Society (Società Sismologica Italiana) was created in 1895; L. Palmieri, T. Bertelli, and G. Mercalli were among its first members. Another national society with great influence in the history of seismology is the Seismological Society of America, which was founded in 1906 as a response to the great San Francisco earthquake, with G. Davidson as its first president. The idea of an international association of seismology was first proposed by G. Gerland, during the sixth International Congress of Geography that was held in London in 1895. In 1904, the International Association of Seismology was finally created, but it was suppressed in 1916. Since 1922, seismology has formed a section of the International Union of Geodesy and Geophysics (IUGG), created in 1919. In 1930, the IUGG was reorganized and included as one of its associations the International Association of Seismology, which finally, in 1951, received its present name of the International Association of Seismology and Physics of the Earth's Interior (IASPEI). One of its commissions is the European Seismological Commission (ESC), which was founded in 1951. There are also active seismology sections of geophysical scientific societies such as the American Geophysical Union, European Geophysical Society and European Union of Geosciences.

Exchange of seismologic data between observatories was carried out in the past through the publication of seismologic bulletins. These bulletins preserve a great wealth of information about earthquakes of the early instrumental period. One of the first publications of epicenter determinations was The Reports of the Seismological Committee of the British Association for the Advancement of Science, which started in 1911 with the determinations for the period 1899–1903. In 1922, this publication became the International Seismological Summary (ISS), its first volume being dedicated to the earthquakes of 1918. Later, in 1963, the publication was continued by the

International Seismological Centre (ISC), Newbury, UK. The Bureau Central International de Séismologie (BCIS) was created in Strasbourg in 1906 and published a bulletin with epicenter determinations from 1904 until 1975. In 1976 the Centre Séismologique Européen Méditerranéen (CSEM) was created by the ESC with the task of determining hypocenters of earthquakes of the Mediterranean region. Other agencies started also to publish epicenter determinations, such as, in North America, the Jesuit Seismological Association that was active between 1925 and 1960 and the United States Coast and Geodetic Survey (USCGS), which later was transferred to the National Earthquake Information Center (NEIC), which was dependent on the United States Geological Survey. Since 1968, its monthly publication Preliminary Determination of Epicenters has included also information on determinations of focal mechanisms for sufficiently large earthquakes. Similar information has also been published since 1977 by Harvard University. At present, there are several world centers of seismologic data including digital seismograms from broad-band stations, such as the IRIS (USA), GEOFON (Germany), and ORFEUS (Holland).

1.6   Books and journals

Among the early treatises on seismology are those of Mallet (1862), *Great Neapolitan Earthquake of 1857: The First Principles of Observational Seismology* (London); Milne (1886), *Earthquakes and Other Earth Movements* (Fig. 1.1) (New York); and Hoernes (1893), *Erdbebenkunde* (Leipzig). At the beginning of this century, several books on seismology were published, among them those by Sieberg (1904), *Handbuch der Erdbebenkunde* (Braunschweig); Hobbs (1908), *Earthquakes. An Introduction to Seismic Geology* (London); Montessus de Ballore (1911), *La sismologie moderne* (Paris); and Galitzin (1914), *Vorlesungen der Seismometrie* (Leipzig).

From 1930, textbooks about seismology that may be considered modern started to be published. Only those of general character will be mentioned (full references are given in the Bibliography): Macelwane and Sohon (1936), *Introduction to Theoretical Seismology. Part I, Geodynamics* and *Part II, Seismometry*; Byerly (1942), *Seismology*; Bullen (1947), *An Introduction to the Theory of Seismology*; Richter (1958), *Elementary Seismology*; Sawarensky and Kirnos (1960), *Elemente der Seismologie und Seismometrie*; and Bath (1973), *Introduction to Seismology*.

More recently, since 1979, several textbooks on general seismology at various levels have been published. Four excellent advanced books are by Pilant (1979), *Elastic Waves in the Earth*; Aki and Richards (1980), *Quantitative Seismology. Theory and Methods*; Ben Menahem and Singh (1981), *Seismic Waves and Sources* and Dahlen and Tromp (1998) *Theoretical Global Seismology. At an introductory level there are books by Bullen and Bolt (1985), An Introduction to the Theory of Seismology*; Bolt (1978), *Earthquakes, a Primer*; Gubbins (1990), *Seismology and Plate Tectonics*; Madariaga and Perrier (1991), *Tremblements de terre*; Lay and Wallace (1995), *Modern Global Seismology*; Doyle (1995), *Seismology*; Gershanik (1995), *Sismologia*; and Udías and Mezcua (1996), *Fundamentos de sismología*.

There are books covering only certain aspects of seismology, such as, for example, wave propagation and free oscillations, by Officer (1958), Ewing *et al.* (1957), Lapwood

THE INTERNATIONAL SCIENTIFIC SERIES

EARTHQUAKES

AND

OTHER EARTH MOVEMENTS

BY

JOHN MILNE

PROFESSOR OF MINING AND GEOLOGY IN THE IMPERIAL COLLEGE OF ENGINEERING,
TOKIO, JAPAN

WITH THIRTY-EIGHT FIGURES

NEW YORK
D. APPLETON AND COMPANY
1, 8, AND 5 BOND STREET
1886

Fig. 1.1. The title page of Milne's book on seismology.

and Usami (1981), Kennett (1983), and Babuska and Cara (1991); source mechanisms, by Kasahara (1981), Kostrov and Das (1988), and Scholz (1990); seismicity, earthquake prediction, and other topics, by Gutenberg and Richter (1954), Kisslinger and Zuzuki (1978), Kulhanek (1990), and Lomnitz (1994). There are excellent collections of review papers such as those by Dziewonski and Boschi (1980), Kanamori and Boschi (1983), and Boschi *et al.* (1996). Entries on seismologic subjects in James' (1989) *The Encyclopedia of Solid Earth Geophysics* are very good short up-to-date presentations.

The first scientific articles about seismology were published in the *Bollettino del vulcanismo italiano* founded by de Rossi in 1874 and in the *Beiträge zur Geophysik*, founded by Gerland in 1887. The first journals exclusively dedicated to seismology were the *Transactions of the Seismological Society of Japan* published from 1880 to 1892 and the *Seismological Journal of Japan* published from 1892 to 1895, both directed by Milne. In 1985 the *Bollettino della Società Sismologica Italiana* was founded by the Italian Seismologic Society and, in 1897, the *Mitteilungen der Erdbeben* was founded by the Vienna Academy of Sciences. In 1907 the publication of the *Bulletin of the Imperial Earthquake Investigation Committee* started in Japan, in 1908, the *Publications du Bureau Central de l'Association International de Sismologie* started to be published in Strasbourg, in 1911, the *Bulletin of the Seismological Society of America*, in 1926, the *Bulletin of the Earthquake Research Institute* of Tokyo University, and in 1929 *Earthquakes Notes*, that changed its name to *Seismological Research Letters* in 1987. In 1997, the publication of the *Journal of Seismology* (Kluwer, Dordrecht) started. The following journals are dedicated to the field of earthquake engineering: *Earthquake Engineering and Spectral Dynamics*, *European Earthquake Engineering*, *Soil Dynamics and Earthquake Engineering*, and *Earthquake Spectron*.

Besides the journals dedicated entirely to seismology, articles on this subject are published in geophysical journals. The list is very long so only the most representative are mentioned, in chronologic order of the first year of publication: *Geophysical Magazine* (1926), *Fizica ziemly* (1937), *Pure and Applied Geophysics* (*Geofisica pura e applicata*) (1939), *Annali di geofisica* (1948), *Journal of Geophysical Research* (1949), *Journal of Physics of the Earth* (1952), *Geophysical Journal International* (1992) (a fusion of the *Geophysical Journal of the Royal Astronomical Society* (1958), the *Zeitschrift für Geophysik* (1924) and the *Annale de géophysique* (1948)), *Reviews of Geophysics* (1963), *Tectonophysics* (1964), *Earth and Planetary Science Letters* (1966), and *Physics of the Earth and Planetary Interiors* (1967).

Actually, the amount of published material in seismology keeps on increasing considerably. Students will find it useful to consult the most important textbooks, where they will find different approaches to the topics treated in this book. Also, they should read some of the classical papers, references to which are given in the Bibliography, and look through the recent issues of seismologic journals to find out about the present topics of research.

2 FUNDAMENTAL EQUATIONS OF AN ELASTIC MEDIUM

2.1 Stress, strain, and displacement

An important part of seismology consists in the study of the generation and propagation of seismic waves, that is, waves produced in the Earth by earthquakes. For this purpose it is necessary to approximate the Earth by a continuous elastic medium to which the equations of mechanics can be applied. A continuous medium is an idealization of a material in which the distance between two contiguous points can be made infinitesimally small. In this idealization, the granular structure of the materials of the Earth and their molecular and atomic nature are not considered. In a continuous medium the term particle is used to mean a geometric point without dimensions. Density and mechanical properties are considered as continuous functions of spatial coordinates and time.

The study of the mechanical behavior of continuous media can be traced back to the discovery in 1660 by Hooke of the linear law that relates stress and strain in an elastic body and its applications formulated by Mariotte around 1680. The first studies on the behavior of elastic materials were those of Bernoulli, Euler, Lagrange, and Coulomb. In 1827, Navier established the general equations for equilibrium and vibration in elastic solids. This work was continued by Cauchy, Poisson, Lamé, Kirchhoff, Lord Kelvin, Lamb, Rayleigh, and Love, among others.

We start by recalling some of the basic ideas of the mechanical behavior of continuous media. Let us consider, in a continuous medium, a region of volume V surrounded by a closed surface S, in such a way that there is matter on both sides of the surface. For each point inside V we can define elastic stresses, strains, and displacements as continuous functions of the spatial coordinates and time (Fig. 2.1).

Stresses at a point inside V are defined as the limits of the quotients of the forces that act at this point per unit surface through a plane with a certain orientation. Both the orientation of the force F and that of the plane ΔS are included in the definition of the stress. The stress through a plane with normal ν is represented by a vector T,

$$T = \lim_{\Delta S \to 0} \frac{F}{\Delta S}$$

Stress exists only if there is material on both sides of the surface. Stress can also be represented by a second-order tensor τ with nine elements τ_{ij} that are the stresses through three orthogonal planes (Appendix 1). In Cartesian coordinates, these planes are normal to the three axes (x_1, x_2 and x_3) (Fig. 2.2). With this tensor we can define the state of stress at a point through a plane of arbitrary orientation. For a plane with a normal given by the unit vector ν, Cauchy's relation between the vector T and the

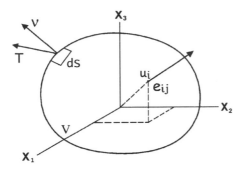

Fig. 2.1. The stress, strain, and displacement in an elastic medium.

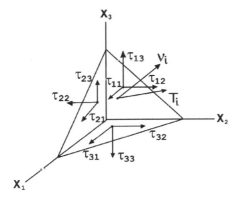

Fig. 2.2. The components of the stress tensor, τ_{ij}, through three orthogonal planes and of the stress vector T_i through a plane normal to the unit vector in ν_i.

tensor τ is

$$T_i = \tau_{ji}\nu_j \tag{2.1}$$

In this equation we have used the index notation and the convention that repeated subindexes are summed through their three values (Appendix 1). It is easy to show that, in the absence of external moments, the tensor τ is symmetric ($\tau_{ij} = \tau_{ji}$) and only six of the nine components are different. With respect to the three planes normal to the coordinates' axes, the components of τ_{ij} with subindex $i = j$ correspond to normal stresses and those with $i \neq j$ correspond to tangential or shear stresses.

Displacements at a point in a continuous medium are given by the vector $\boldsymbol{u}$. If dl is the distance between two points, for a rigid-body displacement, $\boldsymbol{u}$ is the same for all points and dl remains constant. When a body is deformed $\boldsymbol{u}$ is different for different points and the distance dl changes. The strain is defined as the change of this distance. If the distance before deformation is dl and that after deformation is dl', we can write, in a first-order approximation,

$$dl'^2 - dl^2 = 2\frac{\partial u_i}{\partial x_j}\, dx_j\, dx_i \tag{2.2}$$

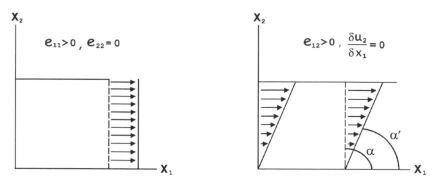

Fig. 2.3. Representations of the longitudinal strain e_{11} and shear strain e_{12}.

The deformation or strain in the body depends on the derivatives of the displacements. Because the strain implies variations in the displacements in the directions of the three spatial coordinates, it must be represented by a tensor. In the case of infinitesimal deformations, the strain can be represented by the Cauchy tensor **e**, which, if $\boldsymbol{u}$ varies continuously and slowly with position, is given by

$$e_{ij} = \frac{1}{2}\left(\frac{\partial u_i}{\partial x_j} + \frac{\partial u_j}{\partial x_i}\right) = \frac{1}{2}(u_{i,j} + u_{j,i}) \tag{2.3}$$

In this expression we have used the commas to represent partial derivatives with respect to the corresponding spatial coordinate (Appendix 1). This definition implies that the increments in the displacements ($\delta\boldsymbol{u}$) are smaller than those in position ($\delta\boldsymbol{x}$). The tensor **e** is, by definition, symmetric ($e_{ij} = e_{ji}$) and only six of its nine components are different. The components of the strain tensor **e** have a simple geometric interpretation. Those with subindexes $i = j$ correspond to longitudinal deformations along the axes of coordinates. For $i \neq j$, they correspond to variations in the angles between the directions of the axes of the deformed and undeformed states; deformations of this type are called shear strains (Fig. 2.3).

Another important tensor related to changes in the displacements in a deformable continuous body is the rotation tensor

$$\omega_{ij} = \tfrac{1}{2}(u_{i,j} - u_{j,i}) \tag{2.4}$$

The rotation tensor **ω** is antisymmetric. Using the tensors *e* and **ω** the partial derivatives of the displacements are given by

$$u_{i,j} = e_{ij} + \omega_{ij} \tag{2.5}$$

The rotation tensor ω_{ij} is related to the curl of the displacements,

$$\boldsymbol{\omega} = \nabla \times \boldsymbol{u}$$

$$\omega_i = e_{ijk} u_{k,j} \tag{2.6}$$

by the equation

$$\omega_i = 2e_{ijk}\omega_{jk} \tag{2.7}$$

where e_{ijk} is the alternating or permutation tensor (Appendix 1). According to (2.5), the partial derivatives of $\boldsymbol{u}$ are totally defined by the two tensors e_{ij} and ω_{ij}. This means that the variations of displacements from one point to another in a deformable medium include both strains and rotations.

2.1.1 Eigenvalues and eigenvectors

The tensors τ_{ij} and e_{ij} can be studied in terms of their eigenvalues and eigenvectors. Since both tensors are represented by 3×3 symmetric matrices, in each case, their three eigenvalues are real and their three eigenvectors mutually orthogonal. The eigenvectors form a system of orthogonal axes such that, when referred to it, all components of the tensor are zero, except those in the main diagonal, where they are the eigenvalues. The reference system formed by the eigenvectors represents the system of principal axes of stress and strain. The eigenvalues are also called the principal values of the stress and strain. Eigenvectors and eigenvalues are found through the equation

$$(\tau_{ij} - \sigma\delta_{ij})\nu_i = 0 \tag{2.8}$$

In this equation δ_{ij} is the Kronecker delta tensor (Appendix 1). The eigenvalues are the three roots of the cubic equation resulting from equating to zero the determinant of (2.8),

$$\mathrm{Det}[\tau_{ij} - \sigma\delta_{ij}] = 0 \tag{2.9}$$

The three eigenvectors ν_i^1, ν_i^2 and ν_i^3 correspond to the three eigenvalues σ_1, σ_2 and σ_3 and are obtained by substituting each of these values into equation (2.8). Thus, referred to its principal axes, the tensor τ_{ij} has the following form:

$$\tau_{ij} = \begin{bmatrix} \sigma_1 & 0 & 0 \\ 0 & \sigma_2 & 0 \\ 0 & 0 & \sigma_3 \end{bmatrix} \tag{2.10}$$

In this system of axes, the tensor τ_{ij} has only normal components. Regarding the planes normal to the axes, there exist only normal stresses (τ_{ij}, $i = j$) and the shear stresses (τ_{ij}, $i \neq j$) are null. The principal values of the stress, σ_1, σ_2 and σ_3, are, usually, ordered so that σ_1 is the largest and σ_3 the smallest.

A similar result is obtained for the strain tensor e_{ij}. When it is referred to its principal axes, the shear components are null and the tensor takes the form

$$e_{ij} = \begin{bmatrix} \varepsilon_1 & 0 & 0 \\ 0 & \varepsilon_2 & 0 \\ 0 & 0 & \varepsilon_3 \end{bmatrix} \tag{2.11}$$

In both cases, for $\boldsymbol{\tau}$ and $\mathbf{e}$, the sum of the elements of the main diagonal is the first invariant of the matrix, so that, for any orientation of the axes,

$$\tau_{11} + \tau_{22} + \tau_{33} = \sigma_1 + \sigma_2 + \sigma_3 = \text{constant}$$

$$e_{11} + e_{22} + e_{33} = \varepsilon_1 + \varepsilon_2 + \varepsilon_3 = \text{constant}$$

In the case of the strain tensor, this sum represents the change in volume per unit volume and it is called the cubic dilation θ. It can be obtained from (2.3) that the cubic dilation is equal to the divergence of the displacements,

$$e_{11} + e_{22} + e_{33} = u_{1,1} + u_{2,2} + u_{3,3} = \theta$$

$$\nabla \cdot \boldsymbol{u} = \theta \qquad (2.12)$$

The tensors τ_{ij} and e_{ij} can be expressed as the sum of two tensors, one isotropic and the other deviatoric:

$$\tau_{ij} = \sigma_0 \delta_{ij} + \tau'_{ij}$$

$$e_{ij} = \varepsilon_0 \delta_{ij} + e'_{ij} \qquad (2.13)$$

where σ_0 and ε_0 are one third the sums of the principal stresses and strains, respectively:

$$\sigma_0 = \tfrac{1}{3}(\sigma_1 + \sigma_2 + \sigma_3)$$

$$\varepsilon_0 = \tfrac{1}{3}(\varepsilon_1 + \varepsilon_2 + \varepsilon_3)$$

The deviatoric stress and strain τ'_{ij} and e'_{ij} are, thus, defined by equations (2.13). In the case that the deformation implies only changes in volume without changes in form, $e'_{ij} = 0$. If there are only changes in form, without changes in volume, $\varepsilon_0 = 0$. In this case the strain is purely deviatoric.

2.2   Elasticity coefficients

The mechanical behavior of a continuous material is defined by the relation between the stress and the strain. For a linear elastic body, this relation is given by Hooke's law, which states that the strain is proportional to the stress. Cauchy's formulation in tensor form of this law is

$$\tau_{ij} = C_{ijkl} e_{kl} \qquad (2.14)$$

This equation states the linear relation between the strain and stress tensors and is the foundation of the theory of linear elasticity. The fourth-order tensor C_{ijkl} is the tensor of the elasticity coefficients or moduli and has 81 components. Owing to the symmetry of τ_{ij} and e_{ij}, only 36 are different. For perfect elasticity, there exists a strain-energy function and it follows that $C_{ijkl} = C_{klij}$ and the number of elasticity coefficients is further reduced to 21 (Malvern, 1969). Equation (2.14) can also be expressed in terms of the derivatives of the displacements, by substituting (2.3) into it:

$$\tau_{ij} = C_{ijkl} u_{k,l} \qquad (2.15)$$

The simplest case for the elasticity coefficients corresponds to an isotropic medium, that is, a medium with the same properties in all directions. For such a medium, all components of C_{ijkl} can be expressed by two, λ and μ, called Lamé's coefficients:

$$C_{ijkl} = \lambda \delta_{ij} \delta_{kl} + \mu (\delta_{ik} \delta_{jl} + \delta_{il} \delta_{jk}) \qquad (2.16)$$

The 21 different components of C_{ijkl} in terms of λ and μ are

$$C_{1111} = C_{2222} = C_{3333} = \lambda + 2\mu$$

$$C_{1122} = C_{1133} = C_{2233} = C_{2211} = C_{3311} = C_{3322} = \lambda$$

$$C_{1212} = C_{2121} = C_{1221} = C_{2112} = C_{1313} = C_{3131} = C_{1331} = C_{3113} = C_{2323}$$

$$= C_{3232} = C_{2332} = C_{3223} = \mu$$

By substitution of (2.16) into (2.14), the relation between the stress and the strain for an isotropic medium is found to be

$$\tau_{ij} = \lambda \delta_{ij} e_{kk} + 2\mu e_{ij} \qquad (2.17)$$

In terms of the derivatives of the displacements

$$\tau_{ij} = \lambda \delta_{ij} u_{k,k} + \mu(u_{i,j} + u_{j,i}) \qquad (2.18)$$

From a different point of view, the mechanical behavior of an isotropic elastic medium can be stated in terms of two coefficients, K, the bulk modulus, that relates the changes in volume without changes in form and G, the shear modulus that relates the changes in form without changes in volume to the stresses that produce them. For the first case,

$$\tau_{11} + \tau_{22} + \tau_{33} = 3K(e_{11} + e_{22} + e_{33}) \qquad (2.19)$$

In the case of hydrostatic pressure, the normal stresses are equal ($\tau_{11} = \tau_{22} = \tau_{33} = -P$) and equation (2.18), taking into account (2.12), results in

$$P = -K\theta \qquad (2.20)$$

The coefficient K represents the quotient relating the pressure to the change in volume it produces ($K = -P/\theta$). For this reason it is called the bulk modulus or incompressibility.

The second elasticity coefficient G relates changes in form without changes in volume to the shear stresses or deviatoric strain and stress:

$$\tau'_{ij} = 2G e'_{ij} \qquad (2.21)$$

The coefficient G is equivalent to μ and is called the shear modulus or rigidity. By substituting equation (2.19) into (2.17), and considering (2.13), we obtain $G = \mu$ and

$$K = \lambda + \tfrac{2}{3}\mu \qquad (2.22)$$

Another elasticity coefficient is Young's modulus E that relates the longitudinal stress and strain in the same direction:

$$E = \tau_{11}/e_{11} \qquad (2.23)$$

The relations of E to λ, μ and K are

$$E = \frac{\mu(3\lambda + 2\mu)}{\lambda + \mu} = \frac{9K\mu}{3K + \mu} \qquad (2.24)$$

In a medium subject only to a longitudinal stress τ_{11} in the direction of the x_1 axis, the quotient relating the strain in a perpendicular direction e_{22} and that in the same direction

Table 2.1. *Correspondences among the values of K, λ, and μ for some values of σ.*

σ	λ	K
0	0	$\frac{2}{3}\mu$
$\frac{1}{8}$	$\frac{1}{3}\mu$	μ
$\frac{1}{4}$	μ	$\frac{5}{3}\mu$
$\frac{1}{2}$	K	λ

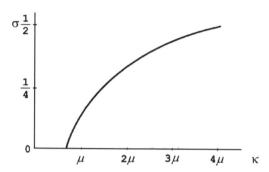

Fig. 2.4. The relation between Poisson's ratio σ and the bulk modulus K in terms of the rigidity μ or shear modulus.

as the stress e_{11} is called Poisson's ratio:

$$\sigma = -e_{22}/e_{11} \tag{2.25}$$

On substituting this into equation (2.17) and considering that $e_{33} = e_{22}$, the following relation is obtained:

$$\sigma = \frac{\lambda}{2(\lambda + \mu)} \tag{2.26}$$

Poisson's ratio σ has values between 0 and $\frac{1}{2}$. The correspondence among the values of K, λ, and μ for some values of σ are given in Table 2.1 and shown in Fig. 2.4.

For $\sigma < \frac{1}{8}$, the material changes volume with greater facility than it changes its form ($K < \mu$). For $\sigma > \frac{1}{8}$, we have the opposite case; the material changes form more easily than it changes in volume ($K > \mu$). When $\mu = 0$, the material cannot support any shear stress, which is the case for a fluid. If $\mu = 0$ and $K = \infty$, the fluid is incompressible. If $K = \infty$ and $\mu \neq 0$, the material is an incompressible solid. The condition $\sigma = \frac{1}{4}$, for which $\lambda = \mu$, is known as Poisson's condition and the elastic behavior of the material (Poisson's solid) is defined by only one parameter. This condition is approximately satisfied for most materials in the Earth's interior (σ varies in the range 0.22–0.35). Solutions of many seismologic problems are greatly simplified by using this condition.

The units used nowadays in seismology are preferentially those of the International System (IS), although sometimes cgs units are still used. For displacements the units are m (IS) and cm (cgs), for stress $\mathrm{N\,m^{-2}}$ or Pa (IS) and $\mathrm{dyn\,cm^{-2}}$ (cgs) and also bars

(1 bar $= 10^6$ dyn cm^{-1}) (10^6 Pa $= 1$ MPa $= 10$ bars). The elastic coefficients K, λ, μ, and E have the same units as stress. Strain has units of m m^{-1} (IS) and cm cm^{-1} (cgs).

2.3   The influence of temperature

In the process of deformation of a continuous elastic medium, temperature changes may be produced. This means that the thermodynamic conditions must be considered. If, during a process, the temperature changes from an initial value T_0 to another T, the relation between the stress and the strain for an isotropic medium is

$$\tau_{ij} = \delta_{ij}\lambda e_{kk} + 2\mu e_{ij} - K\alpha(T - T_0)\delta_{ij} \tag{2.27}$$

where, in the last term, α is the coefficient of thermal expansion. If there are no applied stresses ($\tau_{ij} = 0$), then, considering equation (2.22), we obtain from (2.27)

$$e_{kk} = \theta = \alpha(T - T_0) \tag{2.28}$$

If the temperature increases ($T > T_0$), θ is positive and there is an increase in volume. The coefficient α relates the increase in volume to the increase in temperature at constant pressure:

$$\alpha = \left(\frac{\partial V}{\partial T}\right)_P \tag{2.29}$$

We have defined K as the quotient relating the applied pressure to changes in volume; its inverse $1/K$ represents the change in volume with a change in pressure. Since, in the definition of K, we have not considered changes in temperature, this coefficient is valid for processes with a constant temperature and is called the isothermal bulk modulus:

$$\frac{1}{K} = \left(\frac{\partial V}{\partial P}\right)_T \tag{2.30}$$

In processes with changes in temperature when there is no heat transfer, that is, adiabatic processes, we can define the adiabatic bulk modulus K_a. Its inverse $1/K_a$ represents the change in volume with a change in pressure at constant entropy:

$$\frac{1}{K_a} = \left(\frac{\partial V}{\partial P}\right)_S \tag{2.31}$$

From the basic thermodynamic equations, the increments of heat dQ and enthalpy dH are

$$dQ = \left(\frac{\partial Q}{\partial T}\right)_P dT + \left(\frac{\partial Q}{\partial P}\right)_T dP \tag{2.32}$$

$$dH = T\,dS + V\,dP \tag{2.33}$$

From these two equations and thermodynamic relations, we find that

$$\left(\frac{\partial V}{\partial P}\right)_S = \left(\frac{\partial V}{\partial P}\right)_T + \frac{T}{C_P}\left(\frac{\partial V}{\partial T}\right)_P^2 \tag{2.34}$$

If we substitute (2.29), (2.30), and (2.31) into (2.34), we obtain

$$\frac{1}{K_a} = \frac{1}{K} - \frac{T}{\alpha^2 C_P} \tag{2.35}$$

where

$$C_P = \left(\frac{\partial Q}{\partial T}\right)_P \tag{2.36}$$

(C_P is the specific heat at constant pressure). Equation (2.35) relates isothermal and adiabatic bulk moduli.

In conclusion, the coefficients that define the mechanical behavior of isotropic elastic media are K, μ, and ρ. If there are changes in temperature we have to add α and, for adiabatic processes, substitute K for K_a.

2.4   Work and energy

In an elastic medium stresses that produce deformations result in an amount of work. The work done by each increment of strain per unit volume is

$$dW = \tau_{ij}\, de_{ij} \tag{2.37}$$

If the medium is perfectly elastic there is no dissipation of energy and dW is an exact differential that depends uniquely on the strain:

$$dW = \frac{\partial W}{\partial e_{ij}}\, de_{ij} \tag{2.38}$$

From (2.37) and (2.38) we can derive

$$\tau_{ij} = \frac{\partial W}{\partial e_{ij}} \tag{2.39}$$

In this expression W can be considered as the potential function of elastic energy. If we replace τ_{ij} in (2.37) by its value in terms of e_{ij} (2.14), we obtain

$$dW = C_{ijkl}e_{kl}\, de_{ij} \tag{2.40}$$

If we integrate this expression, the energy per unit volume for an elastic body is given by

$$W = \tfrac{1}{2} C_{ijkl}e_{kl}e_{ij} = \tfrac{1}{2}\tau_{ij}e_{ij} \tag{2.41}$$

Considering the thermodynamic problem, the change in internal energy U in an elastic body subject to changes in work and heat is given by

$$dU = dQ + dW \tag{2.42}$$

On substituting for dW according to (2.37) and dQ in terms of the change in entropy dS, we have that

$$dU = T\, dS + \tau_{ij}\, de_{ij} \tag{2.43}$$

In an adiabatic process, there is no transport of heat and the entropy is constant, so

$$dU = \tau_{ij}\,de_{ij} \tag{2.44}$$

This equation can also be written as

$$\left(\frac{\partial U}{\partial e_{ij}}\right)_S = \tau_{ij} \tag{2.45}$$

On comparing it with equation (2.39), this expression shows that, in an elastic body, under adiabatic conditions, the internal energy is formed only by the elastic potential energy.

In order to consider isothermal processes, we introduce the free energy or Helmholtz's function F defined in the form

$$F = U - TS \tag{2.46}$$

If we take differentials of this equation and substitute for dU from (2.44), since the temperature and entropy are constant, we obtain

$$dF = \tau_{ij}\,de_{ij} \tag{2.47}$$

This expression can also be written as

$$\left(\frac{\partial F}{\partial e_{ij}}\right)_T = \tau_{ij} \tag{2.48}$$

For isothermal processes, the elastic energy potential function is given by the free energy. In conclusion, in a perfectly elastic body there exists an elastic energy potential function of the strain from which we can derive the stress. This function is given by the internal energy in adiabatic processes and by the free energy in isothermal processes.

2.5   Equations of continuity and motion

The fundamental equations that rule the mechanical behavior of an elastic medium are those of continuity and motion. The first is a consequence of the principle of conservation of mass and energy and the second is a consequence of Newton's second law. The application of these two equations to the processes in the Earth derived from the occurrence of earthquakes constitutes the fundamentals of theoretical seismology.

2.5.1   *The equation of continuity*

The mass contained in a volume V of a continuous medium of density $\rho(x, t)$, a function of the spatial coordinates and time, is given by

$$M(t) = \int_V \rho(x, t)\,dV \tag{2.49}$$

The principle of conservation of mass states that mass is conserved in all physical processes. This may be expressed mathematically using the concept of the material derivative with respect to time (D/Dt) that must be zero. For a volume V, contained inside a

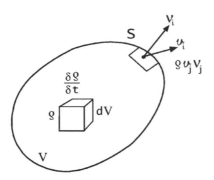

Fig. 2.5. The change with time of the mass inside the volume V and the flux of mass through its external surface S.

closed surface S, the material derivative of the mass is (Malvern, 1969)

$$\frac{D}{Dt}\int_V \rho \, dV = \int_V \frac{\partial \rho}{\partial t} \, dV + \int_S \rho v_j v_j \, dS \tag{2.50}$$

On the right-hand side of (2.50), the first integral represents the change of mass with time inside the volume V and the second the flux of mass through the surface S. For each element of the surface dS with normal ν, the velocity of mass flow is v and the flux per unit surface area is $\boldsymbol{F} = \rho\boldsymbol{v}$ (Fig. 2.5). The conservation of mass implies that expression (2.50) is null. Therefore the variation of mass with time inside the volume V must equal the flux through its surface S. If we apply Gauss's theorem to convert the surface integral into a volume integral in equation (2.50) and equate it to zero, we obtain

$$\int_V \left(\frac{\partial \rho}{\partial t} + \frac{\partial}{\partial x_i}(\rho v_i) \right) dV = 0 \tag{2.51}$$

In differential form this expression is

$$\frac{\partial \rho}{\partial t} + \frac{\partial}{\partial x_i}(\rho v_i) = 0 \tag{2.52}$$

In an analogous way, if $w(\boldsymbol{x}, t)$ is the energy density per unit volume and the flux of energy $\boldsymbol{F} = w\boldsymbol{U}$, then the conservation of energy in a volume V is given by

$$\int_V \left(\frac{\partial w}{\partial t} + \frac{\partial}{\partial x_i}(wU_i) \right) dV = 0 \tag{2.53}$$

where now $\boldsymbol{U}$ is the velocity of the energy flow through the surface S that surrounds V. In differential form the resulting equation is similar to (2.52). The equations of continuity (2.51) and (2.53) show the relation between the change with time of the mass and energy inside a volume V and their flow through a surface S. If the change is negative (the mass or energy in V diminishes), the flow is positive (going out through S) and vice versa.

2.5.2   The equation of motion

The motion at each point inside a volume V is determined by the forces acting in its interior and stresses on its external surface. Newton's second law, that the sum of the

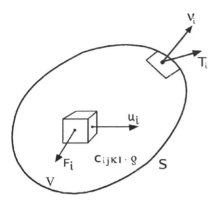

Fig. 2.6. The displacement u, force per unit volume F (the body force), and stress per unit surface area T, acting on an elastic medium of density ρ and elastic coefficients C_{ijkl}.

forces equals the time derivative of the linear momentum, for a continuous body, according to Euler's formulation, is given by

$$\int_V F_i \, dV + \int_S T_i \, dS = \frac{d}{dt} \int_V \rho v_i \, dV \qquad (2.54)$$

where F are forces acting on volume elements dV or body forces, T stresses acting on surface elements dS and v is the velocity at each point of the volume (Fig. 2.6). On substituting for the vector T the tensor τ according to (2.1) and applying Gauss's theorem to the surface integral to convert it in a volume integral, we obtain

$$\int_V \left(F_i + \frac{\partial \tau_{ij}}{\partial x_j} \right) dV = \frac{d}{dt} \int_V \rho v_i \, dV \qquad (2.55)$$

If the density is constant with time, this equation becomes

$$\frac{\partial \tau_{ij}}{\partial x_j} + F_i = \rho \frac{dv_i}{dt}$$

$$\tau_{ij,j} + F_i = \rho \dot{v}_i$$

in differential form. In the second formulation, we use commas for derivatives with respect to spatial coordinates just like in (2.3) and (2.4) and an overdot for the total time derivative.

The total derivative with respect to time of the velocity v can be expressed in the form

$$\frac{dv_i}{dt} = \frac{\partial v_i}{\partial t} + v_j \frac{\partial v_i}{\partial x_j}$$

When using infinitesimal deformations and for very small velocities and changes in velocity with distance, second-order terms can be neglected and the total derivative with respect to time approximated by the partial derivatives. An overdot on a letter will denote now a partial derivative with respect to time. In this type of approximation, there is no difference between the Lagrangian and Eulerian formulations. However, since in seismology we are more interested in the displacement field, the Lagrangian formulation is more appropriate.

If we substitute for the total the partial time derivative in (2.54) and express the velocity in terms of displacement, replacing T by τ (2.1), we obtain,

$$\int_V F_i\,\mathrm{d}V + \int_S \tau_{ij}\nu_j\,\mathrm{d}S = \int_V \rho\ddot{u}_i\,\mathrm{d}V \tag{2.56}$$

For an infinite medium, replacing the surface integral by one over the volume by means of Gauss's theorem, as in (2.55), we obtain

$$\frac{\partial\tau_{ij}}{\partial x_j} + F_i = \rho\frac{\partial^2 u_i}{\partial t^2} \tag{2.57}$$

Equation (2.57) can be written in terms of the strain for an elastic medium by the substitution of (2.14) into it:

$$\frac{\partial}{\partial x_j}(C_{ijkl}e_{kl}) + F_i = \rho\frac{\partial^2 u_i}{\partial t^2} \tag{2.58}$$

In terms of the derivatives of displacements, according to (2.15), for constant elastic coefficients, we obtain

$$C_{ijkl}u_{k,lj} + F_i = \rho\ddot{u}_i \tag{2.59}$$

For an isotropic material, using equation (2.18) and substituting into (2.57), we obtain

$$[\lambda\delta_{ij}u_{k,k} + \mu(u_{i,j} + u_{j,i})]_{,j} + F_i = \rho\ddot{u}_i \tag{2.60}$$

For a homogeneous material, that is, for λ and μ constant, we can write the equation of motion in index and vector notation as

$$(\lambda + \mu)u_{k,ki} + \mu u_{i,jj} + F_i = \rho\ddot{u}_i$$

$$(\lambda + \mu)\nabla(\nabla\cdot\boldsymbol{u}) + \mu\nabla^2\boldsymbol{u} + \boldsymbol{F} = \rho\ddot{\boldsymbol{u}} \tag{2.61}$$

This expression represents the equation of motion in terms of displacements for a continuous, homogeneous, isotropic, infinite, elastic medium. The first term is the gradient of the divergence and the second the Laplacian of the displacements. This is, then, a second-order differential equation for partial derivatives with respect to the spatial coordinates and time with an independent term formed by the body forces. If we specify these forces, the solution of (2.61) gives us the elastic displacement field for the infinite medium. This equation is very important in seismology, since many problems can be solved using this approximation.

We can suppose that an earthquake is generated by processes that can be represented by a system of body forces acting in a certain focal region; then, outside this region, the only body forces are those of gravity ($F = g\rho$). However, except for waves with very large periods ($T > 600$ s), the influence of gravity is very small and is usually neglected (section 3.5). Thus, we can use equation (2.61), approximating the Earth by an infinite medium, for the elastic displacements outside the focal region (Chapter 16).

The equation of motion (2.61) can also be expressed in terms of the cubic dilation θ and the rotation vector ω, whose relations to the displacements are given by (2.12) and (2.6), respectively. For this we use the equation that relates the curl of the curl of a vector to the gradient of the divergence and the Laplacian (Appendix 1; (A1.30) or (A1.31)). On substituting for the Laplacian of u in (2.61) its value in (A1.30), we obtain

$$(\lambda + 2\mu)u_{k,ki} - \mu e_{ijk} e_{kln} u_{n,lj} + F_i = \rho \ddot{u}_i \tag{2.62}$$

Replacing θ and ω according to (2.12) and (2.6) and dividing by ρ results in

$$\alpha^2 \theta_{,i} - \beta^2 e_{ijk} \omega_{k,j} + \frac{F_i}{\rho} = \ddot{u}_i$$

$$\alpha^2 \nabla \theta - \beta^2 \nabla \times \omega + \frac{F}{\rho} = \ddot{u} \tag{2.63}$$

In this equation we have introduced the parameters α and β whose values in terms of the elastic coefficients are

$$\alpha^2 = \frac{\lambda + 2\mu}{\rho} = \frac{K + \frac{4}{3}\mu}{\rho} \tag{2.64}$$

$$\beta^2 = \mu/\rho \tag{2.65}$$

The parameter α is related to θ and, in consequence, to changes in volume, and β is related to ω, that is, to changes in form without changes in volume.

If, in expressions (2.61) and (2.63), we make the forces F null, we obtain the homogeneous equation of motion, also known as Navier's equation:

$$(\lambda + \mu)\nabla(\nabla \cdot u) + \mu \nabla^2 u = \rho \ddot{u} \tag{2.66}$$

$$\alpha^2 \nabla \theta - \beta^2 \nabla \times \omega = \ddot{u} \tag{2.67}$$

Solutions for this equation are the elastic displacements in a medium where there are no forces acting. With this equation we can study elastic perturbations in an infinite medium without considering the effects of any forces. We will see that this equation can be easily transformed into the wave equation.

2.6 Potential functions of displacements and forces

Displacements $u(x,t)$ in an elastic medium form a vector field. We can, therefore, apply Helmholtz's theorem that allows their representation in terms of two potential functions, a scalar potential ϕ and a vector potential ψ:

$$u = \nabla \phi + \nabla \times \psi$$

$$u_i = \phi_{,i} + e_{ijk} \psi_{k,j} \tag{2.68}$$

The vector potential ψ must satisfy the condition that its divergence is zero ($\nabla \cdot \psi = 0$). Using expressions (2.12) and (2.6) it is easy to deduce the relations of the two potentials to the cubic dilation θ and the rotation ω:

$$\theta = \nabla^2 \phi \tag{2.69}$$

$$\omega = -\nabla^2 \psi \tag{2.70}$$

These relations indicate that ϕ is related to changes in volume and ψ to changes in form.

The body forces $\boldsymbol{F}$ can also be represented in a similar form by two potential functions, a scalar potential Φ and a vector potential of zero divergence $\boldsymbol{\Psi}$:

$$\boldsymbol{F} = \nabla\Phi + \nabla \times \boldsymbol{\Psi} \tag{2.71}$$

Equation (2.63) can now be written in terms of the potentials, using (2.69)–(2.71), and can be separated into a scalar equation and a vector equation:

$$\alpha^2 \nabla^2 \phi + \frac{\Phi}{\rho} = \ddot{\phi} \tag{2.72}$$

$$\beta^2 \nabla^2 \psi + \frac{\Psi}{\rho} = \ddot{\psi} \tag{2.73}$$

The vector differential equation of motion (2.61) and (2.63) is now expressed in terms of the potentials by a scalar and a vector equation of a much more simple form. We must remember that the three components of the vector potentials ψ and $\boldsymbol{\Psi}$ are not independent, since they must satisfy the condition that their divergencies are zero. Then (2.72) and (2.73) represent only three independent equations corresponding to the three equations of (2.61) or (2.63). Equation (2.72) is related to the elastic perturbations which imply changes in volume and equation (2.73) is related to perturbations with changes in form only. This separation of the components of the equation of motion is very important, because it greatly facilitates its solution.

2.7   The Green function of elastodynamics

In the Earth, neglecting the forces of gravity, body forces in the equation of motion (2.61) may be used to represent the processes that generate earthquakes. In general, these forces $\boldsymbol{F}(\boldsymbol{x}, t)$, functions of the spatial coordinates and time, may be different for each earthquake and are defined only inside a certain volume. An example of a convenient simple time dependence is the harmonic function

$$\boldsymbol{F}(\boldsymbol{x}, t) = \boldsymbol{F}(\boldsymbol{x})\, e^{i\omega t} \tag{2.74}$$

This form of time dependence simplifies the solution of many problems in seismology. Use of the harmonic function is not in itself a very realistic way to represent the time dependence of the forces that generate earthquakes, but solutions to this problem may be used to find solutions for other time functions by means of the Fourier transform (Appendix 4).

A type of body forces of great importance in the solution of many problems of elastodynamics is that formed by a unit impulsive force in space and time with an arbitrary direction. This force may be represented mathematically by means of Dirac's delta function

$$F_i(x_s, t) = \delta(x_s - \xi_s)\delta(t - \tau)\delta_{in} \tag{2.75}$$

The force is applied at the point of the coordinates ξ_i and the time τ, and is null outside this point and time. Its orientation is given by its three components represented by the

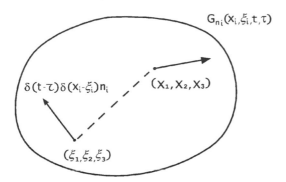

Fig. 2.7. Displacements G_{ni} (Green's function) at x_i produced by an impulsive force in time and space at ξ_i.

subindex n. If we substitute this force into the equation of motion (2.54), its solutions are the elastic displacements for every point of coordinates x for every time t, in a certain volume V surrounded by a surface S. Each component of the displacement (subindex i) depends on the orientation of the force (subindex n) and thus the displacement is given by a second-order tensor with components $G_{ni}(x_s, \xi_s, t, \tau)$, which is a function of the coordinates and time (x_s, t) of each point in V and of the coordinates and time of the point of application of the force (ξ_s, τ) (Fig. 2.7). On replacing the stress in terms of the derivatives of the displacements (2.15) into the equation of motion (2.56), we obtain

$$\int_V \rho \ddot{G}_{ni} \, dV - \int_S C_{ijkl} G_{nk,l} \nu_j \, dS = \int_V \delta(x_s - \xi_s)\delta(t - \tau)\delta_{ni} \, dV \tag{2.76}$$

On taking the reciprocal of the surface integral by using Gauss's theorem as in (2.55), for a homogeneous, infinite medium according to (2.59), we have

$$\rho \ddot{G}_{ni} - C_{ijkl} G_{nk,lj} - \delta(x_s - \xi_s)\delta(t - \tau)\delta_{ni} \tag{2.77}$$

The solutions of equations (2.76) and (2.77) represent the elastic displacements due to a unit impulse force in space and time. For this reason the tensor **G** is called the Green function of elastodynamics or the response of the medium to an impulsive excitation. The form of this function depends on the characteristics of the medium, its elastic coefficients, and its density. In a finite medium (2.76), it depends also on the shape of the volume V and the boundary conditions on its surface S. For each medium there is a different Green function that defines how this medium reacts mechanically to an impulsive excitation force and is, therefore, a proper characteristic of each medium.

2.8 Theorems of reciprocity and representation

As we have seen, displacements in an elastic medium depend on the body forces and stresses acting on it. Let us consider an elastic medium of volume V surrounded by a surface S. For a system of body forces f acting on each volume element dV and stresses $\mathbf{T}^u$ on every element of surface dS, the displacements are u. In the same volume let us

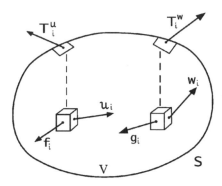

Fig. 2.8. Elastic displacements u and w corresponding to body forces f and g acting inside the volume V and stresses T^u and T^w acting on the surface S.

consider also a different system of forces and stresses g and $\mathbf{T}^w$, to which correspond the displacements w (Fig. 2.8). For each case we can write the equation of motion (2.56) in the form

$$\int_V (f_i - \rho u_i)\, dV + \int_S T_i^u\, dS = 0 \tag{2.78}$$

A similar expression is found for g, $\mathbf{T}^w$ and w. In (2.78) we take the scalar product of each term with w and, in its analog, we do this with u. Since each expression is now a scalar equation equal to zero, we can equate them, resulting in

$$\int_V (f_i - \rho \ddot{u}_i)w_i\, dV + \int_S T_i^u w_i\, dS = \int_V (g_i - \rho \ddot{w}_i)u_i + \int_S T_i^w u_i\, dS \tag{2.79}$$

This expression is known as Betti's reciprocity theorem. It shows the reciprocal relation between the displacements corresponding to two systems of forces and stresses acting in the same volume. If we reorder the terms in (2.79) and integrate each one through all time, we obtain

$$\int_{-\infty}^{\infty} dt \int_V \rho(u_i \ddot{w}_i - w_i \ddot{u}_i)\, dV = \int_{-\infty}^{\infty} dt \int_V (u_i g_i - w_i f_i)\, dV$$

$$+ \int_{-\infty}^{\infty} dt \int_S (u_i T_i^w - w_i T_i^u)\, dS \tag{2.80}$$

In this expression, the products are convolutions in time (Appendix 4). Thus, the first term can be written in full form by changing the order of integration:

$$\int_V dV \int_{-\infty}^{\infty} \rho[u_i(t)\ddot{w}_i(\tau - t) - \ddot{u}_i(t)w_i(\tau - t)]\, dt \tag{2.81}$$

Equation (2.80) is known as the Green–Volterra formula and is a generalization of Betti's reciprocity theorem. This expression relates the displacements and accelerations produced by two systems of forces and stresses on the same volume integrated over space and time.

A particular case results if displacements and velocities are null for all times previous to a given one. This condition implies the causality principle, namely that the medium is at rest until a given time when motion starts. If this time is $t = 0$, it follows that

$$u_i = \dot{u}_i = 0, \qquad t \leq 0$$

$$w_i = \dot{w}_i = 0, \qquad t \leq 0$$

Under these conditions, the integral of time in (2.81) can be written in the form

$$\rho \int_0^\tau \frac{\mathrm{d}}{\mathrm{d}t} [\dot{u}_i(t) w_i(\tau - t) + \dot{w}_i(\tau - t) u_i(t)] \, \mathrm{d}t \tag{2.82}$$

The change in sign is due to the fact that the first derivative with respect to time of **w** is negative. On integrating we obtain

$$\rho[\dot{u}_i(\tau) w_i(0) + \dot{w}_i(0) u_i(\tau) - \dot{u}_i(0) w_i(\tau) - \dot{w}_i(\tau) u_i(0)]$$

Taking into account the conditions specified above for **u**, **w**, and their derivatives at $t = 0$, this expression is zero (Aki and Richards, 1980). Then, equation (2.80) becomes

$$\int_{-\infty}^{\infty} \mathrm{d}t \int_V (u_i g_i - w_i f_i) \, \mathrm{d}V = \int_{-\infty}^{\infty} \mathrm{d}t \int_S (w_i T_i^u - u_i T_i^w) \, \mathrm{d}S \tag{2.83}$$

This is a very important result for seismology, since it allows the representation of the displacements due to a system of forces by those produced by a different system, given that causality conditions are satisfied. We can, then, represent the displacements due to a complicated system of forces in terms of those produced by a simpler one.

We can select, as the simplest system of forces, an impulsive force in space and time. As we have seen above, the displacements corresponding to this type of force are given by Green's function. Thus, we substitute into (2.83) **g** by expression (2.75) and the displacements **w** by Green's tensor **G**. Accordingly, equation (2.83) becomes

$$\int_{-\infty}^{\infty} \mathrm{d}t \int_V [u_i \delta(x_s - \xi_s) \delta(t - \tau) \delta_{in} - G_{ni} f_i] \, \mathrm{d}V$$

$$= \int_{-\infty}^{\infty} \mathrm{d}t \int_S (G_{ni} T_i - u_i C_{ijkl} G_{nk,l} \nu_j) \, \mathrm{d}S \tag{2.84}$$

where we have replaced $\mathbf{T}^u$ by $\mathbf{T}$ and $\mathbf{T}^w$ by its value in terms of the derivatives of the displacements (2.15),

$$T_i^w = \tau_{ij} \nu_j - C_{ijkl} G_{nk,l} \nu_j \tag{2.85}$$

According to the definition of the delta function,

$$\int_{-\infty}^{\infty} f(x) \delta(x - x_0) \, \mathrm{d}x = f(x_0) \tag{2.86}$$

the first integral of (2.83) results in

$$\int_{-\infty}^{\infty} \mathrm{d}t \int_V u_i(x_s, t) \delta(x_s - \xi_s) \delta(t - \tau) \delta_{in} \, \mathrm{d}V = u_n(\xi_s, \tau) \tag{2.87}$$

Since equation (2.84) is symmetric with respect to the variables x, t, ξ, and τ, we can interchange them and, on replacing them in (2.87), we obtain

$$u_n(x_s, t) = \int_{-\infty}^{\infty} \mathrm{d}\tau \int_V f_i G_{ni} \, \mathrm{d}V + \int_{-\infty}^{\infty} \mathrm{d}\tau \int_S [G_{ni} T_i - C_{ijkl} u_i G_{nk,l} \nu_j] \, \mathrm{d}S \tag{2.88}$$

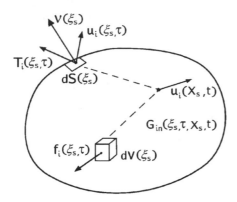

Fig. 2.9. Elastic displacements u at point x and time t, corresponding to a body force f and stress T acting at ξ and τ, in a medium with Green's function G.

This is an important result, as a special case of (2.83), and is generally referred to as the representation theorem. This equation gives us the elastic displacements inside a volume V as the sum of two double integrals over time and space. The first is the volume integral of the body forces multiplied by Green's function. The second is the surface integral of the stress multiplied by Green's function minus the displacements times the derivatives of Green's function. The displacements $u(x, t)$ are given for every point of the volume V at every time. In the two integrals f, T and u are functions of the integration variables ξ_i and τ, and, in general, also of $\nu_i(\xi_s)$, $dV(\xi_s)$, and $dS(\xi_s)$. Green's function $G_{ni}(\xi_s, \tau; x_s, t)$ depends both on x_s and t and on ξ_s and τ (Fig. 2.9).

Equation (2.88) allows us to determine the elastic displacements produced by a system of body forces defined in a volume or by a system of stresses and displacements defined over a surface by means of the Green function, as we will see in Chapter 16. To use this equation we need to have determined Green's function for the medium in question previously. This means that we need only solve the equation of motion once to find Green's function. Once Green's function for a certain medium with specified characteristics is known, we can determine the elastic displacements for any type of system of body forces in a volume and of stresses and displacements on a surface using equation (2.88). In this equation, Green's function acts as a propagator, since it propagates the effects of forces, stresses, or displacements defined on coordinates (ξ, τ) to determine the elastic displacements at points of the coordinates (x, t).

In general, the determination of Green's function is not easy. The difficulty increases with the complexity of the medium. The most simple case is for an infinite, homogeneous, isotropic, elastic medium (Chapter 16). Solutions for other media such as a layered half-space and a radially symmetric sphere have been determined. The advantage of using equation (2.88) is that, for a specified medium, Green's function, which implies the solution of the equation of motion, need only be solved once. As we will see in Chapter 16, equation (2.88) is very important in the determination of the source mechanism of earthquakes.

3 ELASTIC WAVES

3.1 Wave equations for an elastic medium

In an infinite, homogeneous, isotropic, elastic medium, the equation of motion in the absence of body forces is given by Navier's equation, which may be expressed in terms of the displacements (2.66) and the cubic dilation and rotation vector (2.67). These two equations may be easily transformed into wave equations.

In the first place, we apply the divergence operation in equation (2.67). The divergence of the gradient of θ is its Laplacian, that of the curl of ω is null and the divergence of the displacement u is the cubic dilation θ. Thus, we obtain

$$\nabla^2\theta = \frac{1}{\alpha^2}\frac{\partial^2\theta}{\partial t^2} \tag{3.1}$$

To the same equation, (2.67), we apply the curl operation. The curl of the gradient of the scalar function θ is null and that of the displacement u is the rotation vector ω. The curl of the curl of ω is equal to the gradient of the divergence, which is null minus the Laplacian (A1.30). The result is

$$\nabla^2\omega = \frac{1}{\beta^2}\frac{\partial^2\omega}{\partial t^2} \tag{3.2}$$

The same results (3.1) and (3.2) can be obtained by taking the divergence and the curl operations in equation (2.66). Equations (3.1) and (3.2) have the form of wave equations for the scalar function θ and vector function ω. The solutions of both equations represent waves that propagate in the elastic medium and the parameters α and β are their velocities. These velocities are functions of the elastic coefficients λ and μ and the density ρ, according to (2.64) and (2.65). When Poisson's ratio is $\frac{1}{4}$, $\alpha = \sqrt{3}\beta$. Because θ represents changes in volume without changes in shape, solutions of equation (3.1) correspond to compressional and dilational motion, or longitudinal waves. As waves propagate, the elastic material expands and contracts keeping the same form. In seismology, these waves are given the name of P (*prima*) waves, since they are the first to be observed in the seismograms (α is greater than β). Solutions of equation (3.2) represent shear waves that propagate with velocity β. The medium changes in shape, but not in volume, since the divergence of ω is null. In seismology these waves are given the name of S (*secunda*) waves, because they are the second type of prominent waves arriving after P waves. Therefore, in an infinite, homogeneous, isotropic, elastic medium there exist only these two types of waves that are called body waves. This important result was first found by Poisson in 1829.

The same result can be obtained for the potentials ϕ and ψ defined in (2.68). If in equations (2.72) and (2.73) we make null the potentials of the forces Φ and Ψ, we obtain

$$\nabla^2\phi = \frac{1}{\alpha^2}\frac{\partial^2\phi}{\partial t^2} \tag{3.3}$$

$$\nabla^2\psi = \frac{1}{\beta^2}\frac{\partial^2\psi}{\partial t^2} \tag{3.4}$$

Therefore, under conditions of the absence of body forces, the potentials ϕ and ψ are also solutions of the wave equation. Since α and β are the velocities of P and S waves, ϕ is the potential of P waves and ψ that of S waves. The total elastic displacement u is the sum of the displacements of P and S waves and can be written as

$$u = u^P + u^S \tag{3.5}$$

According to equation (2.68),

$$u^P = \nabla\phi \tag{3.6}$$

$$u^S = \nabla \times \psi \tag{3.7}$$

Displacements of the P and S waves can be deduced from the potentials ϕ and ψ, respectively.

Elastic displacements u in the absence of body forces are solutions of equation (2.66), which is not a wave equation. However, if they represent only P or S waves, they are solutions of wave equations. For example, if u depends only on x_1 and has only a component in the direction of x_2, that is, $u_2(x,t)$, corresponding to a transversal displacement, then, on substituting into equation (2.66), we obtain

$$\frac{\partial^2 u_2}{\partial x_1^2} = \frac{1}{\beta^2}\frac{\partial^2 u_2}{\partial t^2} \tag{3.8}$$

The displacement u_2 is in this case the solution of the wave equation. Solutions of (3.8) correspond to S waves that propagate in the x_1 direction with velocity β.

The same result can be obtained for a displacement of only a component $u_1(x_1, t)$ that depends only on x_1. By substitution into (2.66) we obtain

$$\frac{\partial^2 u_1}{\partial x_1^2} = \frac{1}{\alpha^2}\frac{\partial^2 u_1}{\partial t^2} \tag{3.9}$$

Therefore u_1 is, in this case, the solution of the wave equation. The solutions are P waves with displacements and propagation in the same direction x_1 and velocity α.

If, in the wave equation (3.3), the potential ϕ has a harmonic dependence on time with frequency ω,

$$\phi(x, t) = \phi(x)\,e^{-i\omega t}$$

After substitution, we obtain

$$(\nabla^2 + k_\alpha^2)\phi = 0 \tag{3.10}$$

where $k_\alpha = \omega/\alpha$ is the wave number of P waves. This is Helmholtz's equation, from which the time dependence has been eliminated. The same can be done starting from equation (3.4) and we obtain an equation for ψ similar to (3.10), in which $k_\beta = \omega/\beta$ is the wave number of S waves.

3.2   Solutions of the wave equation

From the equation of motion for displacements in an elastic medium in the absence of body forces (2.66), we have derived wave equations for the cubic dilation θ (3.1), the rotation vector ω (3.2) and the scalar ϕ (3.3) and vector ψ (3.4) potentials. The simplest case of a wave equation is that in one dimension:

$$\frac{\partial^2}{\partial x^2} f(x,t) = \frac{1}{c^2} \frac{\partial^2}{\partial t^2} f(x,t) \tag{3.11}$$

where c is the velocity of wave propagation. To solve this equation we apply the method of separation of variables, making $f(x,t) = R(x)T(t)$. By substitution into (3.11), we obtain

$$\frac{c^2}{R} \frac{d^2 R}{dx^2} = \frac{1}{T} \frac{d^2 T}{dt^2} = -\omega^2 \tag{3.12}$$

Where we have introduced ω^2 as the variable separation constant. From (3.12) we obtain two separate equations for R and T:

$$\frac{d^2 R}{dx^2} + k^2 R = 0 \tag{3.13}$$

$$\frac{d^2 T}{dt^2} + \omega^2 T = 0 \tag{3.14}$$

where $k^2 = \omega^2/c^2$. The solutions of both equations are harmonic functions and, for $f(x,t)$, their product is

$$f(x,t) = A\, e^{i(kx - \omega t)} + B\, e^{i(kx + \omega t)} \tag{3.15}$$

From the form of this solution, we can see that the separation constant ω represents the angular frequency and k is the wave number. The velocity of wave propagation is $c = \omega/k$. In conclusion, equation (3.15) represents harmonic waves of frequency ω that propagate with velocity c in the positive and negative x directions.

A more general solution of the wave equation, not restricted to harmonic functions, can be written in the form

$$f(x,t) = f(x - ct) + f(x + ct) \tag{3.16}$$

where, as before, $c = \omega/k$ is the velocity of propagation and, depending on the sign, waves propagate in the positive or negative x direction.

A particular solution, such as that in (3.15), may be given by harmonic functions, represented by imaginary exponentials, sines, and cosines. For one value of ω, they represent monochromatic waves, which, for propagation in the positive x direction,

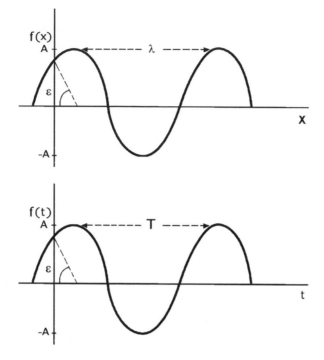

Fig. 3.1. A representation of sinusoidal wave motion as a function of the distance x and time t. A is the amplitude, ε is the initial phase, λ is the wave length, and T is the period.

may be expressed in the following forms:

$$f(x,t) = C\,e^{i(kx-\omega t+\varepsilon)} \tag{3.17}$$

$$f(x,t) = C\cos[k(x-ct)+\varepsilon] \tag{3.18}$$

$$f(x,t) = A\cos(kx-\omega t) + B\sin(kx-\omega t) \tag{3.19}$$

In (3.17) and (3.18), C is the amplitude of the wave and ε is the phase at the origin or the initial phase. In (3.19), A and B include the amplitude and initial phase. In all cases, the solutions have two arbitrary constants, A and B or C and ε. Their relation is

$$C^2 = A^2 + B^2 \tag{3.20}$$

$$\varepsilon = \tan^{-1}(B/A) \tag{3.21}$$

For harmonic waves, $f(x,t)$ has the same values at constant intervals of distance, multiples of the wave length $\lambda = 2\pi/k$ and of time, and multiples of the period $T = 2\pi/\omega$ (Fig. 3.1). In terms of λ and T, equation (3.18) may be written in the form

$$f(x,t) = C\cos\left[2\pi\left(\frac{x}{\lambda} - \frac{t}{T}\right) + \varepsilon\right]$$

The argument ξ of the harmonic function for each value of x and t is the phase

$$\xi = k(x-ct) + \varepsilon \tag{3.22}$$

For a constant value of ξ, if we vary t, x also varies, and this phase value propagates in time and space with a velocity c. For this reason, c is the velocity of propagation of each phase or phase velocity.

3.2.1 Wave fronts and rays

In three dimensions, a solution of the wave equation for monochromatic waves of frequency ω may be written as

$$f(x_j, t) = A \exp\{i[kS(x_j) - \omega t + \varepsilon]\} \tag{3.23}$$

The geometric surface of all points in space for which the phase ξ has the same constant value forms a wave front. According to (3.23), for each time, the function $S(x_j)$ gives us the spatial coordinates of the corresponding wave front. Let us suppose that the phase $\xi = 0$, then the spatial coordinates of the wave front corresponding to this value of the phase for times t_1 and t_2 are given by (Fig. 3.2)

$$S_1(x_j) = \frac{\omega}{k} t_1 - \frac{\varepsilon}{k} \tag{3.24}$$

$$S_2(x_i) = \frac{\omega}{k} t_2 - \frac{\varepsilon}{k} \tag{3.25}$$

The unit vector normal at each point of the wave front at each time defines the ray's direction or direction of propagation of each element of the wave front. The trajectory followed by the normal of the same element of the wave front during a certain time interval defines the ray's trajectory. The orientation of the ray at each time is given by its direction cosines that can be derived from the gradient of S:

$$\nu_i = \frac{\partial S}{\partial x_i} \bigg/ \left| \frac{\partial S}{\partial x_i} \right| \tag{3.26}$$

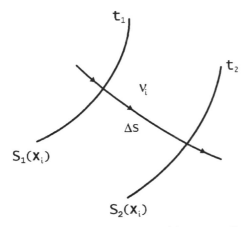

Fig. 3.2. Wave fronts $S_1(x)$ and $S_2(x)$ corresponding to times t_1 and t_2 and the ray's trajectory. Δs is an element of distance along the ray.

Equations of ray trajectories are given by

$$\frac{\mathrm{d}x_1}{\dfrac{\partial S}{\partial x_1}} = \frac{\mathrm{d}x_2}{\dfrac{\partial S}{\partial x_2}} = \frac{\mathrm{d}x_3}{\dfrac{\partial S}{\partial x_3}} \tag{3.27}$$

where $\mathrm{d}s^2 = \mathrm{d}x_1^2 + \mathrm{d}x_2^2 + \mathrm{d}x_3^2$ is the element of distance in the ray's direction. If the wave fronts for the same phase, which correspond to times t and $t + \Delta t$, are, respectively, S_1 and S_2, then, from (3.24) and (3.25), we obtain

$$kS_1 - \omega t = kS_2 - \omega(t + \Delta t) \tag{3.28}$$

If the increment in the direction of propagation between the two wave fronts is $S_2 - S_1 = \Delta S$, then, from (3.28), we obtain

$$\frac{\Delta S}{\Delta t} = \frac{\omega}{k} = c \tag{3.29}$$

where c is, again, the velocity of propagation of the wave front corresponding to the same phase or phase velocity. According to (3.29), with this velocity, the wave front advances a distance ΔS in a time Δt. This velocity may depend on the spatial coordinates and in this case each element of the wave front advances with a different velocity in a different direction.

3.2.2   *Waves of several frequencies*

Up to this point, we have considered only monochromatic harmonic waves, that is, harmonic waves with only one value of the frequency ω. Waves with an arbitrary dependence on time can be represented by the sum or integral of harmonic waves of different frequencies using Fourier's theorem (Appendix 4):

$$f(x_i, t) = \frac{1}{2\pi} \int_{-\infty}^{\infty} F(\omega) \exp\left[\mathrm{i}\left(\frac{\omega}{c(\omega)} S(x_i) - \omega t\right)\right] \mathrm{d}\omega \tag{3.30}$$

where $F(\omega)$ is a complex function that is called the complex spectrum of $f(x, t)$ and can be represented as

$$F(\omega) = R(\omega) + \mathrm{i}I(\omega) = A(\omega)\,\mathrm{e}^{\mathrm{i}\Phi(\omega)} \tag{3.31}$$

where $A(\omega)$ is the amplitude spectrum and $\Phi(\omega)$ is the phase spectrum. These two functions of frequency represent the contributions of amplitude and initial phase by each harmonic component to the resulting function of time. The phase velocity $c(\omega)$ is now a function of frequency, that is, each harmonic component may have a different phase velocity. Motion due to a propagating wave at a point is given by a function of time, $f(t)$. The same information can be represented as a function of frequency, $F(\omega)$, that is obtained by means of the inverse Fourier transform (A4.12):

$$F(\omega) = \int_{-\infty}^{\infty} f(t)\,\mathrm{e}^{-\mathrm{i}\omega t}\,\mathrm{d}t \tag{3.32}$$

$f(t)$ is a real function that represents the amplitudes of waves at each time for a particular point of space. Its Fourier transform $F(\omega)$ is a complex function that can be represented by its amplitudes $A(\omega)$ and initial phases $\Phi(\omega)$ for each frequency. In this way, problems of wave propagation can be studied in the time or frequency domain.

3.3 Displacement, velocity, and acceleration

In the one-dimensional case, the displacement at a point of an elastic medium due to a monochromatic wave that propagates in the positive direction of the x axis with velocity c is given by

$$u(x, t) = A \cos \left[\omega \left(\frac{x}{c} - t \right) + \varepsilon \right] \tag{3.33}$$

At a distance x, the displacement varies with time from A to $-A$ and the maxima correspond to phase values of 0, π, 2π, 3π, etc. (Fig. 3.3). The velocity of this motion, or the particle velocity (not to be mistaken for the phase velocity), is given by the time derivative,

$$v(x, t) = \frac{\partial u}{\partial t} = A\omega \sin \left[\omega \left(\frac{x}{c} - t \right) + \varepsilon \right] \tag{3.34}$$

The acceleration is given by the second derivative,

$$a(x, t) = \frac{\partial^2 u}{\partial t^2} = -A\omega^2 \cos \left[\omega \left(\frac{x}{c} - t \right) + \varepsilon \right] \tag{3.35}$$

The particle velocity is shifted in phase by $\pi/2$ with respect to the displacement and by π with respect to the acceleration (Fig. 3.3). The amplitude of the velocity is multiplied by the frequency and that of the acceleration by its square. The study of ground displacement, velocity, and acceleration is an important subject in seismology. Damage produced by earthquakes is caused by the effects of such motion, especially the effect of its acceleration on buildings and other structures.

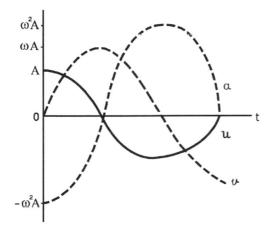

Fig. 3.3. The displacement, velocity, and acceleration for a cosine wave.

3.4   The propagation of energy. The Lagrangian formulation

The solution of the equation of motion for an elastic medium results in the existence of elastic waves in its interior. The wave phenomenon is a way of transporting energy without transport of matter. The propagation of energy is, then, a very important aspect of wave propagation. In mechanics, the formulation of a problem from the point of view of energy is different from that of the equation of motion, which we have considered.

The equation of energy continuity in a continuous medium in differential form, according to (2.53), is given by

$$\frac{\partial w}{\partial t} + \frac{\partial}{\partial x_i}(U_i w) = 0 \tag{3.36}$$

where w is the energy density and U is the velocity of the energy flux or propagation. The energy density in an elastic medium is given by the sum of kinetic T and potential V energies per unit mass:

$$w = T + V \tag{3.37}$$

In terms of energy, the problem must be solved using the Lagrangian formulation. This formulation uses the Lagrangian density function, $L = T - V$. For a discrete system, formed by N particles, this function depends on the generalized coordinates of each particle q_i, their time derivatives $\dot{q}_i$, and time, $L(q_i, \dot{q}_i, t)$. The application of Hamilton's principle of stationary action results in Lagrange's equation (Lanczos, 1986; Lindsay, 1960):

$$\frac{d}{dt}\frac{\partial L}{\partial \dot{q}_i} - \frac{\partial L}{\partial q_i} = 0 \tag{3.38}$$

In a continuous system, which is the case for the problem of elastic waves, the Lagrangian density function depends on the generalized coordinates of each point of the continuous medium $q_i(x, t)$, their derivatives with respect to the spatial coordinates and time, and the independent variables x_i and t:

$$L\left(q_i, \frac{\partial q_i}{\partial t}, \frac{\partial q_i}{\partial x_k}, x_k, t\right)$$

In this case, application of the principle of stationary action leads to Euler's equations (Lindsay, 1960):

$$\sum_k \frac{\partial}{\partial x_k}\left(\frac{\partial L}{\partial\left(\frac{\partial q_i}{\partial x_k}\right)}\right) - \frac{\partial L}{\partial q_i} = 0 \tag{3.39}$$

Equations (3.38) and (3.39) allow the solution of mechanical problems from the point of view of energy as an alternative to the application of Newton's second law.

The energy density w can be expressed in terms of the Lagrangian function L, the generalized coordinates q_i and their derivatives:

$$w = \sum_i \frac{\partial q_i}{\partial t}\frac{\partial L}{\partial q_i} - L \tag{3.40}$$

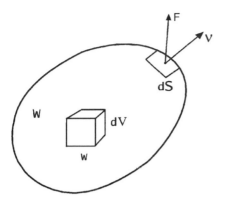

Fig. 3.4. The energy W contained in a volume V surrounded by a surface S. The energy density w and energy flux F through an element of surface dS are shown.

By taking the time derivative of this expression, we obtain

$$\frac{\partial w}{\partial t} = \sum_i \left(\frac{\partial^2 q_i}{\partial t^2} \frac{\partial L}{\partial q_i} + \frac{\partial q_i}{\partial t} \frac{\partial}{\partial t} \frac{\partial L}{\partial q_i} \right) - \frac{\partial L}{\partial t} \tag{3.41}$$

The flux of energy per unit surface and unit time F is given by (Fig. 3.4)

$$\boldsymbol{F} = w\boldsymbol{U} \tag{3.42}$$

where U is the velocity of propagation or transport of energy. The flux of energy can also be derived from the Lagrangian function L (Lindsay, 1960):

$$F_k = \sum_i \frac{\partial q_i}{\partial t} \frac{\partial L}{\partial \left(\dfrac{\partial q_i}{\partial x_k} \right)} \tag{3.43}$$

A simple application of these principles to the propagation of energy in an elastic medium is the case of a shear wave with displacement u and propagation in the x direction. In this case $x_i = x$ and $q_i = u$. The kinetic T and potential V energy densities are given by

$$T = \frac{\rho}{2} \left(\frac{\partial u}{\partial t} \right)^2 \tag{3.44}$$

$$V = \frac{\mu}{2} \left(\frac{\partial u}{\partial x} \right)^2 \tag{3.45}$$

The Lagrangian function is

$$L = \frac{\rho}{2} \left(\frac{\partial u}{\partial t} \right)^2 - \frac{\mu}{2} \left(\frac{\partial u}{\partial x} \right)^2 \tag{3.46}$$

On applying Euler's equation (3.39), we obtain

$$\frac{\partial}{\partial x} \left(-\mu \frac{\partial u}{\partial x} \right) + \frac{\partial}{\partial t} \left(\rho \frac{\partial u}{\partial t} \right) = 0 \tag{3.47}$$

From this equation, we finally derive

$$\frac{\partial^2 u}{\partial x^2} = \frac{\rho}{\mu} \frac{\partial^2 u}{\partial t^2} \tag{3.48}$$

This is the wave equation for a shear displacement (3.8), where $\beta^2 = \mu/\rho$ is the velocity of phase propagation. Here we have derived the wave equation from energy considerations by applying Euler's equation. Previously, we had obtained the same result from the equation of motion.

The energy density w, according to (3.37), is the sum of the kinetic and potential energies:

$$w = \frac{\rho}{2} \left(\frac{\partial u}{\partial t} \right)^2 + \frac{\mu}{2} \left(\frac{\partial u}{\partial x} \right)^2 \tag{3.49}$$

and its time derivative is

$$\frac{\partial w}{\partial t} = \rho \frac{\partial u}{\partial t} \frac{\partial^2 u}{\partial t^2} + \mu \frac{\partial u}{\partial x} \frac{\partial}{\partial t} \left(\frac{\partial u}{\partial x} \right) \tag{3.50}$$

If we replace the second time derivative of u according to (3.48) and change the order of derivatives in (3.50), we obtain

$$\frac{\partial w}{\partial t} = \mu \frac{\partial}{\partial x} \left(\frac{\partial u}{\partial t} \frac{\partial u}{\partial x} \right) \tag{3.51}$$

Substituting this expression into the energy-continuity equation (3.36) results in

$$\frac{\partial}{\partial x} \left(\mu \frac{\partial u}{\partial t} \frac{\partial u}{\partial x} + wU \right) = 0 \tag{3.52}$$

The quantity inside the brackets is independent of x. For a localized wave whose value tends to zero when x tends to infinity, this quantity is zero for all times. Hence, the velocity of the propagation of energy U is given by

$$U = -\frac{\mu}{w} \frac{\partial u}{\partial t} \frac{\partial u}{\partial x} \tag{3.53}$$

Thus, the velocity of energy propagation U does not necessarily coincide with the phase velocity c of the harmonic wave's motion.

3.4.1   Phase and group velocities

We have seen that, for harmonic wave motion, the phase velocity $c = \omega/k$, is not, in general, equal to the energy-transport velocity. For waves with more than one frequency, the phase velocity is a function of the frequency $c(\omega)$ or wave number $c(k)$. This implies that the wave number is a function of frequency $k(\omega)$ and vice versa for $\omega(k)$. In this case, as we will see in Chapter 12, we have the phenomenon of wave dispersion and can define the group velocity as

$$v = \frac{\mathrm{d}\omega}{\mathrm{d}k} \tag{3.54}$$

This is the velocity of the propagation of the packets or groups of waves and we will see that is the velocity corresponding to the frequencies that make the phase stationary (section 12.3). If we substitute $\omega = ck$ into equation (3.54), we obtain the relation between phase and group velocities

$$v = c + k\frac{dc}{dk} \tag{3.55}$$

Since energy is contained in wave packets, the group velocity coincides with the velocity of propagation of energy $v = U$. According to equation (3.55), if the phase velocity c depends on the wave number k, then the phase and group velocities are different, whereas if c is constant they are equal.

For the simple case of a monochromatic shear wave that propagates in the x direction

$$u = A\cos(kx - \omega t) \tag{3.56}$$

The phase velocity is $c = \beta = \omega/k$. According to (3.49), the energy density is

$$w = \mu A^2 k^2 \sin^2(kx - \omega t) \tag{3.57}$$

The energy contained in one wave length λ can be obtained by integration:

$$W = \mu A^2 k^2 \int_0^\lambda \sin^2(kx - \omega t)\, dx \tag{3.58}$$

and this results in

$$W = \pi\mu k A^2 = 2\pi^2 \beta \rho T\left(\frac{A}{T}\right)^2 \tag{3.59}$$

where we have taking into account that $k = 2\pi/(\beta T)$ and $\beta^2 = \mu/\rho$. This result can be applied also to P waves by changing the velocity β into α. Hence, the energy contained in a wave length of a monochromatic wave is proportional to the square of the amplitude or the square of the amplitude divided by the period.

We can also calculate the velocity of energy transport U using equation (3.53) by substitution of w from (3.57) and the derivatives of u from (3.56). The result is $U = \beta$; the velocity of energy propagation is equal, in this case, to the phase velocity. This result is consistent with the fact that the phase velocity is constant and therefore equal to the group velocity.

In conclusion, in an elastic medium, energy is propagated by waves with a velocity that, in general, is equal to the group velocity and different than the phase velocity. In the case that the phase velocity is constant with frequency, the two velocities are equal.

3.5 The effect of gravity on wave propagation

The Earth possesses a gravitational field that affects the propagation of waves in its interior. In a nonrotational spherical Earth, the gravitational potential is $U = GM/r$. For points in its interior, the potential satisfies Poisson's equation:

$$\nabla^2 U = -4\pi G\rho \tag{3.60}$$

With the passage of a wave, the potential at an element of volume is perturbed so that $\delta U = U - U_0$ is the difference between the perturbed and the unperturbed potentials. If the unperturbed and perturbed densities are ρ_0 and ρ, we have

$$\nabla^2(U - U_0) = -4\pi G(\rho - \rho_0) \tag{3.61}$$

In terms of the cubic dilation $(\theta = \delta V / V)$, this equation is

$$\nabla^2(U - U_0) = -4\pi G\rho\theta \tag{3.62}$$

The force due to the change in the gravitational potential is $\boldsymbol{F} = \rho\nabla(U - U_0)$. If we substitute this force into the equation of motion (2.63), we obtain

$$\alpha^2\nabla\theta - \beta^2\nabla \times \boldsymbol{\omega} + \nabla(U - U_0) = \ddot{\boldsymbol{u}} \tag{3.63}$$

by taking the divergence in (3.63) and substituting (3.62) into it, we obtain

$$\alpha^2\nabla^2\theta + 4\pi G\rho\theta = \ddot{\theta} \tag{3.64}$$

If we substitute as a solution $\theta = A\cos(kx - \omega t)$ and solve for $c = \omega/k$, we obtain

$$c = \alpha(1 - \varepsilon)^{1/2} \tag{3.65}$$

$$\varepsilon = \frac{4\pi G\rho}{\alpha^2 k^2}$$

The phase velocity c is now a function of the wave number k and differs from α according to the value of ε. If, for the Earth's mantle, we take $\alpha = 8\,\mathrm{km\,s^{-1}}$ and $\rho = 4.4\,\mathrm{g\,cm^{-3}}$, then, for waves of 1 s period $(k = 0.78)$, $\varepsilon = 10^{-7}$ and the velocity of P waves is not affected by gravity. If we consider waves of 300 s period, the value of ε is 0.01 and the velocity will be decreased by approximately 0.1%. This shows that we need consider the effect of gravity on wave propagation only for waves with very long periods.

3.6   Plane waves

The simplest geometry of the wave front is that of a plane and the corresponding waves are called plane waves. The equation of the wave front in Cartesian coordinates is given by

$$S(x_1, x_2, x_3) = \nu_1 x_1 + \nu_2 x_2 + \nu_3 x_3$$

According to equation (3.26), ν_1, ν_2 and ν_3 are the direction cosines of the normal to the plane wave front which define the direction of propagation or ray trajectory.

In Cartesian coordinates the wave equation is

$$\frac{\partial^2 f}{\partial x_1^2} + \frac{\partial^2 f}{\partial x_2^2} + \frac{\partial^2 f}{\partial x_3^2} = \frac{1}{c^2}\frac{\partial^2 f}{\partial t^2} \tag{3.66}$$

Solutions for monochromatic waves of frequency ω can be written as

$$f(x_1, x_2, x_3, t) = A\exp\{\mathrm{i}[k(x_1\nu_1 + x_2\nu_2 + x_3\nu_3) - \omega t + \varepsilon]\} \tag{3.67}$$

or

$$f(x_j, t) = A\exp[\mathrm{i}(k_j x_j - \omega t + \varepsilon)] \tag{3.68}$$

where $k_j = k\nu_j$ is the wave-number vector. The equation of the wave front for a phase at a given time is

$$k_1 x_1 + k_2 x_2 + k_3 x_3 = \text{constant} \tag{3.69}$$

If we are dealing with waves with plane wave fronts of infinite extension, it is implied that their source is at an infinite distance. This condition does not correspond to any real problem. However, if we are interested only in wave propagation and can assume that the source is at a very large distance, plane waves are a good simplifying approximation.

Plane-wave solutions for the potentials ϕ and ψ are

$$\phi = A \exp[ik_\alpha(\nu_j x_j - \alpha t + \varepsilon)] \tag{3.70}$$

$$\psi_k = B_k \exp[ik_\beta(\nu_j x_j - \beta t + \eta)] \tag{3.71}$$

where $k_\alpha = \omega/\alpha$ and $k_\beta = \omega/\beta$ are the wave numbers of P and S waves. Displacements for P and S waves are obtained by substitution of (3.70) and (3.71) into (3.6) and (3.7):

$$u_k^P = Aik_\alpha(\nu_1, \nu_2, \nu_3) \exp[ik_\alpha(\nu_j x_j - \alpha t + \varepsilon)] \tag{3.72}$$

$$u_k^S = [(B_3\nu_2 - B_2\nu_3), (B_1\nu_3 - B_3\nu_1), (B_2\nu_1 - B_1\nu_2)]ik_\beta \exp[ik_\beta(\nu_j x_j - \beta t + \eta)] \tag{3.73}$$

Cartesian components of amplitudes of displacements of P waves are proportional to the direction cosines of the rays corresponding to longitudinal waves. Displacements of S waves are perpendicular to the direction of propagation, which can be verified by taking the scalar product of u^S and ν, which is always zero. S waves are, therefore, transverse waves.

Another way of showing the properties of displacements of P and S waves is to consider the total elastic displacements in the form of plane waves (with the initial phase null):

$$u_i = C_i \exp[ik(\nu_j x_j - ct)] \tag{3.74}$$

where c is the phase velocity whose value we do not yet know. By substitution of (3.74) into the homogeneous equation of motion (2.72), we obtain three equations for the three components of the displacement that, after dividing by k^2 and the exponential, may be expressed in matrix form as

$$\begin{pmatrix} \nu_1^2 - \dfrac{\rho c^2 - \mu}{\lambda + \mu} & \nu_1\nu_2 & \nu_1\nu_3 \\[2mm] \nu_1\nu_2 & \nu_2^2 - \dfrac{\rho c^2 - \mu}{\lambda + \mu} & \nu_2\nu_3 \\[2mm] \nu_1\nu_3 & \nu_2\nu_3 & \nu_3^2 - \dfrac{\rho c^2 - \mu}{\lambda + \mu} \end{pmatrix} \begin{pmatrix} C_1 \\ C_2 \\ C_3 \end{pmatrix} = 0 \tag{3.75}$$

This is a homogeneous system of equations for C_1, C_2 and C_3, and the condition for the existence of a solution is that the determinant of the system is null. From this condition we obtain a cubic equation for c^2 that has two different solutions that correspond to the already known values of α and β (2.67) and (2.68). This result confirms what we knew from the solution for the potentials ϕ and ψ, namely that, in a homogeneous,

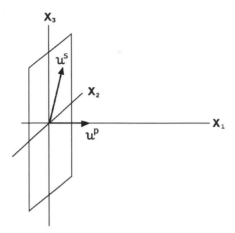

Fig. 3.5. Displacements of P and S plane waves propagating in the x_1 direction.

isotropic, elastic medium, there are only these two types of waves. The components of the displacement for each wave can be obtained by replacing the two values of c in equation (3.75).

As an example, if we have a wave that propagates in the positive direction of the x_1 axis, that is, $\nu_i = (1,0,0)$, equation (3.75) now has the simpler form

$$\begin{pmatrix} 1 - \dfrac{\rho c^2 - \mu}{\lambda + \mu} & 0 & 0 \\[2ex] 0 & \dfrac{\rho c^2 - \mu}{\lambda + \mu} & 0 \\[2ex] 0 & 0 & \dfrac{\rho c^2 - \mu}{\lambda + \mu} \end{pmatrix} \begin{pmatrix} C_1 \\ C_2 \\ C_3 \end{pmatrix} = 0 \tag{3.76}$$

For the value $c = \alpha$, we obtain $C_1 \neq 0$ and $C_2 = C_3 = 0$. Displacements of the P waves are in the same direction as the propagation. For the value $c = \beta$, we obtain $C_1 = 0$, $C_2 \neq 0$, and $C_3 \neq 0$. Displacements of S waves are on the (x_2, x_3) plane, normal to the propagation direction (Fig. 3.5). The equations for displacements of the P and S waves are given by

$$u_k^{\mathrm{P}} = (C_1, 0, 0) \exp[ik_\alpha(x_1 - \alpha t)] \tag{3.77}$$

$$u_k^{\mathrm{S}} = (0, C_2, C_3) \exp[ik_\beta(x_1 - \beta t)] \tag{3.78}$$

In the general case, displacements of P and S plane waves can be written as

$$u_k^{\mathrm{P}} = D_k \exp[ik_\alpha(\nu_j x_j - \alpha t + \varepsilon)] \tag{3.79}$$

$$u_k^{\mathrm{S}} = E_k \exp[ik_\beta(\nu_j x_j - \beta t + \eta)] \tag{3.80}$$

Since displacements of P and S waves are perpendicular to each other, their scalar product must be null; $D_j E_j = 0$. The relations between the amplitudes of displacements (3.79) and (3.80) and those of potentials (3.70) and (3.71), according to (3.72) and

(3.73), are

$$D_j = ik_\alpha \nu_j A \qquad (3.81)$$

$$E_j = ik_\beta e_{jkl} \nu_k B_l \qquad (3.82)$$

If the amplitudes of displacements are given in meters, those of potentials are given in m^2. Their phases are shifted by $\pi/2$.

3.7 The geometry of P and S wave displacements

In seismology, the reference coordinate system generally used is the geographic one (x_1, x_2, x_3), with x_1 and x_2 in the horizontal plane, x_1 positive to the North and x_2 positive to the West, and x_3 in the vertical direction positive up (zenith) (another choice is positive axes in the directions of North, East, and down (nadir)). The positive directions of the axes must satisfy a right-handed system. The direction of propagation of a wave is given by the vector $r(\nu_1, \nu_2, \nu_3)$ and its projection onto the horizontal plane by R. The angle between r and x_3 is the angle of incidence i $(0-\pi/2)$, and its complementary is $e = 90 - i$, the angle of emergence. The angle between R and x_1 is the azimuth α, measured from the North $(0-2\pi)$ (Fig. 3.6). In terms of i and α, the direction cosines of the ray are

$$\nu_1 = \sin i \cos \alpha \qquad (3.83)$$

$$\nu_2 = \sin i \sin \alpha \qquad (3.84)$$

$$\nu_3 = \cos i \qquad (3.85)$$

In reference to the direction of wave propagation or the ray's direction given by the vector r, we can define an orthogonal system of coordinate axes (R, T, Z); R, already

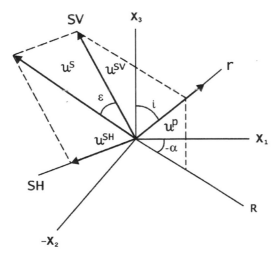

Fig. 3.6. Geometries of P and S wave displacements referred to geographic coordinate axes x_i (North, West, and zenith). Definitions of SH and SV components of the S wave are shown. α is the azimuth, i is the angle of incidence, and ε is the angle of polarization.

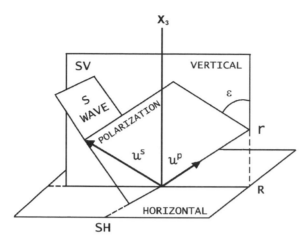

Fig. 3.7. Geometries of P and S wave displacements in reference to vertical or incidence, horizontal, S wave motion and S wave planes of polarization.

defined, is the horizontal component of the direction of the ray (r), T is normal to R in the horizontal plane, and Z is vertical. Z is positive up, R is positive in the direction of propagation and T is positive to the left looking forward in the direction of propagation.

Displacements of P waves have, in general, three components with respect to the geographic axes (u_1^P, u_2^P, u_3^P) and only two with respect to the propagation axes $(u_R^P, 0, u_Z^P)$. Displacements of S waves are contained on a plane normal to the vector r and have, in general, three components in both systems, (u_1^S, u_2^S, u_3^S) and (u_R^S, u_T^S, u_Z^S).

In order to understand the geometry of plane P and S waves in Cartesian coordinates, it helps to define four planes: the vertical plane or plane of incidence; the horizontal plane; the plane of S motion; and the polarization plane of the S wave (Fig. 3.7). The plane of incidence contains the vectors Z, r, and R. The plane of S motion is normal to the vector r. The intersection of this plane with the horizontal plane defines the SH direction and that with the vertical plane defines the SV direction. In consequence, the displacement of the S wave can be divided into two components, namely, SH and SV. The polarization plane of the S wave contains the vectors r and u^S. This plane forms an angle ε, called the polarization angle, with the incidence or vertical plane. In terms of the SH and SV components of S displacement, the angle of polarization is given by

$$\tan \varepsilon = u_{SH}/u_{SV} \tag{3.86}$$

The projection of the angle ε onto the horizontal plane is the angle γ, which is the angle between R and the horizontal component of S motion (u_{II}^S):

$$\tan \varepsilon = \tan \gamma \cos i \tag{3.87}$$

The unit vectors r, SH, and SV form another orthogonal system of axes. With respect to this system, amplitudes of P waves have components $(u_r, 0, 0)$ and those of S waves have $(0, u_{SH}, u_{SV})$. The relations among the different components of the amplitudes of P and S waves on the plane of incidence, plane of S motion, and horizontal plane are represented in Fig. 3.8.

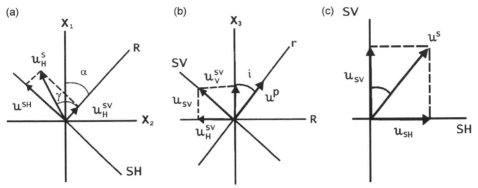

Fig. 3.8. Components of displacements of P and S waves: (a) on the horizontal plane, (b) on the vertical or incidence plane, and (c) on the plane of S wave motion. Planes are defined in Fig. 3.6 and angles are defined in Fig. 3.7.

3.8 Particular forms of the potentials

In many problems of propagation of plane waves in Cartesian coordinates, the relations between displacements and their potentials are simplified if the coordinate axes are selected to be in some particular direction. In general, expressions of the components of the displacements of P and S waves in terms of the scalar ϕ and vector ψ potentials are, according to (2.68),

$$u_1 = \frac{\partial \phi}{\partial x_1} + \frac{\partial \psi_3}{\partial x_2} - \frac{\partial \psi_2}{\partial x_3} = u_1^P + u_1^S \tag{3.88}$$

$$u_2 = \frac{\partial \phi}{\partial x_2} + \frac{\partial \psi_1}{\partial x_3} - \frac{\partial \psi_3}{\partial x_1} = u_2^P + u_2^S \tag{3.89}$$

$$u_3 = \frac{\partial \phi}{\partial x_3} + \frac{\partial \psi_2}{\partial x_1} - \frac{\partial \psi_1}{\partial x_2} = u_3^P + u_3^S \tag{3.90}$$

If we take the axis x_1 as the direction of propagation, the displacements are

$$u_j = A_j \exp\{i[k(x_1 - ct) + \varepsilon]\} \tag{3.91}$$

In this case, the displacements can be derived from three scalar potentials, ϕ, ψ, and Λ. The relations between the last two and the components of the vector potential ψ are $\psi = \psi_2$ and $\Lambda = -\psi_3$. The relations between the components of displacements and the scalar potentials are

$$u_1 = \frac{\partial \phi}{\partial x_1} = u^P \tag{3.92}$$

$$u_2 = \frac{\partial \Lambda}{\partial x_1} = u^{SH} \tag{3.93}$$

$$u_3 = \frac{\partial \psi}{\partial x_1} = u^{SV} \tag{3.94}$$

As we see, given the geometry of the problem, u_1 corresponds to the complete P wave and u_2 and u_3 to the SH and SV components of the S wave.

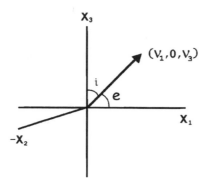

Fig. 3.9. A ray trajectory on the (x_1, x_3) plane, showing the angles of incidence i and emergence e.

In many problems of wave propagation, it is convenient to use as the plane of incidence the plane (x_1, x_3), so that the ray is contained in such a plane (Fig. 3.9). The direction cosines of the ray $(\nu_1, 0, \nu_3)$ in terms of the angles of incidence i and emergence e are

$$\nu_1 = \sin i = \cos e \tag{3.95}$$

$$\nu_3 = \cos i = \sin e \tag{3.96}$$

In this case, the potentials and displacements are functions of x_1 and x_3 only. The components u_1 and u_3 can be derived from two scalar potentials ϕ and ψ, where $\psi = \psi_2$:

$$u_1 = \frac{\partial \phi}{\partial x_1} - \frac{\partial \psi}{\partial x_3} = u_1^{\mathrm{P}} + u_1^{\mathrm{SV}} \tag{3.97}$$

$$u_3 = \frac{\partial \phi}{\partial x_3} + \frac{\partial \psi}{\partial x_1} = u_3^{\mathrm{P}} + u_3^{\mathrm{SV}} \tag{3.98}$$

The u_2 component is usually kept separate, since, in this case, it is a solution of the wave equation (as in (3.8)). However, it can also be derived from a third scalar potential Λ:

$$u_2 = -\frac{\partial \Lambda}{\partial x_1} = u^{\mathrm{SH}} \tag{3.99}$$

The components u_1 and u_3 correspond to P and SV motion and the component u_2 corresponds to SH. The total displacement can be expressed in a single vector equation as a function of the ϕ, ψ, and Λ potentials:

$$\boldsymbol{u} - \nabla\phi + \nabla \times (0, \psi, 0) + \nabla \times (0, 0, \Lambda) \tag{3.100}$$

This equation is a particular form of the more general relation of Eringen and Suhubi (Pilant, 1979) that expresses displacements as functions of three scalar potentials. We must remember that, when we use the vector potential ψ, it must satisfy the condition that its divergence is null. In consequence, only two of its components are independent and it is always possible to use three scalar potentials to represent the displacements.

3.9 Cylindrical waves

Many problems of wave propagation in seismology have axial symmetry and it is convenient to use cylindrical coordinates in their resolution. An example is the propagation of surface waves. Using expressions for the gradient and curl in cylindrical coordinates (r, ϕ, z) (Appendix 2), expressions for the displacement components (u_r, u_ϕ, u_z) (Fig. 3.10) in terms of the scalar potential Φ and vector potential $(\psi_r, \psi_\phi, \psi_z)$ are

$$u_r = \frac{\partial \Phi}{\partial r} + \frac{1}{r}\frac{\partial \psi_z}{\partial \phi} - \frac{\partial \psi_\phi}{\partial z} \tag{3.101}$$

$$u_\phi = \frac{1}{r}\frac{\partial \Phi}{\partial \phi} + \frac{\partial \psi_r}{\partial z} - \frac{\partial \psi_z}{\partial r} \tag{3.102}$$

$$u_z = \frac{\partial \Phi}{\partial z} + \frac{1}{r}\frac{\partial}{\partial r}(r\psi_\phi) - \frac{1}{r}\frac{\partial \psi_r}{\partial \phi} \tag{3.103}$$

The condition for null divergence of the vector potential ψ, according to (A2.12), is

$$\frac{1}{r}\frac{\partial}{\partial r}(r\psi_r) + \frac{1}{r}\frac{\partial \psi_\phi}{\partial \phi} + \frac{\partial \psi_z}{\partial z} = 0 \tag{3.104}$$

In many problems we can assume symmetry with respect to ϕ and the equations are simplified. In this case, instead of the vector potential ψ, we can use two scalar potentials, ψ and Λ, and the components of the displacements are

$$u_r = \frac{\partial \Phi}{\partial r} + \frac{\partial \psi}{\partial z} \tag{3.105}$$

$$u_\phi = \frac{\partial \Lambda}{\partial \phi} \tag{3.106}$$

$$u_z = \frac{\partial \Phi}{\partial z} + \frac{1}{r}\frac{\partial}{\partial r}(r\psi) \tag{3.107}$$

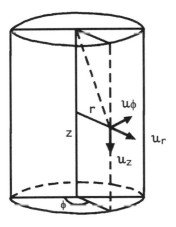

Fig. 3.10. Cylindrical coordinates (r, z, ϕ) and components of the displacement (u_r, u_z, u_ϕ).

If waves propagate in the r direction and z is the vertical axis, u_r corresponds to P waves, u_z to the SV, and u_ϕ to the SH component of the S wave.

To study the solutions of the wave equation in cylindrical coordinates, we will consider first the case of axial symmetry (independence of ϕ). If we assume a harmonic dependence on time, Helmholtz's equation (3.10) for the potential $\Phi(r, z)$ is

$$\frac{\partial^2 \Phi}{\partial r^2} + \frac{1}{r}\frac{\partial \Phi}{\partial r} + \frac{\partial^2 \Phi}{\partial z^2} + k_\alpha^2 \Phi = 0 \tag{3.108}$$

Using the method of separation of variables, $\Phi(r, z) = R(r)Z(z)$ (as in section 3.2), the following equations are found:

$$\frac{d^2 R}{dr^2} + \frac{1}{r}\frac{dR}{dr} + k_{\alpha r}^2 R = 0 \tag{3.109}$$

$$\frac{d^2 Z}{dz^2} + k_{\alpha z}^2 Z = 0 \tag{3.110}$$

where $k_\alpha^2 = k_{\alpha r}^2 + k_{\alpha z}^2$ and $k_{\alpha z}^2$ is the constant of separation. On making in equation (3.109) the change of variable $x = k_{\alpha r} r$, we obtain for $R(x)$ the equation

$$\frac{d^2 R}{dx^2} + \frac{1}{x}\frac{dR}{dx} + R = 0 \tag{3.111}$$

This is a differential Bessel equation for $n = 0$ and its solutions are the Bessel functions of zeroth order $J_0(k_{\alpha r} r)$ (Appendix 3). Equation (3.110) for the dependence of z has solutions that are harmonic functions. Adding the harmonic time dependence, the complete solution of the potential Φ is

$$\Phi(r, z, t) = AJ_0(k_{\alpha r} r)\exp[i(k_{\alpha z} z - \omega t)] \tag{3.112}$$

For large values of r, we can express Bessel functions in an asymptotic form in terms of harmonic functions:

$$\Phi(r, z, t) = \left(\frac{2}{\pi r k_{\alpha r}}\right)^{1/2} A\exp\left[i\left(k_{\alpha r} r + k_{\alpha z} z - \omega t - \frac{\pi}{4}\right)\right] \tag{3.113}$$

In this equation, we find the dependence of the amplitude of the potential on the inverse of the square root of the distance r. In this case, wave fronts are cylindrical surfaces whose areas increase with increasing r. If the energy generated at the source is finite, then, with increasing r, it must be distributed over an increasing area and, in consequence, amplitudes per unit area decrease with distance. This decrease in amplitude with distance is called the geometric spreading. This term has not appeared for plane waves, since their wave fronts are always infinite planes and their source is at an infinite distance, implying infinite energy.

Solutions for the two other scalar potentials ψ and Λ are of similar forms to (3.112) and (3.113), with k_α replaced by k_β. Components of displacements are obtained from the three potentials Φ, ψ, and Λ using equations (3.105)–(3.107).

For the more general case in which the potentials depend on the three coordinates, for example, $\Phi(r, \phi, z)$, Helmholtz's equation is given by

$$\frac{\partial^2 \Phi}{\partial r^2} + \frac{1}{r}\frac{\partial \Phi}{\partial r} + \frac{1}{r^2}\frac{\partial^2 \Phi}{\partial \phi^2} + \frac{\partial^2 \Phi}{\partial z^2} + k_\alpha^2 \Phi = 0 \tag{3.114}$$

Using the method of separation of variables, as before, $\Phi(r, \phi, z) = R(r)Y(\phi)Z(z)$, we obtain the following equations:

$$\frac{d^2 R}{dr^2} + \frac{1}{r}\frac{dR}{dr} + \left(k_{\alpha r}^2 - \frac{n^2}{r^2}\right)R = 0 \qquad (3.115)$$

$$\frac{d^2 Y}{d\phi^2} + n^2 Y = 0 \qquad (3.116)$$

$$\frac{d^2 Z}{dz^2} + k_{\alpha z}^2 Z = 0 \qquad (3.117)$$

The first equation is a differential Bessel equation and its solutions, finite at $r = 0$, are Bessel functions of order n. The other two equations (3.116) and (3.117) have harmonic functions for solutions. The solution for Φ, adding the harmonic time dependence, is

$$\Phi(r, \phi, z, t) = A_n J_n(k_{\alpha r} r) \exp[i(k_{\alpha z} z + n\phi - \omega t)] \qquad (3.118)$$

The solutions for components of the vector potential ψ have similar forms, replacing k_α by k_β. As in the previous case, components of displacements are obtained by using equations (3.101)–(3.103).

3.10 Spherical waves

The problem of propagation of elastic waves generated by a point source, which is commonly encountered in seismology, leads to the consideration of spherical waves. In this problem, we use spherical coordinates (r, θ, ϕ) and the components of displacements are u_r, u_θ, and u_ϕ (Appendix 2). If r is the direction of the ray, θ is measured from the vertical axis and ϕ on the horizontal plane, then u_r corresponds to the P wave, u_θ to SV and u_ϕ to SH (Fig 3.11).

Components of displacements, as functions of potentials, are obtained by replacing the gradient and curl in spherical coordinates in equation (2.68) (Appendix 2). The

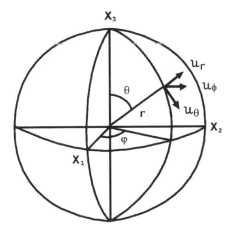

Fig. 3.11. Spherical coordinates (r, θ, ϕ) and components of the displacement (u_r, u_θ, u_ϕ).

vector potential ψ must satisfy the condition of null divergence that, in spherical coordinates, is given by (A2.28):

$$u_r = \frac{\partial \Phi}{\partial r} + \frac{1}{r \sin \theta} \frac{\partial}{\partial \theta} (\psi_\phi \sin \theta) - \frac{1}{r \sin \theta} \frac{\partial \psi_\theta}{\partial \phi} \tag{3.119}$$

$$u_\theta = \frac{1}{r} \frac{\partial \Phi}{\partial \theta} + \frac{1}{r \sin \theta} \frac{\partial \psi_r}{\partial \phi} - \frac{1}{r} \frac{\partial}{\partial r} (r \psi_\phi) \tag{3.120}$$

$$u_\phi = \frac{1}{r \sin \theta} \frac{\partial \Phi}{\partial \phi} + \frac{1}{r} \frac{\partial}{\partial \phi} (r \psi_\phi) - \frac{1}{r} \frac{\partial \psi_r}{\partial \theta} \tag{3.121}$$

For P waves that propagate from a point source with spherical symmetry and harmonic time dependence, we have only the potential $\Phi(r)$ and Helmholtz's equation is given by

$$\frac{d^2 \Phi}{dr^2} + \frac{2}{r} \frac{d\Phi}{dr} + k_\alpha^2 \Phi = 0 \tag{3.122}$$

If we make the substitution $\Phi = F/r$, we obtain

$$\frac{d^2 F}{dr^2} + k_\alpha^2 F = 0 \tag{3.123}$$

whose solution is

$$F = A \exp(\pm i k_\alpha r) \tag{3.124}$$

Adding the term of the harmonic time dependence for waves that propagate only in the positive direction of r, the potential $\Phi(r, t)$ is

$$\Phi(r, t) = \frac{A}{r} \exp[i(k_\alpha r - \omega t)] \tag{3.125}$$

The amplitude of the potential decreases with distance by the factor $1/r$. This is the factor of geometric spreading due to the increase in area of the spherical wave front with increasing distance from the origin.

If the problem has symmetry with respect to ϕ, displacements may be derived from three scalar potentials Φ, ψ, and Λ, which are functions of r, θ, and t:

$$u_r = \frac{\partial \Phi}{\partial r} + \frac{1}{\sin \theta} \frac{\partial}{\partial \theta} (\psi \sin \theta) \tag{3.126}$$

$$u_\theta = \frac{1}{r} \frac{\partial \Phi}{\partial \theta} - \frac{1}{r} \frac{\partial}{\partial r} (r \psi) \tag{3.127}$$

$$u_\phi = -\frac{\partial \Lambda}{\partial \theta} \tag{3.128}$$

Assuming a harmonic dependence on time, the potentials must satisfy Helmholtz's equation, which, for a potential $\Phi(r, \theta)$, is given by

$$\frac{1}{r^2} \frac{\partial}{\partial r} \left(r^2 \frac{\partial \Phi}{\partial r} \right) + \frac{1}{r^2 \sin \theta} \frac{\partial}{\partial \theta} \left(\sin \theta \frac{\partial \Phi}{\partial \theta} \right) + k_\alpha^2 \Phi = 0 \tag{3.129}$$

Applying the method of separation of variables, $\Phi(r, \theta) = R(r) Y(\theta)$, we obtain the following differential equations:

$$r^2 \frac{d^2 R}{dr^2} + 2r \frac{dR}{dr} + [k_\alpha^2 r^2 - n(n+1)] R = 0 \tag{3.130}$$

$$\frac{1}{r^2} \frac{d^2 Y}{d\theta^2} + \cot\theta \frac{dY}{d\theta} + n(n+1) Y = 0 \tag{3.131}$$

where $n(n+1)$ is the constant of separation. By the substitution of $R = S/\sqrt{r}$ in equation (3.130), we obtain

$$\frac{d^2 S}{dr^2} + \frac{1}{r} \frac{dS}{dr} + \left(k_\alpha^2 - \frac{(n+\frac{1}{2})^2}{r^2} \right) S = 0 \tag{3.132}$$

If, in this equation, we substitute $x = k_\alpha r$, we obtain a differential Bessel equation whose solutions are Bessel functions of the type $J_{n+1/2}(k_\alpha r)$. These functions may be expressed in terms of spherical Bessel functions $j_n(k_\alpha r)$ (Appendix 3):

$$J_{n+1/2}(k_\alpha r) = \left(\frac{2k_\alpha r}{\pi} \right)^{1/2} j_n(k_\alpha r) \tag{3.133}$$

For complex solutions spherical Hankel functions, $h_n(z) - j_n(z) + i n_n(z)$, are used (Appendix 3). Then, the solution for the function $R(r)$ is given by

$$R(r) = A_n \left(\frac{2k_\alpha}{\pi} \right)^{1/2} j_n(k_\alpha r) \tag{3.134}$$

Making in equation (3.131) the change of variable $u = \cos\theta$, we obtain

$$(1 - u^2) \frac{d^2 Y}{du^2} - 2u \frac{dY}{du} + n(n+1) Y = 0 \tag{3.135}$$

This is a differential Legendre equation and its solutions are Legendre functions, $P_n(u)$ and $Q_n(u)$ (Appendix 3).

The complete solution for the potential $\Phi(r, \theta, t)$, adding the harmonic time dependence and using only Legendre functions of the first kind and Hankel spherical functions, is given by

$$\Phi(r, \theta, t) = A_n \left(\frac{2k_\alpha}{\pi} \right)^{1/2} h_n(k_\alpha r) P_n(\cos\theta) e^{-i\omega t} \tag{3.136}$$

For large values of r, using the asymptotic expression for $h_n(k_\alpha r)$ in terms of harmonic functions, the solution may be written as

$$\Phi = A_n \frac{1}{r} \left(\frac{2}{k_\alpha \pi} \right)^{1/2} P_n(\cos\theta) \exp\left[i\left(k_\alpha r - \omega t - \frac{1}{2}(n+1)\pi \right) \right] \tag{3.137}$$

For each value of n, there is a particular solution and the general solution is given by their sum. Solutions for the potentials ψ and Λ have the same form, with k_α replaced by k_β.

For the most general case in which there is no symmetry with respect to ϕ, displacements must be derived from scalar Φ and vector $(\psi_r, \psi_\theta, \psi_\phi)$ potentials that are functions of the three coordinates r, θ, and ϕ. For the scalar potential $\Phi(r,\theta,\phi)$, Helmholtz's equation is given by

$$\frac{1}{r^2}\frac{\partial}{\partial r}\left(r^2\frac{\partial\Phi}{\partial r}\right) + \frac{1}{r^2\sin\theta}\frac{\partial}{\partial\theta}\left(\sin\theta\frac{\partial\Phi}{\partial\theta}\right) + \frac{1}{r^2\sin^2\theta}\frac{\partial^2\Phi}{\partial\phi^2} + k_\alpha^2\Phi = 0 \qquad (3.138)$$

Using the method of separation of variables, $\Phi(r,\theta,\phi) = R(r)N(\theta)L(\phi)$, we obtain the following differential equations:

$$\frac{1}{r^2}\frac{d}{dr}\left(r^2\frac{dR}{dr}\right) + \left(k_\alpha^2 - \frac{n(n+1)}{r^2}\right)R = 0 \qquad (3.139)$$

$$\frac{1}{\sin\theta}\frac{d}{d\theta}\left(\sin\theta\frac{dN}{d\theta}\right) + \left(n(n+1) - \frac{m^2}{\sin^2\theta}\right)N = 0 \qquad (3.140)$$

$$\frac{d^2L}{d\phi^2} + m^2L = 0 \qquad (3.141)$$

where $n(n+1)$ and m^2 are the constants of separation. The solutions of equation (3.139), as in the previous case, may be expressed by Bessel or Hankel spherical functions. Solutions of equation (3.140) are associate Legendre functions (Appendix 3) and those of (3.141) are harmonic functions. The potential $\Phi(r,\theta,\phi,t)$ is given by the product of the three solutions and the harmonic function of time:

$$\Phi(r,\theta,\phi,t) = A_{nm}\left(\frac{2k_\alpha}{\pi}\right)^{1/2}h_n(k_\alpha r)P_n^m(\cos\theta)\,e^{i(\pm m\phi - \omega t)} \qquad (3.142)$$

As in the previous case, the general solution is the sum for all values of n. For each value of n, there are values of m from $m = -n$ to $m = n$. Then, for each value of n, there are $2n + 1$ solutions corresponding to different values of m. Solutions for the components of the vector potential $(\psi_r, \psi_\theta, \psi_\phi)$ are of similar form, with k_α replaced by k_β. The components of displacements u_r, u_θ, and u_ϕ can be obtained from the potentials by using equations (3.119)–(3.121).

An important relation between the formulations in spherical coordinates (r,θ,ϕ) and cylindrical coordinates (ρ,z,ϕ), when there is symmetry with respect to ϕ is

$$\frac{1}{r}\exp(ik_r r) = \int_0^\infty J_0(k\rho)\exp(ik_z z)\frac{k}{k_z}\,dk \qquad (3.143)$$

where

$$r^2 = \rho^2 + z^2$$
$$k_r^2 = k^2 + k_z^2$$

This equation, known as Sommerfeld's relation, allows one to relate problems of spherical and cylindrical symmetries.

4 NORMAL MODE THEORY

4.1 Standing waves and modes of vibration

In Chapter 3 we saw that perturbations produced in an unbounded elastic body have the form of waves that propagate in its interior. If the medium is perfectly elastic, homogeneous, and isotropic, there are only two types of waves (P and S) that propagate with constant velocities (α and β) that depend on the elastic coefficients and density with no conditions imposed on their frequency. The Earth has finite dimensions and is bounded by a free surface, therefore it can not be considered an infinite medium. For this reason, we have to consider the elastic behavior of an elastic body of finite dimensions. This consideration leads us into normal mode theory. In this chapter we will give the fundamentals of this theory that will be applied to the Earth in Chapter 13.

As a first approximation, we start with the results of Chapter 3 and consider the phenomenon of standing waves. Let us assume waves in one dimension with the same frequency and amplitude that propagate in both directions. For sinusoidal waves, according to (3.18),

$$u(x,t) = A[\sin(kx + \omega t + \phi_1) + \sin(kx - \omega t + \phi_2)] \tag{4.1}$$

where ϕ_1 and ϕ_2 are the initial phases. Applying the relation $\sin(a + b) + \sin(a - b) = 2 \sin a \cos b$, we can write (4.1) in the form

$$u(x,t) = 2A \sin(kx + \phi_1') \cos(\omega t + \phi_2') \tag{4.2}$$

where $\phi_1' = \phi_1/2 + \phi_2/2$ and $\phi_2' = \phi_1/2 - \phi_2/2$. Since in (4.2) we do not have the propagating term $k(x \pm ct)$, this expression corresponds to standing waves such that the dependences of x and t are separated. For each value of t, $u(x)$ is a sine function of x with wave length $\lambda = 2\pi/k$ and, for each value of x, $u(t)$ is a cosine function of t with period $T = 2\pi/\omega$.

If we impose the condition that, for $x = 0$, the amplitude of the standing wave is zero for all values of t, we obtain

$$u(x,t) = 2A \sin(kx) \cos(\omega t + \phi) \tag{4.3}$$

where $\phi = \phi_1$. If we further impose that, for another value, $x = L$, the amplitude is also zero for all t, then we have

$$2A \sin(kL) = 0 \quad \text{and} \quad kL = (n+1)\pi, \quad n = 0, 1, 2, 3, \ldots \tag{4.4}$$

Hence, in order to fulfill both conditions, the wave number k must have certain discrete values, namely,

$$k_n = \frac{(n+1)\pi}{L}; \qquad n = 0, 1, 2, 3, \ldots \tag{4.5}$$

Since the velocity of the waves that travel in opposite directions has the same constant value c (4.1), frequencies of standing waves that satisfy the two imposed conditions are also limited to certain discrete values,

$$\omega_n = \frac{(n+1)\pi c}{L}; \qquad n = 0, 1, 2, 3, \ldots \tag{4.6}$$

Finally, standing waves that satisfy both conditions are given by

$$u_n(x, t) = 2A \sin\left(\frac{(n+1)\pi x}{L}\right) \cos(\omega_n t + \phi); \qquad n = 0, 1, 2, \ldots \tag{4.7}$$

Since, for each value of n, equation (4.7) is a solution, the general solution of the problem is given by their sum:

$$u(x, t) = \sum_{n=0}^{\infty} 2A \sin\left(\frac{(n+1)\pi x}{L}\right) \cos(\omega_n t + \phi)$$

Standing waves generated by waves traveling in opposite directions with the same velocity, when they are forced to have certain values at two points, limit their frequencies and wave numbers to multiples of the inverse of the distance between the two points ((4.6) and (4.5)). Each solution in (4.7) is called a mode of vibration, the lowest, that for $n = 0$, is called the fundamental mode and the rest are higher modes, harmonics, or overtones. The largest wave length and period correspond to the fundamental mode, $\lambda_0 = 2L$ and $T_0 = 2L/c$. The period of the fundamental mode corresponds to the time it takes for progressive waves to travel in both directions between 0 and L. For higher modes, wave lengths and periods are fractions of those of the fundamental mode ($\lambda_n = 2L/(n+1)$ and $T_n = 2L/(n+1)c$). Total motion of standing waves is the sum of all modes.

4.2   Vibrations of an elastic string of finite length

A mechanical problem concerning vibrations of an elastic medium with finite dimensions is that of the vibrations of an elastic string of length L fixed at both ends (Fig. 4.1). The transverse motion of each point of the string $y(x, t)$ depends only on the tension T in the direction of the string that is supposed to be constant. If ρ is the mass per unit length, the equation of transverse motion for a point of the string is

$$\sum_y T_y = \rho \, dx \frac{\partial^2 y}{\partial t^2} \tag{4.8}$$

The forces T_y, the components of $\mathbf{T}$ in the directions y between a point x and another $x + dx$, are given by $T_y = T \sin\theta$, and $T_y' = T \sin\theta'$ (Fig. 4.2). For small angles we

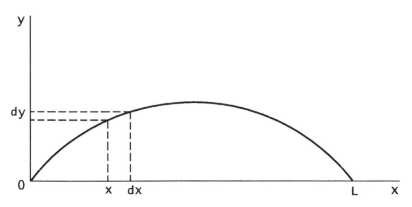

Fig. 4.1. A vibrating elastic string with both ends fixed.

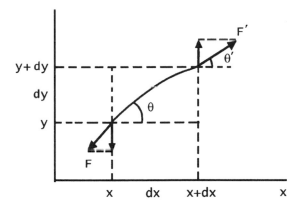

Fig. 4.2. Tensions **F** and **F′** acting on an element of an elastic string.

can approximate sines by tangents and equation (4.8) becomes

$$T(\tan\theta' - \tan\theta) = \rho\,dx\frac{\partial^2 y}{\partial t^2} \tag{4.9}$$

However, $\tan\theta$ and $\tan\theta'$ are the slopes of the curve formed by the string at the two points, and, therefore, for point x, $\tan\theta = \partial y/\partial x$. For the point $x + dx$, the slope has changed and it is given by

$$\tan\theta' = \frac{\partial y}{\partial x} + \frac{\partial}{\partial x}\left(\frac{\partial y}{\partial x}\right)dx \tag{4.10}$$

By substitution into (4.9), we obtain

$$\frac{\partial^2 y}{\partial x^2} = \frac{\rho}{T}\frac{\partial^2 y}{\partial t^2} \tag{4.11}$$

This equation has the same form as the wave equation (3.11) for a displacement y that propagates in the x direction. To solve equation (4.11), we apply the method of separation of variables, as in section 3.2. On making the substitutions $y(x,t) = u(x)v(t)$ and

$a^2 = T/\rho$, we obtain the following equations:

$$\frac{d^2u}{dx^2} + \frac{\omega^2}{a^2}u = 0 \tag{4.12}$$

$$\frac{d^2v}{dt^2} + \omega^2 v = 0 \tag{4.13}$$

where $-\omega^2$ is the constant of separation of variables. The solution of (4.11) is the product of the solutions of (4.12) and (4.13) and can be written as

$$y(x, t) = \left[A \cos\left(\frac{\omega x}{a}\right) + B \sin\left(\frac{\omega x}{a}\right) \right] [C \cos(\omega t) + D \sin(\omega t)] \tag{4.14}$$

This equation is equivalent to (4.2), where ω is the frequency of transverse displacements of the string. The boundary conditions at both ends of the string are that they must remain fixed for all times, that is, $y(0, t) = 0$ and $y(L, t) = 0$. From the first condition results that $A = 0$ and from the second

$$y(L, t) = 0$$

which implies that

$$\sin\left(\frac{\omega}{a}L\right) = 0$$

In consequence, the argument of the sine function must have the values

$$\omega L/a = (n + 1)\pi, \qquad n = 0, 1, 2, 3, \ldots \tag{4.15}$$

where we have not considered the value of the argument equal to zero, because for that value $y(x, t) = 0$ for all values of x. Then, the frequency can only have certain values, namely,

$$\omega_n = a\pi(n + 1)/L, \qquad n = 0, 1, 2, 3, \ldots \tag{4.16}$$

Since for each value of ω_n there exists a solution of (4.11), there is an infinite number of solutions that, according to (4.14), can be written as

$$y_n(x, t) = B_n \sin\left(\frac{(n + 1)\pi a x}{L}\right) [C_n \cos(\omega_n t) + D_n \sin(\omega_n t)] \tag{4.17}$$

If we add the condition that, for $t = 0$, the string is at rest, $(\partial y/\partial t = 0)$, and has an initial configuration $y(x, 0) = f(x)$, then equation (4.17) can be written as

$$y_n(x, t) = F_n \sin(k_n x) \cos(\omega_n t) \tag{4.18}$$

where $F_n = f(x)/\sin[\pi(n + 1)x/L]$ and $k_n = (n + 1)\pi/L$, where ω_n are the frequencies of vibration and k_n are the corresponding wave numbers. Equation (4.18) is analogous to (4.7) in the problem of standing waves with null values at $x = 0$ and $x = L$. The general solution is given by the sum of solution (4.17) or (4.18) for all values of n. Therefore, the final solution of the problem of the vibrating string is a sum of an infinite number of functions $y_n(x, t)$, each representing a harmonic motion of a different frequency ω_n, that is, a mode of vibration:

$$y(x, t) = \sum_{n=0}^{\infty} F_n \sin(k_n x) \cos(\omega_n t) \qquad (4.19)$$

The mode corresponding to the lowest frequency ($n = 0$) is the fundamental mode,

$$\omega_0 = \frac{\pi}{L} \left(\frac{T}{\rho} \right)^{1/2} \qquad (4.20)$$

and its period is

$$T_0 = 2L \left(\frac{\rho}{T} \right)^{1/2} \qquad (4.21)$$

The other values of n correspond to the higher modes, harmonics, or overtones whose frequencies are multiplied by $n + 1$ and whose periods are divided by $n + 1$.

Regarding the form that the string takes in its vibrations, it is easy to prove that, for the fundamental mode, there is no value of x in the interval $(0, L)$ for which $y(x, t)$ is zero; that is, there are no nodes of motion. For each higher mode there is a number of nodes equal to the order number of the mode (Fig. 4.3). The x coordinate of the position of the nodes for each mode of order n is given by

$$x_n^m = \frac{mL}{n + 1}; \qquad m = 1, 2, 3, \ldots, n \qquad (4.22)$$

In conclusion, the problem of the vibration of an elastic string results in a solution with an infinite number of modes of vibration (4.18). The solutions $y_n(x, t)$ corresponding to each mode of vibration are the eigenfunctions of the system and the frequencies ω_n are the eigenvalues (eigenfrequencies). The complete solution of the problem is given by the sum of all modes (4.19).

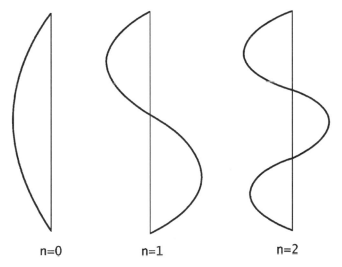

n=0   n=1   n=2

Fig. 4.3. The shape of a vibrating elastic string for the fundamental ($n = 0$) and first two higher modes ($n = 1, 2$).

4.3   Vibrations of an elastic rod

Another example of motion in a finite body is the vibration of an elastic rod of finite length. This problem introduces us, in a simple way, to the vibrations of an elastic medium of finite dimensions that can be applied to the Earth. Let us consider a cylindrical rod of radius a and length L ($L \gg a$) and use cylindrical coordinates (r, z, ϕ) (Appendix 2) (Fig. 4.4).

4.3.1   *Longitudinal vibrations*

If we apply to the elastic rod a force in the direction of its axis (z), it will start to vibrate longitudinally along its axis and will continue vibrating after the force has ceased to act. Let us consider these vibrations after the force has been removed. If the material is isotropic, the relation between the stress and the derivatives of displacement (2.18), assuming that there are displacements only along the axis ($u_r = u_\phi = 0$ and $u_z(z, t)$), is

$$\tau_{zz} = (\lambda + 2\mu)\frac{\partial u_z}{\partial z} \tag{4.23}$$

The equation of motion in the absence of forces (2.66) is

$$\frac{\partial^2 u_z}{\partial z^2} = \frac{1}{\alpha^2}\frac{\partial^2 u_z}{\partial t^2} \tag{4.24}$$

where $\alpha^2 = (\lambda + 2\mu)/\rho$ (2.64) is the velocity of longitudinal waves (P waves). The solution of this equation in the form of (4.3) may be written in two ways:

$$u_z = A\sin(kz)\cos(\omega t + \varepsilon) \tag{4.25}$$

$$u_z = A\cos(kz)\sin(\omega t + \varepsilon) \tag{4.26}$$

First, we will consider that both ends of the rod are fixed to a rigid material, in such a way that displacements at them are null ($u_z(0, t) = u_z(L, t) = 0$) (Fig. 4.5(a)). In this case, we use the solution (4.25), since it satisfies $u_z(0, t) = 0$. For the other end, the condition $u(L, t) = 0$ gives

$$A\sin(kL) = 0; \qquad kL = (n + 1)\pi, \qquad n = 0, 1, 2, 3, \ldots \tag{4.27}$$

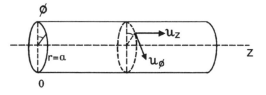

Fig. 4.4. A vibrating elastic rod: cylindrical coordinates and components of the displacements u_z and u_ϕ are shown.

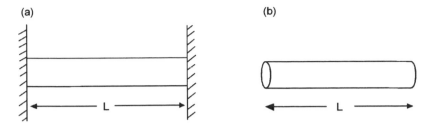

Fig. 4.5. A vibrating elastic rod: (a) with both ends fixed to a rigid medium, and (b) with both ends free.

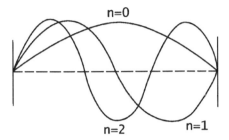

Fig. 4.6. Longitudinal vibrations of an elastic rod with both ends fixed, for the fundamental mode and the first two higher modes.

In consequence, as in the case of the vibrating string (4.18), the solution is given in terms of modes:

$$u_z^n(z, t) = A_n \sin\left(\frac{(n+1)\pi z}{L}\right) \cos(\omega_n t + \varepsilon_n); \qquad n = 0, 1, 2, \ldots \tag{4.28}$$

where, just like in (4.6),

$$\omega_n = \frac{(n+1)\pi \alpha}{L}; \qquad n = 0, 1, 2, \ldots \tag{4.29}$$

For each value of n, the solution (4.28) represents a mode of vibration. In this case, because of the type of motion, they are called longitudinal modes. The total motion is given by the sum of all modes. The lowest order mode, corresponding to $n = 0$, is the fundamental mode, whose frequency is $\omega_0 = \pi \alpha/L$; its wave length is $\lambda_0 = 2L$ and its period is $T_0 = 2L/\alpha$ (the time that it takes a longitudinal wave to travel along the rod in both directions). Frequencies of higher modes are multiples and periods of higher modes are fractions of those of the fundamental mode. Since we have not considered the applied force, this solution corresponds to free longitudinal vibrations of the elastic rod.

For the fundamental mode there is no value of z along the rod where u_z is null. For each higher mode of order n, as in the case of the string, there exist n values of z, between 0 and L, where u_z is zero. These are given by the same equation (4.22) (Fig. 4.6).

Another possibility is that the two ends of the rod are free (Fig. 4.5(b)). The boundary conditions for free surfaces are that stresses through them (tractions) are null. In this case, we consider only normal components of the stress ($\tau_{zz}(0) = \tau_{zz}(L) = 0$). Now we

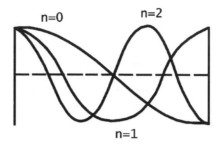

Fig. 4.7. Longitudinal vibrations of an elastic rod with both ends free, for the fundamental mode and the first two higher modes.

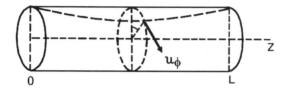

Fig. 4.8. Torsional vibrations of an elastic rod.

take solution (4.26) and normal stress (if $e_{rr} = e_{\phi\phi} = 0$), according to (2.18) and (2.64), is given by

$$\tau_{zz} = -\rho\alpha^2 Ak \cos(kz) \sin(\omega t + \varepsilon) \tag{4.30}$$

The condition of null stress at $z = 0$ and $z = L$ leads to the same values of the wave numbers k_n and frequencies ω_n as in the previous case, (4.27) and (4.29).

According to (4.26), for all modes, the amplitude of displacements at each end of the rod is always a maximum (for the fundamental mode, $n = 0$, $u_z(0) = A$ and $u_z(L) = -A$). The relation for the position of nodes along the rod's axis for different modes is now given (Fig. 4.7) by

$$z_n^m = \frac{(2m+1)L}{2(n+1)}; \qquad m = 0, 1, 2, 3, \ldots, n \tag{4.31}$$

For the fundamental mode there exists a node at $z = L/2$. The number of nodes for each mode of order n is $n + 1$.

4.3.2   *Torsional vibrations*

Another way of generating vibrations in an elastic rod is to apply a torsional moment at a tangent to its surface and normal to its axis. The result is a torsion of the rod with deformations in the direction of the angle ϕ, corresponding to displacements u_ϕ (Fig. 4.8). If u_ϕ varies only with z along the rod, the equation of motion is given by

$$\frac{\partial^2 u_\phi}{\partial z^2} = \frac{1}{\beta^2} \frac{\partial^2 u_\phi}{\partial t^2} \tag{4.32}$$

where, just like in (2.65), $\beta^2 = \mu/\rho$ is the velocity of transverse waves (S waves). If the two ends of the rod are fixed to a rigid body so that its displacements are null $(u_\phi(0, t) = u_\phi(L, t) = 0)$, then the solution of equation (4.32), after imposing the boundary conditions, as in the previous case, is given by

$$u_\phi^n(z, t) = A_n \sin\left(\frac{(n+1)\pi z}{L}\right) \cos(\omega_n t + \varepsilon_n); \qquad n = 0, 1, 2, 3, \ldots \tag{4.33}$$

where

$$\omega_n = \frac{(n+1)\pi\beta}{L}; \qquad n = 0, 1, 2, 3, \ldots \tag{4.34}$$

For each value of n, there exists a mode of vibration with frequency ω_n. The value $n = 0$ corresponds to the fundamental mode and the others correspond to higher modes. Since the elastic deformation is of torsional character, these modes are called torsional modes.

In conclusion, free vibrations of an isotropic elastic rod of finite length, with the simplifying assumptions that we have used, are formed by modes of two types, longitudinal and torsional ones, which are related to longitudinal (P) and transversal (S) waves. Just like the infinite medium, a finite isotropic elastic body has different responses to longitudinal and torsional stresses. The lowest frequency corresponds to the fundamental mode and the frequencies for higher modes are multiples of it. The total motion is given by the sum of all modes.

4.4   The general problem. The Sturm–Liouville equation

The general problem of the vibration of an elastic body of finite dimensions with arbitrary shape and boundary conditions does not have such simple solutions as those we have seen in this chapter for the string and rod. The solutions depend on the geometry of the shape of the body and the boundary conditions on its external surface. In many cases, however, the problem can be reduced to the Sturm–Liouville equation. This differential equation for a function $f(z)$ has the general form

$$\frac{\mathrm{d}}{\mathrm{d}z}\left(p(z)\frac{\mathrm{d}f(z)}{\mathrm{d}z}\right) + [q(z) + ns(z)]f(z) = 0 \tag{4.35}$$

where $p(z)$, $q(z)$, and $s(z)$ are algebraic functions with finite numbers of zeros and poles. The solutions of $f(z)$ for specific $p(z)$, $q(z)$, and $s(z)$ depend on the values of the parameter n and on the boundary conditions imposed on the problem. The most general properties of this problem are the following. For each value of n there is a solution $f_n(z)$, which is called an eigenfunction, and the general solution is the sum of these functions:

$$f(z) = \sum_{n}^{\infty} f_n(z) \tag{4.36}$$

There is a minimum value of n, but not a maximum value. As n increases, zeros of $f_n(z)$ correspond to values of z that are more similar to each other. When, for the same value of n, there are several different eigenfunctions $f_n(z)$, this is known as a degenerate case. The eigenfunctions $f_n(z)$ for different values of n form a complete set of orthogonal functions. If the functions are normalized, it follows that

$$\int_{-\infty}^{\infty} f_n(z) f_m(z)\, \mathrm{d}z = \begin{cases} 0, & m \neq n \\ 1, & m = n \end{cases} \tag{4.37}$$

In the cases that we have considered for the elastic string and rod, equations (4.11), (4.24), and (4.32), applying the separation of variables as in (4.12) or assuming a harmonic dependence on time results in equations like

$$c^2 \frac{\mathrm{d}^2 f}{\mathrm{d}z^2} + \omega^2 f = 0 \tag{4.38}$$

This is the simplest form of the Sturm–Liouville equation, in which $p(z) = c^2$, $q(z) = 0$, and $ns(z) = \omega^2$. The solutions f_n resulting from the application of boundary conditions are the eigenfunctions of the problem and the frequencies ω_n are its eigenvalues. As we have seen, the eigenvalues are real and the eigenfunctions, which are harmonic functions, are orthogonal and form a complete set.

The application of this approach to free oscillations or vibrations of the Earth will be handled in Chapter 13. The first approximation is that of a homogeneous elastic sphere with the same radius, density, and elastic properties as the average values for the Earth. Further problems will introduce variations of density and elastic properties with the radius, more heterogeneous conditions, a lack of sphericity, and anelastic properties. From this point of view, elastic displacements generated by earthquakes can be obtained as the sum of normal modes or eigenvibrations. This is an alternative to the problem of wave propagation.

5 REFLECTION AND REFRACTION

5.1 Snell's law

In Chapter 3 we considered the propagation of elastic waves in homogeneous media. We know that materials in the Earth are not homogeneous, rather their elastic properties vary with depth and from one region to another. This variation may be gradual, but there are also discontinuities that separate media with different densities and elastic coefficients. Let us consider now the phenomena that takes place when waves propagate from one medium to another with different properties. When waves fall upon the surface separating the two media, part of the energy is reflected back into the first medium and part is transmitted or refracted into the second medium. Reflection of waves also occurs when there is a free surface. These problems are very important in seismology, since in the Earth there is a free surface and several discontinuities that separate media with different densities and elastic coefficients. In some cases, such as for the crust and mantle, and for the mantle and core, the contrast between materials is large and produces notable phenomena of reflection and refraction of waves. The theory of reflection and refraction of seismic waves was first developed by Zöppritz and Knott. In this chapter, we will consider the problems of reflection and refraction using plane geometry and waves.

Fermat's principle in mechanics and optics states that waves follow a trajectory for which the duration of the journey is stationary and minimum. From this principle, there follow two well-known consequences for the reflection and refraction of waves on a plane surface that separates two media. First, incident, reflected, and refracted rays are in the same plane, normal to the plane of separation of the two media, which is called the plane of incidence. Second, the trajectories of incident, reflected, and refracted rays follow Snell's law. For two media in which the waves' velocities are v and v', if i is the angle between the incident ray in the medium of velocity v and the normal to the plane and i' is that of the refracted ray in the medium of velocity v', this law establishes that

$$\frac{\sin i}{v} = \frac{\sin i'}{v'} = \frac{1}{c} = p \tag{5.1}$$

where p is called the ray parameter and $c = v/\sin i = v'/\sin i'$ is the component of the velocity in the direction parallel to the plane separating the two media. Also c is the apparent velocity of propagation of the points of intersection of wave fronts with the plane of separation. In Fig. 5.1, the point P travels a distance OP with velocity α during the same time as that in which the point Q travels a distance QO with velocity c. For reflection, the angles of incident and reflected rays are obviously the same, for both rays are in the same medium.

63

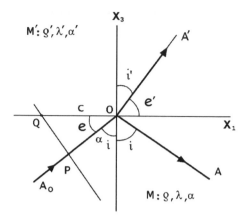

Fig. 5.1. Reflection and refraction of waves in two liquid media (M and M′) for waves incident from medium M.

5.2   Reflection and refraction in two liquid media

We will start with the problem of the reflection and refraction of waves in two liquid media of densities ρ and ρ' and bulk moduli K and K'. In a liquid medium there are no shear waves, only longitudinal or acoustic waves whose displacements are defined by only one potential, ϕ. Let us consider that the two media are separated by the plane (x_1, x_2). The potentials ϕ and ϕ' in each medium are solutions of the wave equation (3.3) with velocities of propagation α and α'. For a liquid $\alpha^2 = \lambda/\rho$ and $\lambda = K$. If we take the plane (x_1, x_3) as the incidence plane (Fig. 5.1) and express the direction cosines in terms of the angle of incidence i and angle of refraction i' ((3.95) and (3.96)), then the potentials in each medium for plane waves of frequency ω are

$$\phi = A_0 \exp[ik_\alpha(x_1 \sin i + x_3 \cos i - \alpha t)]$$
$$+ A \exp[ik_\alpha(x_1 \sin i - x_3 \cos i - \alpha t)] \tag{5.2}$$
$$\phi' = A' \exp[ik_{\alpha'}(x_1 \sin i' + x_3 \cos i' - \alpha' t)] \tag{5.3}$$

where the wave numbers in each medium are

$$k_\alpha = \omega/\alpha \tag{5.4}$$
$$k_{\alpha'} = \omega/\alpha' \tag{5.5}$$

The potential ϕ represents the sum of the incident and reflected waves in the medium M and the potential ϕ' is for a wave refracted or transmitted in medium M′ (Fig. 5.1). The amplitudes of the potentials of the incident, reflected, and refracted waves are A_0, A, and A'. Equations (5.2) and (5.3) can be expressed in terms of the tangent of the angles of emergence ($e = \pi/2 - i$) in the form

$$\phi = A_0 \exp[ik(x_1 + x_3 \tan e - ct)] + A \exp[ik(x_1 - x_3 \tan e - ct)] \tag{5.6}$$
$$\phi' = A' \exp[ik(x_1 + x_3 \tan e' - ct)] \tag{5.7}$$

where, according to Snell's law,

$$k = k_\alpha \cos e = k_{\alpha'} \cos e' \tag{5.8}$$

$$c = \frac{\alpha}{\cos e} = \frac{\alpha'}{\cos e'} \tag{5.9}$$

where k is the wave number associated with the apparent velocity c ($k = \omega/c$). From equation (5.9), we can derive

$$\tan e = (c^2/\alpha^2 - 1)^{1/2} = r \tag{5.10}$$

$$\tan e' = (c^2/\alpha'^2 - 1)^{1/2} = r' \tag{5.11}$$

Equations (5.6) and (5.7) can be expressed using (5.10) and (5.11) in terms of the velocities α, α', and c. When there are reflected and refracted waves, c is larger than α and α', and in consequence r and r' are real. Equations (5.6) and (5.7) for ϕ and ϕ' are of a very convenient form since, for $x_3 = 0$, all exponentials are equal.

The relations between the amplitude A_0 of the potential of the incident wave and A and A' of reflected and refracted waves are obtained by applying boundary conditions at the surface separating the two media ($x_3 = 0$). For two liquids, the boundary conditions are the continuity of the normal component of stress τ_{33} and of the displacement u_3:

$$\tau_{33} = \tau'_{33} \tag{5.12}$$

$$u_3 = u'_3 \tag{5.13}$$

In general, for the potential of a ray (each term of (5.6) and (5.7)), if we express the stress in terms of the strain (2.17), the strain in terms of derivatives of displacements (2.18), taking into account that for a liquid $\mu = 0$, and, finally, displacements as functions of the potential ϕ (2.68), we obtain

$$\tau_{33} = \lambda(\phi_{,11} + \phi_{,33}) = \rho\omega^2\phi \tag{5.14}$$

$$u_3 = \phi_{,3} = ik \tan e\phi \tag{5.15}$$

where we have used the fact that, according to the wave equation, $\lambda\nabla^2\phi = \rho\omega^2\phi$. By substituting (5.14) and (5.15) into (5.12) and (5.13), with the potentials of each ray from equations (5.6) and (5.7), taking into account (5.10) and (5.11), we finally obtain for $x_3 = 0$

$$A_0 + A = \frac{\rho'}{\rho}A' \tag{5.16}$$

$$A_0 - A = \frac{r'}{r}A' \tag{5.17}$$

These two equations can be written in matrix form:

$$A_0 \begin{bmatrix} 1 \\ 1 \end{bmatrix} = \begin{bmatrix} -1 & \rho'/\rho \\ 1 & r'/r \end{bmatrix} \begin{bmatrix} A \\ A' \end{bmatrix} \tag{5.18}$$

The reflection coefficients V and refraction or transmission coefficients W are defined as the quotients of reflected and refracted or transmitted amplitudes with respect to the incident amplitude:

$$V = \frac{A}{A_0} = \frac{\rho' r - \rho r'}{\rho' r + \rho r'} \tag{5.19}$$

$$W = \frac{A'}{A_0} = \frac{2\rho r}{\rho' r + \rho r'} \tag{5.20}$$

If, instead of using equations (5.6) and (5.7), we use (5.2) and (5.3), we obtain

$$V = \frac{\rho' \alpha' \cos i - \rho \alpha \cos i'}{\rho' \alpha' \cos i + \rho \alpha \cos i'} \tag{5.21}$$

$$W = \frac{2\rho \alpha' \cos i}{\rho' \alpha' \cos i + \rho \alpha \cos i'} \tag{5.22}$$

These two equations can also be expressed in terms of the refractive index $n = \alpha/\alpha'$, the contrast $m = \rho'/\rho$ of densities between the two media, and angles of emergence e and e':

$$V = \frac{m \sin e - n \sin e'}{m \sin e + n \sin e'} \tag{5.23}$$

$$W = \frac{2 \sin e}{m \sin e + n \sin e'} \tag{5.24}$$

According to these equations, the relation between W and V is

$$W = \frac{1}{m}(1 + V) \tag{5.25}$$

The mechanical impedance Z is defined as the quotient relating the stress and particle velocity in a particular orientation. In the x_3 direction, for a liquid, according to (5.14) and (5.15), Z is given by

$$Z = \frac{\tau_{33}}{\dot{u}_3} = \frac{\rho \alpha}{\sin e} \tag{5.26}$$

Equations (5.23) and (5.24) can be written in terms of Z as

$$V = \frac{Z' - Z}{Z' + Z} \tag{5.27}$$

$$W = \frac{1}{m}\left(\frac{2Z'}{Z' + Z}\right) \tag{5.28}$$

If waves travel from a medium M' to medium M, the reflection and refraction coefficients V' and W', according to (5.19) and (5.20), are

$$V' = \frac{A'}{A_0'} = \frac{\rho r' - \rho' r}{\rho r' + \rho' r} \tag{5.29}$$

$$W' = \frac{A}{A_0'} = \frac{2r' \rho'}{\rho r' + \rho' r} \tag{5.30}$$

The complete problem with waves traveling from both media in both directions is completely described by the four coefficients V, V', W, and W'. The matrix formed

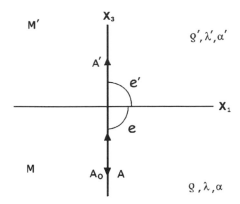

Fig. 5.2. Normal incidence in two liquids.

by these four coefficients is called the dispersion matrix; it is

$$\mathbf{D} = \begin{pmatrix} V & V' \\ W & W' \end{pmatrix}$$

The first column represents the partition of amplitudes between reflected and refracted waves for waves traveling from medium M to medium M' and the second represents the partition for waves traveling from M' to M. It is important to notice that V and W represent the partition of amplitudes of potentials but not the partition of energy between reflected and transmitted waves. For this reason their sum is not unity.

5.2.1 *Normal incidence*

A particular case of reflection and refraction corresponds to normal incidence, that is, when rays travel perpendicular to the surface separating the two media ($i = 0$, or $e = \pi/2$) (Fig. 5.2). The reflection and transmission coefficients are given by

$$V = \frac{\alpha'\rho' - \alpha\rho}{\alpha'\rho' + \alpha\rho} \tag{5.31}$$

$$W = \frac{2\rho\alpha'}{\alpha'\rho' + \alpha\rho} \tag{5.32}$$

We can see from these two equations that, if the contrast between the densities and velocities of the two media is small, V is very small and W is nearly unity. Waves are transmitted into the second medium with little reflection. For the opposite case, if the contrast is large, V is nearly unity and W is small. Waves are reflected back into the same medium with little transmission. The larger the contrast between the properties of the two media the larger the reflections. From equations (4.31) and (4.32) we can easily verify that the sum of V and W is not unity, for they do not represent the partition of energy between the reflected and refracted waves.

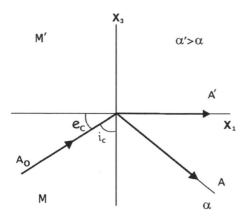

Fig. 5.3. Critical incidence in two liquids (e_c and i_c are the critical angles).

5.2.2   *Critical incidence*

According to Snell's law, if $\alpha' > \alpha$, there is an angle i_c or e_c, called the critical angle of incidence or emergence, which corresponds to rays refracted in the medium of velocity α' with angle $e' = 0$, or $i' = \pi/2$. Then, according to (5.1),

$$\frac{\sin i_c}{\alpha} = \frac{\cos e_c}{\alpha} = \frac{1}{\alpha'} = \frac{1}{c} \tag{5.33}$$

and, therefore,

$$\cos e_c = \sin i_c = \alpha/\alpha' = n \tag{5.34}$$

If we substitute $e' = 0$ into equations (5.23) and (5.24) we obtain $V = 1$ and $W = 2\rho/\rho'$. Waves refracted in medium M' propagate along the surface separating the two media and therefore $c = \alpha'$ (Fig. 5.3). Such a wave is called a critically refracted or head wave. Its potential is given by

$$\phi' = 2A_0\rho/\rho' \exp[ik_{\alpha'}(x_1 - \alpha't)] \tag{5.35}$$

Rays with angles of emergence $e < e_c$ or angles of incidence $i > i_c$ are not refracted into medium M' and their energy is completely reflected back into medium M. This situation is called total reflection. Thus, from the point of view of the angle of incidence, we have subcritical ($i < i_c$) and supercritical ($i > i_c$) reflections. Supercritical reflected waves carry more energy than do subcritical waves since, in the former case, no energy is transmitted to the second medium. For supercritical waves, according to (5.9) and (5.33), $\cos e'$ would have a value larger than unity, which cannot correspond to a real angle, and $c < \alpha'$. Hence, according to (5.11), $\tan e'$ is an imaginary number:

$$\tan e' = i(1 - c^2/\alpha'^2)^{1/2} = i\bar{r}' \tag{5.36}$$

The reflection coefficient V (5.19) is now a complex number with modulus equal to unity:

$$V = \frac{\rho'r - i\rho\bar{r}'}{\rho'r + i\rho\bar{r}'} = e^{-i\theta} \tag{5.37}$$

where

$$\tan\left(\frac{\theta}{2}\right) = \frac{\rho\bar{r}'}{\rho'r} = \frac{\rho(1 - c^2/\alpha'^2)^{1/2}}{\rho'(c^2/\alpha^2 - 1)^{1/2}} \tag{5.38}$$

On putting $A = A_0 V$ and replacing V from (5.37), the expression for the potential of supercritical reflected waves is given by

$$\phi = A_0 \exp[ik(x_1 - rx_3 - ct - \theta/k)] \tag{5.39}$$

Supercritical reflected waves have the same amplitude as incident waves, but they have a phase shift of θ.

5.2.3   *Inhomogeneous waves*

As we have seen, waves with angles of incidence greater than the critical angle ($i > i_c$, or $e < e_c$) are not refracted into the second medium; all of the energy is reflected back. However, we can verify from (5.20) that W is not zero but rather has a complex value, since r' is imaginary. Therefore, there is a potential ϕ' that corresponds to some type of elastic perturbation in the medium M'. This perturbation is formed not by normal transmitted waves but by waves of a special type. If we substitute the imaginary value of r' in expression (5.7) for ϕ', we obtain

$$\phi' = A_0 W \exp[-k\bar{r}'x_3 + ik(x_1 - ct)] \tag{5.40}$$

where W is a complex number and $\bar{r}'$ is real. This potential represents waves that propagate in medium M' in the direction of x_1 with velocity c with values $\alpha < c < \alpha'$ and amplitudes decreasing exponentially with the distance x_3 from the surface separating the two media. This potential corresponds to an elastic perturbation in medium M' that exists only near the surface separating the two media. Owing to the special characteristics of these waves whose amplitudes decrease with the distance x_3 along the same wave front, they are called inhomogeneous or evanescent waves.

Displacements of inhomogeneous waves also have different properties than those of normal transmitted waves. If we derive the displacements from the potential ϕ' in equation (5.40) according to (3.97) and (3.98), keeping only the real part, we obtain

$$u_1' = -A_0|W|k\,e^{-k\bar{r}'x_3} \sin[k(x_1 - ct) - \varepsilon] \tag{5.41}$$

$$u_3' = -A_0|W|k\bar{r}'\,e^{-k\bar{r}'x_3} \cos[k(x_1 - ct) - \varepsilon] \tag{5.42}$$

where the phase angle $\varepsilon = \tan^{-1}[\rho\bar{r}'/(\rho'r)]$.

Components u_1 and u_3 of displacements are shifted in phase by $\pi/2$ and in consequence the particles' motion is elliptical with its major axis in the direction of x_1 ($\bar{r}' < 1$) and of retrograde sense (Fig. 5.4). These are not, therefore, normal longitudinal waves of the type that exist in liquids, but waves of a special type with elliptical motion, similar in some ways to waves generated by wind on the surface of a liquid. Inhomogeneous waves are a phenomenon characteristic of supercritical reflections and we will see that they are related to surface waves.

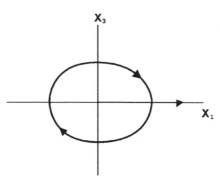

Fig. 5.4. Particle motion for an inhomogeneous wave produced by an incident wave with angle of incidence $e < e_c$.

5.2.4   *Reflected and transmitted energy*

We have already mentioned that coefficients of reflection V and transmission W do not represent partition of energy and for this reason their sum is not unity. If we want to find the partition of the incident wave energy between reflected and refracted waves we have to consider the energy that arrives at the surface of separation. For plane waves the energy contained in an element of volume $\delta V = \delta s\,\delta x$, where δs is an element of the wave front and δx is the distance traveled in the direction of propagation in an increment of time δt, according to (3.49), is the sum of kinetic and potential energies

$$\delta E = \frac{1}{2}\left[\rho\left(\frac{\partial u}{\partial t}\right)^2 + \alpha^2\rho\left(\frac{\partial u}{\partial x}\right)^2\right]\delta s\,\delta x \tag{5.43}$$

For a unidimensional plane wave in a liquid of velocity α, $u = B\cos(kx - \omega t)$, and $\alpha = \omega/k$,

$$\frac{\partial u}{\partial x} = -\frac{1}{\alpha}\frac{\partial u}{\partial t} \tag{5.44}$$

Since $\delta x = \alpha\,\delta t$, substituting into (5.43) gives

$$\delta E = \rho\left(\frac{\partial u}{\partial t}\right)^2 \alpha\,\delta s\,\delta t \tag{5.45}$$

We define the intensity I as the energy per unit surface of the wave front and per unit time. From (5.45) the intensity is given by

$$I = \rho\alpha\left(\frac{\partial u}{\partial t}\right)^2 \tag{5.46}$$

If we derive the displacements from a potential ϕ,

$$\phi = A\sin(kx - \omega t) \tag{5.47}$$

then, on substituting into (5.46), we can find the intensity in terms of the amplitude of the potential,

$$I = \alpha\rho k^2\omega^2 A^2 = \frac{\rho\omega^4}{\alpha}A^2 \tag{5.48}$$

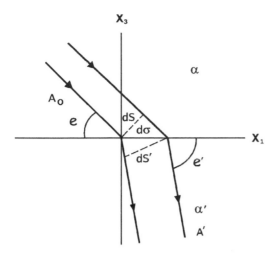

Fig. 5.5. The transmission of energy from a liquid medium to another by a beam of incident rays with element of wave-front area dS. The element of wave-front area of transmitted rays is dS'.

For normal incidence, an element of the wave front is equal to an element of the surface separating the two media and we can write directly the partition of energy on such a surface:

$$I_{inc} = I_{refl} + I_{trans} \tag{5.49}$$

If we substitute values of intensities according to (5.48), knowing that the amplitudes of incident, reflected, and refracted potentials are, respectively, A_0, VA_0, and WA_0, we obtain, after dividing by A_0,

$$1 = V^2 + \frac{\rho'\alpha}{\rho\alpha'} W^2 \tag{5.50}$$

This equation represents the partition of incident energy between reflected and refracted waves for normal incidence. If we substitute expressions for V and W from (5.31) and (5.32), we can easily verify this relation.

In the general case, we have to take into consideration the angle of incidence. Let us consider the energy contained in a ray bundle that occupies an element of the wave front of area ds that strikes the surface separating the two media with an angle of emergence e (Fig. 5.5). The ray bundle of transmitted or refracted waves leaves the surface with an angle e' and occupies an element of the wave front of area ds', while that of reflected waves occupies an element ds equal to that of incident waves. The element of area of the surface separating the two media dσ occupied by the three ray bundles (incident, reflected, and refracted) must be the same. The relations among ds, ds', and dσ are

$$ds = d\sigma \sin e \tag{5.51}$$

$$ds' = d\sigma \sin e' \tag{5.52}$$

The energy incident on the element of area dσ must be divided between the energies of

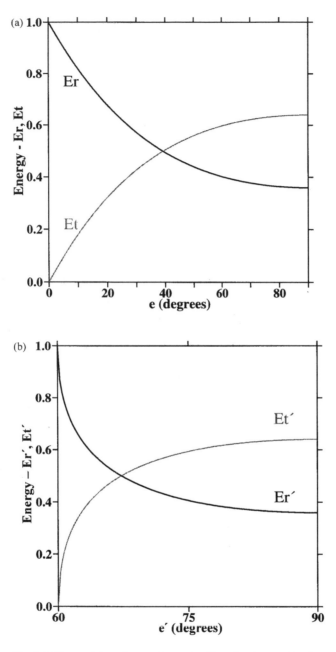

Fig. 5.6. The partition of energy between reflected and transmitted waves in two liquids of velocities $\alpha' = \alpha/2$ and densities $\rho' = \rho/2$ as a function of the angle of emergence e. (a) Waves travel from medium M to medium M'. (b) Waves travel from medium M' to medium M; for $e' < 60°$ all energy is reflected. (c) The phase shift between incident and reflected waves for angles of emergence less than the critical one ($e' < 60°$).

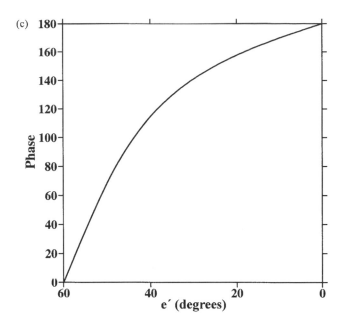

the refracted and reflected waves:

$$I_{inc}\, ds = I_{refl}\, ds + I_{trans}\, ds'$$　　　　　　(5.53)

On substituting equations (5.51) and (5.52) and intensities in terms of the amplitudes of incident, reflected, and refracted potentials (5.48), we finally obtain

$$1 = V^2 + \frac{\rho' \alpha \sin e'}{\rho \alpha' \sin e}\, W^2$$　　　　　　(5.54)

This equation represents the partition of incident energy between reflected and refracted waves for an arbitrary angle of incidence.

Figure 5.6 shows an example of the partition of energy between reflected and refracted waves in two liquids with velocities $\alpha' = \alpha/2$ and densities $\rho' = \rho/2$ as a function of the angle of emergence. In Fig. 5.6(a), waves travel from medium M to medium M'; this represents waves traveling from a medium of larger velocity to one of smaller velocity. For angles $e < 39°$, the reflected energy is larger than the refracted energy, whereas for angles $e > 39°$, the transmitted energy is larger. Figure 5.6(b) shows the case for waves traveling from medium M' to medium M, that is, from a medium of smaller velocity to one of larger velocity. The critical angle is $e'_c = 60°$; for smaller angles waves are not transmitted and all the energy is reflected into medium M'. For angles larger than $e' = 68°$, there is more energy transmitted than there is reflected. Figure 5.6(c) shows the phase shift of the reflected waves for supercritical incidence ($e' < e'_c$).

5.2.5　Reflection on a free surface

The case of the reflection of waves on a free surface (Fig. 5.7) is of special interest. This is a particular case in which the second medium is the vacuum. The

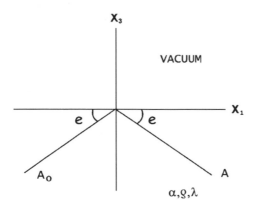

Fig. 5.7. Incident and reflected waves on a free surface of a liquid.

potential for incident and reflected waves is given by equation (5.6):

$$\phi = A_0 \exp[ik(x_1 + rx - ct)] + A \exp[ik(x_1 - rx_3 - ct)] \qquad (5.55)$$

The boundary conditions on a free surface are that normal stresses or tractions are null. In our case, for $x_3 = 0$, $\tau_{33} = 0$. According to equation (5.14), this condition in terms of potentials leads to

$$\lambda(\phi_{,11} + \phi_{,33}) = 0 \qquad (5.56)$$

By substitution of ϕ from (5.55), we obtain

$$A = -A_0 \qquad (5.57)$$

The potential of reflected waves has the same amplitude as that of the incident wave but with the opposite sign; that is, there is a phase shift of π. The total displacement at the free surface is given by the sum of those of the incident and reflected waves, $U = u_{\text{inc}} + u_{\text{ref}}$. Using the potential ϕ from (5.55), taking derivatives according to (3.97) and (3.98) and using (5.57), we obtain for the real part that $U_1 = 0$ and U_3 is given by

$$U_3 = -2rkA_0 \sin[k(x_1 - ct)] \qquad (5.58)$$

As we expected, the motion of the free surface of a liquid produced by the arrival of acoustic waves has only a vertical component and its amplitude depends on the angle of incidence of the waves.

5.3   Reflection and refraction in elastic media

Since in elastic media there are two types of waves, P and S, the problem of the reflection and refraction of waves is more complicated than that for two liquids. For incident P and S waves there are, generally, reflected and refracted P and S waves. This is due to Huyghens' principle that states that, when waves reach the surface separating two media, its points become sources of waves that, in an elastic medium, are of both types. This problem was first solved by Knott in 1899 and Zöppritz in 1919. We take the plane (x_1, x_2) as that separating the two media and the plane (x_1, x_3) as the plane of incidence. The components of elastic displacements u_1 and u_3 can be derived from the scalar potentials ϕ and ψ, according to equations (3.97) and (3.98), and the component u_2 is kept apart. This way the potential ϕ represents P waves, whereas the potential ψ represents the SV and u_2 represents the SH component of S waves (section 3.7).

If e and f are the angles of emergence of P and S incident and reflected waves in medium M of velocities α and β and e' and f' are those of waves refracted or transmitted in medium M$'$ with velocities α' and β', then Snell's law (5.1) is given by

$$\frac{\cos e}{\alpha} = \frac{\cos f}{\beta} = \frac{\cos e'}{\alpha'} = \frac{\cos f'}{\beta'} = \frac{1}{c} = p \tag{5.59}$$

where p and c have already been defined. As functions of the velocities, the tangents of e and f are

$$\tan e = (c^2/\alpha^2 - 1)^{1/2} = r \tag{5.60}$$

$$\tan f = (c^2/\beta^2 - 1)^{1/2} = s \tag{5.61}$$

and similarly for $\tan e' = r'$ and $\tan f' = s'$. The boundary conditions at the surface separating two elastic bodies $(x_3 = 0)$ are continuity of the three components of the displacement u_i and continuity of the three components of stresses τ_{3i} across the surface (x_1, x_2):

$$u_i = u_i' \tag{5.62}$$

$$\tau_{3i} = \tau_{3i}' \tag{5.63}$$

These conditions imply that the two media are welded together. We will treat separately the cases for incident SH (S waves with an SH component only), P, and SV waves.

5.3.1 Incident SH waves

Incident SH waves, owing to the selection of the orientation of the reference coordinate axes, have only the displacement component u_2. It is easy to verify that, for only SH incident waves, the boundary conditions for u_1, u_3, τ_{31}, and τ_{33} result in homogeneous equations for the amplitudes of the reflected and refracted potentials ϕ, ψ, ϕ', and ψ' that admit only a zero solution. This means that, if the incident S wave has only an SH component, then there are only reflected and refracted SH waves (Fig. 5.8). The problem is reduced to that of the displacement component u_2, which is given in both media by

$$u_2 = C_0 \exp[ik(x_1 + sx_3 - ct)] + C \exp[ik(x_1 - sx_3 - ct)] \tag{5.64}$$

$$u_2' = C' \exp[ik(x_1 + s'x_3 - ct)] \tag{5.65}$$

where C_0, C, and C' are the amplitudes of the incident, reflected, and refracted waves.

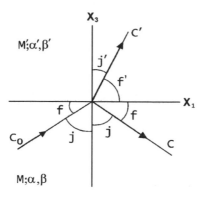

Fig. 5.8. Reflection and refraction in two elastic media for incident SH waves.

The boundary conditions at the surface separating the two media $(x_3 = 0)$ are

$$u_2 = u_2'$$ (5.66)

$$\tau_{32} = \tau_{32}'$$ (5.67)

By expressing the stress τ_{32} in terms of the strain e_{32} (2.17), and this in terms of the derivatives of the displacement u_2 (2.18), and substituting the expressions (5.64) and (5.65) into the boundary conditions (5.66) and (5.67), we obtain the equations for the amplitudes in matrix form (just like in (5.18)):

$$C_0 \begin{pmatrix} 1 \\ 1 \end{pmatrix} = \begin{pmatrix} -1 & 1 \\ 1 & \dfrac{s'\mu'}{s\mu} \end{pmatrix} \begin{pmatrix} C \\ C' \end{pmatrix}$$ (5.68)

On solving for the amplitudes' quotients C/C_0 and C'/C_0 we obtain

$$V_{\text{SH}} = \frac{C}{C_0} = \frac{\mu s - \mu' s'}{\mu s + \mu' s'}$$ (5.69)

$$W_{\text{SH}} = \frac{C'}{C_0} = \frac{2\mu s}{\mu s + \mu' s'}$$ (5.70)

where V_{SH} is the reflection coefficient and W_{SH} is the refraction or transmission coefficient for SH waves. In this case, the coefficients are quotients of displacement amplitudes. Expressions (5.69) and (5.70) are similar to (5.19) and (5.20) obtained for the two liquids. Here, also, the reflected energy is larger for a large contrast between the shear moduli μ and μ' of the two media.

If SH waves travel from medium M' into medium M, the reflection and transmission coefficients are

$$V_{\text{SH}}' = \frac{C'}{C_0'} = \frac{\mu' s' - \mu s}{\mu' s' + \mu s}$$ (5.71)

$$W_{\text{SH}}' = \frac{C}{C_0'} = \frac{2\mu' s'}{\mu' s' + \mu s}$$ (5.72)

The matrix for the total dispersion of SH motion between two elastic media is given by

$$\begin{pmatrix} V_{\mathrm{SH}} & V'_{\mathrm{SH}} \\ W_{\mathrm{SH}} & W'_{\mathrm{SH}} \end{pmatrix}$$

In a similar way to what we did with the case of two liquids, we can find the partition of energy between the reflected and refracted waves for an arbitrary angle of emergence f:

$$1 = V_{\mathrm{SH}}^2 + \frac{\rho' \beta \sin f'}{\rho \beta' \sin f} W_{\mathrm{SH}}^2 \tag{5.73}$$

As in the case of two liquids, this equation shows the partition of the energy of the incident wave between the energy that is reflected back into the same medium and that which is transmitted into the second medium.

5.3.2 Critical incidence and inhomogeneous waves

When for the two media $\beta' > \beta$, then, for waves that travel from medium M to M', there is a critical value for the angle of incidence $f_c = \cos^{-1}(\beta/\beta')$, for which waves are refracted into medium M' with the angle $f' = 0$ and propagate along the surface separating the two media. For angles of incidence $f < f_c$, no waves are transmitted into the second medium and all the energy is reflected back into medium M (total reflection). For those values of f, s' is imaginary, since $c < \beta'$ and V_{SH} is a complex quantity of modulus unity. Reflected supercritical waves present a phase shift θ with respect to incident waves:

$$\tan\left(\frac{\theta}{2}\right) = \frac{\mu' \bar{s}'}{\mu s} \tag{5.74}$$

where

$$\bar{s}' = (1 - c^2/\beta'^2)^{1/2} \tag{5.75}$$

For supercritical reflections ($f < f_c$) we find also the phenomenon of the generation of inhomogeneous waves in medium M', whose displacements are given by

$$u'_2 = C_0 W_{\mathrm{SH}} \exp[-k\bar{s}' x_3 + ik(x_1 - ct)] \tag{5.76}$$

These waves have the same characteristics as SH waves with only a u_2 component of displacement; they propagate in the x_1 direction with a velocity c ($\beta < c < \beta'$) that depends on f and their amplitude on the same wave front decreases exponentially with the distance (x_3) from the surface separating the two media.

5.3.3 Incident P and SV waves

For incident P waves, we have both reflected and refracted P and S waves (Fig. 5.9). Incident P waves that travel from medium M are represented by the potential ϕ with the amplitude A_0. Reflected P and S waves represented by potentials ϕ (P) and ψ (SV) with amplitudes A and B, and a displacement component u_2 (SH) of

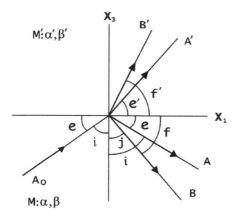

Fig. 5.9. Reflection and refraction in two elastic media for incident P waves. Reflected and refracted P and SV waves.

amplitude C. Waves transmitted in medium M' are represented by potentials ϕ' (P) and ψ' (SV) with amplitudes A' and B' and a displacement component u'_2 (SH) with amplitude C'.

The relations between the amplitudes of incident P waves and those of reflected and refracted P and S waves are found by applying the boundary conditions of the continuity of stress and displacement at the surface separating the two media ($x_3 = 0$). From the conditions $u_2 = u'_2$ and $\tau_{32} = \tau'_{32}$ it is easily found that C and C' are zero. This means that, for an incident P wave, reflected and refracted S waves have only SV components. The other four boundary conditions are

$$u_1 = u'_1 \tag{5.77}$$

$$u_3 = u'_3 \tag{5.78}$$

$$\tau_{31} = \tau'_{31} \tag{5.79}$$

$$\tau_{33} = \tau'_{33} \tag{5.80}$$

The potentials for P and SV motion in both media corresponding to incident, reflected, and refracted waves are

$$\phi = A_0 \exp[ik(x_1 + rx_3 - ct)] + A \exp[ik(x_1 - rx_3 - ct)] \tag{5.81}$$

$$\psi = B_0 \exp[ik(x_1 + sx_3 - ct)] + B \exp[ik(x_1 - rx_3 - ct)] \tag{5.82}$$

$$\phi' = A' \exp[ik(x_1 + r'x_3 - ct)] \tag{5.83}$$

$$\psi' = B' \exp[ik(x_1 + s'x_3 - ct)] \tag{5.84}$$

On substituting in equations (5.77)–(5.80), and taking into account the relations between the stress and strain (2.17), strain and derivatives of displacements (2.18), and displacements and potentials (3.97) and (3.98), we obtain for the potentials the following equations:

$$\phi_{,1} - \psi_{,3} = \phi'_{,1} - \psi'_{,3} \tag{5.85}$$

$$\phi_{,3} + \psi_{,1} = \phi'_{,3} + \psi'_{,1} \tag{5.86}$$

$$\mu(2\phi_{,13} + \psi_{,33} + \psi_{,11}) = \mu'(2\phi'_{,13} - \psi'_{,33} + \psi_{,11}) \tag{5.87}$$

$$\lambda(\phi_{,33} + \phi_{,11}) + 2\mu\psi_{,13} = \lambda'(\phi'_{,33} + \phi'_{,11}) + 2\mu'\psi'_{,13} \tag{5.88}$$

By substitution into equations (5.85)–(5.88) of the expressions for the potentials (5.81)–(5.84) and writing the resulting equations in matrix form, we obtain

$$A_0 \begin{pmatrix} 1 \\ r \\ 2r\mu \\ \lambda(1+r^2) \end{pmatrix} + B_0 \begin{pmatrix} -s \\ 1 \\ 1+s^2 \\ 2\mu s \end{pmatrix}$$

$$= \begin{pmatrix} -1 & -s & 1 & -s' \\ -r & -1 & -r' & 1 \\ -2r\mu & -(1-s^2) & 2\mu'r' & \mu'(1-s'^2) \\ -\lambda(1+r^2) & -2\mu s & \lambda'(1+r^2) & 2\mu's' \end{pmatrix} \begin{pmatrix} A \\ B \\ A' \\ B' \end{pmatrix} \tag{5.89}$$

For an incident P wave, $B_0 = 0$, and for an incident SV wave, $A_0 = 0$. The resulting expressions are known as Zöppritz's equations. For incident P waves, by solving the resulting equations from (5.89) with $B_0 = 0$, we determine the reflection and transmission coefficients. Now we have four coefficients corresponding to reflected and transmitted P and SV waves:

$$V_{PP} = \frac{A}{A_0}; \qquad V_{PS} = \frac{B}{A_0}; \qquad W_{PP} = \frac{A'}{A_0}; \qquad W_{PS} = \frac{B'}{A_0}$$

Reflected and transmitted SV waves from an incident P wave are often called converted waves.

For incident SV waves, the problem is similar. The reflection and transmission coefficients are derived from equation (5.89) by putting $A_0 = 0$. We have four coefficients and the converted reflected and transmitted waves are P waves:

$$V_{SS} = \frac{B}{B_0}; \qquad V_{SP} = \frac{A}{B_0}; \qquad W_{SS} = \frac{B'}{B_0}; \qquad W_{SP} = \frac{A'}{B_0}$$

Reflection and transmission coefficients for incident P and SV waves in isotropic media are functions of the elastic coefficients λ and μ, the densities ρ of the two media and the angles of incidence e and f. The resulting expressions are simplified if we assume the value of Poisson's ratio $\sigma = 0.25$ ($\lambda = \mu$) (Section 2.2).

The dispersion matrix for the general P–SV motion generated in two elastic media in contact with incident waves traveling in both directions has 16 elements, eight for incident P and SV waves traveling from medium M and another eight for those traveling from medium $\mathbf{M}'$:

$$
\begin{pmatrix}
V_{PP} & V_{SP} & V'_{PP} & V'_{SP} \\
V_{PS} & V_{SS} & V'_{PS} & V'_{SS} \\
W_{PP} & W_{SP} & W'_{PP} & W'_{SP} \\
W_{PS} & W_{SS} & W'_{PS} & W'_{SS}
\end{pmatrix}
$$

The partition of incident energy between the reflected and transmitted waves, P and SV, is deduced in a similar way to that which was presented for two liquids (5.54) and for the case of two solids for SH waves (5.73). For incident P waves we obtain

$$
V_{PP}^2 + \frac{\alpha \sin f}{\beta \sin e} V_{PS}^2 + \frac{\rho' \alpha \sin e'}{\rho \alpha' \sin e} W_{PP}^2 + \frac{\rho' \alpha \sin f'}{\rho \beta' \sin e} W_{PS}^2 = 1 \tag{5.90}
$$

For incident SV waves,

$$
V_{SS}^2 + \frac{\beta \sin e}{\alpha \sin f} V_{SP}^2 + \frac{\rho' \beta \sin f'}{\rho \beta' \sin f} W_{SS}^2 + \frac{\rho' \beta \sin e'}{\rho \alpha' \sin f} W_{SP}^2 = 1 \tag{5.91}
$$

In both cases, the energy of an incident P or SV wave is divided into that of reflected and transmitted P and SV waves.

Just like incident SH waves, incident P and SV waves traveling from a medium of lesser velocity to one of greater velocity ($\alpha > \alpha'$) present the phenomenon of total reflection with critical angles given by $\cos e_c = \alpha/\alpha'$ and $\cos f_c = \beta/\beta'$. For supercritical angles we have also the generation of inhomogeneous waves. Since $\beta < \alpha$, there is also a critical angle for incident SV waves for which there is no reflected P wave: $\cos \bar{f}_c = \beta/\alpha$.

5.4   Reflection on a free surface

The problem of the reflection of elastic waves on a free surface is of particular interest in seismology since it represents the situation of the arrival of seismic waves at the Earth's surface. The boundary conditions on a free surface, as we saw for a liquid, are that the components of stress across the surface are zero. We will treat first incident SH waves and then incident P and SV waves.

5.4.1   *Incident SH waves*

Let us consider an elastic half-space bounded by the free surface formed by the plane (x_1, x_2). It is easy to show that, for incident SH waves, there are no reflected P or SV waves, but only reflected SH waves. The displacement u_2 of incident and reflected SH waves is given (Fig. 5.10) by

$$
u_2 = C_0 \exp[ik(x_1 + sx_3 - ct)] + C \exp[ik(x_1 - sx_3 - ct)] \tag{5.92}
$$

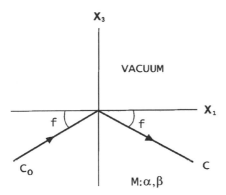

Fig. 5.10. SH waves incident on and reflected from a free surface of an elastic medium.

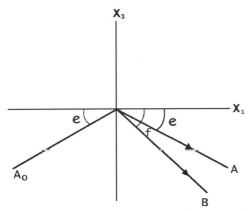

Fig. 5.11. Incident P and reflected P and SV waves for a free surface of an elastic medium.

where C_0 is the amplitude of incident waves and C is that of reflected waves. On applying the boundary condition for $x_3 = 0$, we obtain

$$\tau_{32} = \mu u_{2,3} = 0 \tag{5.93}$$

On substituting (5.92) we obtain $C = C_0$ and that the reflection coefficient $V_{\text{SH}} = 1$. The reflected SH wave has the same amplitude as the incident wave.

5.4.2   Incident P waves

For an incident P wave there are reflected P and SV waves (Fig. 5.11). The non-existence of reflected SH waves is also easily shown from the boundary condition $\tau_{32} = 0$. The potentials of incident P and reflected P and SV waves are

$$\phi = A_0 \exp[ik(x_1 + rx_3 - ct)] + A \exp[ik(x_1 - rx_3 - ct)] \tag{5.94}$$

$$\psi = B \exp[ik(x_1 - sx_3 - ct)] \tag{5.95}$$

The two pertinent boundary conditions at $x_3 = 0$ are that the stress components τ_{31} and τ_{33} are null. Writing the stress in terms of the strain for an isotropic material (2.17), the

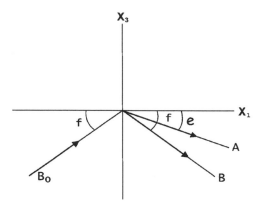

Fig. 5.12. Incident SV and reflected P and SV waves for a free surface of an elastic medium.

strain in terms of the derivatives of displacements (2.18), and, finally, displacements in terms of the potentials (3.97) and (3.98), we obtain

$$\tau_{31} = \mu(2\phi_{,13} - \psi_{,33} + \psi_{,11}) \tag{5.96}$$

$$\tau_{33} = \lambda\phi_{,11} + (\lambda + 2\mu)\psi_{,33} + \psi_{,13} \tag{5.97}$$

For Poisson's relation ($\lambda = \mu$), by substituting the potentials (5.94) and (5.95) into equations (5.96) and (5.97), we obtain

$$(3r^2 + 1)(A_0 + A) - 2sB = 0 \tag{5.98}$$

$$2r(A_0 - A) + B(1 - s^2) = 0 \tag{5.99}$$

The coefficients of reflection are

$$V_{PP} = \frac{A}{A_0} = \frac{4rs - (1 + 3r^2)^2}{4rs + (1 + 3r^2)^2} \tag{5.100}$$

$$V_{PS} = \frac{B}{A_0} = \frac{4r(1 + 3r^2)}{4rs + (1 + 3r^2)^2} \tag{5.101}$$

These two coefficients represent the ratios of the amplitudes of the potentials of reflected P and SV waves with respect to that of incident P waves.

5.4.3   *Incident SV waves*

For incident SV waves, the problem is very similar. Again there are reflected SV and P waves, but no reflected SH waves (Fig. 5.12). The potentials of the incident and reflected waves are

$$\psi = B_0 \exp[ik(x_1 + sx_3 - ct)] + B \exp[ik(x_1 - sx_3 - ct)] \tag{5.102}$$

$$\phi = A \exp[ik(x_1 - rx_3 - ct)] \tag{5.103}$$

By applying the boundary conditions as in the previous case, (5.96) and (5.97), and with the condition $\lambda = \mu$, we obtain

$$(1 - s^2)(B_0 + B) - 2rA = 0 \tag{5.104}$$

$$2s(B_0 - B) + (1 + 3r^2)A = 0 \tag{5.105}$$

The reflection coefficients are

$$V_{SS} = \frac{B}{B_0} = \frac{4rs - (1 + 3r^2)^2}{4rs + (1 + 3r^2)^2} \tag{5.106}$$

$$V_{SP} = \frac{A}{B_0} = \frac{-4s(1 + 3r^2)}{4rs + (1 + 3r^2)^2} \tag{5.107}$$

Equations (5.98), (5.99), (5.104), and (5.105) can be expressed in matrix form by

$$\begin{pmatrix} 1 + 3r^2 \\ 2r \end{pmatrix} A_0 + \begin{pmatrix} 2s \\ 1 - s^2 \end{pmatrix} B_0 = \begin{pmatrix} -(1 + 3r^2) & 2s \\ 2r & -(1 - s^2) \end{pmatrix} \begin{pmatrix} A \\ B \end{pmatrix} \tag{5.108}$$

From this equation, we can derive the particular cases for incident P or SV waves by letting $B_0 = 0$ or $A_0 = 0$.

5.4.4   Critical reflection of SV waves

According to Snell's law (5.59) for SV waves incident on a free surface, since $\alpha > \beta$, there exists a value of the angle of incidence f for which the angle e of reflected P waves is zero. This angle is called the critical angle f_c and is given by

$$\cos f_c = \beta/\alpha \tag{5.109}$$

Incident SV waves with angles $f < f_c$ (supercritical SV waves) produce no reflected P waves, but only reflected SV waves. For these waves, we have $\beta < c < \alpha$ and therefore r is imaginary. In consequence, the reflection coefficient V_{SS} is a complex number:

$$V_{SS} = \frac{(1 - 3\bar{r}^2)^2 - i4\bar{r}s}{(1 - 3\bar{r}^2)^2 + i4\bar{r}s} = e^{-i\theta} \tag{5.110}$$

where

$$\bar{r} = (1 - c^2/\alpha^2)^{1/2} \tag{5.111}$$

$$\tan\left(\frac{\theta}{2}\right) = \frac{4\bar{r}s}{(1 - 3\bar{r}^2)^2} \tag{5.112}$$

Reflected SV waves have the same amplitude as incident waves but with a phase shift θ. There are no reflected P waves, but there is a reflected potential ϕ that corresponds to an elastic perturbation in the form of inhomogeneous waves of P type that travel in the x_1 direction with velocity $c < \alpha$:

$$\phi = B_0 V_{SP} \exp[k\bar{r}x_3 + ik(x_1 - ct)] \tag{5.113}$$

Since $r = i\bar{r}$ is imaginary, V_{SP} has complex values (5.107). This situation is similar to that found for supercritical incident waves in two elastic media when they travel from a medium of lesser velocity to one of larger velocity. Inhomogeneous P waves present in a half-space travel in the direction parallel to its surface and their amplitudes are attenuated exponentially with the distance from this surface. This is an elastic perturbation that exists near the free surface and we will see that it is related to the generation of surface waves.

5.4.5   The partition of energy

In the same way as for the reflection and refraction of waves in two elastic media, we can derive the partition of energy between P and SV waves reflected on a free surface. For incident P waves, the energy brought to the surface is divided between those of reflected P and SV waves (Fig. 5.13). In terms of energy intensities as in (5.53) we have the relation

$$I_{inc}^{P} \, ds = I_{refl}^{P} \, ds + I_{refl}^{SV} \, ds' \tag{5.114}$$

where $ds/\sin e = ds'/\sin f = d\sigma$. On substituting ds and ds' in terms of $d\sigma$, the energy intensities as functions of the amplitudes of the potentials, as in (5.48), and those in terms of the coefficients of reflection for P and SV waves, the partition of energy is given by

$$V_{PP}^{2} + \frac{\alpha \sin f}{\beta \sin e} V_{PS}^{2} = 1 \tag{5.115}$$

For an incident SV wave, the relation is

$$V_{SS}^{2} + \frac{\beta \sin e}{\alpha \sin f} V_{SP}^{2} = 1 \tag{5.116}$$

Figures 5.14 and 5.15(a) show the partition of energy between reflected P and SV waves for incident P and SV waves with $\lambda = \mu$. For an incident P wave, $V_{PP} = 0$, for

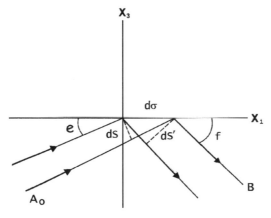

Fig. 5.13. An incident ray beam of P waves and a reflected ray beam for SV waves at a free surface of an elastic medium.

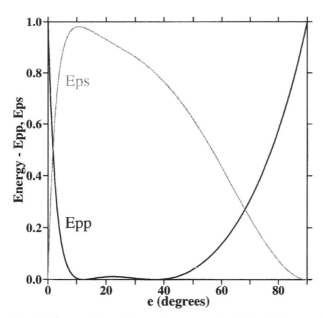

Fig. 5.14. The partition of energy between reflected P and SV waves for a P wave incident on a free surface of an isotropic elastic medium with $\lambda = \mu$.

angles of incidence $e = 30°$ and $12.47°$; and $V_{PP} = 1$ for $e = 0$ and $90°$. For an incident SV wave, for angles of incidence less than the critical angle ($f_c = 54.73°$), $V_{SS} = 1$ and there are no reflected P waves. For normal incidence, the reflected waves are of the same kind as the incident ones. Figure 5.15(b) shows the phase shift of the supercritical reflected SV waves.

5.5   Motion at the free surface

Seismologic observations are carried out at the Earth's surface, thus seismograms are recordings of the total motion of this surface upon the arrival of seismic waves, not only of the incident waves. Damage to buildings produced by earthquakes is also caused by the total motion of the Earth's surface. It is important, then, to study the total motion of the free surface of an elastic medium upon the arrival of waves that were traveling in its interior. The total displacement of points on a free surface is the sum of those of incident and reflected P and S waves. We will treat separately the motions of the surface upon the arrival of incident P and S waves.

5.5.1   Incident P waves

For an incident P wave, the total displacement of the free surface (U^P) is the sum of the displacements of the incident wave (u^{Pi}) and of the reflected P (u^{Pr}) and SV (u^{Sr}) waves (Fig. 5.16):

$$U^P = u^{Pi} + u^{Pr} + u^{Sr} \tag{5.117}$$

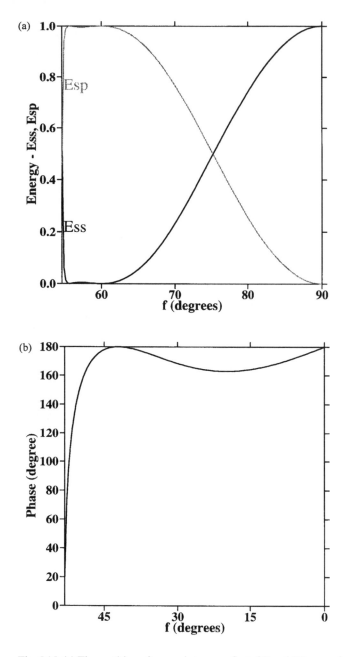

Fig. 5.15. (a) The partition of energy between reflected P and SV waves for an SV wave incident on a free surface of an isotropic elastic medium with $\lambda = \mu$; for $f < f_c$ there are no reflected P waves. (b) The phase shift between incident and reflected SV waves for supercritical angles of incidence ($f < f_c$).

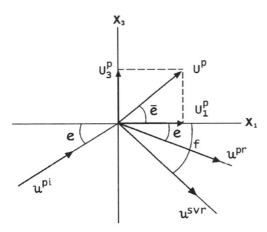

Fig. 5.16. The total motion of the free surface of an elastic medium for an incident P wave. The apparent angle of emergence is $\bar{e}$.

If (x_3, x_1) is the plane of incidence, A_0 is the amplitude of the potential ϕ of incident P waves, and A and B are those of the potentials ϕ and ψ of reflected P and SV waves, (5.94) and (5.95), then the amplitudes of the horizontal and vertical components of the displacement at the free surface, according to (3.97) and (3.98), are

$$U_1^{\mathrm{P}} = k(A_0 + A + B \tan f) \tag{5.118}$$

$$U_3^{\mathrm{P}} = k(A_0 \tan e - A \tan e + B) \tag{5.119}$$

On substituting for the values of A and B in terms of A_0 and the reflection coefficients V_{PP} and V_{PS}, according to equations (5.100) and (5.101), we obtain

$$U_1^{\mathrm{P}} = \frac{12 A_0 k}{D} \tan e \tan^2 f \sec e \tag{5.120}$$

$$U_3^{\mathrm{P}} = \frac{6 A_0 k}{D} (1 + 3 \tan^2 e) \tan e \sec^2 e \tag{5.121}$$

$$D = 4 \tan e \tan f + (1 + 3 \tan^2 e)^2 \tag{5.122}$$

5.5.2 Incident S waves

For an incident S wave with SH and SV components, such that, according to the orientation of the coordinates axes, U_2^{S} is the SH component, and U_1^{S} and U_3^{S} are the horizontal and vertical components of SV motion (Fig. 5.17),

$$U_1^{\mathrm{S}} = u_1^{\mathrm{Si}} + u_1^{\mathrm{Sr}} + u_1^{\mathrm{Pr}} \tag{5.123}$$

$$U_2^{\mathrm{S}} = u_2^{\mathrm{Si}} + u_2^{\mathrm{Sr}} \tag{5.124}$$

$$U_3^{\mathrm{S}} = u_3^{\mathrm{Si}} + u_3^{\mathrm{Sr}} + u_3^{\mathrm{Pr}} \tag{5.125}$$

On substituting into these equations the amplitudes of SH displacement (C_0 incident and C reflected) and of the potential ψ of SV displacement (B_0 incident and B reflected) and A

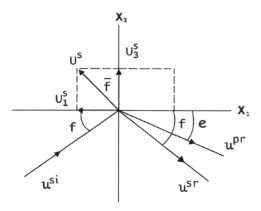

Fig. 5.17. The total motion of the free surface of an elastic medium for an incident SV wave. The apparent angle of emergence is $\bar{f}$.

of the potential ϕ of reflected P waves, according to equations (5.92), (5.102), and (5.103), and the relations between the displacements and the potentials (3.97) and (3.98), we obtain

$$U_1^S = k(B_0 \tan f + A - B \tan f) \tag{5.126}$$

$$U_2^S = C_0 + C \tag{5.127}$$

$$U_3^S = -k(-B_0 + A \tan e + B) \tag{5.128}$$

Just like in the previous case, using the relations between the reflected and incident amplitudes in terms of reflection coefficients, according to (5.106) and (5.107), and $C = C_0$, we obtain

$$U_1^S = \frac{6B_0 k}{D}(1 + 3\tan^2 e)\tan f \sec^2 e \tag{5.129}$$

$$U_2^S = 2C_0 \tag{5.130}$$

$$U_3^S = -\frac{12B_0 k}{D}\tan e \tan f \sec^2 e \tag{5.131}$$

The amplitudes of the total displacement of the free surface depend on the amplitudes and angles of incidence of the incident waves. For incident P waves, the variables are A_0 and e; for incident S waves, they are B_0, C_0, and f. Equations (5.120)–(5.122) and (5.129)–(5.131) allow also the determination of incident wave displacement amplitudes from the total motion observed at the free surface.

5.5.3   *Apparent angles of incidence and polarization*

Displacements observed at the Earth's surface correspond to the total motion of a free surface and thus include incident and reflected waves. If we want to calculate the incidence and polarization angles from amplitudes observed at the surface we have to

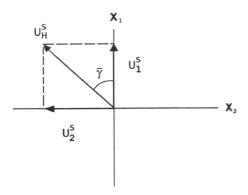

Fig. 5.18. The apparent polarization angle $\bar{\gamma}$ measured from the horizontal transverse (U_2) and radial (U_1) displacements at the free surface of an elastic medium for an incident S wave.

take this fact into account. The angles of incidence $\bar{e}$ and $\bar{f}$ derived directly from the observed total motion for arriving P and S waves are called apparent angles of incidence (Figs. 5.16 and 5.17). If we substitute for incident P waves equations (5.120) and (5.121), and for incident S waves (5.129) and (5.131), we obtain

$$\tan \bar{e} = \frac{U_3^{\mathrm{P}}}{U_1^{\mathrm{P}}} = \frac{1 + 3 \tan^2 e}{2 \tan f} \tag{5.132}$$

$$\tan \bar{f} - \frac{U_1^{\mathrm{S}}}{U_3^{\mathrm{S}}} = -\frac{1 + 3 \tan^2 e}{2 \tan e} \tag{5.133}$$

The apparent angle of polarization $\bar{\gamma}$ (3.87) is measured from the radial component, in our case U_1^{S}, and the transverse component, U_2^{S}, of the observed motion at the surface for an arriving S wave (Fig. 5.18):

$$\tan \bar{\gamma} = \frac{U_2^{\mathrm{S}}}{U_1^{\mathrm{S}}} = \frac{C_0 [4 \tan e \tan f + (1 + 3 \tan^2 e)^2]}{3 B_0 k \tan f (1 + 3 \tan^2 e) \sec^2 e} \tag{5.134}$$

According to the definition of the polarization angle ε (3.86),

$$\tan \varepsilon = \frac{u_2^{\mathrm{S}}}{(u_1^{\mathrm{S2}} + u_3^{\mathrm{S2}})^{1/2}} = \frac{C_0 \cos f}{B_0 k} \tag{5.135}$$

On substituting (5.135) into (5.134) we obtain the relation between $\bar{\gamma}$ and ε:

$$\tan \bar{\gamma} = \frac{D}{3 \sin f \sec^2 e (1 + 3 \tan^2 e)} \tan \varepsilon \tag{5.136}$$

where D is given in (5.122). By replacing equation (3.87) into (5.136) we obtain the relation between $\bar{\gamma}$ and γ. Equation (5.136) is valid only for angles of incidence less than the critical one, that is, for $f > f_{\mathrm{c}}$. When $f < f_{\mathrm{c}}$, there are no reflected P waves, $\tan e$ is an imaginary quantity and the factor in (5.136) is a complex quantity. There is a phase shift between the incident and reflected SV, but not for the SH waves, and the total motion of the free surface is not linear.

In conclusion, the total motion of a free surface upon the arrival of P and S waves is complicated by the presence of reflected waves. For incident S waves and angles of incidence greater than the critical angle, the total motion is not linear. Thus, it is not always easy to find the amplitudes of incident waves.

6 RAY THEORY. MEDIA OF CONSTANT VELOCITY

6.1 Ray theory. The eikonal equation

The differential equation of motion in a continuous medium is given in terms of the divergence of the stress tensor (2.57). For an elastic body, we substitute for the stress in terms of the strain and the elastic coefficients (2.58). If the medium is homogeneous and isotropic with constant elastic coefficients, we obtain equation (2.61) for displacements. If there are no body forces, Navier's equation (2.66) and (2.67) can easily be reduced to the wave equations (3.1)–(3.4), where the velocities of propagation α and β are also constant.

If the medium is not homogeneous, the elastic coefficients and density are functions of the spatial coordinates. The equation of motion (2.58) can be written in terms of the displacements:

$$\frac{\partial}{\partial x_j}(C_{ijkl}u_{k,l}) + F_i = \rho \frac{\partial^2 u_i}{\partial t^2} \tag{6.1}$$

For a liquid, for which $C_{ijkl}u_{k,l} = Ku_{k,k}$, if K is a function of the coordinates, equation (6.1), for no body forces, becomes

$$\frac{\partial K}{\partial x_i}\frac{\partial u_k}{\partial x_k} + K\frac{\partial}{\partial x_i}\left(\frac{\partial u_k}{\partial x_k}\right) = \rho \frac{\partial^2 u_i}{\partial t^2} \tag{6.2}$$

This equation can not be reduced to the wave equation either for the cubic dilation θ or for the displacement potential ϕ, as we did in the homogeneous case. Taking the divergence in (6.2), we obtain

$$K\frac{\partial^2 \theta}{\partial x_i\,\partial x_i} - \rho\frac{\partial^2 \theta}{\partial t^2} = \frac{\partial \rho}{\partial x_i}\frac{\partial^2 u_i}{\partial t^2} - \theta\frac{\partial^2 K}{\partial x_i\,\partial x_i} - 2\frac{\partial K}{\partial x_i}\frac{\partial \theta}{\partial x_i}$$

The three terms on the right-hand side are introduced by the dependences of K and ρ on the coordinates and the equation is difficult to solve. For this reason, for the problem of the propagation of waves in inhomogeneous media or media of variable velocity, we use a different approach, called the ray theory. As we saw in Section 3.2, the trajectories followed by the propagation of waves are given by the trajectories of rays or normals to wave fronts. In many problems of seismology, we are primarily interested in these trajectories, their traveling times, and the behavior of their amplitudes. In many of these problems it is not necessary to find a complete solution of the wave equation; rather, it is sufficient to solve for the behavior of rays using ray theory.

91

We will treat the problem of the propagation of P waves that are represented by only one scalar potential ϕ. Let us consider a medium in which the velocity $v(x_i)$ varies continuously with the coordinates. If we call v_0 an initially constant value of the velocity or the value for a reference point or region, then the wave equation for such a reference region is

$$\nabla^2\phi = \frac{1}{v_0^2}\frac{\partial^2\phi}{\partial t^2} \tag{6.3}$$

The solution for this equation for harmonic waves of frequency ω with constant amplitude and zero initial phase can be written just like (3.23):

$$\phi = A\exp\{i[k_0 S(x_i) - \omega t]\} \tag{6.4}$$

where $k_0 = \omega/v_0$ is the wave number corresponding to the velocity v_0 and $S(x_i) = $ constant, for the equations of wave fronts. Let us consider the phase $\xi = k_0 S(x_i) - \omega t$, with the same value at two different times t_1 and t_2 separated by a time interval $\Delta t = t_2 - t_1$, and wave fronts separated by a distance $\Delta S = S_2 - S_1$; then we can write

$$k_0 S_1 - \omega t_1 = k_0 S_2 - \omega t_2 \tag{6.5}$$

From these relations we can deduce that

$$v_0 = \frac{\omega}{k_0} = \frac{\Delta S}{\Delta t} \tag{6.6}$$

In another region of the medium, where the velocity has a different value, v, the relation between two wave fronts, corresponding to the same phase, separated by a time interval Δt and distance ΔS, gives

$$v = \frac{\omega}{k} = \frac{\Delta S}{\Delta t} \tag{6.7}$$

Let us consider now the propagation of wave fronts through a medium of smoothly continuous variable velocity $v(x_i)$. For each wave front corresponding to a constant value of the phase ξ, its time derivative must be zero:

$$\frac{d\xi}{dt} = \frac{\partial\xi}{\partial t} + v_i\frac{\partial\xi}{\partial x_i} = 0 \tag{6.8}$$

and, therefore,

$$\frac{\partial\xi}{\partial t} = -v_i\frac{\partial\xi}{\partial x_i} \tag{6.9}$$

where the vector component v_i is the velocity in the direction normal to the wave front. The gradient of ξ can be substituted by the derivative in the direction of the normal to the wave front $(\boldsymbol{n})$ and the vector $\boldsymbol{v}$ by the scalar v, the velocity in the same direction:

$$\frac{\partial\xi}{\partial t} = -v\frac{\partial\xi}{\partial n} \tag{6.10}$$

According to (6.4), the gradient of ξ can be replaced by the gradient of S. Then its derivatives with respect to the normal to the wave front and with respect to time are given by

$$\frac{\partial \xi}{\partial n} = k_0 \frac{\partial S}{\partial n} \tag{6.11}$$

$$\frac{\partial \xi}{\partial t} = -\omega \tag{6.12}$$

By substituting (6.11) and (6.12) into (6.10), we obtain

$$v k_0 \frac{\partial S}{\partial n} = \omega \tag{6.13}$$

Taking the square of (6.13) and replacing $k_0 = \omega/v_0$ and $(\partial S/\partial n)^2 = (\partial S/\partial x_1)^2 + (\partial S/\partial x_2)^2 + (\partial S/\partial x_3)^2$, we finally obtain

$$\left(\frac{\partial S}{\partial x_1}\right)^2 + \left(\frac{\partial S}{\partial x_2}\right)^2 + \left(\frac{\partial S}{\partial x_3}\right)^2 = \left(\frac{v_0}{v}\right)^2 \tag{6.14}$$

This equation is known as the eikonal equation (from the Greek word *eikon* for image) or the equation of the characteristic functions of Hamilton. The equation relates the value of the square of the gradient of the wave front at a point of a medium of variable velocity to the quotient of the velocity v_0 at a reference point or the initial velocity and v at any other point. These velocities are in the direction normal to the wave fronts or along the ray's direction. The quotient $v/v_0 = n$ is the refractive index at each point. Since the gradient of the wave front represents the direction of propagation or the ray's trajectory, equation (6.14) shows how this trajectory changes as wave fronts propagate through a medium of variable velocity. If we know how v varies in the medium, the eikonal equation shows us how the trajectories of rays change.

Let us consider now the wave equation for an arbitrary point of the medium of variable velocity v, as in (6.3):

$$\nabla^2 \phi = \frac{1}{v^2} \frac{\partial^2 \phi}{\partial t^2} \tag{6.15}$$

For a harmonic time dependence of ϕ, we can write Helmholtz's equation (3.10), in which the wave number $k = \omega/v$ is a variable. This wave number can be expressed in terms of that of the reference region or the initial value $k_0 = \omega/v_0$, in the form $k = k_0 v_0/v$; thus, we obtain

$$\left(\nabla^2 + k_0^2 \frac{v_0^2}{v^2}\right)\phi = 0 \tag{6.16}$$

If we substitute into this equation a solution given by (6.4), we obtain

$$i\frac{\partial^2 S}{\partial x_i \partial x_i} - k_0 \frac{\partial S}{\partial x_i}\frac{\partial S}{\partial x_i} + k_0 \frac{v_0^2}{v^2} = 0 \tag{6.17}$$

This equation reduces to the eikonal equation (6.14) if

$$\left|\frac{\partial^2 S}{\partial x_i \partial x_i}\right| \ll k_0 \left|\frac{\partial S}{\partial x_i}\frac{\partial S}{\partial x_i}\right| \tag{6.18}$$

Therefore, if this condition is satisfied, for a medium of variable velocity, solutions of the eikonal equation (6.14) and, in consequence, the principles of ray theory are good approximations to the solutions of the wave equation (6.15).

6.1.1   *The condition of validity*

The condition of validity for the approximation of ray theory in a medium of variable velocity is given by equation (6.18). To see the significance of this condition, we start with the expression for the direction cosines of the ray (3.26):

$$\frac{\partial S}{\partial x_i} = \nu_i \left| \frac{\partial S}{\partial x_i} \right| \tag{6.19}$$

If we substitute for the gradient of the wave front in (6.19) the derivatives in the direction of its normal **n** and then substitute these values into (6.18), we obtain

$$\left| \frac{\partial}{\partial x_i} \left(\nu_i \frac{\partial S}{\partial n} \right) \right| \ll k_0 \left| \nu_i \frac{\partial S}{\partial n} \right|^2 \tag{6.20}$$

where we have used

$$\frac{\partial S}{\partial x_i} = \nu_i \frac{\partial S}{\partial n} \tag{6.21}$$

By replacing equation (6.14) into the second term of (6.20), we obtain

$$\left| \frac{\partial}{\partial x_i} \left(\nu_i \frac{\partial S}{\partial n} \right) \right| \ll \omega \frac{v_0}{v^2} \tag{6.22}$$

The first term of (6.22) can be written using equation (6.13) in the form

$$\frac{\partial}{\partial x_i} \left(\nu_i \frac{\partial S}{\partial n} \right) = \nu_i \frac{\partial}{\partial x_i} \left(\frac{v_0}{v} \right) + \left(\frac{v_0}{v} \right) \frac{\partial \nu_i}{\partial x_i} \tag{6.23}$$

On putting (6.23) into condition (6.22), this is satisfied if each term of the sum in (6.23) is much smaller than the right-hand term of (6.22). Taking the derivatives of v_0/v, the validity condition can be expressed by two conditions:

$$\left| \frac{\partial v}{\partial n} \right| \ll \omega \tag{6.24}$$

$$\left| \frac{\partial \nu_i}{\partial x_i} \right| \ll \frac{\omega}{v} \tag{6.25}$$

In terms of the wave length λ, these conditions can be written as

$$\frac{1}{v} \left| \frac{\partial v}{\partial n} \right| \ll \frac{1}{\lambda} \tag{6.26}$$

$$\left| \frac{\partial \nu_i}{\partial x_i} \right| \ll \frac{1}{\lambda} \tag{6.27}$$

The first condition (6.24) implies that the change in velocity in the direction normal to the wave front must be small compared with the angular frequency, or, according to

(6.26), this change per unit velocity must be small compared with the inverse of the wave length. In both cases, this condition is satisfied if the frequency of the wave is large or its wave length is small. The second condition implies that the variation of the normal to the wave front (the ray's direction) must be small compared with the inverse of the wave length (6.27). This means that wave lengths must be small in comparison with the distance along which the ray changes significantly in direction. In conclusion, ray theory is a good approximation that is valid for high frequencies or short wave lengths.

The theory we have presented is known as the classical ray theory. The presence of discontinuities and strong velocity gradients in the medium, such as are found in the Earth's interior, leads to serious inadequacies and limitations of this theory. An improvement in ray theory for inhomogeneous media is obtained by using the WKBJ approximation (the initials are those of Wenzel, Kramers, Brillouin, and Jeffreys). A presentation of this theory can be found in brief form in Bullen and Bolt (1985) and in a more complete form in Aki and Richards (1980).

6.2   Ray trajectories

According to the eikonal equation (6.14), the trajectories of rays or ray paths in a medium of variable velocity change depending on the variations of velocity. As wave fronts advance through the medium, they change in shape and consequently change the directions of their normals or the direction of rays. If we select a point on a wave front and, as it propagates, follow the changes in the direction of its normal, we have the trajectory of the ray or ray path corresponding to that point of the wave front.

The trajectory or line of propagation of a ray is given by equation (3.27). The direction cosines of rays can be written from (6.19) and (6.14) as

$$\nu_i = \frac{v}{v_0} \frac{\partial S}{\partial x_i} \tag{6.28}$$

where v_0 is a reference velocity and $v_0/v = n$ is the refraction coefficient at each point of the medium.

If we consider the trajectory for a particular ray given by the curve s (Fig. 6.1), the direction cosines of the ray at each point can be written as

$$(\nu_1, \nu_2, \nu_3) = \left(\frac{dx_1}{ds}, \frac{dx_2}{ds}, \frac{dx_3}{ds} \right) \tag{6.29}$$

By substitution into (6.28), we obtain

$$\frac{\partial S}{\partial x_i} = \frac{v_0}{v} \frac{dx_i}{ds} \tag{6.30}$$

By substituting for $\partial S/\partial x_i$ from equation (6.28) and using $d/ds = \nu_i \, \partial/\partial x_i$ we obtain an equation for the refraction coefficient n:

$$\frac{\partial n}{\partial x_i} = \frac{d}{ds} \left(n \frac{dx_i}{ds} \right) \tag{6.31}$$

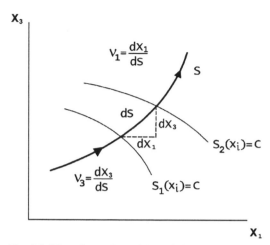

Fig. 6.1. Wave fronts S_1 and S_2 and the ray trajectory.

This equation can be considered to be a generalized form of Snell's law. It shows how the direction of the ray changes along its trajectory for a medium of variable velocity. In a medium of constant velocity, $v = v_0$ and $n = 1$, from equation (6.31) we obtain that the direction cosines dx_i/ds are constant along the ray's trajectory; that is, rays are straight lines.

6.3   Propagation in the (x, z) plane

Many problems of ray theory in two dimensions can be solved in a more convenient way if we take the (x, z) plane as the plane that contains the rays. In this plane x represents the horizontal direction and z the vertical (positive downward). For a ray that forms an angle i (the angle of incidence) with the z axis (Fig. 6.2), the potential of P waves of a monochromatic harmonic wave of frequency ω can be expressed according to (5.2) as

$$\phi = A \exp[ik_\alpha(\sin i\, x + \cos i\, z - \alpha t)] \tag{6.32}$$

Two other ways of writing this are

$$\phi = A \exp[i(k_x x + k_z z - \omega t)] \tag{6.33}$$

$$\phi = A \exp\left[i\omega\left(\frac{x}{c_x} + \frac{z}{c_z} - t\right)\right] \tag{6.34}$$

where c_x and c_z are the apparent velocities of the wave front in the directions of the axes x and z, and k_x and k_z are the corresponding wave numbers. Thus, we have the relations

$$k_x^2 + k_z^2 = k_\alpha^2 \tag{6.35}$$

$$\frac{1}{c_x^2} + \frac{1}{c_z^2} = \frac{1}{\alpha^2} \tag{6.36}$$

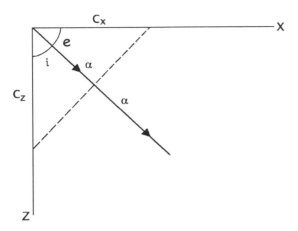

Fig. 6.2. A ray path or trajectory. The true velocity of propagation is α and the apparent velocities are c_x and c_z.

According to (6.34) and (6.32),

$$c_x = \frac{\alpha}{\sin i}; \qquad c_z = \frac{\alpha}{\cos i} \tag{6.37}$$

In many problems, for example when α is a function of z, it is convenient to separate the propagation in the z direction from that in the x direction. For this we call c and k the apparent velocity and wave number in the x direction. If we have waves propagating in positive and negative z directions (downward and upward), they can be expressed in the following way:

$$\phi = [A \exp(ikrz) + B\exp(-ikrz)] \exp[ik(x - ct)] \tag{6.38}$$

where $r = \tan e = \cot i$ from (5.10), and also in the form

$$\phi = [A \exp(i\omega qz) + B\exp(-i\omega qz)] \exp[i\omega(px - t)] \tag{6.39}$$

where $p = 1/c = \sin i/\alpha$ and $q = 1/c_z = \cos i/\alpha$ are the inverses of the apparent velocities in the x and z directions, also called the slowness components in the two directions. As functions of c and α, r and q are given by

$$r = \left(\frac{c^2}{\alpha^2} - 1\right)^{1/2} \tag{6.40}$$

$$q = \frac{1}{c}\left(\frac{c^2}{\alpha^2} - 1\right)^{1/2} = \left(\frac{1}{\alpha^2} - \frac{1}{c^2}\right)^{1/2} = pr \tag{6.41}$$

In order to have real waves propagating in the z direction, r and q must be real, or $c > \alpha$. If $c < \alpha$, r and q are imaginary, and, according to (6.38) and (6.39), waves propagate in the x direction and their amplitudes decrease or increase exponentially in the z direction. The exponential increase has no physical meaning and hence we must put $B = 0$. The exponential decrease corresponds to inhomogeneous waves, as we saw in the study of supercritical incidence (sections 5.2 and 5.3). The same procedure can be applied to S waves by replacing α by β. When the two types of waves are treated together, the following notation is used: for P waves, r, p_α, and q_α; and for S waves, s, p_β, and q_β. Then, we have the relations $kr = \omega q_\alpha$ and $ks = \omega q_\beta$.

6.4   Ray trajectories and travel times. A homogeneous half-space

We start by considering the propagation of rays in a homogeneous half-space. If the free surface is the horizontal plane, this is a simple way to represent a model of the Earth of planar geometry. A flat Earth is a sufficiently good approximation for small distances (less than 1000 km), for which effects of its curvature can be neglected. For the two-dimensional problem, we choose the coordinates (x, z) as defined in the previous section with x along the horizontal plane and z in the vertical direction positive downward. The origin of rays is at a point (focus or hypocenter) situated at a certain depth h from the surface and observation points are situated along the x axis on the surface. The origin of coordinates is located at the projection of the focus onto the surface, or epicenter (Fig. 6.3). The depth of the focus introduces a certain spatial dimension; therefore, the application of ray theory requires that the wave lengths be much smaller than the depth. We are interested in ray trajectories or paths and travel times, that is, the times which waves take to propagate from the focus to each observation point as a function of the horizontal distance $t(x)$. For a medium of constant velocity the ray trajectories are straight lines (6.31) contained in the (x, z) plane.

Times taken for rays to travel from the focus F to an observation point P for a medium of velocity v (v represents the velocity either of P or of S waves) are given (Fig. 6.3) by

$$t = \frac{(x^2 + h^2)^{1/2}}{v} \tag{6.42}$$

Both the travel time t and the distance x can also be expressed in terms of the angle of incidence at the focus i:

$$t = \frac{h}{v \cos i} \tag{6.43}$$

$$x = h \tan i \tag{6.44}$$

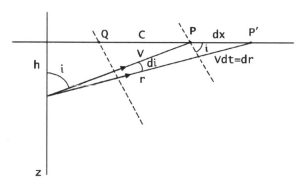

Fig. 6.3. Trajectories of two near rays in a half-space of constant velocity. The wave front advances from P to P' in time dt.

Equation (6.42) gives the travel times directly as a function of the horizontal distances $t(x)$, whereas equations (6.43) and (6.44) are parametric equations for times and distances $t(i)$ and $x(i)$. In this case, the parameter used is the take-off angle at the focus. Only for cases of simple geometry is it possible to write explicit equations for $t(x)$. However, no matter how complicated the geometry, we can write, using Snell's law, equations of parametric form. We will see that this holds also for media of variable velocity.

As we saw already, if v is the velocity of propagation along the ray and c is the apparent velocity of the intersection of the wave front with the x axis (Fig. 6.3), then

$$c = \frac{v}{\sin i} \tag{6.45}$$

For two rays that arrive at distances x and $x + dx$ with traveling times t and $t + dt$, we can deduce easily from Fig. 6.3 ($\sin i = v \, dt/dx$) that

$$\frac{dt}{dx} = \frac{\sin i}{v} = \frac{1}{c} = p \tag{6.46}$$

where p is the ray parameter, which, according to Snell's law, is constant along each ray. This is an important relation between the slope of the curve of travel times at a distance x and the parameter of the ray that arrives at that distance. From (6.46), we can express $\sin i$, $\cos i$ and $\tan i$ in terms of v and p, and $\eta = 1/v$, the slowness or inverse of the velocity:

$$\sin i = vp = p/\eta$$

$$\cos i = (1 - v^2 p^2)^{1/2} = v(\eta^2 - p^2)^{1/2}$$

$$\tan i = \frac{vp}{(1 - v^2 p^2)^{1/2}} = \frac{p}{(\eta^2 - p^2)^{1/2}}$$

On substituting these equations into (6.43) and (6.44), we obtain

$$t = \frac{h}{v^2 (\eta^2 - p^2)^{1/2}} \tag{6.47}$$

$$x = \frac{ph}{(\eta^2 - p^2)^{1/2}} \tag{6.48}$$

These are another pair of parametric equations for t and x, in terms now of the ray parameter and the slowness. These are very useful equations, as we will see in the next chapter.

The travel time curve $t(x)$, according to (6.42), is a hyperbola that tends asymptotically toward a straight line of slope $1/v$ (Fig. 6.4). If the focus is at the surface ($h = 0$), $t(x)$ is a straight line with slope $1/v$ that passes through the origin.

At every distance x there arrives a ray of a different parameter p, so the curve $p(x)$ is also characteristic of each medium. For the homogeneous half-space we find from (6.48) that

$$p = \frac{\eta x}{(h^2 + x^2)^{1/2}} \tag{6.49}$$

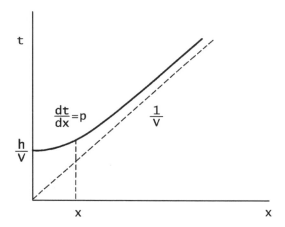

Fig. 6.4. The traveling time curve of rays in a half-space with constant velocity v for a focus at depth h.

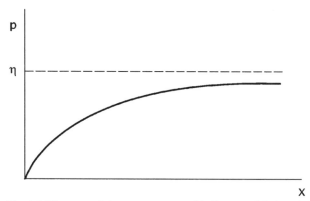

Fig. 6.5. The curve of the ray parameter with distance $p(x)$, for a half-space with constant velocity v.

The curve $p(x)$ (Fig. 6.5) starts with $p = 0$ for $x = 0$, corresponding to the vertical ray upward ($i = 0$). For very large values of x the curve tends asymptotically toward $p = \eta = 1/v$. This straight line corresponds to a ray that travels horizontally ($i = 90°$), which naturally exists only for $h = 0$.

Another important relation is the change in distance x with the ray parameter p. Taking the derivative with respect to p in (6.48), we obtain

$$\frac{dx}{dp} = \frac{h}{(\eta^2 - p^2)^{1/2}} + \frac{hp^2}{(\eta^2 - p^2)^{3/2}} \tag{6.50}$$

According to this equation, dx/dp is always positive; that is, if p increases, x increases too (Fig. 6.6). For $p = 0$, dx/dp has a constant value (hv) and, for the maximum value of p ($p = \eta$), tends to infinity. For small values of p, rays are nearly vertical ($i \simeq 0°$) and x varies little for changes in the angle i, whereas for values of p near $p = \eta$, rays are nearly horizontal ($i = \pi/2$) and small changes in i produce large changes in x.

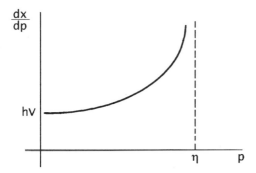

Fig. 6.6. The curve of the variation of distance with the ray parameter dx/dp, in a half-space with constant velocity v.

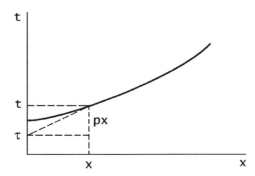

Fig. 6.7. The traveling time curve $t(x)$ and the meaning of the reduced time $\tau(p)$ (the *tau* function) for a half-space with constant velocity v.

Another important relation in the study of ray propagation is that of reduced times with the ray parameter or *tau* function $\tau(p)$, defined by

$$\tau(p) = t(p) - px(p) \tag{6.51}$$

If we take derivatives with respect to p, we obtain

$$\frac{d\tau}{dp} = \frac{dt}{dp} - x - p\frac{dx}{dp} \tag{6.52}$$

On substituting,

$$\frac{dt}{dp} = \frac{dt}{dx}\frac{dx}{dp} = p\frac{dx}{dp}$$

we obtain

$$\frac{d\tau}{dp} = -x \tag{6.53}$$

The slope of the curve $\tau(p)$ gives us the distance x which corresponds to the ray with parameter p. From Fig. 6.7 we can see that τ is the intersection of the tangent to the curve $t(x)$ with the t axis at a point at a distance x and with a slope p. An important

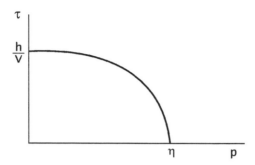

Fig. 6.8. The curve of the *tau* function (the reduced time with the ray parameter) $\tau(p)$, in a half-space with constant velocity v.

property of $\tau(p)$ is that it is always a single-valued function even for cases in which $t(x)$ is a multiple-valued function.

For a homogeneous half-space, on replacing (6.47) and (6.48) in (6.51), we obtain

$$\tau(p) = h(\eta^2 - p^2)^{1/2} \tag{6.54}$$

Figure 6.8 shows the curve $\tau(p)$. For $p = 0$, τ represents the traveling time of the vertical ray and, in the upper limit of p ($p = \eta$), it corresponds to a horizontal ray, $\tau = 0$.

This exercise has shown the behavior of ray propagation in a homogeneous half-space and has served to define some important concepts such as the ray's trajectory, horizontal distance, travel time, ray parameter, apparent horizontal velocity, and reduced time. We have also seen the curves $t(x), p(x), \tau(p)$, and $dx/dp(p)$ that describe the characteristics of ray propagation.

6.5   A layer over a half-space

Another example of ray propagation in media of constant velocity is that of a layer of thickness H and velocity v' over a half-space of velocity v. This situation allows the study of ray trajectories and travel times due to the presence of a surface of contact between the layer and the half-space. The layer's thickness is a parameter that gives a spatial dimension to the model. For this reason, the application of ray theory is valid if the wave lengths are much smaller than the thickness of the layer. This example represents, in a simplified way, the situation of the Earth's crust over the upper mantle for small distances, for which the flat-Earth approximation is valid.

We start with the velocity of the layer smaller than that of the half-space ($v' < v$) and the focus at the surface ($h = 0$) (Fig. 6.9). At a point P at a distance x from the focus we have three types of rays: (a) direct rays, traveling from F to P; (b) rays reflected on the surface between the layer and the half-space (FCP); and (c) critically refracted rays or head waves, that is, rays with a critical angle of incidence i_c at the contact surface that propagate a certain distance horizontally through the half-space and come back to the free surface with the same angle (FBDP) (see section 5.2 and equation (5.33)). Travel

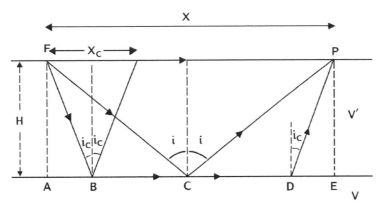

Fig. 6.9. Ray paths in a layer over a half-space ($v' < v$), for a surface focus. The direct ray is FP, the reflected ray is FCP, and the critically refracted ray is FBDP.

times for the three types of ray are

$$t_a = x/v' \tag{6.55}$$

$$t_b = \frac{2(H^2 + x^2/4)^{1/2}}{v'} \tag{6.56}$$

$$t_c = \frac{2H}{v' \cos i_c} + \frac{x - 2H \tan i_c}{v} \tag{6.57}$$

The equation for head waves can be written as a function of the velocities, taking into account that $\sin i_c = v'/v$:

$$t_c = \frac{x}{v} + \frac{2H(v^2 - v'^2)^{1/2}}{vv'} \tag{6.58}$$

The second term in (6.58) is called the 'delay time' and is the difference between the times taken to travel the distances FB with velocity v' and AB with velocity v (Fig. 6.9). Head waves reach the free surface from a minimum distance called the 'critical distance' x_c, which corresponds to the reflected ray with the critical angle (Fig. 6.9). As a function of the velocities, the critical distance is given by

$$x_c = \frac{2Hv'}{(v^2 - v'^2)^{1/2}} \tag{6.59}$$

Travel time curves of the three types of ray are given in Fig. 6.10. For direct rays (a) and head waves (c), they are straight lines with slopes $1/v'$ and $1/v$, respectively. For reflected rays they are hyperbolas that tend asymptotically toward the straight line of the direct rays. Reflected rays can be divided into subcritical and supercritical rays according to the value of their angle of incidence. Supercritical reflected rays are total reflections and thus carry more energy than do subcritical reflections (Chapter 5). First arrivals correspond to direct rays up to a certain distance x', and, for greater distances, to critically refracted rays. The distance x', in terms of velocities, is given by

$$x' = 2H \left(\frac{v + v'}{v - v'} \right)^{1/2} \tag{6.60}$$

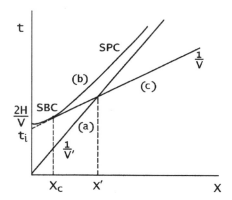

Fig. 6.10. Travel times of rays in a layer over a half-space: (a) direct rays; (b) reflected rays, SBC, subcritical and SPC, supercritical; and (c) critically refracted rays (head waves).

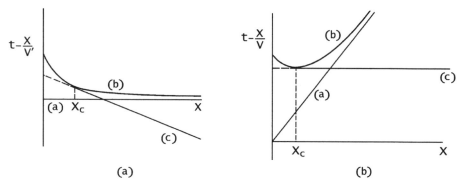

Fig. 6.11. Reduced travel time curves for a layer over a half-space: (a) the reduced velocity v' (the layer velocity), and (b) the reduced velocity v (the half-space velocity).

A useful representation is given by the reduced travel times, the ordinate representing times reduced with respect to a certain velocity v_R in the form $t - x/v_R$. Figure 6.11(a) shows reduced travel times with $v_R = v'$, and Fig. 6.11(b) shows those with $v_R = v$. In both cases, this type of representation increases the differences between the slopes of the lines for the travel times of the rays. The curve corresponding to rays with velocity equal to v_R is parallel to the x axis, those corresponding to lower velocities have positive slopes, and those corresponding to higher velocities have negative slopes.

In order to study travel times of reflected rays, it is also useful to represent square times and distances (t^2, x^2). In this representation, travel times of reflected rays are straight lines with slopes $1/v'^2$ (Fig. 6.12).

For each of the three types of ray (direct, head waves, and reflected) there is a curve $p(x)$ (Fig. 6.13). For direct rays (a), it is a straight line, since the value of p is constant; $p = 1/v'$. For reflected rays (b), the curve is a hyperbola that starts at the origin and tends asymptotically toward the straight line of slope $1/v'$. Finally, for head waves (c), the curve is a straight line with $p = 1/v$.

The relation between travel times and the parameters of the model (H, v and v') makes their determination easy. The layer velocity v' is the inverse of the slope of the travel

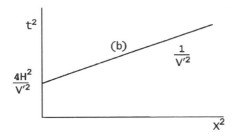

Fig. 6.12. The square of the traveling time curve (t^2, x^2) for reflected rays in a layer over a half-space.

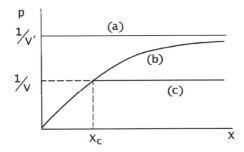

Fig. 6.13. A plot of the ray parameter versus distance $p(x)$ for a layer over a half-space: (a) direct rays, (b) reflected rays, and (c) critically refracted rays (head waves). The critical distance is x_c.

times corresponding to direct rays and the velocity of the half-space v is that of critically refracted rays. Once the velocities are known, the thickness of the layer can be determined from the value of the critical distance x_c using (6.59) or from the time of intersection of head waves ($x = 0$) using (6.58). The velocity and thickness of a layer can also be determined from the slope and intercept time of reflected rays in the (x^2, t^2) representation. This methodology, explained only for the case of there being one layer, can be applied to the determination of velocities and thicknesses of layered media in seismic refraction profiles for the study of the Earth's crustal structure (Chapter 9). The one-layer model may be used as a first approximation to the study of sediments over a rock basement or the crust over the upper mantle.

If the focus is not at the surface (Fig. 6.14), but rather at a depth h, equations of traveling times for direct rays (a), reflected rays (b), and critically refracted rays (c) are modified with respect to (6.55), (6.56), and (6.58) to the forms

$$t_a = \frac{(x^2 + h^2)^{1/2}}{v'} \tag{6.61}$$

$$t_b = \frac{[x^2 + (2H - h)^2]^{1/2}}{v'} \tag{6.62}$$

$$t_c = \frac{x}{v} + \frac{(2H - h)(v^2 - v'^2)^{1/2}}{vv'} \tag{6.63}$$

The travel time curves now have the following forms (Fig. 6.15): the curve for direct rays is a hyperbola tending asymptotically toward the line of slope $1/v'$ and does not pass

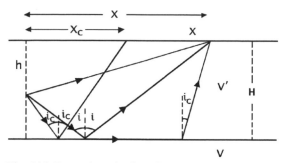

Fig. 6.14. Ray trajectories for a focus at depth h in a layer over a half-space: direct, reflected, and critically refracted rays.

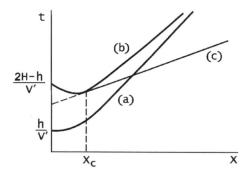

Fig. 6.15. Traveling times for a focus at depth h in a layer over a half-space: (a) the direct ray, (b) the reflected ray, and (c) the critically refracted ray (head wave).

through the origin; the curve for critically refracted rays is a straight line of slope $1/v$, but its time intersection depends also on the focal depth; and that of reflected rays is a hyperbola as in the previous case. In these two examples we have considered that the velocity of the layer is smaller than that of the half-space. In the opposite case, there are no critically reflected rays, but only direct and reflected rays.

6.6　The dipping layer

Let us consider now that the surface of the contact between the layer and the half-space is not parallel to the free surface, but has a dip angle θ. This is known as the case of a dipping layer. The thickness H of the layer is now measured perpendicularly from the base of the layer under the focus to the free surface. We have two different cases if we observe arrivals of rays in down-dip or up-dip manner (Fig. 6.16). Travel times of critically refracted and reflected rays are affected by the dip of the layer. For critically refracted rays and down-dip observations of a surface focus, using the results from (6.57) and that, according to Fig. 6.16(a), $FC = x\cos\theta$, $EC = x\sin\theta$, $E'C = x\sin\theta\tan i_c$ and $EE' = x\sin\theta/\cos i_c$, the travel times t^- are

$$t^- = \frac{x\cos\theta}{v} - \frac{x\sin\theta\tan i_c}{v} + \frac{2H\cos i_c}{v'} + \frac{x\sin\theta}{v'} \tag{6.64}$$

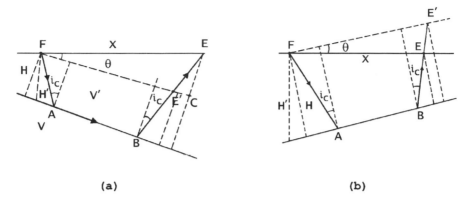

Fig. 6.16. Trajectories of critically refracted rays (head waves) for a dipping layer: (a) down-dip observations, and (b) up-dip observations.

By replacing $v = v'/\sin i_c$ into the first and second terms and combining them, we obtain

$$t^- = \frac{x \sin(i_c + \theta)}{v'} + \frac{2H \cos i_c}{v'} \tag{6.65}$$

and, as a function of the velocities,

$$t^- = \frac{x \sin(i_c + \theta)}{v'} + \frac{2H(v^2 - v'^2)^{1/2}}{vv'} \tag{6.66}$$

Because H is measured normal to the base of the layer, the depth of the layer under the focus is $H' = H/\cos \theta$.

For up-dip observations (the layer is inclined upward from the focus) (Fig. 6.16(b)), traveling times of critically refracted rays are obtained in a similar way:

$$t^+ = \frac{x \sin(i_c - \theta)}{v'} + \frac{2H \cos i_c}{v'} \tag{6.67}$$

If we compare equations (6.65) and (6.67) with (6.58) we find that, on putting $v' = v \sin i_c$ in the first two, the first terms are multiplied by $\sin(i_c + \theta)/\sin i_c$ and $\sin(i_c - \theta)/\sin i_c$, respectively. If $\theta = 0$, both factors are unity and equations (6.65) and (6.67) are equal to (6.58). These factors contain the influence of the layer's dip on the traveling times.

Travel time curves for these rays are shown in Fig. 6.17, in which direct rays are also drawn. The slopes of straight lines corresponding to down- and up-dip traveling times of head waves according to (6.65) and (6.67) are

$$S^- = \frac{\sin(i_c + \theta)}{v \sin i_c} \tag{6.68}$$

$$S^+ = \frac{\sin(i_c - \theta)}{v \sin i_c} \tag{6.69}$$

For down-dip observations, the slope is larger than that for a flat layer of the same velocity, whereas for up-dip observations, it is smaller. If we determine the velocity of

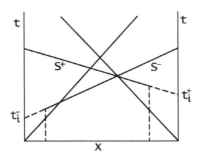

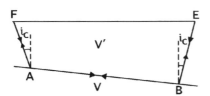

Fig. 6.17. Travel time curves in both directions (direct and inverse) for direct and critically refracted rays in a dipping layer. S^- and S^+ are the slopes of the curves of the critically refracted rays for down- and up-dip observations.

the half-space from the inverse of the traveling time curve without considering the dip of the layer, the result is larger or smaller than the real velocity. If we have observations in one direction only, we can not detect the dip of the layer and in consequence the slope of the curve of head waves gives us only an apparent velocity that may be smaller or larger than the real one, depending on the direction of the dip of the layer.

If we have observations in both directions, down- and up-dip ones, we can determine the dip of the layer and the true velocity of the half-space. The dip θ can be determined from the slopes of the critically reflected rays S^+ and S^-, using (6.68) and (6.69), and making the substitution $v' = v \sin i_c$:

$$\theta = \tfrac{1}{2}[\sin^{-1}(v'S^-) - \sin^{-1}(v'S^+)] \qquad (6.70)$$

The velocity v' of the layer is deduced from the slope of the traveling time curve of direct waves. Once the dip θ is known, the true velocity of the half-space is given by

$$v = \frac{2\cos\theta}{S^+ + S^-} \qquad (6.71)$$

The depth along the vertical from the focus to the base of the layer can be determined from the intersection times of travel time curves of down- and up-dip head waves t_I^+ and t_I^-:

$$H'^- = \frac{t_i^- v'}{\cos i_c \cos\theta} \qquad (6.72)$$

$$H'^+ = \frac{t_i^+ v'}{\cos i_c \cos\theta} \qquad (6.73)$$

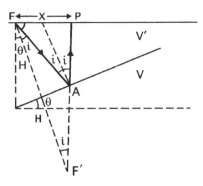

Fig. 6.18. The trajectory of a reflected ray in a dipping layer.

The determination of dips of layers and true velocities from observations of traveling times of head waves is possible only if we have observations along the same line in both directions. In the context of seismic refraction profiles these are known as direct and inverse profiles.

The equation for travel times of the reflected rays in a dipping layer can easily be found by using the image method of optics (FAP = F'AP) (Fig. 6.18):

$$t = \frac{1}{v'}(x^2 + 4H^2 \pm 4xH \sin \theta)^{1/2} \tag{6.74}$$

The sign depends on whether the observations are down-dip (+) or up-dip (−). The dipping layer problem can be used as a first approximation to the study of dips in the Earth's crust.

6.7 A plane layered medium

A plane layered medium consists of N flat parallel layers of different thicknesses H_i and different constant velocities v_i over a half-space. For a surface focus, if the velocity increases with the depth in all layers, the traveling times of critically refracted rays at the interface at the top of layer k can be written from equation (6.58) as

$$t_k = \frac{x}{v_k} + \sum_{i=1}^{k} \frac{2H_k(v_k^2 - v_i^2)^{1/2}}{v_k v_i} \tag{6.75}$$

The second term corresponds to the sum of the delay times from layer 1 to layer k. The critical distances and intersection times are given by

$$x_c^k = \sum_{i=1}^{k} \frac{2H_i v_i}{(v_k^2 - v_i^2)^{1/2}} \tag{6.76}$$

$$t_I^k = \sum_{i=1}^{k} \frac{2H_i(v_k^2 - v_i^2)^{1/2}}{v_k v_i} \tag{6.77}$$

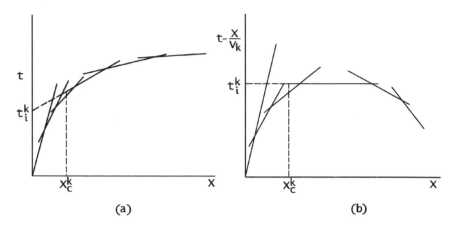

(a)   (b)

Fig. 6.19. Traveling time curves for direct and critically refracted rays (head waves) for a plane-layered medium: (a) normal traveling times, and (b) reduced traveling times (reduced velocity v_k).

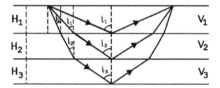

Fig. 6.20. Trajectories of reflected rays in a plane-layered medium.

Travel time curves for these rays have the form of N segments each with a slope equal to $1/v_k$ (Fig. 6.19(a)). The slopes have decreasing values and the critical distances and inter-section times have increasing values. The reduced traveling times with the reference velocity v_k are line segments with positive slopes for rays refracted at layers above layer k (smaller velocities) and negative slopes for rays refracted under it (higher velocities) (Fig. 6.19(b)).

If a layer has a smaller velocity than the one on top of it, this is called a low-velocity layer. At the top of this layer there are no head waves. Head waves that arrive at the surface after traveling through the low-velocity layer are refracted at a layer with a velocity higher than the one at the top of this layer. These rays are delayed by the time spent traveling through the low-velocity layers.

Travel times and distances of reflected rays at each layer interface in parametric form are given (Fig. 6.20) by

$$t_k = 2 \sum_{i=1}^{k} \frac{H_i}{v_i \cos i_i} \tag{6.78}$$

$$x_k = 2 \sum_{i=1}^{k} H_i \tan i_i \tag{6.79}$$

According to (6.47) and (6.48), these two equations can also be written in terms of the ray parameter p_k and slowness in each layer η_i:

$$t_k = 2 \sum_{i=1}^{k} \frac{H_i}{v_i(\eta_i^2 - p_k^2)^{1/2}} \tag{6.80}$$

$$x_k = \sum_{i=1}^{k} \frac{p_k H_i}{(\eta_i^2 - p_k^2)^{1/2}} \tag{6.81}$$

The angles of incidence of rays that pass from one layer to another, according to Snell's law, satisfy the relation

$$\frac{\sin i_i}{v_i} = \frac{\sin i_{i+1}}{v_{i+1}} \tag{6.82}$$

Using this relation, equations (6.78) and (6.79) can be written solely in terms of the take-off angle at the focus. Therefore, for each reflected ray, we can find the traveling time and corresponding distance in terms solely of the take-off angle at the focus or of the ray parameter.

The parametric expressions for reflected waves (6.78)–(6.81) can also be generalized to other geometries of layers or blocks of different constant velocities. These expressions can also be written for cases in which the focus is not at the surface. This analysis is part of the technique called ray tracing in which travel times and amplitudes are calculated by following the trajectories of the rays. Once the take-off angle of the ray at the focus has been fixed, its trajectory is specified by Snell's law and travel times and distances can be determined. However, this is not possible for critically refracted rays or head waves, since rays that leave the focus with the same take-off angle arrive at different distances.

7 RAY THEORY. MEDIA OF VARIABLE VELOCITY

7.1 A variable velocity with depth

Although in the general case of a variable velocity its dependence is on the three spatial coordinates, for many seismologic problems it is sufficient to consider only the variation of the velocity with depth. For relatively short distances ($\Delta < 1000\,\text{km}$), plane geometry is a good approximation and the Earth may be considered as a half-space limited by a free surface. The vertical direction is represented by the z coordinate positive downward and the horizontal direction is represented by x. Rays are contained in the (x, z) plane and problems are reduced to two dimensions (section 6.3). The velocity v_0 and angle of incidence i_0 at the surface are taken as reference values. The direction cosines for an arbitrary point on the trajectory of a ray are given (Fig. 7.1) by

$$\nu_x = \frac{\mathrm{d}x}{\mathrm{d}s} = \sin i \tag{7.1}$$

$$\nu_z = \frac{\mathrm{d}z}{\mathrm{d}s} = \cos i \tag{7.2}$$

where $\mathrm{d}s$ is an increment of the distance along the ray. Taking the x component in equation (6.31), and replacing (7.1), we obtain

$$\frac{\mathrm{d}}{\mathrm{d}s}\left(\frac{v_0}{v}\sin i\right) = 0 \tag{7.3}$$

where $v(z)$ and $i(z)$ are arbitrary functions of the depth. This equation indicates that $(v_0/v)\sin i$ has a constant value along the ray. Since, for $z = 0$, this constant is $\sin i_0$, equation (7.3) represents Snell's law for media with velocities that are variable with depth:

$$\frac{\sin i(z)}{v(z)} = \frac{\sin i_0}{v_0} = p \tag{7.4}$$

where, as we have already seen, p is the ray parameter, which has a constant value for each ray along its complete trajectory. If we take derivatives in equation (7.4) with respect to s (the ray's direction), then

$$\frac{\mathrm{d}}{\mathrm{d}s}(\sin i) = \frac{\mathrm{d}}{\mathrm{d}s}(pv) \tag{7.5}$$

and substituting in (7.2), we have

$$\frac{\mathrm{d}i}{\mathrm{d}s} = p\frac{\mathrm{d}v}{\mathrm{d}z} \tag{7.6}$$

112

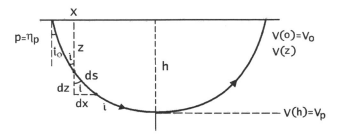

Fig. 7.1. A ray path or trajectory in a medium with velocity increasing with depth.

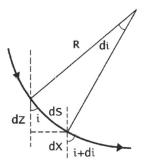

Fig. 7.2. The radius of curvature of a ray trajectory.

This equation relates changes in the angle of incidence $i(z)$ along the ray, that is changes in the ray trajectory, to changes in velocity.

According to (7.6), the curvature of the ray changes along its trajectory. At a point of this trajectory, the radius of curvature (Fig. 7.2) is given by

$$R = \frac{\mathrm{d}s}{\mathrm{d}i} \qquad (7.7)$$

Substituting equation (7.6) into this gives

$$R = \frac{1}{p\dfrac{\mathrm{d}v}{\mathrm{d}z}} \qquad (7.8)$$

If the velocity is constant, the radius of curvature is infinite and ray trajectories are straight lines. If the gradient of the velocity is constant, R is also constant and the ray trajectory is a circle. The curvature of a ray changes along its trajectory according to the changes in the velocity gradient. The curvatures of different rays vary according to their p values.

For two contiguous rays with parameters p and $p + \mathrm{d}p$, and for two wave fronts at times t and $t + \mathrm{d}t$, separated by a horizontal distance $\mathrm{d}x$ (Fig. 7.3), we can easily deduce (from $\sin i = \mathrm{d}s/\mathrm{d}x$) that

$$\frac{\mathrm{d}t}{\mathrm{d}x} = \frac{\sin i}{v} = p \qquad (7.9)$$

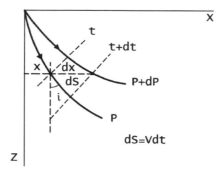

Fig. 7.3. The relation between the wave fronts and trajectories of two rays with parameters p and $p + dp$.

This result is similar to that obtained for a medium of constant velocity (6.46) and is known as the Benndorf relation.

As we have seen for the propagation of rays in media of constant velocity (Chapter 6), in the study of rays in media with variable velocities it is important to obtain expressions for the horizontal distance x, the distance traveled along the ray's trajectory s, and the travel time t. For media in which the velocity varies with depth, for a ray that starts at the surface and arrives at a certain depth z (Fig. 7.1), x, s, and t are given by

$$x = \int_0^z \tan i \, dz \tag{7.10}$$

$$s = \int_0^z \frac{dz}{\cos i} \tag{7.11}$$

$$t = \int_0^z \frac{dz}{v \cos i} \tag{7.12}$$

If the velocity increases with depth, dv/dz is positive and the radius of curvature is also positive; that is, rays are concave upward. If rays start at the surface, they will also end at the surface, turning at a maximum depth h (Fig. 7.1). The horizontal distance x, distance traveled along the ray s, and travel time t are given by twice the integrals (7.10)–(7.12), putting $z = h$. In a similar manner to what we did with equations (6.47) and (6.48), expressions (7.10)–(7.12) for rays that start and end at the surface, turning at a maximum depth h, can be written in terms of the ray parameter p and slowness η ($\eta = 1/v$) as

$$x = 2 \int_0^h \frac{p \, dz}{(\eta^2 - p^2)^{1/2}} \tag{7.13}$$

$$s = 2 \int_0^h \frac{\eta \, dz}{(\eta^2 - p^2)^{1/2}} \tag{7.14}$$

$$t = 2 \int_0^h \frac{\eta^2 \, dz}{(\eta^2 - p^2)^{1/2}} \tag{7.15}$$

We can also derive an expression for reduced times or the *tau* function $\tau(p)$, by substituting (7.13) and (7.15) into (6.51):

$$\tau(p) = 2 \int_0^h (\eta^2 - p^2)^{1/2} \, dz \tag{7.16}$$

We notice that expressions (7.13) and (7.15) for x and t are similar to those derived for a half-space of constant velocity (6.47) and (6.48). If the velocity increases with depth, then for each value of p there is a ray that arrives at a distance x in a time t, reaching a maximum depth h. Since, for $z = h$, $i = \pi/2$, the ray parameter $p = \eta_p = 1/v_p$, where v_p is the velocity at depth h. Along the ray $\eta \geq p$, and there is a singularity in expressions (7.13)–(7.15) at the limit $z = h$. According to (7.9), the slope of the travel time curve $t(x)$ equals the ray parameter at each value of x and corresponds to the inverse of the velocity at the turning depth h ($dt/dx = 1/v_p$). This is an important relation between travel time curves and velocity distributions. Depending on the form of $v(z)$, integrals for x and t may have analytic solutions. Only in some simple cases is it possible to find an explicit analytic function for $t(x)$.

7.2 The generalized formulation

The analysis of ray propagation in media with velocities that depend solely on depth allows a simplified approach to the generalized ray theory in terms of a Lagrangian formulation and Hamilton's canonical equations. We present only a brief discussion of the problem. The generalized formulation is very useful in the solution of problems in which the velocity varies with more than one coordinate. The travel time can be expressed, in a similar way to that in (7.12), in terms of the horizontal distance x and the differential ds along the ray (Cisternas, 1982):

$$t = \int_0^x \frac{ds}{v} \tag{7.17}$$

The differential along the ray ds is given (Fig. 7.1) by

$$ds = (dx^2 + dz^2)^{1/2} = (1 + z'^2)^{1/2} \, dx \tag{7.18}$$

where $z' = dz/dx$ and the independent variable is x, so that v is a function of z and z is a function of x. Equation (7.17) can be written as

$$t = \int_0^x \frac{(1 + z'^2)^{1/2}}{v} \, dx = \int_0^x L(z, z') \, dx \tag{7.19}$$

where $L(z, z')$ may be considered the Lagrangian function of the ray,

$$L(z, z') = \frac{(1 + z'^2)^{1/2}}{v} \tag{7.20}$$

Since (7.19) represents the travel time, according to Fermat's principle, this must be stationary, and, therefore,

$$\delta \int_0^x L(z, z') \, dx = 0 \tag{7.21}$$

This condition is satisfied if L is a solution of Lagrange's equation,

$$\frac{d}{dx} \left(\frac{\partial L}{\partial z'} \right) - \frac{\partial L}{\partial z} = 0 \tag{7.22}$$

If now we define the moment $P = \partial L / \partial z'$, then from equation (7.22) we deduce that

$$\frac{dP}{dx} = P' = \frac{\partial L}{\partial z} \tag{7.23}$$

We can now define the Hamiltonian $H(z, P)$:

$$H(z, P) = Pz' - L \tag{7.24}$$

The differential dH is given by

$$dH = z' \, dP - P' \, dz \tag{7.25}$$

Since dH is a perfect differential,

$$z' = \frac{\partial H}{\partial P} \tag{7.26}$$

$$P' = -\frac{\partial H}{\partial z} \tag{7.27}$$

This is a system of canonical equations of first order of Hamilton. Since, in our problem, the independent variable is x, we can determine dH/dx and, using (7.26) and (7.27), we have

$$\frac{dH}{dx} = \frac{\partial H}{\partial z} z' + \frac{\partial H}{\partial P} P' = 0 \tag{7.28}$$

In consequence, H is constant with x. If we substitute the Lagrangian L from (7.20) into equation (7.24), we obtain for H

$$H = \frac{-1}{v(1 + z'^2)^{1/2}} = p \tag{7.29}$$

where, according to (7.28), $p(x)$ is a constant. From equation (7.16) we have (Fig. 7.1) that

$$\frac{1}{(1 + z'^2)^{1/2}} = \frac{dx}{ds} = \sin i \tag{7.30}$$

On substituting (7.30) into (7.29), we find that p is the ray parameter:

$$\frac{\sin i}{v} = \frac{1}{c} = p \tag{7.31}$$

where c is the apparent velocity of the wave front in the direction of x (section 6.3, Fig. 6.2). The condition $H(x) = $ constant for each ray is a formulation of Snell's law.

This is a different formulation of ray theory that has led to the same results as those we already knew. For more general problems this formulation is very useful.

7.3 The change of distance with the ray parameter

For a given depth distribution of the velocity, at a distance x there arrives a ray with parameter p. In order to understand the behavior of ray trajectories in media of different velocity distributions, it is important to consider the change in distance with the change in ray parameter. Equation (7.13) with p as a variable represents distances corresponding to rays of different parameters. If we take derivatives with respect to p, we obtain

$$\frac{dx}{dp} = 2 \int_0^h \frac{dz}{(\eta^2 - p^2)^{1/2}} + 2p \frac{d}{dp} \int_0^h \frac{dz}{(\eta^2 - p^2)^{1/2}} \tag{7.32}$$

Changing the integral over z to one over η, putting $dz = f(\eta)\, d\eta$ and changing the limits of integration accordingly (for $z = 0$, to $\eta = \eta_0$; and for $z = h$, to $\eta = \eta_p$), we have

$$\frac{dx}{dp} = 2 \int_{\eta_0}^{\eta_p} \frac{f(\eta)\, d\eta}{(\eta^2 - p^2)^{1/2}} + 2p \frac{d}{dp} \int_{\eta_0}^{\eta_p} \frac{f(\eta)\, d\eta}{(\eta^2 - p^2)^{1/2}} \tag{7.33}$$

Integrating by parts in the second term gives

$$2p \frac{d}{dp} \left[f(\eta_0) \cosh^{-1} \left(\frac{\eta_0}{p} \right) \int_{\eta_0}^{\eta_p} \cosh^{-1} \left(\frac{\eta}{p} \right) f'(\eta)\, d\eta \right] \tag{7.34}$$

Derivation with respect to p in this expression results in

$$\frac{\eta_0 f(\eta_0)}{p(\eta_0^2 - p^2)^{1/2}} + \int_{\eta_0}^{\eta_p} \frac{\eta f'(\eta)\, d\eta}{p(\eta^2 - p^2)^{1/2}} \tag{7.35}$$

Substituting (7.35) into (7.33) gives

$$\frac{dx}{dp} = \frac{2\eta_0 f(\eta_0)}{(\eta_0^2 - p^2)^{1/2}} + 2 \int_{\eta_0}^{\eta_p} \frac{[f(\eta) + \eta f'(\eta)]\, d\eta}{(\eta^2 - p^2)^{1/2}} \tag{7.36}$$

Now we introduce the variable ζ, which is the gradient of the velocity per unit of velocity:

$$\zeta = \frac{1}{v} \frac{dv}{dz} = -v \frac{d\eta}{dz} \tag{7.37}$$

Since $f(\eta) = dz/d\eta$, $\eta f(\eta) = -1/\zeta$ and

$$f(\eta) + \eta f'(\eta) = \frac{d}{d\eta} [\eta f(\eta)] = \frac{1}{\zeta^2} \frac{d\zeta}{d\eta} \tag{7.38}$$

By substitution into (7.36) and changing the integral over η to one over z, we finally obtain

$$\frac{dx}{dp} = \frac{-2}{\zeta_0(\eta_0^2 - p^2)^{1/2}} + 2 \int_0^h \frac{\frac{d\zeta}{dz}\, dz}{\zeta^2(\eta^2 - p^2)^{1/2}} \tag{7.39}$$

where, as a function of the gradient of the velocity with depth,

$$\frac{d\zeta}{dz} = -\frac{1}{v^2}\left(\frac{dv}{dz}\right)^2 + \frac{1}{v}\frac{d^2v}{dz^2} \tag{7.40}$$

Equation (7.39) represents the change in horizontal distance x when we pass from one ray of parameter p to another of parameter $p + dp$. This is an important relation for studying the behavior of rays in a medium with a certain distribution of velocity with depth, $v(z)$. The first term of (7.39) depends on the gradient of the velocity at the surface ζ_0 and the second depends on the gradient at each depth ζ and its change with depth, $d\zeta/dz$. If changes in the gradient of the velocity are small, the first term is dominant and dx/dp is negative. This means that rays with smaller values of p arrive at larger distances. If there are large changes in the gradient of the velocity, dx/dp may become positive and there is an inversion, which means that rays with smaller values of p arrive at smaller distances.

7.4   The velocity distribution with ζ constant

A first application of equation (7.39) corresponds to the case with ζ constant and, thus, the second term is zero:

$$\frac{dx}{dp} = \frac{-2}{\zeta_0(\eta_0^2 - p^2)^{1/2}} \tag{7.41}$$

This case is of no direct interest in seismology, but constitutes a good exercise to help the reader understand the applications of this equation. If we assign a constant value $\zeta = \alpha$ (a small positive value), then, from equation (7.37), we obtain

$$\alpha\,dz = \frac{dv}{v}$$

By integration of this equation with the condition that, for $z = 0$, $v = v_0$, we obtain an exponential distribution of velocity with depth:

$$v = v_0\,e^{\alpha z} \tag{7.42}$$

From equation (7.41), integrating with respect to p between η_0 and η_p, we obtain for x

$$x = \frac{1}{\alpha}\int_{\eta_0}^{\eta_p} \frac{-2\,dp}{(\eta_0^2 - p^2)^{1/2}} = \frac{2}{\alpha}\cos^{-1}\left(\frac{\eta_p}{\eta_0}\right) \tag{7.43}$$

Since $\eta_p = p$, the ray parameter for the distance x, on solving for $p(x)$ we have

$$p = \frac{1}{v_0}\cos\left(\frac{\alpha x}{2}\right) \tag{7.44}$$

Since $p = dt/dx$, by integrating (7.44) between 0 and x, we obtain the travel time $t(x)$:

$$t = \frac{2}{\alpha v_0}\sin\left(\frac{\alpha x}{2}\right) \tag{7.45}$$

In this distribution, both the velocity and its gradient increase exponentially with depth, which is not a property applicable to real media. However, this simple exercise has shown us how to obtain travel times $t(x)$ from the expression for dx/dp for a specified distribution of velocity with depth. We will see later how we can find certain characteristics of traveling time curves, using equation (7.39), without solving the integral.

7.5 A linear increase of velocity with depth

A distribution of velocity that is very useful in seismology is that with a linear increase with depth. This distribution is applicable in many cases to the Earth's crust and upper mantle. It is

$$v(z) = v_0 + kz \tag{7.46}$$

where v_0 is the velocity at the surface $(z = 0)$ and k is the constant velocity gradient $(dv/dz = k)$. A ray that arrives at a horizontal distance x penetrates to a maximum depth h (Fig. 7.4). The ray parameter p is, then, given by

$$p = \frac{1}{v_0 + kh} \tag{7.47}$$

According to (7.8), the radius of curvature is

$$R = \frac{v_0}{k} + h \tag{7.48}$$

Since R is constant, the ray's trajectory is circular with its center at a distance v_0/k above the surface (Fig. 7.4). According to ray geometry, the radius of curvature can also be written as a function of the horizontal distance:

$$R = \left[\left(\frac{x}{2} \right)^2 + \left(\frac{v_0}{k} \right)^2 \right]^{1/2} \tag{7.49}$$

From (7.48) and (7.49) we can deduce that the velocity at the maximum ray depth is

$$v_m = k \left[\left(\frac{x}{2} \right)^2 + \left(\frac{v_0}{k} \right)^2 \right]^{1/2} \tag{7.50}$$

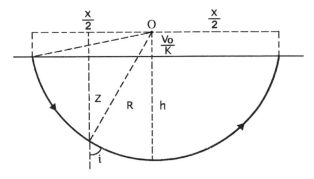

Fig. 7.4. The circular path of a ray in a medium with a linear increase of the velocity with depth.

According to Snell's law,

$$\sin i = \frac{v_0 + kz}{v_0 + kh} \tag{7.51}$$

The travel time can be determined directly by integration; on substituting (7.51) into (7.15), we obtain

$$t = \frac{2}{k} \int_0^h \frac{k \, dz}{(v_0 + kz)\left[1 - \left(\dfrac{v_0 + kz}{v_m}\right)^2\right]^{1/2}} \tag{7.52}$$

On substituting for v_m, we obtain

$$t = \frac{2}{k} \cosh^{-1}\left(\frac{v_m}{v_0}\right) \tag{7.53}$$

and, finally, using (7.50), we get the travel time as a function of the horizontal distance $t(x)$:

$$t = \frac{2}{k} \sinh^{-1}\left(\frac{kx}{2v_0}\right) \tag{7.54}$$

This expression can also be obtained by using equation (7.39) for dx/dp, as we did in section 7.4. According to (7.37), for a linear increase in velocity, ζ and $d\zeta/dz$ are given by

$$\zeta = \frac{k}{v_0 + kz} \tag{7.55}$$

$$\frac{d\zeta}{dz} = \frac{-k^2}{(v_0 + kz)^2} \tag{7.56}$$

On substituting into (7.39), we obtain

$$\frac{dx}{dp} = \frac{-2}{kp^2(1 - v_0^2 p^2)^{1/2}} \tag{7.57}$$

By integration with respect to p between η_0 and η_p (the slowness at the ray's deepest point) and remembering that for each ray $\eta_p = p$, we obtain for $x(p)$

$$x = \frac{2}{kp}(1 - v_0^2 p^2)^{1/2} \tag{7.58}$$

Solving for p as a function of x gives

$$p = \frac{1}{\left[\left(\dfrac{kx}{2}\right)^2 + v_0^2\right]^{1/2}} \tag{7.59}$$

Since $p = \mathrm{d}t/\mathrm{d}x$, $t(x)$ can be obtained by integration of (7.59) between 0 and x:

$$t = \frac{1}{v_0} \int_0^x \frac{\mathrm{d}x}{\left[\left(\dfrac{kx}{2v_0} \right)^2 + 1 \right]^{1/2}} \tag{7.60}$$

The result of this integral is again expression (7.54).

For a medium with a linear increase in velocity with depth, that is, one with a constant velocity gradient (Fig. 7.5(a)), rays are circular (Fig. 7.5(b)). The curve $\mathrm{d}x/\mathrm{d}p$ (Fig. 7.5(c)) always has negative values and singularities at $p = 0$ and $p = \eta_0$. The ray parameter p decreases when x increases, with a maximum value η_0 for $x = 0$, and tends to zero for x infinite (Fig. 7.5(d)). The travel time curve $t(x)$ has a positive slope with value $\mathrm{d}t/\mathrm{d}x = \eta_0$, for $x = 0$, that decreases monotonically as x increases (Fig. 7.5(e)).

7.6 Distributions of velocity with depth

Equation (7.39) may be used to find the behavior of rays for different distributions of velocity with depth without solving completely for expressions of $p(x)$ and $t(x)$. This equation provides us with information on rays' trajectories and travel times for general types of velocity distributions. The first step is to relate the characteristics of the velocity distribution $v(z)$ and its gradient $\mathrm{d}v/\mathrm{d}z$ to values of ζ and $\mathrm{d}\zeta/\mathrm{d}z$, and thus find the general form of the curve $\mathrm{d}x/\mathrm{d}p$ as a function of p. Integration of this equation gives us the curve for $p(x)$, that is, the behavior of the ray parameter with distance. Since $p = \mathrm{d}t/\mathrm{d}x$, a new integration gives us the travel time curve $t(x)$. Also, from knowledge of the velocity gradient, we obtain the radius of curvature (7.8) and the general characteristics of rays' trajectories. In this way, even if we do not know the analytic expression for the velocity distribution with depth, knowing some of its characteristics we can find some properties of rays' trajectories and travel times. We will study three cases corresponding to velocity distributions with a gradual increase, rapid increase, and decrease of velocity with depth.

7.6.1 *A gradual increase of velocity*

A useful distribution of velocity with depth applicable to the Earth is a gradual increase with a small gradient that changes very slowly. This distribution is known in seismology as the normal distribution of velocity. Since $\mathrm{d}v/\mathrm{d}z$ is small and positive, ζ has the same characteristics (7.37). Since there are only very small changes in velocity gradient, according to (7.40), $\mathrm{d}\zeta/\mathrm{d}z$ is very small. Therefore, in (7.39), the first term is the most important and $\mathrm{d}x/\mathrm{d}p$ is negative. In this aspect, this case is similar to that of constant ζ (section 7.4). The velocity distribution with constant gradient of section 7.5, for small values of gradient, is a particular case of the normal distribution.

The most important properties of this distribution are represented in Fig. 7.6 for a surface focus. The velocity $v(z)$ increases gradually with depth (Fig. 7.6(a)) from a value v_0 at the surface; the gradient of velocity changes very slowly with depth. The slowness η decreases with depth from a value η_0 at the surface (Fig. 7.6(b)). The rays

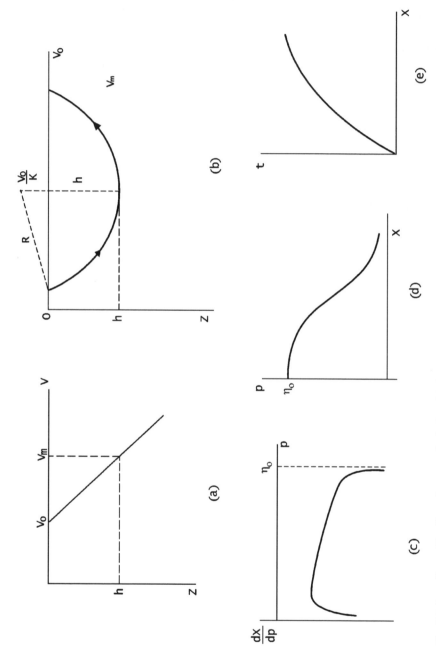

Fig. 7.5. A medium with a linear increase of velocity with depth. (a) The distribution of the velocity with depth. (b) A ray trajectory. (c) The variation of the ray parameter dx/dp. (d) The variation of the ray parameter with distance. (e) The travel time curve.

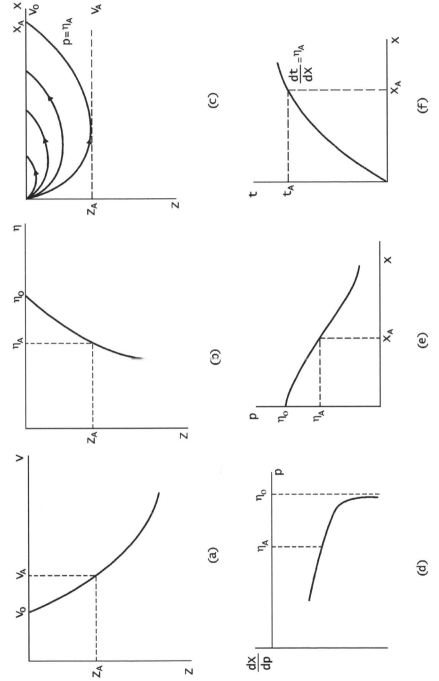

Fig. 7.6. A medium with a normal distribution of velocity with depth. (a) The distribution of velocity with depth. (b) The distribution of slowness (the inverse of velocity) with depth. (c) Ray trajectories. (d) The variation of the distance with the ray parameter dx/dp. (e) The variation of the ray parameter with distance. (f) The travel time curve.

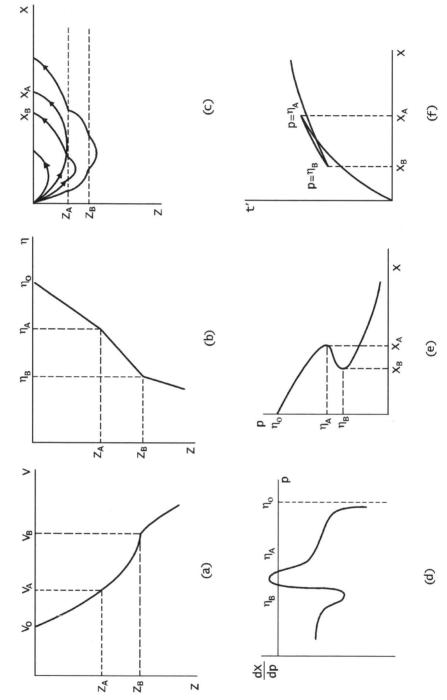

Fig. 7.7. A medium with a zone of a rapid increase of velocity with depth. (a) The distribution of velocity with depth. (b) The distribution of slowness (the inverse of velocity) with depth. (c) Ray trajectories. (d) The variation of distance with the ray parameter dx/dp. (e) The variation of the ray parameter with distance. (f) The travel time curve.

124

are curved, concave upward since the radii of curvature (7.8) are positive and large, dv/dz being small (Fig. 7.6(c)). For each distance, the ray parameter p equals the slowness at each deepest point $p = \eta_p$. A generic ray that arrives at x_A, turns at its deepest point z_A, and has a parameter $p = \eta_A$, has a velocity at that point of $v(z_A) = v_A = 1/\eta_A$.

dx/dp is always negative and varies with p from the singularity at $p = \eta_0$ that corresponds to $x = 0$, to a limit value for very small values of p that correspond to distances that tend to infinity (Fig. 7.6(d)). The limit, $p = 0$, can not be considered since it corresponds to a vertical ray that does not arrive at the surface (or arrives at an infinite distance). For all other values of p, rays arrive at a finite distance. The integral of this curve gives us the relation between p and x (Fig. 7.6(e)). For distances between zero and infinity, p varies between η_0 and 0. Since $p = dt/dx$, by integration of the curve $p(x)$ we obtain the travel time curve $t(x)$ (Fig. 7.6(f)). At the origin the slope of $t(x)$ is $dt/dx = \eta_0$, inverse of the surface velocity v_0. As x increases, the slope of $t(x)$ decreases monotonically. The travel time curve has, then, the form of a curve with a continuously decreasing slope, that is, d^2t/dx^2 is negative.

7.6.2   A rapid increase of velocity

Let us consider a distribution of velocity such that, at a certain depth, the velocity passes from a gradual increase to a rapid increase and, at a further depth, returns to a gradual increase (Fig. 7.7(a)). The zone of rapid increase is located between depths z_A and z_B and there the velocity varies from v_A to v_B (the slowness varies from η_A to η_B). The gradient in this zone is larger than those for depths above and below, so that it changes rapidly at both ends (Figs. 7.7(a) and (b)). From the origin to a distance x_A, rays do not penetrate the zone of rapid increase and have the properties of a normal distribution.

Since the radius of curvature depends on the inverse of the velocity gradient (7.8), rays inside the zone of a rapid increase in velocity have smaller radii (Fig. 7.7(c)). Owing to this change, a ray that turns there may arrive at a distance x_B smaller than x_A. To understand this situation, let us consider the curve dx/dp. If, for a range of values of p near the value η_A, the change in the velocity gradient is sufficiently large that $d\zeta/dz$ is large and positive, according to (7.39), then dx/dp becomes positive (Fig. 7.7(d)). The curve for $p(x)$ (Fig. 7.7(e)) now has a range of values of p between η_A and η_B, for which x decreases when p decreases. Rays with these values of p that penetrate deeper arrive at shorter distances (Figs. 7.7(c) and (e)). For rays that turn at greater depth ($z > z_B$), where the velocity distribution is again normal, p decreases with increasing x.

The travel time curve $t(x)$ has its first part (between 0 and x_A) just like that in the form of the normal distribution (Fig. 7.7(f)). For distances corresponding to rays with values of p for which dx/dp is positive (between η_A and η_B), distances decrease from x_A to x_B and dt/dx increases with x (d^2t/dx^2 is positive). These rays form a second branch of the travel time curve for which the curvature has changed sign; now it is concave upward. Rays that penetrate to greater depths have negative values of dx/dp and arrive at greater distances as p decreases. On the travel time curve (Fig. 7.7(f)), these rays correspond to a third branch with negative curvature, concave downward, like the first. According

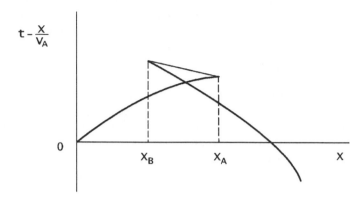

Fig. 7.8. The reduced travel time curve for a medium with a zone of a rapid increase of velocity with depth. The reduced velocity v_A corresponds to the beginning of the rapid increase.

to the $p(x)$ curve (Fig. 7.7(e)), for distances between x_B and x_A there exist three values of p for each distance x. This means that three different rays arrive at each distance. This results in the triplication of the travel time curves with cusps at their ends, which represent a focusing of rays known as a caustic. There is a branch of positive curvature (upward concavity), called a retrograde branch, since rays that penetrate deeper arrive at shorter distances $(dp/dx > 0)$. The presence of this triplication of travel times is a typical feature of velocity distributions for which there is a zone of rapid increase in velocity with depth with a pronounced change in the velocity gradient. If the change in velocity gradient is not very pronounced, dp/dx does not become positive and there is no retrograde branch in the travel time curve. In this case, there is at a certain distance a change in the slope of the travel time curve which is formed by two branches, both with downward concavity.

In the reduced travel time representation, the curve has the same characteristics (Fig. 7.8). If the reduced velocity v_A corresponds to the region above the zone of rapid increase, the first branch of the curve corresponds to rays that penetrate to a depth z_A, the second for distances between x_B and x_A is the retrograde branch with positive values of dp/dx and upward concavity, and the third is a normal branch for rays that penetrate deeper than z_B. In this representation, differences among the three branches are accentuated.

We have seen that, in this case, the travel time $t(x)$ is not a single-valued function (there is a zone of the curve for which, for each value of x, there are three values of t). However, for the same case, the *tau* function $\tau(p)$ (6.51) is a single-valued function (for each value of p there is one value of τ). According to (6.51), for $x = 0$, $t = 0$, $p = \eta_0$, and $\tau = 0$. The slope of the curve $d\tau/dp$ is always negative (6.53) and τ increases when p decreases (Fig. 7.9). The change in the slope is given by $d^2\tau/dp^2$ which, according to (6.53), is $-dx/dp$. Thus, for values of p for which dx/dp is negative, the curve is concave upward, and, when dx/dp is positive, the curve is concave downward (Fig. 7.9). In the $\tau(p)$ curve the rapid change in velocity gradient results in a change in the sign of the curvature corresponding to the triplication of the $t(x)$ curve.

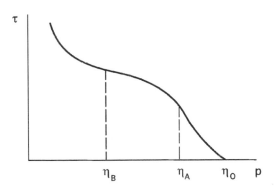

Fig. 7.9. The curve of the *tau* function (reduced time) with the ray parameter, for a medium with a zone of a rapid increase of velocity with depth.

7.6.3   *A decrease of velocity. A low-velocity layer*

We consider now a velocity distribution such that the velocity decreases at a certain depth and, further down, increases again (Fig. 7.10(a)). Between the surface and a depth z_A, the velocity increases just like it does in the normal distribution. Between z_A and z_B, the velocity decreases with depth. From z_B downward the velocity increases again. There are, then, two depths, z_A and z_C, at which the velocity has the same value v_A. The zone where the velocity is less than v_A is called a low velocity layer. Rays that penetrate to depths less than z_A behave just like they do in the normal distribution, reaching a maximum distance x_A (Fig. 7.10(c)). Rays that penetrate into the zone where the velocity decreases with depth, since dv/dz is negative, change their curvature, which is now concave downward. Those that penetrate below z_B have again an upward concavity. According to Snell's law, rays can not turn upwards until a depth z_C where the velocity starts to be larger than v_A.

Values of dx/dp are negative for ray parameters p between η_0 and η_A, that is, for those rays that do not penetrate into the low-velocity layer. For depths between z_A and z_C, the ray parameter p is not defined, since there are no rays that turn at those depths. The curve for dx/dp is discontinuous at $p = \eta_A$ (Fig. 7.10(d)). Values of p less than η_A correspond to rays that turn upward at depths below z_C. Since, in the integral of (7.39), for these rays there are two zones (z_A and z_B) where the change in velocity gradient is large, $d\zeta/dz$ is also large and dx/dp may become positive. In the integral of (7.39) for rays that penetrate further down, this contribution is smaller and dx/dp becomes negative again. In the $p(x)$ curve (Fig. 7.10(e)), the discontinuity of dx/dp at $p = \eta_A$ results in the existence of two rays for the same value of p, those that turn upward at z_A and arrive at x_A, and those that turn upward at z_C and arrive at a larger distance x_C. Since, for $p = \eta_A$, dx/dp is positive, x decreases as p decreases, corresponding to rays that arrive at distances from x_C to x_B. When dx/dp becomes negative again, x increases as p decreases (Fig. 7.10(e)). This change happens at a distance x_B that is always larger than x_A.

Depending on the thickness of the low-velocity zone and its magnitude, the distance interval from x_A to x_B is larger or smaller. Since, at these distances, there are no values of

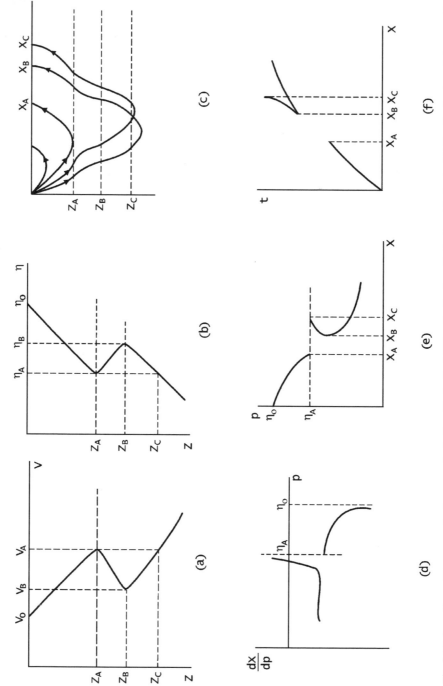

Fig. 7.10. A medium with a zone of a decrease of velocity with depth. (a) The distribution of velocity with depth. (b) The distribution of slowness (the inverse of velocity) with depth. (c) Ray trajectories. (d) The variation of the distance with the ray parameter dx/dp. (e) The variation of the distance with the ray parameter with distance. (f) The travel time curve.

p, this means that no rays arrive and the interval is called a 'shadow zone'. Behind the shadow zone, between x_B and x_C, there are two values of p for each value of x. Therefore, two rays arrive at each of these distances (Fig. 7.10(e)). For distances greater than x_C, the situation becomes normal again.

The travel time curve $t(x)$ (Fig. 7.10(f)) has, for distances up to x_A, the same form as that in the normal velocity distribution. Between the distances x_A and x_B, there is a shadow zone where no rays arrive and the curve is discontinuous. At the distance x_B two branches correspond to the two values of p for each value of x in the $p(x)$ curve (Fig. 7.10(e)). The first branch starts at x_B, has the form corresponding to a normal distribution, concave downward, and continues indefinitely. The second branch starts at x_B with slope η_A, the same as that at x_A before the shadow zone, has positive curvature (concave upward) corresponding to positive values of dx/dp, and ends at x_C. This is a retrograde branch since it corresponds to the part of the $p(x)$ curve for which x decreases as p also decreases (Fig. 7.10(c)). A caustic appears at the cusp of the two branches. The time corresponding to the distance x_B, after the shadow zone, has a delay with respect to the prolongation of the curve from the point before it. In conclusion, travel time curves for cases in which there are low-velocity zones exhibit a shadow zone, a duplication of rays, and a delay of arrivals after the shadow zone.

The *tau* function curve $\tau(p)$ is single-valued; its slope $d\tau/dp$ increases for decreasing values of p from η_0 to η_A. At this point there is a discontinuity and, in the interval from η_A to η_B, the slope of the curve decreases with decreasing p. For values of p smaller than η_B the slope increases again.

7.7 Travel times for deep foci

Let us consider now travel times corresponding to deep foci in a medium with a normal distribution of velocity with depth. The travel time curve starts at a value of time corresponding to that of the vertical ray from the focus to the surface. For short distances, rays travel from the focus upward $(i_h > \pi/2)$, at a certain distance the ray leaves the focus horizontally $(i_h = \pi/2)$, and for greater distances rays leave the focus in the downward direction $(i_h < \pi/2)$ (Fig. 7.11(a)).

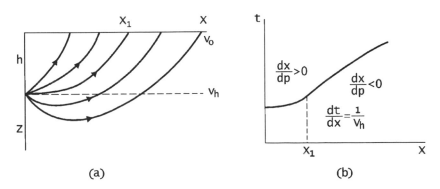

Fig. 7.11. A deep focus. (a) Ray paths. (b) The travel time curve.

The travel time curve has a first part that is concave upward corresponding to upward rays ($dx/dp > 0$). The inflection point corresponds to the distance for the ray with $i_h = \pi/2$. For larger distances, the curvature of the travel time curve is concave downward ($dx/dp < 0$) (see Fig. 7.13(b) later). At the inflection point, the slope of the travel time curve corresponds to the inverse of the velocity at the focus ($dt/dx = 1/v_h$). This property can be used to find the velocity at focal depth.

7.8   The determination of the velocity distribution

We have seen how the form of travel time curves depends on the velocity's distribution with depth. Therefore, we may be able to obtain velocity distributions from travel time curves. This is one of the classical inverse problems in seismology. The direct problem, that of how to calculate travel time curves from velocity distributions, is always soluble, with greater or lesser difficulty. The inverse problem involves a greater difficulty and need not always be soluble in an exact and unique way.

Let us suppose that we know the continuous travel time curve $t(x)$, in such a way that we can calculate its slope $dt/dx = p$ at each point. The distance x at which a ray of parameter p arrives, according to (7.13), changing the integral over z to one over η, is given by

$$x = 2p \int_{\eta_p}^{\eta_0} \frac{\dfrac{dz}{d\eta}\, d\eta}{(\eta^2 - p^2)^{1/2}} \tag{7.61}$$

The limits of the integral are η_0, the slowness for $z = 0$, and η_p that corresponds to $z = h$, the depth of the turning point of the ray that arrives at a distance x with the parameter $p = \eta_p$. Expression (7.61) represents an integral along a single ray of parameter p. Let us consider now all rays that arrive at distances from $x = 0$ to $x = x_1$, corresponding to values of p between p_0 and p_1. For all these rays we may write the integral

$$\int_{p_1}^{p_0} \frac{dp}{(p^2 - \eta_1^2)^{1/2}}$$

where p is a variable.

If we apply this operation to both sides of (7.61), we obtain

$$\int_{p_1}^{p_0} \frac{x\, dp}{(p^2 - \eta_1^2)^{1/2}} = \int_{p_1}^{p_0} \int_{\eta_p}^{\eta_0} \frac{2p \dfrac{dz}{d\eta}\, d\eta\, dp}{[(p^2 - \eta_1^2)(\eta^2 - p^2)]^{1/2}} \tag{7.62}$$

In this expression x and p are variables that take values between 0 and x_1 and between p_0 and p_1, respectively. If we consider the relation between η and p, in the right-hand-side term of (7.62), the first integral between η_p and η_0 corresponds to a single ray that arrives at a distance x. In the second integral over p from p_1 to p_0, η_p is a variable and the operation corresponds to integration over all rays that arrive at distances from 0 to x_1. Considering the plane (η, p) (Fig. 7.12), the first integral from η_0 to η_p corresponds to a vertical strip. The second integral from p_1 to p_0 covers horizontally the triangle

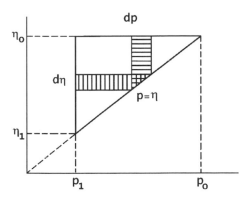

Fig. 7.12. The relation of the integrals over p and over η on the plane (p, η).

with vertices (p_0, η_0), (p_1, η_0), and (p_1, η_1). Then, we can cover the same triangle by changing the order of integration, integrating first over p along a horizontal strip from p_1 to p and then vertically over η from η_1 to η_0. The double integral is now

$$\int_{\eta_1}^{\eta_0} \int_{p_1}^{p} \frac{p \, dp}{[(p^2 - \eta_1^2)(\eta^2 - p^2)]^{1/2}} \frac{dz}{d\eta} \, d\eta \tag{7.63}$$

The integral over p, taking into account that $p = \eta$ and $p_1 = \eta_1$ at the limits, is a definite integral whose value is π:

$$\int_a^b \frac{x \, dx}{[(x^2 - a^2)(b^2 - x^2)]^{1/2}} = \pi$$

The integral over η can be transformed into an integral over z from 0 to z_1 and its value is z_1. The value of the double integral (7.63) is finally πz_1. In the integral of the left-hand-side term of (7.62), changing the limits to $p_0 = \eta_0$ and $p_1 = \eta_1$ and integrating by parts, we obtain

$$\int_{\eta_1}^{\eta_0} \frac{x \, dp}{(p^2 - \eta_1^2)^{1/2}} = \left[x \cosh^{-1} \left(\frac{p}{\eta_1} \right) \right]_{\eta_1}^{\eta_0} - \int_{\eta_1}^{\eta_0} \cosh^{-1} \left(\frac{p}{\eta_1} \right) \frac{dx}{dp} \, dp \tag{7.64}$$

The first term is zero, since, for $\eta = \eta_0$, $x = 0$, and, for $\eta = \eta_1$, the hyperbolic arccosine is zero. The second term can be written as an integral over x, from 0 to x_1. The argument of the hyperbolic arccosine can be rewritten, since $p = \eta_p = 1/v_p$ and $\eta_1 = 1/v_1$, and finally we obtain

$$z_1 = \frac{1}{\pi} \int_0^{x_1} \cosh^{-1} \left(\frac{v_1}{v_p} \right) dx \tag{7.65}$$

This equation is known as the Herglotz–Wiechert integral for a plane medium. Using this expression we can calculate the velocity's distribution with depth from the travel time curve. The application of equation (7.65) consists in the following steps. The depth z_1 is the turning point (maximum depth) of the ray which arrives at a distance x_1. At this depth the velocity is v_1, which can be obtained from the slope of the $t(x)$

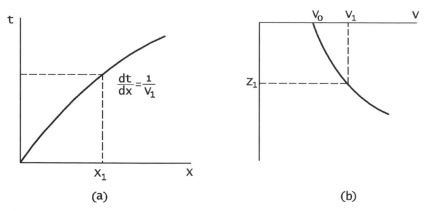

Fig. 7.13. The application of the Herglotz–Wiechert integral formula to obtain the velocity distribution from the travel time curve. (a) The travel time curve. (b) The distribution of velocity with depth.

curve (at x_1, $dt/dx = \eta_1 = 1/v_1$). To obtain the value of z_1 we have to calculate the integral of (7.65). This can be done numerically, since we can determine v_p from the slope of $t(x)$ for each value of x between 0 and x_1 (Fig. 7.13(a)). In this way we obtain v_1 and z_1, that is, one point of the curve $v(z)$ (Fig. 7.13(b)). We obtain the other points by changing the value of x_1.

This method is valid if we know $t(x)$ and dt/dx for each value of x and dt/dx is a monotonically decreasing function of x. This limits the method to normal velocity distributions. The method can not be applied directly if there are velocity discontinuities, zones of very rapid increase, or low-velocity layers. However, there are modifications of the method that allow its application to such distributions. Another limitation is the need to know the travel time curve in a continuous fashion, which is not possible for observed curves that are formed by discreet values of x and t. This limitation is overcome by interpolation between observed values. For plane geometry, this method can be applied to find velocity distributions for regions of the crust or upper mantle where the above conditions are assumed to be satisfied.

7.9   The energy propagated by ray beams. Geometric spreading

From the point of view of ray theory, energy is propagated along a bundle or beam of nearby rays. Since rays may converge or diverge during their propagation due to the velocity distribution, this fact must be taken into account in order to determine the energy that arrives at a certain distance. Let us consider a point source of seismic waves from which an energy E is emitted homogeneously in all directions per unit time. The energy emitted per unit solid angle is $P = dE/d\Omega$ and the total energy per unit time is $E = 4\pi P$. For a focus at depth h, an element of solid angle $d\Omega$ sustained by an element of take-off angle di_h corresponding to a ray beam with take-off angles between i_h and $i_h + di_h$ is given by

$$d\Omega = 2\pi \sin i_h \, di_h \qquad (7.66)$$

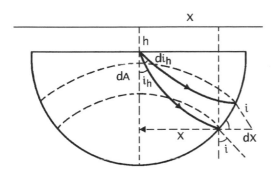

Fig. 7.14. The circular area covered by a wave front of a ray beam that leaves the focus with take-off angle i_h for a distance x.

At a horizontal distance x, the surface element of the wave front dA swept by a ray beam of solid angle $d\Omega$ (Fig. 7.14) is

$$dA = 2\pi x \cos i \, dx \tag{7.67}$$

where i is the angle of incidence of the ray at a distance x. The energy intensity or energy per unit of wave-front area is $I = dE/dA$ or, in terms of P, $I = P \, d\Omega/dA$. On substituting (7.66) and (7.67) into this, we obtain

$$I = \frac{P \sin i_h \, di_h}{x \cos i \, dx} \tag{7.68}$$

Because the distance x reached by a ray depends on the take-off angle at the focus i_h, the distance interval can be expressed by the interval in angle of incidence $dx = (dx/di_h) \, di_h$ and, by substitution into (7.68),

$$I = \frac{P \sin i_h}{x \cos i_h \dfrac{dx}{di_h}} \tag{7.69}$$

If we differentiate with respect to x in Snell's law (6.46), we obtain

$$\frac{d}{dx} \left(\frac{\sin i_h}{v_h} \right) = \frac{dp}{dx} = \frac{d^2 t}{dx^2} \tag{7.70}$$

From (7.70) we deduced that

$$\frac{dx}{di_h} = \frac{\cos i_h}{v_h \dfrac{d^2 t}{dx^2}} \tag{7.71}$$

where v_h is the velocity at the focus. By substituting in (7.69), we can express (7.69) in terms of the second derivative of the traveling time:

$$I = \frac{P v_h \tan i_h}{x \cos i} \frac{d^2 t}{dx^2} \tag{7.72}$$

If rays arrive at the surface with angle of incidence i_0, then, by substitution into (7.68) and (7.72), we obtain the energy per unit time and per unit of wave-front area that

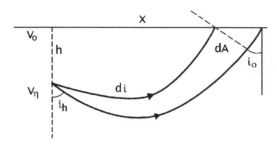

Fig. 7.15. The arrival at the surface of a wave front corresponding to a ray beam that leaves the focus with take-off angle i_h and arrives at the surface with angle of incidence i_0.

arrives at a distance x (Fig. 7.15):

$$I = \frac{P \sin i_h}{x \cos i_0 \dfrac{\mathrm{d}x}{\mathrm{d}i_h}} = \frac{P v_h \tan i_h}{x \cos i_0} \frac{\mathrm{d}^2 t}{\mathrm{d}x^2} \tag{7.73}$$

If the focus is at the surface, $i_h = i_0$.

In expression (7.73), we can see that the energy intensity decreases with the inverse of the distance. This dependence on the factor $1/x$ is called geometric spreading and is due to the increase in the wave-front area comprised by a ray beam as rays propagate from the focus. According to the first term of (7.73), the energy intensity depends inversely on $\mathrm{d}x/\mathrm{d}i_h$. When $\mathrm{d}x/\mathrm{d}i_h$ is small this means that x varies little with i_h and therefore the ray density at an interval of distance $\mathrm{d}x$ is large. If $\mathrm{d}x/\mathrm{d}i_h$ is large, x varies greatly with i_h and the ray density for an interval of distance is small. According to the second term of (7.73), the variation in energy density with distance is related to the second derivative of the travel time. If the travel time has a uniform gradient, then the distribution of the energy density with distance is also uniform. When there are large changes in the gradient, at the cusps, we find an accumulation of energy at the corresponding distances. This may occur when there is a zone of rapid increase in velocity with depth, or also when there is a low-velocity zone (Figs. 7.7 and 7.10).

8 RAY PROPAGATION IN A SPHERICAL MEDIUM

8.1 The geometry of ray trajectories and displacements

In the study of seismic waves for large distances ($x > 1000\,\mathrm{km}$), the flat-Earth approximation is no longer valid and the spherical shape of the Earth must be considered. The theory of propagation of seismic rays was developed by Rudzki, Benndorf, Zöppritz, Geiger, and Wiechert. Ray propagation inside a sphere requires certain modifications to the notions treated in Chapter 7. Let us consider a sphere of radius r_0 and velocity depending on the radius $v(r)$. Rays propagate from a focus F on the surface to a point P that is also on the surface (Fig. 8.1). If the velocity depends solely on the radius, rays are contained on a plane that includes the center of the sphere. If the velocity increases toward the center, ray trajectories are concave toward the surface and they reach a point on the surface. Distances between two points on the surface (from F to P in Fig. 8.1) are represented by the angle Δ at the center of the sphere (the distance on the surface is $r_0 \Delta$). The angle of incidence, i, is measured from the ray trajectory to the radius. The radial distance from the center to the point where a ray turns upward is r_p, corresponding to the velocity v_p (Fig. 8.1).

If the spherical medium is formed by concentric spherical layers of constant velocity (Fig. 8.2), then according to Snell's law (5.1), a ray that passes from a layer of velocity v_1 to another of velocity v_2 must satisfy

$$\frac{\sin i_1}{v_1} = \frac{\sin f}{v_2} \tag{8.1}$$

We want to relate the angles i_1 at the base of layer 1 to i_2 at the base of layer 2. Considering the triangles POR and QOR, we can write the relation

$$r_2 \sin i_2 = r_1 \sin f \tag{8.2}$$

By substitution in (8.1), we obtain a new formulation of Snell's law,

$$\frac{r_1 \sin i_1}{v_1} = \frac{r_2 \sin i_2}{v_2} = p \tag{8.3}$$

where p is the ray parameter and the variables r, i, and v are referred to the same point on the ray's trajectory. The units of the ray parameter p are now those of time (seconds), whereas in the plane case they were of inverse velocity ($\mathrm{s\,m^{-1}}$). In a spherical medium, where the velocity varies continuously with the radius, we can write for each point on a ray

$$\frac{r \sin i}{v} = p \tag{8.4}$$

135

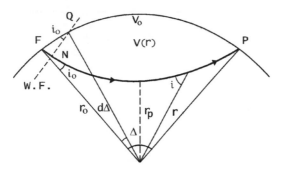

Fig. 8.1. A ray path in a sphere where the velocity increases with depth. The wave front advances from F to N in time dt.

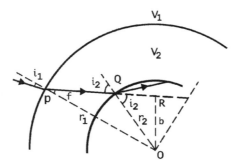

Fig. 8.2. The trajectory of a ray in three concentric spherical regions of constant velocities $v_1 < v_2 < v_3$ (Snell's law).

For a wave front that advances from the focus F, within a time interval dt, a distance along the ray FN $= v_0$ dt, we find (Fig 8.1)

$$\sin i_0 = \frac{\text{FN}}{\text{FQ}} = \frac{v_0}{r_0}\frac{\text{d}t}{\text{d}\varDelta} \tag{8.5}$$

From (8.4) and (8.5) we obtain the Benndorf relation

$$p = \frac{\text{d}t}{\text{d}\varDelta} \tag{8.6}$$

At the point of the ray's trajectory where it turns upward, $i = \pi/2$, and, from (8.4), $r_p/v_p = p$, where r_p and v_p are the radial distance and velocity at that point. The slowness η is now defined as $\eta = r/v$ and, in consequence, $p = \eta_p$.

To study wave propagation in a sphere, we use the spherical coordinates (r, θ, ϕ) (Appendix 2). If θ has its origin in the radius that passes through the focus and ϕ on a plane that contains the ray, components of the displacements of P, SV, and SH waves, in the directions of the unit vectors e_r, e_θ and e_ϕ are (Fig. 8.3)

$$(P_r, P_\theta, 0); \qquad (\text{SV}_r, \text{SV}_\theta, 0); \qquad (0, 0, \text{SH})$$

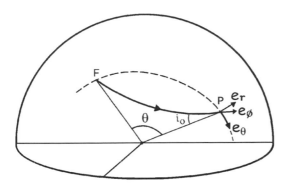

Fig. 8.3. A ray trajectory in a sphere and displacement components.

PNG

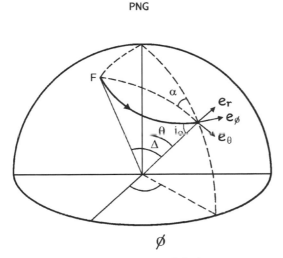

Fig. 8.4. A ray path on a sphere and displacement components at its surface referred to geocentric geographic coordinates r, θ, and ϕ.

In terms of the angle of incidence i_0 at the observation point (Fig. 8.3),

$$P_r = P \cos i_0; \qquad P_\theta = P \sin i_0$$

$$SV_r = SV \sin i_0; \qquad SV_\theta = SV \cos i_0$$

If the sphere represents the Earth, with a system of geographically geocentric coordinates (r, θ, ϕ), with origins at the center (r), North Pole of the rotation axis (θ) and Greenwich meridian on the equatorial plane (ϕ), then the components of P, SV, and SH waves in the directions of the unit vectors e_r, e_θ and e_ϕ positive in the directions of the zenith, South and East (Fig. 8.4) are

$$(P_r, P_\theta, P_\phi); \qquad (SV_r, SV_\theta, SV_\phi); \qquad (0, SH_\theta, SH_\phi)$$

As functions of the angle of incidence i_0 and azimuth α at the observation point P (section 3.7) (Fig. 8.4),

$$P_r = P\cos i_0; \qquad P_\theta = P\sin i_0 \cos\alpha; \qquad P_\phi = P\sin i_0 \sin\alpha$$

$$SV_r = SV\sin i_0; \qquad SV_\theta = SV\cos i_0 \cos\alpha; \qquad SV_\phi = SV\cos i_0 \sin\alpha$$

$$SH_\theta = SH\sin\alpha; \qquad SH_\phi = SH\cos\alpha$$

P and SV displacements have three components whereas SH has only two horizontal components. In relation to the geographic components (x, y, z) in the directions of (North, West, zenith) defined in section 3.7, $x = -e_\theta$, $y = -e_\phi$, and $z = e_r$.

8.2   A sphere of constant velocity

In a sphere of radius R with constant velocity V, rays propagate in straight lines and the traveling time for an angular distance Δ is given by (Fig. 8.5(a))

$$t = \frac{2R}{V}\sin\left(\frac{\Delta}{2}\right) \tag{8.7}$$

The travel time curve is not a straight line, like that in the plane case, but a sine function (Fig. 8.5(b)). The ray parameter p varies with the distance Δ in the form (Fig. 8.5(c))

$$p = \frac{\mathrm{d}t}{\mathrm{d}\Delta} = \frac{R}{V}\cos\left(\frac{\Delta}{2}\right) \tag{8.8}$$

Thus, p equals the radial distance to the center of the ray divided by the velocity (Fig. 8.5(a)). From this expression we obtain

$$\sin\left(\frac{\Delta}{2}\right) = \frac{1}{\eta}(\eta^2 - p^2)^{1/2} \tag{8.9}$$

where $\eta = R/V$. Using (8.9) we derive the expression for the change of Δ with p,

$$\frac{\mathrm{d}\Delta}{\mathrm{d}p} = \frac{-2}{(\eta^2 - p^2)^{1/2}} \tag{8.10}$$

In consequence, $\mathrm{d}\Delta/\mathrm{d}p$ is always negative (Δ increases when p decreases) and has a singularity for $\Delta = 0$ ($p = \eta$) (Fig. 8.5(d)).

8.3   A sphere with a velocity that is variable with the radius

In a sphere in which the velocity varies with the radius, such as in the plane case, the radius of curvature of the ray is given by $R = \mathrm{d}s/\mathrm{d}i$ (7.7). Using Snell's law, we deduce

$$\mathrm{d}i = \frac{p\,\mathrm{d}v}{\cos i} \tag{8.11}$$

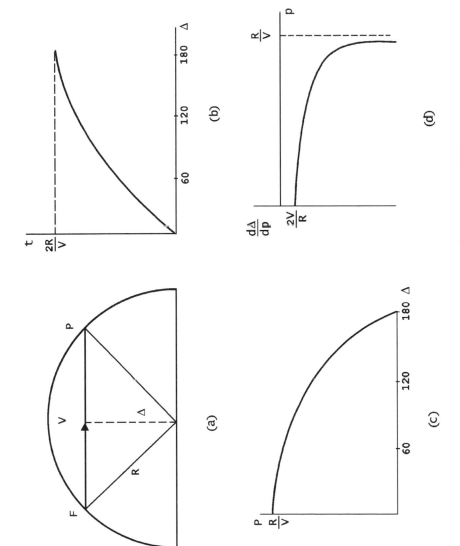

Fig. 8.5. A sphere of constant velocity. (a) A ray trajectory. (b) The travel time curve $t(\Delta)$. (c) A plot of the ray parameter with the angular distance $p(\Delta)$. (d) The variation of $d\Delta/dt$ with p.

139

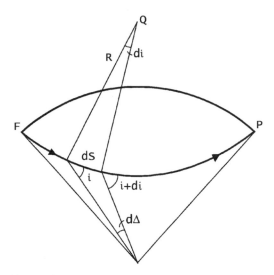

Fig. 8.6. A ray trajectory in a sphere where the velocity increases with depth and radius of curvature.

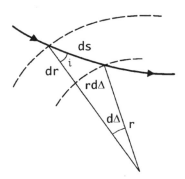

Fig. 8.7. A ray trajectory in a sphere with a velocity that depends on the radius, showing the components dr and $r\,d\Delta$ of an element of distance ds along the ray.

Since $ds = dr/\cos i$ (Fig. 8.6), we obtain

$$R = \frac{r}{p\dfrac{dv}{dr}} \tag{8.12}$$

To deduce the expressions for the angular distance Δ, distance along the ray trajectory s, and travel time t, we use the geometry of the ray (Fig. 8.7):

$$\sin i = \frac{r\,d\Delta}{ds} \tag{8.13}$$

$$ds^2 = dr^2 + r^2\,d\Delta^2 \tag{8.14}$$

By substitution of (8.13) into Snell's law (8.4), we obtain

$$\frac{r^2}{v}\frac{d\Delta}{ds} = p \tag{8.15}$$

From equation (8.14), dividing by ds^2, using (8.15), and making the substitution $\eta = r/v$, we obtain

$$\frac{ds}{dr} = \frac{\eta}{(\eta^2 - p^2)^{1/2}} \tag{8.16}$$

We divide (8.14) by $d\Delta^2$, and in a similar manner to what we did for the derivation of (8.16), obtain

$$\frac{d\Delta}{dr} = \frac{p}{r(\eta^2 - p^2)^{1/2}} \tag{8.17}$$

By integrating equations (8.16) and (8.17), we obtain the expressions for the angular distance Δ and distance along the ray s. The travel time t is obtained by integration of ds/v. If the focus is at the surface we obtain

$$\Delta = 2 \int_{r_p}^{r_0} \frac{p \, dr}{r(\eta^2 - p^2)^{1/2}} \tag{8.18}$$

$$s = 2 \int_{r_p}^{r_0} \frac{\eta \, dr}{(\eta^2 - p^2)^{1/2}} \tag{8.19}$$

$$t = 2 \int_{r_p}^{r_0} \frac{\eta^2 \, dr}{r(\eta^2 - p^2)^{1/2}} \tag{8.20}$$

If the focus is at a depth h from the surface the expressions have two integrals, the first from r_p to $r_0 - h$ and the second from r_p to r_0.

In a similar way to that in the plane case, we can deduce the expression for the reduced time or *tau* function $\tau(p)$ (7.16), which results in

$$\tau(p) = \int_{r_p}^{r_0} \frac{1}{r} (\eta^2 - p^2)^{1/2} \, dr \tag{8.21}$$

These expressions are very similar to those deduced for plane geometry, (7.13)–(7.16). We must bear in mind the different form of Snell's law (8.4), the use of the angular distance, and the definition of slowness.

8.3.1 The change of distance with the ray parameter

The derivation of the expression for $d\Delta/dp$ for a spherical medium is similar to that of dx/dp in plane geometry (section 7.4). By differentiating with respect to p in (8.18), we obtain

$$\frac{d\Delta}{dp} = 2 \frac{d}{dp} \int_{r_p}^{r_0} \frac{p \, dr}{r(\eta^2 - p^2)^{1/2}} \tag{8.22}$$

On making the substitution $f(\eta) = (1/r) \, dr/d\eta$ and changing the integral over r to one over η, we have

$$\frac{d\Delta}{dp} = 2 \int_{\eta_p}^{\eta_0} \frac{f(\eta) \, d\eta}{(\eta^2 - p^2)^{1/2}} + 2p \frac{d}{dp} \int_{\eta_p}^{\eta_0} \frac{f(\eta) \, d\eta}{(\eta^2 - p^2)^{1/2}} \tag{8.23}$$

For the second term on the right-hand side of (8.23), we integrate by parts before taking the derivative:

$$2p\frac{\mathrm{d}}{\mathrm{d}p}\left[f(\eta_0)\cosh^{-1}\left(\frac{\eta_0}{p}\right) - \int_{\eta_p}^{\eta_0}\cos^{-1}\left(\frac{\eta}{p}\right)f'(\eta)\,\mathrm{d}\eta\right] \tag{8.24}$$

After taking the derivative, we obtain

$$\frac{-2\eta_0 f(\eta_0)}{(\eta_0^2 - p^2)^{1/2}} + 2\int_{\eta_p}^{\eta_0}\frac{\eta f'(\eta)\lambda\,\mathrm{d}\eta}{(\eta^2 - p^2)^{1/2}} \tag{8.25}$$

By adding the first term of (8.23), we finally obtain

$$\frac{\mathrm{d}\Delta}{\mathrm{d}p} = \frac{-2\eta_0 f(\eta_0)}{(\eta_0^2 - p^2)^{1/2}} + 2\int_{\eta_p}^{\eta_0}\frac{\frac{\mathrm{d}}{\mathrm{d}\eta}[\eta f(\eta)]\,\mathrm{d}\eta}{(\eta^2 - p^2)^{1/2}} \tag{8.26}$$

On introducing the variable ζ related to the gradient of the velocity,

$$\zeta = \frac{r}{v}\frac{\mathrm{d}v}{\mathrm{d}r} \tag{8.27}$$

it is easy to find that $\eta f(\eta) = (1 - \zeta)^{-1}$ and

$$\frac{\mathrm{d}}{\mathrm{d}\eta}[\eta f(\eta)] = \frac{1}{(1 - \zeta)^2}\frac{\mathrm{d}\zeta}{\mathrm{d}\eta} \tag{8.28}$$

By substitution into (8.26) and changing the integral over η to one over r, we obtain

$$\frac{\mathrm{d}\Delta}{\mathrm{d}p} = \frac{-2}{(1 - \zeta_0)(\eta_0^2 - p^2)^{1/2}} + 2\int_{r_p}^{r_0}\frac{\frac{\mathrm{d}\zeta}{\mathrm{d}r}\,\mathrm{d}r}{(1 - \zeta)^2(\eta^2 - p^2)^{1/2}} \tag{8.29}$$

This expression is similar to that of the plane case (7.39). On making the substitution $\xi = 2/(1 - \zeta)$, equation (8.29) can be written in the form

$$\frac{\mathrm{d}\Delta}{\mathrm{d}p} = \frac{-1}{\xi_0(\eta_0^2 - p^2)^{1/2}} + \int_{r_p}^{r_0}\frac{\frac{\mathrm{d}\xi}{\mathrm{d}r}\,\mathrm{d}r}{(\eta^2 - p^2)^{1/2}} \tag{8.30}$$

Equations (8.29) and (8.30) are very useful when one wants to study the propagation of rays in a spherical medium for different distributions of velocity with the radius.

8.4   A velocity distribution with ζ constant

Just like in the plane case, a velocity distribution with the radius in a sphere corresponding to ζ constant allows an easy derivation of expressions for the relations of the ray parameter to the distance $p(\Delta)$ and travel time $t(\Delta)$. If the velocity increases with depth, according to the definition of ζ (8.27), then, putting $\zeta = -\alpha$, we obtain

$$v = v_0\left(\frac{r_0}{r}\right)^\alpha \tag{8.31}$$

where r_0 and v_0 are the values at the surface. This distribution gives infinite velocity for the center of the sphere, so it can not be used for very small values of r. The gradient of the velocity is negative, that is, it also increases with depth:

$$\frac{dv}{dr} = -\frac{\alpha v_0}{r_0}\left(\frac{r_0}{r}\right)^{\alpha+1} \tag{8.32}$$

Since $d\zeta/dr$ is zero, the second term of (8.29) is null and the expression for $d\Delta/dp$ is

$$\frac{d\Delta}{dp} = \frac{-2}{(1+\alpha)(\eta^2 - p^2)^{1/2}} \tag{8.33}$$

By integration we obtain for $\Delta(p)$ the expression

$$\Delta = \frac{2}{1+\alpha}\cos^{-1}\left(\frac{p}{\eta_0}\right) \tag{8.34}$$

and for $p(\Delta)$

$$p = \eta_0 \cos\left(\frac{(1+\alpha)\Delta}{2}\right) \tag{8.35}$$

Since $p = dt/d\Delta$, by integration of (8.35) we obtain the expression for the travel time $t(\Delta)$:

$$t = \frac{2\eta_0}{1+\alpha}\sin\left(\frac{(1+\alpha)\Delta}{2}\right) \tag{8.36}$$

Equation (8.36) is not valid for rays passing through the center of the sphere, $(1+\alpha)\Delta/2 = \pi/2$, since there the velocity becomes infinite. This velocity distribution is known as Mohorovičić's law and may be used to approximate the velocity distributions for certain regions inside the Earth away from its center.

8.5   Rays of circular trajectories

Another special distribution of increasing velocity with depth is that corresponding to rays with circular trajectories. The radius of curvature (8.12) must be constant, which condition is satisfied if the velocity depends on the radius in the form

$$v = a - br^2 \tag{8.37}$$

where a and b are constants. According to (8.12), the radius of curvature is now $R = 1/(2pb)$ (Fig. 8.8). In terms of the radius of the sphere and surface velocity, the velocity is

$$v = v_0 + b(r_0^2 - r^2) \tag{8.38}$$

The velocity has a finite value at the center. The velocity gradient decreases with depth according to

$$\frac{dv}{dr} = -2br \tag{8.39}$$

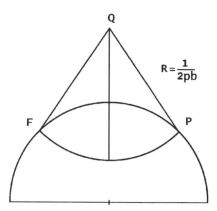

Fig. 8.8. The circular trajectory of a ray in a sphere.

The radius of curvature for a ray with take-off angle at the surface i_0 is given by

$$R = \frac{v_0}{2r_0 b \sin i_0} \tag{8.40}$$

As we have seen, the velocity's distribution with the radius for rays of circular trajectories is different than that for the plane case (7.6) and does not correspond to a constant gradient. It is possible to derive an analytic expression for travel times $t(\Delta)$, but it has a complicated form (Bullen and Bolt, 1985).

8.6   Distributions of the velocity with the radius

We have seen two distributions of the velocity with the radius inside a sphere for which it is possible to find explicit expressions for $p(\Delta)$ and $t(\Delta)$. In general this is not the case. However, we can deduce many characteristics of ray propagation for certain general distributions using the expression for $d\Delta/dp$, in a similar way to what we did in the plane case. For a sphere we must remember that the slowness is defined as $\eta = r/v$ and the ray parameter is $p = \eta \sin i$. Therefore we must take into account the changes with depth of v and r. Just like in the plane case (section 7.6), we will consider a distribution for which the velocity increases very slowly with depth and its gradient decreases or has a normal distribution. To this situation we add a region where the velocity increases rapidly or decreases with depth. The latter case results in the presence of a low-velocity layer. The description of the characteristics of these velocity distributions is very similar to that of section 6.6 for the plane case, so we refer the reader to that section for details.

8.6.1   A normal distribution

In the normal distribution the velocity increases slowly with depth (decreasing r) and its gradient dv/dr, which is negative and small, changes very slowly (Fig. 8.9(a)). Therefore, ζ is negative (its absolute value is small and generally less than 0.25) and $d\zeta/dr$ is very small.

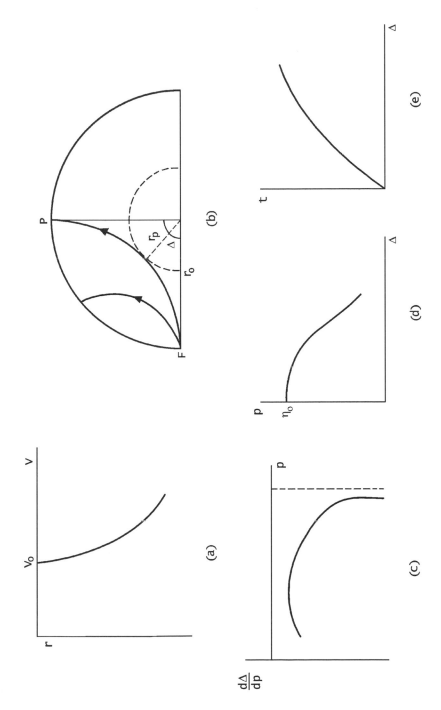

Fig. 8.9. A sphere with a normal distribution of velocity with depth. (a) The distribution of velocity with depth. (b) Ray trajectories. (c) The variation of the ray parameter $d\Delta/dp$. (d) The variation of the ray parameter with distance $p(\Delta)$. (e) The travel time curve $t(\Delta)$.

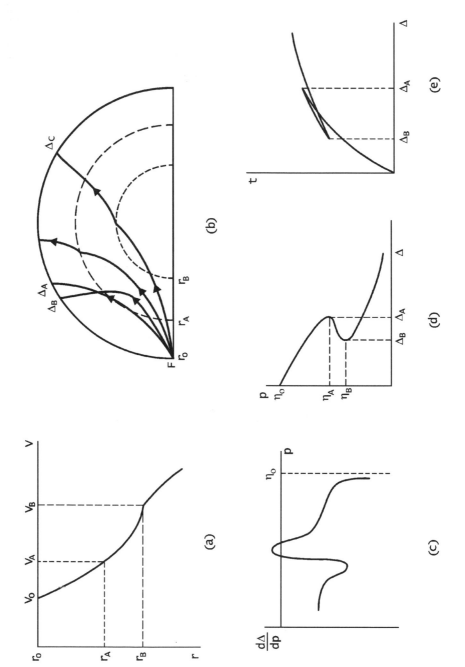

Fig. 8.10. A sphere with a zone of a rapid increase of velocity with depth. (a) The distribution of velocity with depth. (b) Ray trajectories. (c) The variation of the ray parameter $p(\Delta)$. (d) The variation of the angular distance with the ray parameter $d\Delta/dp$. (e) The travel time curve $t(\Delta)$.

Ray trajectories are curved concave upward (Fig. 8.9(b)). The curve for $d\Delta/dp$ (Fig. 8.9(c)) has negative values and a singularity for $p = \eta_0$. The curve of $p(\Delta)$ exhibits a gradual decrease of p with distance (Fig. 8.9(d)), from a value η_0 for $\Delta = 0$. The travel time curve $t(\Delta)$ exhibits a gradual increase of time with distance with a slow decrease of its gradient (Fig. 8.9(e)). A velocity distribution with a small inverse power of the radius (8.36), corresponding to constant ζ, is a particular case of the normal distribution.

8.6.2   A rapid increase in velocity

Just like in the plane case (section 7.6), we consider a region between the radii r_A and r_B, where the velocity increases rapidly with depth, while above and below this region the velocity distribution is normal (Fig. 8.10(a)). When rays penetrate into this region their radii of curvature decrease (8.12), since dv/dr is large (Fig. 8.10(b)). If the change in gradient is sufficiently large that the second term in (8.30) is larger than the first, then the curve of $d\Delta/dp$ has a positive part (Fig. 8.10(c)). In consequence, the curve of $p(\Delta)$ has a part for which p decreases with decreasing Δ (Fig. 8.10(d)). Then, within a certain range of values of distance, say between Δ_A and Δ_B, there are three values of p for each value of Δ, that is, three rays arrive at the same distance. Finally, the travel time curve $t(\Delta)$ (Fig. 8.10(e)) has a triplication with a concave upward branch that corresponds to the rays for which p decreases with decreasing Δ, with two cusps due to the caustic. These characteristics are similar to those explained for the plane case (section 7.6).

8.6.3   A decrease in velocity. A low-velocity layer

In a region bounded by the radii r_A and r_B, the velocity decreases with depth while outside of this region it increases just like it does in the normal distribution (Fig. 8.11(a)). In this region, rays change curvature from concave upward to concave downward (8.12), due to the change in sign of dv/dr (Fig. 8.11(b)). According to Snell's law, rays can not turn upward until they reach the depth where η has again the same value as that at r_A, that is, under r_C. The curve of $d\Delta/dp$ has a discontinuity and the contribution of the large change in gradient leads to the fact that, for certain values of p, $d\Delta/dp$ becomes positive (Fig. 8.11(c)). The curve of $p(\Delta)$ is also discontinuous (Fig. 8.11(d)). Up to a distance Δ_A, rays do not penetrate into the low-velocity zone and the curve corresponds to a normal distribution of velocity. At Δ_A there is a discontinuity and, for the same value of $p = \eta_A$, a ray arrives also at the distance Δ_B. Rays that arrive at distances between Δ_B and Δ_C correspond to a retrograde branch of the curve with positive values of $d\Delta/dp$, that is, p decreases with decreasing Δ (Fig. 8.11(d)). From Δ_C there starts a second normal branch for which p decreases with increasing Δ (Fig. 8.11(d)). There is, then, a range of distances, between Δ_A and Δ_C, for which no rays arrive (there exist no values of p), namely a shadow zone, and another range of distances from Δ_C to Δ_B, for which two rays arrive at the same distance.

The travel time curve corresponds to these characteristics (Fig. 8.11(e)). From zero to Δ_A, the curve corresponds to a normal distribution. Between Δ_A and Δ_C, there is a shadow zone where no rays arrive. Behind the shadow zone, the curve separates into

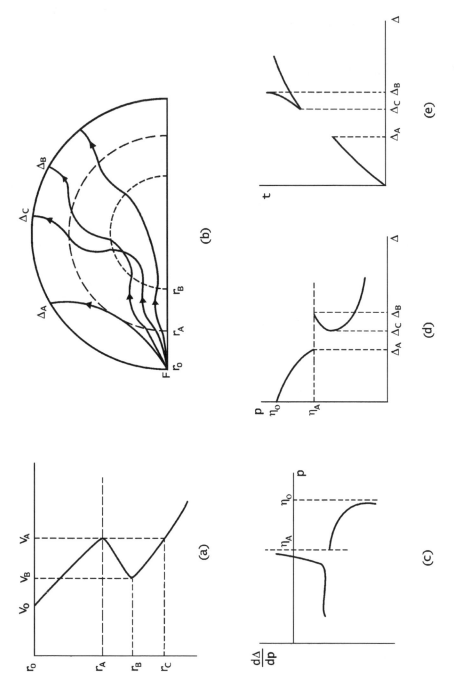

Fig. 8.11. A sphere with a zone of a decrease of velocity with depth. (a) The distribution of velocity with depth. (b) Ray trajectories. (c) The variation of the ray parameter $d\Delta/dp$. (d) The variation of the angular distance with the ray parameter $p(\Delta)$. (e) The travel time curve $t(\Delta)$.

two branches that come together at Δ_C with a cusp, the first with the properties of the normal distribution that continues to the limit of Δ, and the second from Δ_C to Δ_B, for which $dt/d\Delta$ decreases with decreasing Δ (a retrograde concave upward branch). From Δ_C to Δ_B, two rays arrive at different times at each distance. The time for Δ_C is delayed with respect to that corresponding to a prolongation of the first part of the curve. This delay is due to propagation in the low-velocity layer. The main characteristics of the traveling time curve are, then, the existence of a shadow zone and, behind it, a time delay and a duplication of arrivals.

8.7 The determination of velocity distribution

We can derive a relation to determine the velocity distribution with the radius $v(r)$ from the travel times $t(\Delta)$ for a spherical medium in a similar form to that in the plane case (section 7.8). We start with equation (8.18), which, on changing the integral over r to one over η, can be written as

$$\Delta = 2 \int_{\eta_p}^{\eta_0} \frac{p \dfrac{dr}{d\eta} \, d\eta}{r(\eta^2 - p^2)^{1/2}} \tag{8.41}$$

This is an integral along a ray that arrives at a distance Δ, has a parameter p, and turns upward at $\eta = \eta_p$. Consider now all the rays that arrive at distances between $\Delta = 0$ and $\Delta = \Delta_1$, and correspond to values of p between p_0 and p_1. For these rays we can define the integral

$$\int_{p_1}^{p_0} \frac{dp}{(p^2 - \eta_1^2)^{1/2}}$$

By applying this operation to both sides of (8.41) we obtain

$$\int_{p_1}^{p_0} \frac{\Delta \, dp}{(p^2 - \eta_1^2)^{1/2}} = \int_{p_1}^{p_0} \int_{\eta_p}^{\eta_0} \frac{2p \dfrac{dr}{d\eta} \, d\eta}{r(\eta^2 - p^2)^{1/2}(p^2 - \eta_1^2)^{1/2}} \, dp \tag{8.42}$$

The first integral on the right-hand side is over a generic ray of parameter $p = \eta_p$ and the second is over all rays of parameters from p_1 to p_0, where p and η_p are variables. In the integral on the left-hand side Δ and p are variables that take all values from 0 to Δ_1 and from p_1 to p_0, respectively.

In the double integral on the right-hand side of (8.42), similarly to what we did in the plane case (7.63), we can change the order of integration. We integrate first over p from $p = \eta$ to $p = \eta_1$ and the result has the value of π. The integral over η from η_1 to η_0 is changed into an integral over r from r_1 to r_0, resulting in

$$\int_{r_1}^{r_0} \int_{\eta_1}^{\eta} \frac{2p \, dp}{[(\eta^2 - p^2)(p^2 - \eta_1^2)]^{1/2}} \frac{dr}{r} = \pi \ln \left(\frac{r_0}{r_1} \right) \tag{8.43}$$

The term on the left-hand side of (8.42) can be integrated by parts, as we did in (7.64). Taking into account that, for $\eta = \eta_0$, $\Delta = 0$, and, for $\eta = \eta_1$, $\cosh^{-1}(1) = 0$, and

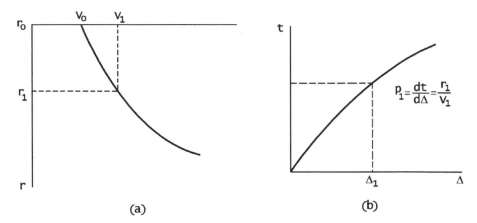

Fig. 8.12. The application of the Herglotz–Wiechert integral to obtain the distribution of velocity with depth in a spherical medium from the travel time curve. (a) The distribution of velocity with depth. (b) The travel time curve.

changing the integral over p to one over Δ, we finally obtain

$$\int_0^{\Delta_1} \cosh^{-1}\left(\frac{p}{\eta_1}\right) d\Delta = \pi \ln\left(\frac{r_0}{r_1}\right) \tag{8.44}$$

This is the integral equation of Herglotz–Wiechert that allows the determination of the velocity's distribution with the radius from the travel time. Just like in the plane case, the equation is valid only for velocity distributions in which the velocity increases monotonically with depth. It can not be directly applied when there are rapid changes in gradient or low-velocity layers (although, as was mentioned in the plane case, there are modifications that can handle those cases).

Equation (8.44) is applied in the following manner. For a fixed distance Δ_1, η_1 is known from the value of the slope of the curve $t(\Delta)$ for such a distance. To find v_1, we need to know r_1. The value of r_1 corresponds to the turning point of the ray that arrives at a distance Δ_1, and is obtained from equation (8.44). The integral can be calculated numerically, since, from each value of Δ, from $\Delta = 0$ to $\Delta = \Delta_1$, p is the slope of $t(\Delta)$ (Fig. 8.12(b)). For each pair of values v_1 and r_1, we have a point of the velocity distribution (Fig 8.12(a)). By repeating the process for different distances, we obtain values of the velocity for different values of r. Since usually $t(\Delta)$ are discrete observed data, we have to interpolate in order to find the values of $dt/d\Delta$. This method can be applied to obtain velocity distributions in the lower mantle of the Earth where the conditions of validity of equation (8.44) are fairly well satisfied.

8.8 Energy propagation by ray beams. Geometric spreading

The energy propagated by a ray beam inside a sphere follows the same treatment as that in the plane case (section 7.9) with a few modifications. From the focus, an energy E is emitted homogeneously per unit time. The energy per unit solid angle is $P = dE/d\Omega$, whence $E = 4\pi P$.

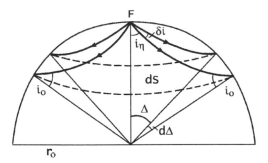

Fig. 8.13. The area on the surface of a sphere covered by the wave front corresponding to a ray beam that leaves the focus with take-off angle i_h at a distance Δ.

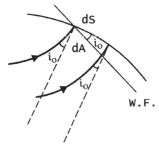

Fig. 8.14. The arrival at the surface of a sphere of a wave front corresponding to a ray beam that leaves the focus with take-off angle i_h and arrives with angle of incidence i_0. The area of the free surface is ds. The area of the wave-front surface is dA.

If the focus is at a depth h from the surface, then the take-off angle at the focus is i_h. Rays that leave the focus between angles i_h and $i_h + di_h$ arrive at the surface at distances between Δ and $\Delta + d\Delta$, with angles of incidence between i_0 and $i_0 + di_0$ (Fig. 8.13) (if the focus is at the surface, $i_h = i_0$). The element of solid angle $d\Omega$ around the focus is

$$d\Omega = 2\pi \sin i_h \, di_h \tag{8.45}$$

The area dS over the surface of the sphere defined by the intersection of the element of the solid angle $d\Omega$, at an angular distance Δ, is given (Fig. 8.13) by

$$dS = 2\pi r_0^2 \sin \Delta \, d\Delta \tag{8.46}$$

The area dA defined on the wave-front surface (Fig. 8.14) is $dA = dS \cos i_0$. The energy intensity, $I = dE/dA$, is the energy that arrives at a distance Δ per unit wave-front surface per unit time. As a function of the energy that leaves the focus per unit solid angle, $I = P \, d\Omega/dA$. By substitution of (8.45) and (8.46), we obtain

$$I = \frac{P \sin i_h}{r_0^2 \cos i_0 \sin \Delta} \frac{di_h}{d\Delta} \tag{8.47}$$

The derivative $di_h/d\Delta$ can be expressed in terms of the slope of the travel time $t(\Delta)$ by using Snell's law:

$$\frac{d}{d\Delta}\left(\frac{r_h \sin i_h}{v_h}\right) = \frac{dp}{d\Delta} = \frac{d^2 t}{d\Delta^2} \tag{8.48}$$

and, therefore,

$$\frac{di_h}{d\Delta} = \frac{v_h}{r_h \cos i_h} \frac{d^2 t}{d\Delta^2} \tag{8.49}$$

By substituting (8.49) into (8.47) we find

$$I = \frac{P v_h \sin i_h}{r_0^2 r_h \cos i_h \cos i_0 \sin \Delta} \frac{d^2 t}{d\Delta^2} \tag{8.50}$$

If the focus is at the surface, $i_h = i_0$ and $r_h = r_0$. Hence, equations (8.47) and (8.50) are simplified.

Equations (8.47) and (8.50) represent the geometric spreading of energy propagated inside a sphere. The energy that arrives at a certain distance Δ decreases with the factor $1/\sin \Delta$ and depends on the value of $di/d\Delta$ or, equivalently, that of $d^2 t/d\Delta^2$. If $di/d\Delta$ is large, its inverse $d\Delta/di$ is small and Δ varies little with changes in i_h. This means that the corresponding interval of distances $d\Delta$ is small, the density of rays is large, and the energy per unit surface area is also large. The opposite happens when $di/d\Delta$ is small. Ray beams spread themselves over a larger surface so that $d\Delta$ is large and the energy density small. Because $di/d\Delta$ is proportional to $dp/d\Delta$, for distances at which $dp/d\Delta$ is large we will have concentrations of energy. This happens when there are large changes in the velocity gradient. For those situations there are also changes in the slope of the travel time curve ($d^2 t/d\Delta^2$ is large). Hence we expect large concentrations of energy at distances corresponding to cusps of $t(\Delta)$ when we have rapid increases in velocity or low-velocity layers.

9 TRAVEL TIMES AND THE STRUCTURE OF THE EARTH

9.1 Observations and methods

In Chapters 6–8 we have seen how travel time curves depend on the characteristics of the media through which the seismic waves propagate and that the velocity's distribution with depth can be deduced from them. In this chapter we will apply these results to observations regarding the Earth and discuss the results concerning its internal structure obtained. For short distances (less than 1000 km) we can use the flat-Earth approximation and plane geometry. Seismic waves for that range of distances give information on depths of about 100 km, that is, on the crust and part of the upper mantle. For this range of distances we can apply the theory derived in Chapters 6 and 7. For greater distances the spherical shape of the Earth must be considered, so the results of Chapter 8 must be applied. The effects due to the deviations of the form of the Earth from a sphere, that is, mainly its flatness, can be taken into account by using corrections to the spherical model. In seismology these effects are not very important.

The first seismic waves used for the study of the Earth's structure were those produced by earthquakes. Even today this is the main source of information, especially for the deep interior. Among the first tables and curves of travel times of seismic waves were those of Oldham, who in 1906 deduced the existence of the Earth's core. These tables were completed by Zöppritz and Turner and later, in 1914, by Gutenberg. In 1940, Jeffreys and Bullen published their tables of travel times that are very widely used even today. In 1968 Herrin published tables of travel times for P waves only. More recently, on a recommendation from the IASPEI, travel time tables have been derived from a spherically symmetric velocity model known as iasp91 based on modern data (the ISC catalog 1964–88) (Kennett and Engdahl, 1991). For shorter distances, the pioneering work was done by Mohorovičić, who in 1909 discovered the discontinuity between the Earth's crust and mantle that has been given his name. Observations of traveling times of P and S waves generated by earthquakes continue to provide important information about the internal structure of the Earth (Bolt, 1982).

Observations of seismic waves generated by earthquakes have the limitation of the lack of control over the exact place and time of their origin. For this reason, for the study of the shallowest layers of the Earth, seismic waves generated artificially by explosions and other methods are used. These studies, which were started around 1920, are the basis of the methods of seismic prospecting using what are known by the generic name of seismic profiles or seismic soundings. These were developed first by the oil industry and consist in the analysis of reflected and refracted waves. These measurements are called deep seismic profiles or soundings when they are applied to

153

the determination of the structure of the Earth's crust and upper mantle. They are divided into two types, namely, refraction and wide-angle reflection and vertical reflection. The first method uses waves refracted and reflected with large angles of incidence at long distances and the second uses waves reflected vertically at very short distances. In both cases, information consists in traveling times and amplitudes of waves observed along linear profiles, but quite different treatments are applied to the data.

Another, more recent, method developed to study the interior of the Earth is that of seismic tomography. This method, based on techniques borrowed from other fields, especially medicine, allows the mapping of inhomogeneities existing in the medium traversed by seismic waves. The method consists in the observation of a very large number of seismic rays that cross a given region in different directions. Applications of this method have resulted in the establishment of three-dimensional models of the Earth's interior.

The first complete velocity models of the Earth's interior based on seismological observations were developed around 1930 by Jeffreys, Bullen, Gutenberg, Macelwane, and Richter, among others. These models have radial symmetry separating the Earth's interior into three regions: the crust (0–30 km), mantle (30–2900 km), and core (2900–6370 km). They have very similar general characteristics, although they differ in terms of the transition zones, especially those between the mantle and the core and between the outer and inner cores. The widely used Jeffreys–Bullen (JB) model was recognized to have an average offset of 1.8 s in the travel times. More accurate velocity models with spherical symmetry have been presented. They include the preliminary Earth model (PREM) of Dziewonski and Anderson (1981) and CAL8 of Bolt and Uhrhammer (1981) (Fig. 9.1, Appendix 6). More refined models using the ISC catalog for 1964–88 with a better fit to the observed travel time data have recently been developed, namely, the already mentioned iasp91 (Kennett and Engdahl, 1991) that has been

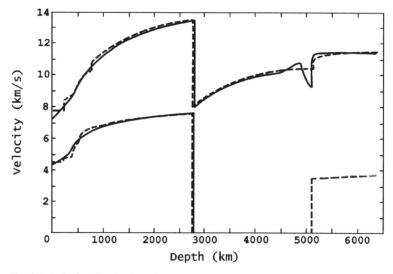

Fig. 9.1. Velocity distribution of P and S waves inside the Earth according to the Jeffreys–Bullen (continuous line) and PREM (dashed line) models (modified from Bullen and Bolt (1985)).

improved for the core phases by the model ak135 (Kennett *et al.*, 1995) and the model SP6 with slightly higher velocities in the upper mantle than iasp91 (Morelli and Dziewonski, 1993). These models have radial symmetry and are used as reference models for the study of lateral inhomogeneities. Actually, more detailed velocity models of the Earth interior are three-dimensional and include also inelastic properties and anisotropy (Boschi *et al.*, 1996).

9.1.1 Refraction and wide-angle reflection

The first methods based on the observation of seismic waves generated by explosions are those employing what are called refraction and wide-angle reflection profiles. These methods were in use in the oil industry during the 1920s. Their use for the study of the structure of the crust has been very extensive, since approximately 1940, especially after the development of portable seismograph stations with magnetic recording in 1960. Since the penetration of seismic rays depends on the distance of observation, long profiles are used to reach depths down to the base of the crust and layers of the upper mantle. This implies the use of large explosions and recording at great distances. The size of explosions establishes the practical limit on the depth that can be investigated with this method.

The general outline of the method consists in recording along lines of portable seismographic stations of lengths from tens of kilometers to nearly 1000 km at intervals of 1–5 km from explosions whose size may vary from tens of kilograms up to five metric tons of explosives (Fig. 9.2). Explosions are generated in holes drilled underground or by the use of underwater charges at various depths (60–200 m) in the sea. Basically, the method is based on the analysis of travel times and amplitudes of recorded waves according to the theory developed in Chapters 6 and 7, for layered media of constant velocities and continuous distributions of velocity with depth with different velocity gradients. The forms of travel time curves, critical distances of reflected waves, and slopes of refracted waves, together with distributions of amplitudes, allow the determination of velocity models for the Earth's crust. When possible, profiles are recorded in both directions along the same line, with direct and inverse profiles, in order to detect dipping layers and true velocities (section 6.6). By obtaining profiles with a high density of recording stations, detailed models with lateral heterogeneity of layer thicknesses and velocities can be obtained.

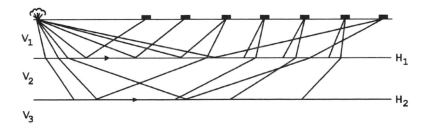

Fig. 9.2. An explosion and the line of recording stations for a refraction and wide-angle-reflection seismic profile in a two layer medium.

A very common representation of observations in seismic profiles is that of record sections, whereby complete seismograms are represented versus distance instead of only travel times so that the information includes also amplitudes of waves. Examples of record sections are shown later in Fig. 9.7 for a theoretical model of the crust and in Fig. 9.8 for an observed profile. A recent technique of interpretation is the comparison of observed and theoretical record sections calculated by the method of ray tracing (see Fig. 9.7 later). This method provides synthetic seismograms at given distances from models of the crust with velocity gradients and lateral changes in velocity and layers' thicknesses. Seismic refraction and wide-angle reflection profiles have provided abundant observations on the structure of the crust and upper mantle during the last 30 years.

9.1.2   *Vertical reflection*

Since 1950, workers involved in industrial seismic exploration have abandoned the use of methods of seismic refraction for those of vertical reflection. These methods are based on observation of waves reflected almost vertically at short distances from the source. The source of energy can be small explosions or mechanical vibrators coupled to the surface. These methods allow one to identify with great detail by using complex processes of data reduction, the presence of reflectors at various depths, and provide an accurate image of the structures present underground (Sheriff and Geldart, 1982). Around 1965, these methods developed by the oil industry started to be used for the study of the crust in order to obtain clear reflections from its base. Since 1970, many programs of extensive deep vertical reflections have been started in many countries, such as COCORP in the USA, BIRPS in the UK, ECORS in France, DEKORP in Germany, and IBERCORS in Spain.

Vertical seismic reflection is based on the generation of reflections in layers of the crust's interior by rays of near-vertical incidence using short distances from the source to observing stations (Fig. 9.3). Assuming that the crust is formed by layers of constant velocity, the equations for the times and distances of reflected waves are

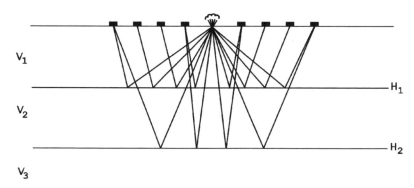

Fig. 9.3. An explosion and the recording stations for a vertical reflection seismic profile in a two-layer medium.

given by (6.78) and (6.79). For vertical reflections, the traveling time of a reflection at layer k is

$$t_k = 2 \sum_{i=1}^{k-1} \frac{H_i}{v_i} \qquad (9.1)$$

This is called the two-way time for normal incidence at layer k, that is, the time taken for the wave to travel from the surface to the top of layer k and back to the surface. Layers with clear reflections are called reflectors. The depth H, at which each reflector is located, is the sum of the thicknesses H_i of all the layers on top of the reflector. Since the observation is t_k, to obtain H we need to know the velocities v_i of all layers on top, or at least a mean value of the velocity from the surface to the reflector. Owing to the difficulty of determining velocities, depths to reflectors are often expressed in terms of two-way times.

To increase the signal-to-noise ratio of reflections from a given reflector, a series of techniques developed by the oil industry is used. A common technique is the stacking or sum of amplitudes of near records into a single trace. Each record is corrected for the differences among two-way times due to the differences in distance by using what is called the 'normal move out' (NMO) correction given by

$$\Delta t = \frac{x^2}{2v^2 t} \qquad (9.2)$$

where x is the distance from each trace to the point where stacking is performed and t is the two-way time for that point. This correction permits the summation of several amplitudes as if they were located at a single point and produced by a common point of reflection (the common depth point; CDP). Owing to the fact that layers are not flat but rather have different dips, vertical displacements and irregularities, the traveling times of reflected waves do not have the simple form of equation (9.1). Besides, we have the presence of diffracted waves and multiple reflections. To eliminate the influences on recordings of these undesirable effects that may conceal the real position of reflectors, a technique called migration is used. There are many methods for performing migration that involve, in general, complex data processing. The results are migrated sections from which diffractions and multiple reflections have been removed and in which non-flat reflectors are correctly located.

Another problem to consider in the resolution of vertical reflections is the influence of the frequency of signals. As we saw in Chapter 6, ray theory, on which all these methods (refraction and reflection) are based, is valid only for high frequencies. There is, then, a limit into the resolution of lateral dimensions or vertical irregularities of reflectors that can be detected with waves of a certain frequency. Approximately, for vertical dimensions to be detectable they must be greater than a quarter of the wave length of the signal. For example, for waves of 20 Hz and velocity 6 km s^{-1}, the minimum size of detectable bodies is about 30 m. For horizontal dimensions of reflectors, we must also take into account the Fresnel zone, that is, the zone from which energy is reflected by half a cycle of a spherical wave incident on a plane surface. Inside the crust, for waves of 20 Hz, minimum detectable horizontal dimensions are of the order of kilometers. Vertical reflection profiles allow correlation of reflectors to the continuation in depth of geologic structures at the surface. Even though this method does not give distributions of velocities, taken together with refraction profiles, it has contributed to the establishment of detailed models of the crustal structure.

9.1.3   *Seismic tomography*

The method of seismic tomography applied to the study of the Earth's interior was developed from the mid-1970s (Aki *et al.*, 1976). This method is based on the study of the time residuals of the arrival of seismic waves with respect to those expected from an initial model. The foundations of the general theory of tomography can be found in the work of the mathematician J. Radon in 1917 and its applications presented by A. Cormack in 1963. Tomography has been applied to other fields, for example in medicine, to obtain detailed images of the human body from x-rays crossing in many directions. To study the Earth's interior, a large number of seismic rays that cross a particular region in many directions is used. From observations of the arrival times of these rays, we can find variations in velocity in the studied region with respect to a reference model. In this way, the method permits the deduction of three-dimensional models of inhomogeneities in the Earth's interior from the crust to the nucleus, such as have been shown to exist by the work of Clayton, Dziewonski, Nakanishi, and Nolet (Nolet, 1987), among others.

A commonly used method of seismic tomography is based on time residuals of arrival times, that is, the differences between observed travel times and those calculated from a reference model, usually with a distribution of velocity with depth ($\Delta t = t_{obs} - t_{theo}$). Time residuals are interpreted in terms of anomalous structures of velocity with respect to the reference velocity model. Reference models of the Earth are taken from those already established with distributions of velocity with the radius (Jeffreys–Bullen, PREM, etc.). If the observed traveling time at one station differs from that of the reference model for the same trajectory, the residual is assigned to a difference in velocity in some region along the ray. The aim of tomography is to explain residuals of travel times in terms of variations in the velocity distribution inside the Earth or velocity anomalies. If we have a large number of ray trajectories in many directions, the anomalous regions can be defined well in terms of the extent and the magnitude of the anomaly.

Since the travel times of waves depend on the inverse of the velocity (the slowness, $\eta = 1/v$), this variable is the one that generally is used. If the anomaly in velocity in a certain region is $\Delta v = v' - v$, the anomaly in slowness is

$$\Delta \eta = \frac{1}{v + \Delta v} - \frac{1}{v} \simeq -\frac{\Delta v}{v^2} \tag{9.3}$$

Once the time residuals Δt have been corrected for other factors, they can be assigned to anomalies in velocity (slowness) in a given region (Fig. 9.4). If the length of a ray's trajectory through the anomalous region is l, then

$$\Delta t = l \, \Delta \eta = -\frac{l \, \Delta v}{v^2} \tag{9.4}$$

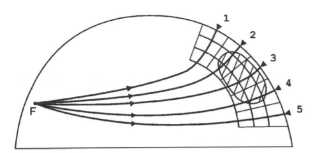

Fig. 9.4. Ray paths that cross an anomalous region (shaded) inside the Earth in a study of seismic tomography. Stations 2, 3, and 4 present traveling time residuals whereas stations 1 and 5 do not. Cells crossed by rays are shown.

In general, the region of the Earth that is being studied is divided into cells of constant dimensions with a velocity that differs from that of the reference model by a quantity Δv that must be determined. The time residual Δt at a station is assigned to the sum of time increments, positive or negative, due to velocity anomalies in each cell:

$$\Delta t = \sum_{i=1}^{N} \frac{l_i \, \Delta v_i}{v_i^2} \qquad (9.5)$$

If a particular ray crosses N cells where the velocities of the reference model are v_i, the relation between Δt and Δv_i is given by equation (9.5). The length across the cells can be written in terms of the vertical dimensions of each cell h, which usually are the same for all cells, and the angles of incidence of the ray at those points, $l_i = h/\cos i_i$. Equation (9.5) has N unknowns (Δv_i), the velocity anomalies of cells crossed by one ray. The complete model has a total number of M cells, of which each ray crosses only a limited number ($N < M$). The problem consists in finding Δv_i for each of the M cells. This is possible only if we have many rays crossing each of the M cells in different directions. The problem requires the solution of a system of many linear equations of the type of (9.5) for the M unknowns (Δv_i). The solution is generally found by a least squares method. For the solution to be well conditioned, the total number of equations must be much larger than M and each cell must be crossed in different directions by many rays. This implies that we have a dense distribution of stations that received rays from a large number of sources. Owing to the large number of observations, complex inversion algorithms and statistical methods are used.

This explanation has shown only the fundamentals of the method, without mentioning many other important aspects, such as the influence of observation errors, errors in hypocenters, and ray trajectories. All of them must be considered if one is to obtain results that are representative of the velocity structure of the region being studied. Seismic tomography studies provide the distributions of zones of positive or negative velocity anomalies with respect to those of the reference model of the Earth. Positive anomalies represent zones of higher velocities and are associated with regions where the material is more rigid and colder, whereas negative anomalies corresponding to lower velocities are associated with less rigid and warmer material.

9.2   Distributions of velocity, elasticity coefficients, and density

The methods we have briefly described provide knowledge of velocity distributions of body waves inside the Earth. In most cases these methods use P waves only, and the Earth's structure is given as distributions of their velocities. Observations of S waves are less commonly used and the distributions of their velocities are less accurate. This is due, in part, to the fact that S waves are not first arrivals and they are not efficiently generated by explosions. If the velocities α and β are known, we can deduce the distribution of Poisson's ratio σ. The quotient β/α can be expressed in terms of σ by using (2.26), (2.64), and (2.65):

$$\frac{\beta}{\alpha} = \left(\frac{2\sigma - 1}{2(\sigma - 1)} \right)^{1/2} \tag{9.6}$$

Also, σ can be written in terms of α and β as

$$\sigma = \frac{\alpha^2 - 2\beta^2}{2(\alpha^2 - \beta^2)} \tag{9.7}$$

In this form, if we know independently the distributions of α and β with depth, we can deduce that of σ.

The elastic coefficients μ (the rigidity modulus), K (the bulk modulus), and λ (the Lamé coefficient) can be expressed in terms of the velocities of elastic waves α and β, and the density ρ:

$$\mu = \rho\beta^2 \tag{9.8}$$

$$K = \rho(\alpha^2 - \tfrac{4}{3}\beta^2) \tag{9.9}$$

$$\lambda = \rho(\alpha^2 - 2\beta^2) \tag{9.10}$$

The variation of the density with depth inside the Earth may be determined from the velocities of seismic waves α and β under certain conditions. In a simplified form, let us assume that a material is homogeneous, in hydrostatic equilibrium, and existing under adiabatic conditions. In this case, the variation of the pressure P with the radius inside the Earth is

$$\frac{\mathrm{d}P}{\mathrm{d}r} = -g\rho \tag{9.11}$$

where g is the gravitational attraction of the mass inside a given value of r (for a spherical Earth $g = -Gm/r^2$). The variation of the density ρ with the radius for hydrostatic equilibrium is

$$\frac{\mathrm{d}\rho}{\mathrm{d}r} = \frac{\mathrm{d}\rho}{\mathrm{d}P} \frac{\mathrm{d}P}{\mathrm{d}r} \tag{9.12}$$

since, by the definition of K, for a homogeneous material at constant volume and adiabatic changes (2.30),

$$\frac{1}{K} = \frac{1}{\rho} \frac{\mathrm{d}\rho}{\mathrm{d}P} \tag{9.13}$$

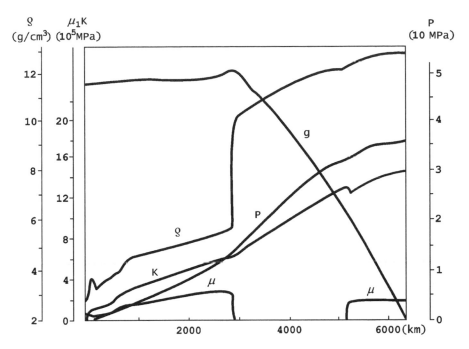

Fig. 9.5. Distributions in the Earth's interior of gravity (g), density (ρ), pressure (P), the bulk modulus (K), and the rigidity modulus (μ).

by substituting (9.11) and (9.13) into (9.12) and expressing K in terms of α and β (9.9), we obtain

$$\frac{d\rho}{dr} = \frac{Gm\rho}{r^2(\alpha^2 - \frac{4}{3}\beta^2)} \tag{9.14}$$

This expression is known as the Adams–Williamson equation. The gradient of the density with the radius for a value of r can be determined from values of α and β and the mass of the Earth inside such a value of r. This equation can be applied to obtain the distribution of density inside the Earth if there are no discontinuities in the values of ρ and consequently also none in those of α and β. Once the distribution of ρ is known, we can find those of the elastic coefficients K and μ, from values of α and β, according to (9.9) and (9.8). For a more general treatment see Bullen and Bolt (1985).

The distributions of μ, K, and ρ inside the Earth are shown in Fig. 9.5. We can see that the density increases gradually with depth from $3.4\,\mathrm{g\,cm}^{-3}$ in the mantle and there is a sharp increase from $5.57\,\mathrm{g\,cm}^{-3}$ in the base of the mantle to 10–$13\,\mathrm{g\,cm}^{-3}$ in the core. Owing to this high density, the largest contribution to gravity is from the mass of the core. The pressure increases gradually with depth in the mantle and core. The bulk modulus K increases slowly in the mantle and more rapidly in the core. The rigidity modulus μ increases very slowly in the mantle, is null inside the outer core, which is in a fluid state, and has a finite value in the inner core.

9.3   The crust

The outermost part of the Earth is called the crust. From the seismologic point of view, the first observations that led to the discovery of the existence of a sharp contrast between the material of the crust and that of the mantle under it were made in 1909 by Mohorovičić, studying the travel times of earthquakes in central Europe. He observed that, for an epicentral distance of about 150 km, there was a change in the slope of traveling times. This change in slope corresponds to a velocity discontinuity at about 30 km depth that he identified as the base of the crust. This discontinuity today bears his name or is called, in abbreviated form, the Moho. A second discontinuity inside the crust was discovered by V. Conrad in 1923 (it is called the Conrad discontinuity) and H. Jeffreys in 1926.

The thickness of the crust is not homogeneous; in shields or stable continental zones it is about 30 km, whereas under the oceans it is only 8–15 km and in regions of high mountains it can be as much as 60 km (Fig. 9.6). The simplest models of the continental crust have two layers with constant velocity or with a small gradient, covered by a thin layer of sediments. These two layers were known as granitic and basaltic layers due to their assumed compositions and today are more generally referred to as the upper and lower crust (Fig. 9.6). Approximate values of P wave velocities for the crust are 2.5–5 km s^{-1} in sediments, 5.7–6.3 km s^{-1} in the upper crust, 6.6–7.3 km s^{-1} in the lower crust, and 7.8–8.3 km s^{-1} for the top of the upper mantle.

The nomenclature used to designate rays traveling through the crust is the following: rays in the upper crust are denoted P_g, waves reflected in the Moho are denoted $P_M P$, and waves refracted in the upper mantle are denoted P_n. Rays refracted and reflected in an internal discontinuity of the crust (the discontinuity between the upper and lower crust) are called P_c and $P_c P$, or P_I and $P_I P$. For observations of earthquakes waves refracted in the lower crust are also called P^*. For S waves the nomenclature is similar, that is, S_g, S_n, S^*, $S_M S$, etc. For reflected waves we must also consider converted

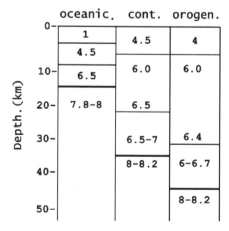

Fig. 9.6. Typical values of P wave velocity (km s^{-1}) in oceanic material, continental shield, and orogenic crust.

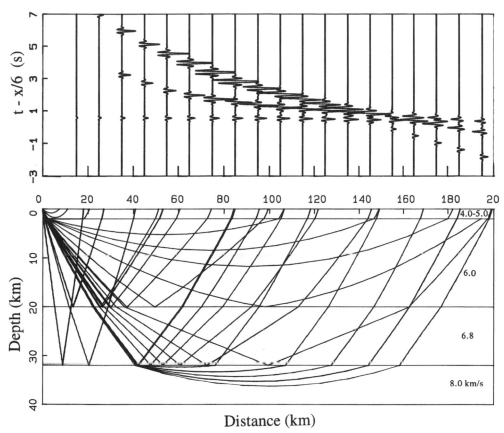

Fig. 9.7. Ray trajectories and a synthetic seismic record section for a two-layer crust with small velocity gradients (courtesy of J. Tellez).

waves, that is, incident P or S waves that are reflected as S or P waves. For rays reflected in the Moho, they are denoted $P_M S$ and $S_M P$.

Rays traveling in a two-layer crust with a sedimentary layer are shown in Fig. 9.7 with the corresponding theoretical reduced record section (reduced velocity 6 km s^{-1}) (section 6.5). For distances less than 140 km, first arrivals correspond to P_g and, for greater distances, to P_n, which have smaller amplitudes. Depending on the thickness of the crust, this distance may vary in the range 90–160 km. A prominent arrival belongs to $P_M P$, with large amplitudes, especially near the critical distance (80–140 km). The critical distance of $P_M P$ can be used to determine the total thickness of the crust (section 6.5). In seismograms of local earthquakes, we observe the same phases but P_g is observed with large amplitudes beyond the critical distance and S phases have large amplitudes in the horizontal components (Fig. 9.8). Rays refracted or reflected at internal discontinuities of the crust, P_I or P^* and S_I or S^*, have, in general, small amplitudes. They are observed better in seismograms of earthquakes than they are in seismic profiles.

Low-velocity layers have often been observed in the crust. Their presence in the upper and lower crust produces delays in rays reflected and refracted at discontinuities below them. In general, models of the crust with layers of constant velocity reproduce many

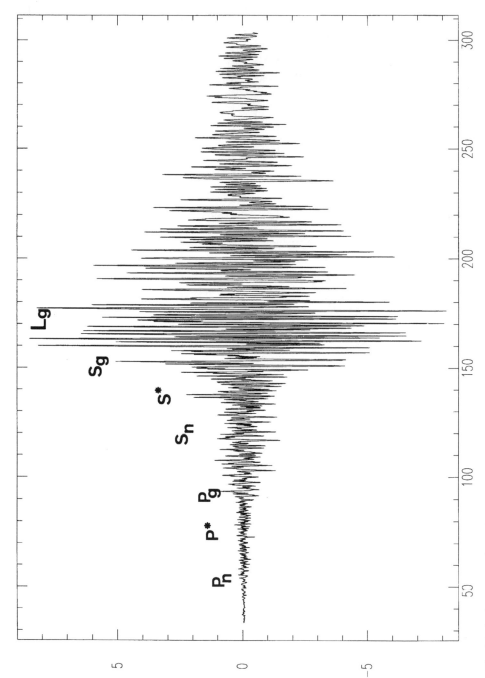

Fig. 9.8. A seismogram (the vertical component) of a local earthquake in Faro, Portugal (1 May 1997) recorded at the SFUC BB station (Cadiz, Spain) at 138 km distance, showing crustal phases.

observed characteristics of travel times and record sections. Better results are obtained upon introducing velocity gradients in the layers. A simple model of this type is that shown in Fig. 9.7 with two layers with small gradients. Rays have curved trajectories and there is a maximum distance for those traveling inside the upper crust (P_g) and those reflected in the mantle (P_MP). The real situation is, in general, more complex than those predicted by simple models, as can be seen from Fig. 9.9. The main phases, however, correspond, in general, to those of the two-layer model (Fig. 9.7).

In oceanic regions, the structure of the crust is different than that explained for a stable continental crust. First of all, its thickness is smaller, 6–10 km, and is formed by a thin layer of sediments underlain by a single layer with P velocities in the range 6.6–7 km s^{-1}, corresponding to what has been called the basaltic layer or lower crust under continents. Underneath, in the upper mantle, P wave velocities are in general lower than they are in continental regions, 7.6–7.9 km s^{-1} (Fig. 9.6). This is manifested by the travel times' smaller values for critical distances of P_MP waves and there being lower slopes of rays refracted inside the crust than those for continents. This holds for a typical oceanic crust corresponding to oceanic abyssal planes (of about 4000 m depth). For shallow seas and near the coast, the crust is thicker and velocities are a little lower.

The structure of the crust below continents is not homogeneous. In regions of mountain ranges, also called orogenic zones, the thickness is usually larger than that observed for shields or stable regions. In zones of large mountains such as in the Andes and Himalayas, the crustal thickness may be as great as 60 km, that is, nearly double that of stable regions. The first seismologic observations that revealed this effect were due to Gutenberg for the Alps and Byerly for the Sierra Nevada, California. This thickening of the crust or roots under mountains agrees with the isostatic models proposed by G. B. Airy in 1856. The structure of the crust in mountain regions may differ also from that of stable regions with more homogeneous characteristics and in some cases there is no discontinuity between the upper and lower crust. The transition from continental crust to oceanic crust in coastal regions is gradual, with decreasing thicknesses and the disappearance of the granitic layer. The surface of the Moho has, in consequence, a complex topography that resembles an inverse image of the free-surface topography (Fig. 9.10).

The crustal structure, which is relatively constant in continental shields or stable regions, varies rapidly from one place to another in geologically complex regions (Meissner, 1986; Taylor and McLennan, 1985). Laterally homogeneous models of the crust are, then, not very representative. The base of the crust and the intermediate layers inside have dips and gradients that vary from place to place. Detailed refraction studies have revealed these lateral inhomogeneities. Vertical reflection surveys have shown the presence of dipping reflectors, offsets, and other complexities. These studies have also shown that the upper crust is relatively transparent to vertical rays whereas the lower crust presents multiple reflections. This has been interpreted as a thin layered or laminated structure in the form of multiple lenses of small dimensions in the lower crust (Fig. 9.11). The upper crust is, then, formed by a more rigid or brittle material and the lower crust by one that is more ductile, laminated, and possibly with greater fluid concentrations. This result agrees with the hypothesis of a seismogenetic layer (a layer where earthquakes are generated) that, more or less, coincides with the upper crust, as we will see in Chapter 20.

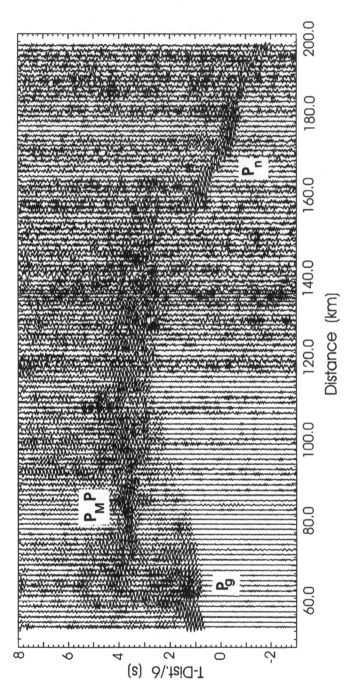

Fig. 9.9. An observed seismic record section of a refraction seismic profile offshore of Galicia, Spain. Pg, Pn, and P$_M$P phases are clearly recorded (courtesy of D. Cordoba).

166

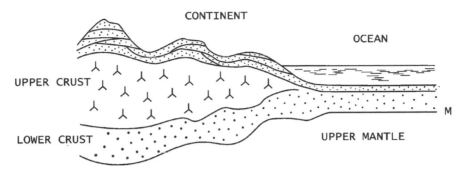

Fig. 9.10. The variation of the crust's thickness on going from oceanic to continental structure.

Profile ESCI on land

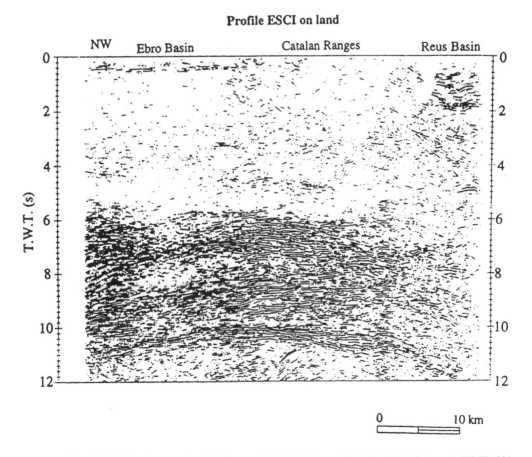

Fig. 9.11. A seismic record section for a vertical reflection profile in the Valencia trough (ESCI 1991, 1992). On the vertical scale are two-way times in seconds. The transparent upper crust and laminated lower crust are shown (Gallart *et al.*, 1995) (with permission from the Sociedad Geológica de España).

Table 9.1. *The compositions of the most common igneous rocks*

	Granite	Basalt	Peridotite
SiO_2	72	51	43
Al_2O_3	14	14	4
Fe_nO_m	3	12	12
MgO	1	6	34
CaO	2	10	4
Na_2O	3	2	1

9.3.1   *The mineralogical composition of the crust*

Seismic observations give us directly only information on the distribution of velocities of body waves inside the Earth and indirectly the density and elastic coefficients. The mineralogic compositions of materials in the interior of the Earth can be derived from those of rocks on the surface, laboratory experiments, and indirect evidence. Surface rocks can be separated according to their origin into sedimentary and igneous ones. Briefly, sedimentary rocks are formed by the compaction of sediments deposited mainly on the sea floor. The most abundant are limestones ($CaCO_3$) and sandstones (SiO_2). The seismic velocities in rocks vary according to the degree of compactness and depend on their age. Igneous rocks are formed by solidification from fused magma. The most representative are granite, basalt, and peridotite. These three types of rocks are examples of acid, basic or mafic, and ultrabasic or ultramafic rocks. This classification refers to the content of silica (SiO_2), which is greater than 52% for acid rocks, 45–52% for basic rocks, and less than 45% for ultrabasic rocks. Other classifications give for acid rocks a silica content of over 66% and call intermediate those with silica contents in the range 52–66%. The proportions of components of the three basic igneous rocks are given in Table 9.1.

Most commonly accepted results give for the global composition of the continental crust a proportion of silica (SiO_2) of approximately 60%, with 64% in the upper crust and 58% in the lower crust. The content of Al_2O_3 is 16% and that of CaO is 7%. The contents of FeO and MgO are 9% and 5%, with greater proportions in the lower crust. The old denomination of granitic and basaltic layers for the upper and lower crust is, then, only an approximation to their mineralogic compositions. The composition of the oceanic crust is 49% SiO_2, 16% Al_2O_3, 11% CaO, 11% FeO, and 8% MgO. This composition is similar to that of the lower continental crust and to those of basaltic rocks.

9.4   The upper mantle and lithosphere

From the point of view of seismology, material under the crust down to a depth of 700 km forms the upper mantle. The most important characteristics of the distribution of velocities of P and S waves in its interior are the following. Velocities increase with depth from values under the Moho of approximately 7.8–8.3 km s^{-1} for P and

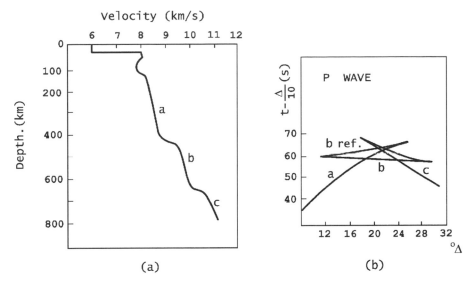

Fig. 9.12. (a) The velocity distribution of P waves in the upper mantle. (b) The reduced travel time curve, showing branches of waves refracted in a, reflected in b, refracted in b, reflected in c, and refracted in c.

3.8–4.1 km s^{-1} for S waves to about 10.7 km s^{-1} for P and 5.9 km s^{-1} for S waves. Approximately between 60 and 220 km depth, a low-velocity layer, which is more pronounced for S waves, is found. The depth, thickness, and decrease in velocity of this layer vary from region to region and in some regions it may even not be present. This low-velocity layer is responsible for the shadow zone observed around 20° distance. Under the low-velocity layer, from approximately 220 km depth, velocities both for P and for S waves increase slowly. At depths of 450 and 670 km there are two zones of rapid increases in velocity (Fig. 9.12(a)). Travel time curves of rays that penetrate to those depths have properties corresponding to rapid increases in velocity with normal and retrograde branches (Fig. 9.12(b)). These two zones are sometimes considered to be discontinuities in the velocity distribution.

The material of the upper mantle, according to A. E. Ringwood (1975), is a composite of magnesium and iron silicates called pyrolite, formed from basalt and olivine. With increasing depth, there are changes of phases in the minerals, for example, that of plagioclase to form garnets. Changes in mineral phases with the same composition are also used to interpret the rapid increases in velocity at the 450 and 670 km discontinuities; for the first, a change from olivine to spinel, and, for the second, a change from spinel to perovskite.

9.4.1 The lithosphere and the astenosphere

The crust and upper mantle are the parts active in tectonic processes. For this reason, other concepts related to tectonics must also be introduced here. In plate tectonics an important role is played by the lithosphere, a rigid layer that forms the units of plate motion and includes the crust and part of the upper mantle. Under the

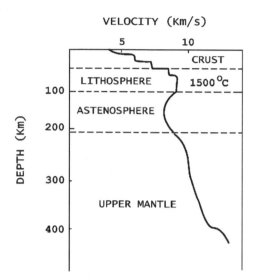

Fig. 9.13. P wave velocity profile in the lithosphere, astenosphere, and mantle.

lithosphere, the material is less rigid, has plastic characteristics, and is in a state of partial fusion, forming a weak layer called the astenosphere (Fig. 9.13). The existence within the Earth of an upper rigid layer (the lithosphere) divided into plates lying over a plastic or viscous layer (the astenosphere) that allows its horizontal motion is necessary in all models of plate tectonics (section 20.5).

From the seismologic point of view, the lithosphere can be identified with the layer that includes the crust and the lid of the upper mantle, where the velocities of P and S waves are relatively high. The thickness found for the lithosphere varies on going from oceans to continents, with values in the range 60–120 km. The astenosphere is identified with the low-velocity layer both for P and for S waves which extends from depths of about 60 km under the oceans and 120 km under the continents to about 200 km down. Its thickness varies from region to region and its lower limit is not well defined. From the thermal point of view, the lithosphere is the relatively cold layer with a large temperature gradient and its lower boundary is located at the isothermal of 1500 °C. These three definitions of the lithosphere (tectonic, seismologic, and thermal) do not always coincide and the term itself has no unique meaning.

The lower limit of the upper mantle at about 700 km depth coincides with the maximum depth of earthquakes and approximately with the 2000 °C isotherm. This depth corresponds also to the depth of material involved in tectonic processes and, for this reason, the whole layer down to this depth is sometimes called the tectonosphere. In relation to the upper mantle, subduction zones have a depth in some cases of as much as 700 km. These zones were first manifested by the occurrence of deep earthquakes or Wadati–Benioff zones where the oceanic lithosphere is introduced under the continental lithosphere in regions of collisions between plates (sections 20.1 and 20.5). These and other heterogeneities in the upper mantle have recently been detected by seismic tomography. Inside the upper mantle, regions of positive (higher than normal velocities) and negative (lower than normal velocities) anomalies with respect to the models of

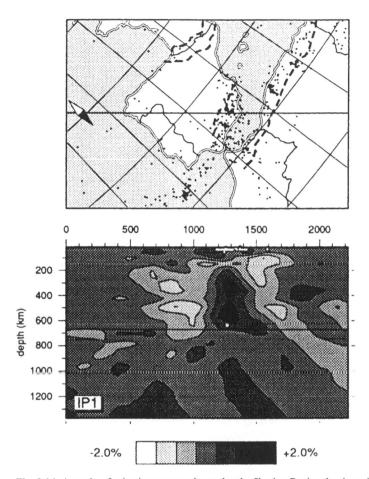

Fig. 9.14. A study of seismic tomography under the Iberian Peninsula. A positive-velocity anomalous region at depths between 200 and 600 km is shown (Blanco and Spakman, 1993) (with permission from Elsevier Science).

radial symmetry with values up to 5% have been found for P and S waves (Fig. 9.14). These regions are interpreted as zones in the upper mantle where material is colder (positive anomalies) and warmer (negative anomalies). Positive anomalies are associated with subduction zones and negative anomalies are associated with the material under rift or extension zones and are related to thermal convection currents (Woodward and Master, 1991). Workers using seismic tomography have also discovered structures in the lower mantle, sometimes down to the core, that may show that tectonic processes are really more deeply rooted.

9.5   The lower mantle

The lower mantle extends in depth from 700 km to the boundary of the core at 2900 km (the core–mantle boundary, CMB). The seismic characteristics of the lower

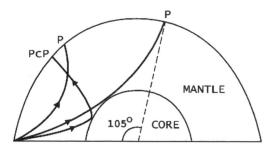

Fig. 9.15. Ray trajectories for P waves inside the mantle for direct rays and rays that are reflected at the core.

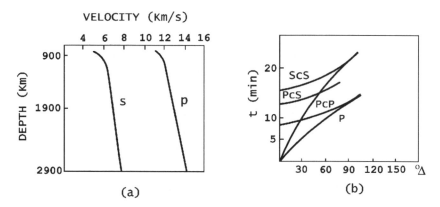

Fig. 9.16. (a) Velocity distributions of P and S waves in the lower mantle. (b) Travel times curves of P and S rays, direct and reflected at the core.

mantle are very uniform, with gradual increases in P and S wave velocities. P velocities increase from 10.75 to 13.72 km s^{-1} and S velocities increase from 5.95 to 7.26 km s^{-1}. The density also increases in a gradual way from 4.38 to 5.57 g cm^{-3}. The trajectories of seismic rays correspond to a normal distribution of velocities (section 8.6). In the lower mantle we find direct P and S rays and rays that are reflected on the free surface and on the surface of the core.

Direct P and S rays propagated in the lower mantle arrive at distances between 35° and 105° (Fig. 9.15). At 105° there arrive rays that turn upward immediately above the core's boundary. At somewhat larger distances, between 105° and 115°, we find arrivals of P and S waves that have been diffracted at the core's surface. Traveling time curves for direct P and S waves have characteristics corresponding to a normal distribution of velocities such that the velocity increases slowly and its gradient changes very little (section 8.6) (Fig. 9.16).

The strong velocity contrast at the CMB (transitions from 13.72 to 8.06 km s^{-1} for P waves and from 8 to 0 km s^{-1} for S waves) produces reflections of P and S waves that are called PcP and ScS waves (c for core) and also converted reflections, PcS and ScP (Fig. 9.16(b)). Travel time curves corresponding to reflected waves are concave upward and tangential to direct branches at their maximum distance (105°) (Fig. 9.16).

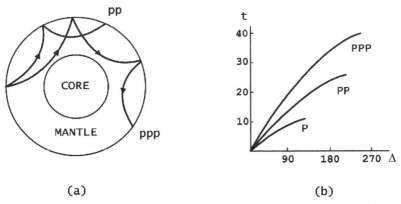

Fig. 9.17. (a) Ray trajectories for P waves reflected at the free surface. (b) Travel time curves.

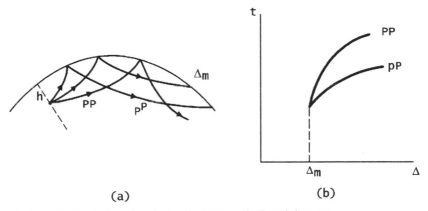

Fig. 9.18. (a) Ray trajectories of pP and PP waves. (b) Travel time curves.

For surface foci, intersection times for PcP and ScS are $8'30''$ and $15'36''$. Diffracted P and S waves arriving at distances greater than $105°$ can also be considered to be grazing reflected PcP and ScS waves at the core's surface.

Inside the lower mantle, we find also P and S waves that are reflected once or several times at the free surface (Fig. 9.17(a)). These waves are designated PP, PPP, etc. and SS, SSS, etc. When there are phase conversions in the reflections we have PS, SP, PSS, PSP, PPS, SSP, etc. Rays reflected several times arrive at later times than do direct rays and those reflected fewer times (Fig. 9.17(b)). One also observes waves reflected at the discontinuities in the upper mantle, that is, at depths of 260 (the lower limit of the low-velocity layer), 450, and 670 km. These reflected rays are designated PdP, where d is the depth of the reflecting layer (e.g. P450P). Owing to the small contrasts in velocity at these discontinuities, the energies of these waves are not large and they are difficult to detect on seismograms.

For deep-focus earthquakes, two types of rays reflected at the free surface can arrive at the same distance, corresponding to rays that travel upward or downward from the focus (Fig. 9.18(a)). In the first case, the rays are designated pP and they arrive at a given distance before the PP rays that correspond to rays that travel downward

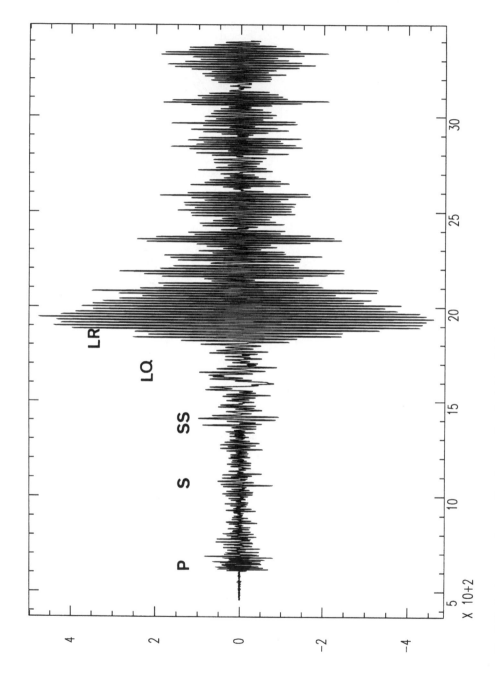

Fig. 9.19. A seismogram for a surface-focus earthquake in Peru (21 February 1996) recorded at the ANMO BB station ($\Delta = 51°$), showing direct and reflected P and S phases and surface waves.

(Fig. 9.18(b)). The ray with take-off angle $i_h = \pi/2$ arrives at the minimum distance Δ_m (for $h = 300$ km, $\Delta_m = 24°$) where the two branches of traveling times for pP and PP rays come together. Since the distance from the focus to the surface is small compared with the distance to the observation point, pP rays arrive a little later than do direct P rays. Hence the time interval $\Delta t = t(pP) - t(P)$ depends on the focal depth and is used in its determination.

Rays that travel through the lower mantle can be observed on seismograms recorded at distances from 30° to 100°. For a surface focus, the most prominent body-wave phases correspond to direct P and S waves; the other phases that are observable are reflected PP and SS waves (Fig. 9.19). On a seismogram of a deep-focus earthquake, we can also observe pP and sS phases and waves reflected at the core, PcP and ScS (Fig. 9.20). Usually, seismograms for deep-focus earthquakes provide clearer recordings of body waves because of the poor generation of surface waves. On seismograms of surface shocks the large amplitudes of surface waves come after S waves (Fig. 9.19).

Seismic tomography has been applied to the study of velocity heterogeneities in the lower mantle (Inoue *et al.*, 1990, Dziewonski, 1996). Regions with higher and lower velocities than those of the reference model have been found. High velocities (positive anomalies) are interpreted as corresponding to colder material and lower velocities (negative anomalies) are thought to correspond to warmer material. These results have been related to the presence in the lower mantle of convection currents with upward (warm) and downward (cold) movements (Fig. 9.21).

The composition of the lower mantle is considered to be very homogeneous and predominantly formed by silicates of magnesium and iron such as perovskite and oxides of magnesium and iron such as magnesiovskite. The gradual increases in velocity and density inside the lower mantle are attributed to crystalline structures which become more compact with depth due to the increase in pressure. Inside the lower mantle the pressure increases from 10^4 to 10^5 MPa and temperatures increase from 1000 to 3000 °C.

9.6   The core

The first evidence about the existence of the Earth's core from the analysis of traveling times of seismic waves was presented in 1906 by Oldham, who deduced that its velocity is lower than that of the mantle. In 1912 Gutenberg fixed the depth of the core at 2900 km from the study of reflected waves. In 1926 Jeffreys discovered its fluid nature from the absence of S waves. The existence of a solid inner core was first proposed in 1936 by Lehmann. The Earth's core is, then, formed by two regions, an outer core (K) of 3486 km radius and an inner core (I) of 1216 km radius (Fig. 9.22).

S waves do not propagate in the outer core, which indicates that its material is fluid enough not to allow the existence of shear stress. The P wave velocity decreases sharply from 13.72 km s^{-1} at the base of the lower mantle to 8.06 km s^{-1} at the surface of the outer core (Fig. 9.22). According to section 8.6, we expect to find a shadow zone in travel time curves and, after it, a duplication with two branches, one normal and another retrograde at a later time (Fig. 8.11). For the Earth, without considering the inner core, the shadow zone extends from 105° to 143°. Rays that penetrate into the core are called PKP rays (K from *Kern*, German for core). Owing to the rapid decrease in velocity, rays

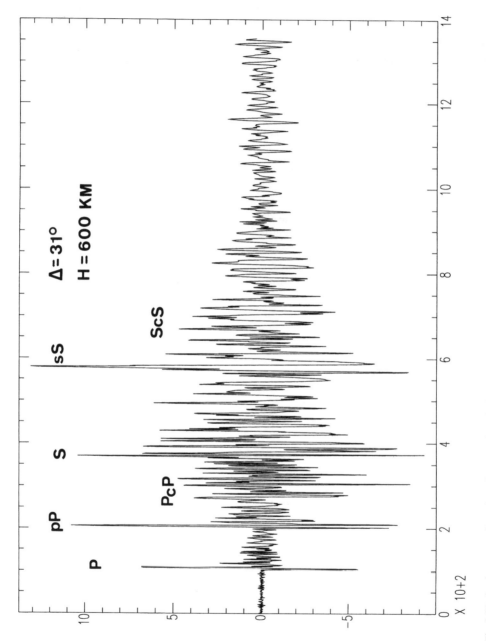

Fig. 9.20. A seismogram for a deep-focus earthquake in Peru (10 January 1994) recorded at the SJG BB station, showing direct P and S phases, and P and S phases reflected at the free surface and at the core.

176

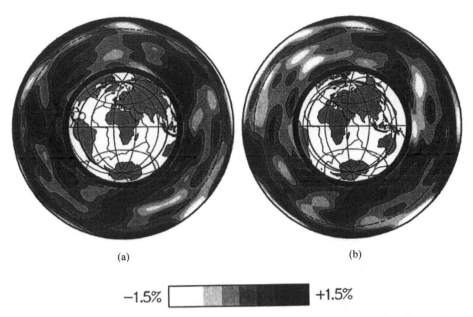

Fig. 9.21. Seismic tomography of the lower mantle for S waves. (a) A polar section along the meridian 108°. (b) A polar section along the meridian 144°. The dashed line circle shows the 670 km discontinuity and the innermost thick line shows the CMB (Dziewonski, 1996) (with permission from the Istituto Nazionale di Geofisica).

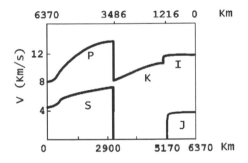

Fig. 9.22. Velocity distributions of P and S waves in the mantle, and in the outer and inner core.

incident on the core with large angles of incidence are refracted to distances of about 170°. With decreasing angle of incidence, rays arrive at shorter distances, forming the retrograde branch PKP$_2$, which is concave upward. The minimum distance for these rays is 143°. At this distance there starts the normal branch PKP$_1$ that continues to 180° (Fig. 9.23). Travel times in the absence of diffracted waves and inner-core reflections present a shadow zone from 105° to 143°. From 143° we find the duplication of branches PKP$_1$ and PKP$_2$ with a time delay of 3 min with respect to direct P waves propagated in the lower mantle (Fig. 9.24(a)).

The presence of the solid inner core with a greater P wave velocity than that of the outer core (Fig. 9.22) modifies the traveling time curve. First of all, there are rays reflected on the surface of the inner core (designated PKiKP) that start to arrive

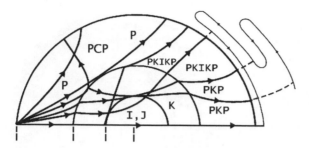

Fig. 9.23. Ray trajectories of P waves through the mantle, and of P waves refracted in and reflected from the outer core.

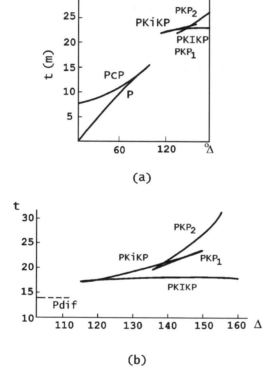

(a)

(b)

Fig. 9.24. (a) Traveling time curves for P waves in the mantle and core. (b) Detail of traveling times for P waves in the outer and inner core.

at a distance of about 120°, in the shadow zone (Fig. 9.24(a)). Arrivals of these waves form the main evidence for the existence of the inner core. Rays refracted in the inner core, PKIKP, arrive ahead of the PKP_1 rays. The travel time curve is, then, formed by four branches, PKP_1, PKP_2, PKiKP, and PKIKP (Fig. 9.24(b)). PKP_1 rays arrive from 143° to 156°, PKP_2 (concave upward) rays from 143° to 170°, PKiPK (concave upward) rays from 120° to 156°, and PKIKP rays from 120° to 180° (Fig. 9.24(b)).

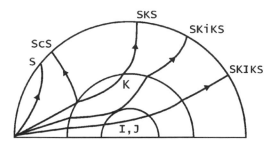

Fig. 9.25. Ray trajectories of S waves in the mantle and of S waves refracted in and reflected from the core.

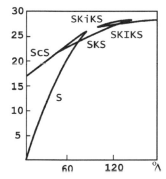

Fig. 9.26. Travel times for S waves in the mantle and for S waves refracted in and reflected from the outer and inner core.

As has been mentioned, the fluid nature of the outer core blocks the propagation of S waves. However, S waves propagated in the lower mantle are converted into P waves in the outer core that are again converted into S waves in the mantle. Such rays, when they travel through the outer core only, are designated SKS rays; if they penetrate the inner core, they are called SKIKS rays. SKiKS rays are those reflected on the surface of the inner core (Fig. 9.25). Since, for rays passing from S (the lower mantle) to K (the outer core) and from K to I (the inner core), the velocity always increases, the travel times have characteristics associated with media with zones of rapid increases in velocity (section 8.6, Fig. 8.10). Direct S rays arrive at distances of up to 105°. Rays refracted in the outer core, SKS rays, start to arrive at 62°, a point common to the branch of reflected ScS rays, and they are observed up to 133°. Rays that are refracted into the inner core, SKIKS rays, arrive at distances from 99° to 180°. Rays reflected from the inner core, SKiKS rays, arrive at distances from 99° to 133°, forming a concave upward branch (Fig. 9.26). Since the inner core is solid, S waves also propagate in its interior. S (J) waves in the inner core are converted from P (K) waves in the outer core. They are called SKJKS rays, and their travel time branch is parallel to the SKIKS branch, but delayed in time. These waves have little energy and are difficult to observe.

On seismograms, for epicentral distances greater than 105°, we observe arrivals of rays that have traveled through the core. Owing to the complexity of arrivals of rays refracted and reflected in the outer and inner core, seismograms vary rapidly in appearance on

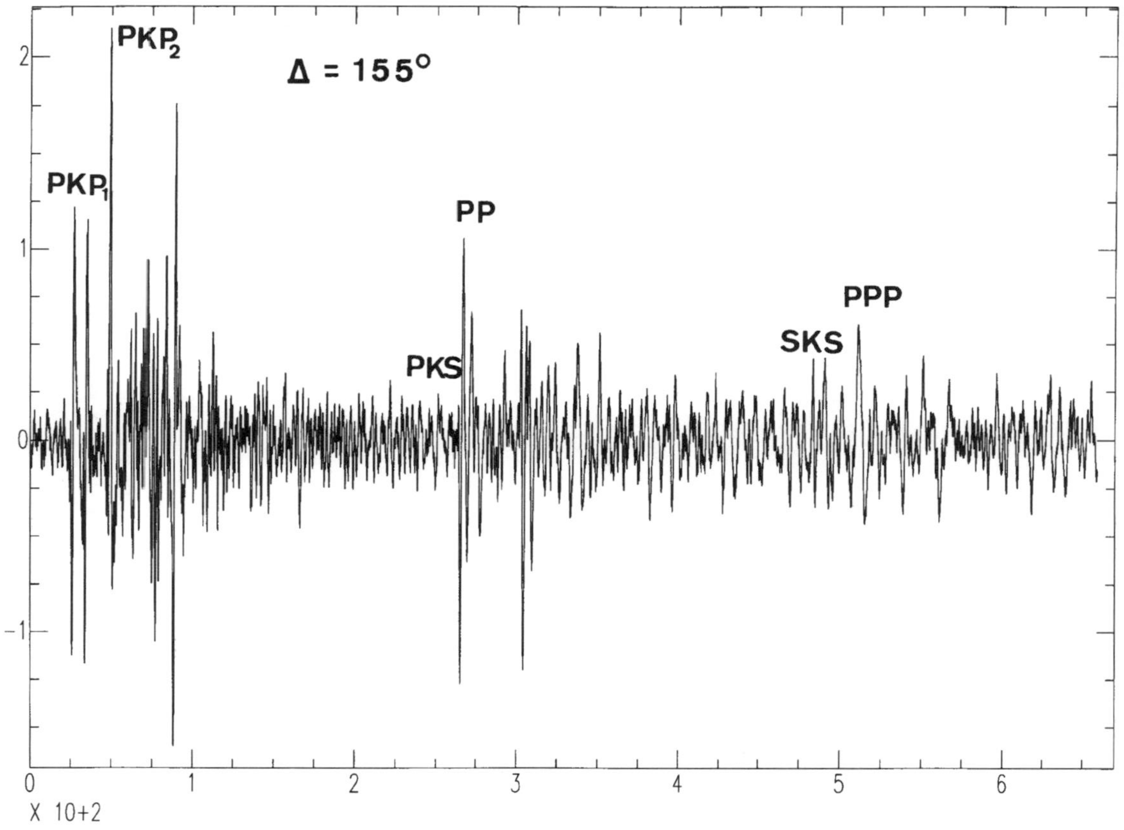

Fig. 9.27. A seismogram of an earthquake in Peru (18 March 1993), recorded at the APR BB station ($\Delta = 155°$), showing the core's phases.

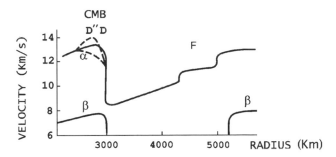

Fig. 9.28. Velocity distributions of P and S waves in the inner and outer core. Transition zones between the core and the mantle (D and D″) (the CMB) and between the inner and the outer core (F) are shown.

going from one distance to another. PKiKP rays start to arrive at about 120°, PKIKP rays at 130°, and PKP_1 and PKP_2 rays at 142° (Fig. 9.27).

Velocity models of the core agree in their general characteristics with that shown in Fig. 9.28 (Bolt, 1982; Jeanloz, 1990). The velocity of P waves decreases sharply from $13.5\,km\,s^{-1}$ at the base of the lower mantle to $8.06\,km\,s^{-1}$ in the core and increases slowly in the outer core from 8.06 to $10.36\,km\,s^{-1}$, where the velocity of S waves is zero. There are several models for the boundary between the outer and inner core. A simple model consists in a decrease in the velocity gradient followed by a sharp increase in velocity from 10.36 to $11.03\,km\,s^{-1}$. In the inner core, the velocity is practically constant or increases very slowly from 11.03 to $11.26\,km\,s^{-1}$.

The density increases rapidly from $5.57\,g\,cm^{-3}$ at the base of the lower mantle to $9.9\,g\,cm^{-3}$ in the outer core. Inside the core, the density increases gradually, with a small increment at the boundary of the inner core, reaching a value of $13.1\,g\,cm^{-3}$. Owing to its large density, the mass of the core is approximately 35% of the mass of the Earth. The composition of the core is basically iron with a small proportion (about 10%) of lighter elements such as silicon, nickel, sulfur, carbon, and oxygen. The presence of some of these elements is necessary in order to explain the values of the density and bulk modulus in the core, both of which are lower than those expected for pure iron at high pressures and temperatures. The iron composition is confirmed by the presence of this mineral in meteorites and the existence of the Earth's internal magnetic field. The generation of this field is explained by models of autoexcited dynamos produced by motion of fluid conductive material of the outer core. The material of the core seems to be very homogeneous and the difference between the outer and inner core is due only to its fluid or solid state. Some authors, however, propose also a difference in composition, with iron and nickel in the inner core and iron and sulfur in the outer core.

Transition zones between the mantle and the outer core, CMB, called by Bullen the D zone, and between the outer and inner cores, or the F zone, are still a subject of discussion. The first models of the CMB posited a sharp decrease in velocity whereas more modern ones favor a previous increase in velocity at a depth of about 2600 km followed by the decrease (Young and Lay, 1987). The CMB marks a sharp contrast in physical and chemical characteristics, from solid material of magnesium and iron silicates to a fluid made basically of iron. Above this boundary, there is an anomalous layer in the mantle known as the D'' zone of up to 300 km thickness and great lateral inhomogeneities. The characteristics of the outer–inner core boundary or F zone are less well known. Models present a gradual increase in velocity from about 4600 km depth (Fig. 9.28). The transition zone may extend further, from 4500 to 5200 km depth (Song and Helmberger, 1992). Modern seismic tomography studies have revealed the existence of lateral velocity anomalies that may be related to thermal convection currents present in the fluid material in the outer core. These currents are related to the generation of the magnetic field. Velocity anomalies and a strong axial anisotropy have also been found in the solid inner core.

10 SURFACE WAVES

10.1 Rayleigh waves in a half-space

The presence of a free surface on an elastic medium introduces a series of phenomena that must be considered in the study of wave propagation. First of all, as we have seen in section 5.4, there are body-wave reflections. Under certain conditions (supercritical incidence of S waves) the generation of inhomogeneous or evanescent waves occurs (sections 5.3 and 5.4). These are body waves that propagate along a direction parallel to the free surface and whose amplitudes decrease with the distance from the free surface. A different phenomenon is the generation of surface waves from constructive interference of body waves in connection with a free surface.

Surface waves are defined as those produced in media with a free surface which propagate parallel to the surface and whose amplitudes decrease with the distance from the surface. Surface waves are generated by energy brought to the free surface by incident body waves. Their existence is related to the presence of a free surface, although they are affected by other surfaces of contact between layers of different elastic properties. There are also waves of similar characteristics, but not related to a free surface, called Stoneley waves, which are associated with an interface between two media in contact.

The first problem is that of determining whether, in an elastic, homogeneous half-space limited by a plane $x_3 = 0$ (x_3 being positive upward), there exist surface waves that propagate in the direction of x_1 with velocity c and whose amplitudes decrease with depth $(-x_3)$ (Fig. 10.1). According to (3.97) and (3.98), the components of displacement u_1 and u_3 can be expressed in terms of the scalar potentials ϕ and ψ, and u_2 is kept apart:

$$u_1 = \phi_{,1} - \psi_{,3} \tag{10.1}$$

$$u_2 = u_2 \tag{10.2}$$

$$u_3 = \phi_{,3} + \psi_{,1} \tag{10.3}$$

For waves of frequency ω that propagate in the positive x_1 direction with velocity c, the potentials, ϕ and ψ, and transversal displacement, u_2, are given by

$$\phi = f(x_3) \exp[ik(x_1 - ct)] \tag{10.4}$$

$$\psi = g(x_3) \exp[ik(x_1 - ct)] \tag{10.5}$$

$$u_2 = h(x_3) \exp[ik(x_1 - ct)] \tag{10.6}$$

where $f(x_3)$, $g(x_3)$, and $h(x_3)$ express the amplitude's dependence on depth and $k = \omega/c$ is the wave number. By substitution into the wave equations for ϕ, ψ, and u_2 (3.66),

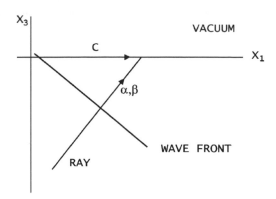

Fig. 10.1. A ray incident on a free surface and the component of velocity along the surface.

we obtain

$$f''(x_3) + kr^2 f(x_3) = 0 \tag{10.7}$$

$$g''(x_3) + ks^2 g(x_3) = 0 \tag{10.8}$$

$$h''(x_3) + ks^2 h(x_3) = 0 \tag{10.9}$$

where, just like in (5.60) and (5.61),

$$r = \left(\frac{c^2}{\alpha^2} - 1 \right)^{1/2} \tag{10.10}$$

$$s = \left(\frac{c^2}{\beta^2} - 1 \right)^{1/2} \tag{10.11}$$

The solutions of equations (10.7)–(10.9) are

$$f(x_3) = A' \exp(ikrx_3) + A \exp(-ikrx_3) \tag{10.12}$$

and expressions of the same form for $g(x_3)$ and $h(x_3)$, with r replaced by s.

Surface waves must have amplitudes that decrease with depth, $(-x_3)$. Hence r and s must be imaginary and positive and consequently $A' = B' = C' = 0$.

According to (10.10) and (10.11), if r and s are imaginary, $c < \beta < \alpha$, that is, the velocity of surface waves is smaller than that of S waves. For incident body waves on the free surface, c is the apparent velocity in the direction of x_1, r and s are real, and $c > \alpha > \beta$ (section 5.4). Also $r = \tan e$ and $s = \tan f$, (5.60 and 5.61), where e and f are the angles of emergence of incident and reflected P and S waves. If $\alpha > c > \beta$, there are incident and reflected S waves and inhomogeneous P waves (r is imaginary). For surface waves, we have both r and s imaginary and $c < \beta < \alpha$. The problem is thus related to the incidence of waves on a free surface.

By substituting $f(x_3)$ from (10.12) and the similar expressions for $g(x_3)$ and $h(x_3)$ into (10.4)–(10.6), with the conditions specified above, we finally obtain

$$\phi = A \exp[-ikrx_3 + ik(x_1 - ct)] \tag{10.13}$$

$$\psi = B \exp[-iksx_3 + ik(x_1 - ct)] \tag{10.14}$$

$$u_2 = C \exp[-iksx_3 + ik(x_1 - ct)] \tag{10.15}$$

To evaluate the arbitrary constants A, B, and C, we apply the boundary conditions at the free surface, that is, null stresses across the surface: for $x_3 = 0$, $\tau_{31} = \tau_{32} = \tau_{33} = 0$. For an isotropic medium, the stresses in terms of the displacements, according to (2.18), are

$$\tau_{31} = \mu(u_{3,1} + u_{1,3}) \tag{10.16}$$

$$\tau_{32} = \mu(u_{3,2} + u_{2,3}) \tag{10.17}$$

$$\tau_{33} = (\lambda + 2\mu)u_{3,3} + \lambda(u_{1,1} + u_{2,2}) \tag{10.18}$$

By replacing the displacements u_1 and u_3 in terms of the potentials (10.1) and (10.3), and considering that they are independent of x_2, for $x_3 = 0$, we obtain the following equations:

$$2\phi_{,31} + \psi_{,11} - \psi_{,33} = 0 \tag{10.19}$$

$$u_{2,3} = 0 \tag{10.20}$$

$$(\lambda + 2\mu)\phi_{,33} + \lambda\phi_{,11} + 2\mu\psi_{,13} = 0 \tag{10.21}$$

By substitution of (10.15) into equation (10.20), we obtain that $C = 0$. Therefore, surface waves in a half-space do not have a transverse component of displacement. By substituting (10.13) and (10.14) into (10.19) and (10.21), we obtain for $x_3 = 0$ that

$$2rA - (1 - s^2)B = 0 \tag{10.22}$$

$$[\alpha^2(r^2 + 1) - 2\beta^2]A - 2\beta^2 sB = 0 \tag{10.23}$$

This is a homogeneous system of equations; therefore, the condition for the existence of a solution, apart from the trivial one $A = B = 0$, is that the determinant of the system be null:

$$\begin{vmatrix} 2r & -(1 - s^2) \\ \alpha^2(r^2 + 1) - 2\beta^2 & -2\beta^2 s \end{vmatrix} = 0 \tag{10.24}$$

Solving the determinant gives

$$[\alpha^2(r^2 + 1) - 2\beta^2](1 - s^2) - 4rs\beta^2 = 0 \tag{10.25}$$

On replacing the values of r and s from (10.10) and (10.11), and taking into account that they are imaginary, we have

$$\left(2 - \frac{c^2}{\beta^2}\right)^2 = 4\left(1 - \frac{c^2}{\alpha^2}\right)^{1/2}\left(1 - \frac{c^2}{\beta^2}\right)^{1/2} \tag{10.26}$$

This equation is known as Rayleigh's equation in honor of John W. Strutt, Lord Rayleigh, who solved this problem for the first time in 1887. In order to study the solutions of Rayleigh's equation, we make the changes of variables $\xi = (c/\beta)^2$ and $q = (\beta/\alpha)^2$. The value of q is related to Poisson's ratio (section 2.2) and depends on the elastic properties of the medium. On taking the square of (10.26) and eliminating the solution $\xi = 0$, we obtain a cubic equation for ξ:

$$\xi^3 - 8\xi^2 + 8(3 + 2q)\xi + 16(q - 1) = 0 \tag{10.27}$$

Table 10.1. *Values of ξ and c_R corresponding to certain values of σ, together with the relations among the values of q, λ, and μ.*

σ	q	λ	μ	ξ	c_R
0	$\frac{1}{2}$	0	$\frac{3}{2}K$	0.7640	0.8741β
$\frac{1}{8}$	$\frac{3}{7}$	$\frac{1}{3}\mu$	K	0.8059	0.8977β
$\frac{1}{4}$	$\frac{1}{3}$	μ	$\frac{3}{5}K$	0.8453	0.9194β
$\frac{1}{2}$	0	K	0	0.9128	0.0

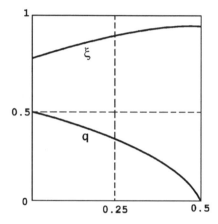

Fig. 10.2. The variations of q (β/α) and ξ (c/β) with Poisson's ratio.

The solutions of this equation depend on the values of q. Since Poisson's ratio has values between 0 and $\frac{1}{2}$ (section 2.2), q varies between $\frac{1}{2}$ and 0. For each value of q, equation (10.27) has three roots. From them only those for values of $\xi < 1$ correspond to surface waves, since they must satisfy the condition that $c < \beta$. Waves corresponding to these solutions are called Rayleigh waves and their velocity c_R is a fraction of the shear-wave velocity in the half-space. In Table 10.1, values of ξ and c_R corresponding to certain values of σ are given, together with the relations among the values of q, λ, and μ.

Values of ξ and q versus σ are shown in Fig. 10.2. For all possible values of σ, ξ varies very little, between 0.7640 and 0.9128. For $\sigma = \frac{1}{2}$, the medium is a liquid, μ and β are null, and the root $\xi = 0.9128$ corresponds to a zero value of c_R. Hence, Rayleigh waves do not exist in a liquid half-space.

A special case is that for $\sigma = \frac{1}{4}$ $(\lambda = \mu)$ and $q = \frac{1}{3}$, which condition is approximately satisfied for the Earth's materials. For this case, we derived in section 5.4 expressions for the reflection coefficients of incident P and S waves reflected on the free surface of an elastic medium, V_{SS} and V_{PP}. Both (5.100) and (5.106) have the same numerator. If we put this numerator equal to zero, we obtain the same equation (10.26) for $q = \frac{1}{3}$. This shows again that the generation of Rayleigh waves is related to the reflection of P and S waves on the free surface of an elastic half-space. For this particular case, equation (10.27) results in

$$\xi^3 - 8\xi^2 + \tfrac{56}{3}\xi - \tfrac{32}{3} = 0 \tag{10.28}$$

The roots of this equation are 4, $2 + 2/\sqrt{3}$ and $2 - 2/\sqrt{3}$. The first two ($\xi > 1$), correspond to cases of wave reflection for which $c > \alpha > \beta$. For incident P waves, these values of ξ correspond to angles of incidence for which there are no reflected P waves, since $V_{PP} = 0$. Because $\cos e = \alpha/c$, these angles are $e = 30°$ and $e = 47°47'$. For an incident S wave, they correspond to $f = 60°$ and $f = 55°44'$ ($V_{SS} = 0$). The third root is $\xi = 0.8453$ and corresponds to Rayleigh waves.

In conclusion, in an elastic half-space, Rayleigh surface waves that propagate parallel to the free surface with a velocity $c_R = \sqrt{\xi}\beta$, where ξ are the roots with values less than unity of equation (10.27), are generated. The amplitudes of these waves decrease exponentially with depth. In a certain way, these waves may be considered to be generated by energy brought to the surface by incident P and S waves that produce no reflections.

10.1.1 Displacements of Rayleigh waves

Displacements of Rayleigh waves are obtained by substituting expressions (10.13) and (10.14) for the potentials ϕ and ψ into (10.1) and (10.3). For the particular case of $\sigma = \tfrac{1}{4}$, we have $c_R = 0.9194\beta$, $r = 0.85i$, and $s = 0.39i$. By first substituting these values into (10.22), we obtain $B = -1.47iA$. Finally, taking only the real part, the displacements are given by

$$u_1 = -Ak(e^{0.85kx_3} - 0.58\,e^{0.39kx_3})\sin[k(x_1 - c_R t)] \tag{10.29}$$

$$u_3 = -Ak(-0.85\,e^{0.85kx_3} + 1.47\,e^{0.39kx_3})\cos[k(x_1 - c_R t)] \tag{10.30}$$

We must remember that A is the potential amplitude (in units of m^2) and Ak is the displacement amplitude (in units of meters). For points at the free surface ($x_3 = 0$), letting $-Ak = a$, we find that

$$u_1 = 0.42a\sin[k(x_1 - c_R t)] \tag{10.31}$$

$$u_3 = 0.62a\cos[k(x_1 - c_R t)] \tag{10.32}$$

Since Rayleigh waves have no transverse component, they are polarized in the vertical plane. The horizontal and vertical components are shifted in phase by $\pi/2$ and thus the motion is elliptical. If we substitute into (10.31) and (10.32) the values of t during a complete cycle (0 to T, where $T = 2\pi/\omega$ is the period), we obtain for the particle's motion an ellipse with a vertical major axis and retrograde motion (opposite to that of wave propagation) (Fig. 10.3).

The dependences of the displacement components u_1 and u_3 on depth ($-x_3$) are given by (10.29) and (10.30). There is a value of x_3 for which u_1 is null, $x_3 = -0.19\lambda$ ($\lambda = 2\pi/k$ is the wave length), whereas u_3 is never null. At the depth where u_1 is null, its amplitude changes sign. For greater depths the particle's motion is prograde (Fig. 10.4). With increasing depth, the amplitudes of u_1 and u_3 decrease exponentially, with u_3 always larger than u_1.

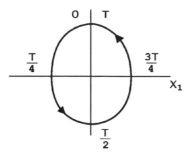

Fig. 10.3. A diagram of a particle's motion on the vertical plane for a Rayleigh wave at the free surface.

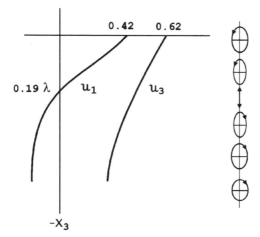

Fig. 10.4. A plot of the amplitudes of the displacement components u_1 and u_3 of Rayleigh waves in a half-space versus depth.

10.2   A liquid layer over a rigid half-space. Guided waves

The existence of one or more layers of finite thickness and different properties lying over a half-space modified the results we have obtained for surface waves. As an introduction to this problem, we consider a liquid layer of thickness H, density ρ, and velocity α lying over a rigid half-space (Fig. 10.5). Since there is no propagation of waves in a rigid medium, we have no surface waves, only waves contained in the liquid layer.

The motion in the liquid layer can be expressed in terms of a scalar potential ϕ. For waves that propagate in the positive x_1 direction with velocity c, the potential is given by

$$\phi = (A\,e^{ikrx_3} + B\,e^{-ikrx_3})\,e^{ik(x_1 - ct)} \tag{10.33}$$

where r is given by (10.10). Since x_3 can vary only between 0 and H, we can not yet impose any condition on the real or imaginary character of r. The boundary condition on the free surface ($x_3 = H$) is that the normal component of stress across the surface be null, $\tau_{33} = 0$. At the base of the layer ($x_3 = 0$), the vertical component of

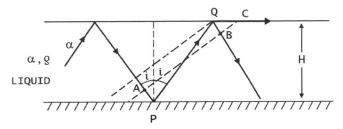

Fig. 10.5. Rays in a liquid layer over a rigid medium, showing constructive interference on the wave front AB.

the displacement is null, $u_3 = 0$. In terms of the potential ϕ, these two conditions are

$$x_3 = H: \qquad \phi_{,11} + \phi_{,33} = 0 \tag{10.34}$$

$$x_3 = 0: \qquad \phi_{,3} - 0 \tag{10.35}$$

The first condition, from the relation among the cubic dilation, the Laplacian of ϕ (2.13), and the wave equation (3.3), can be written as $\rho\omega^2\phi = 0$. Using the potential ϕ according to (10.3), conditions (10.34) and (10.35) give

$$A\,e^{ikrH} + B\,e^{-ikrH} = 0 \tag{10.36}$$

$$A - B = 0 \tag{10.37}$$

Since this is a homogeneous system of equations for A and B, the existence of a solution implies that its determinant is null:

$$e^{ikrH} + e^{-ikrH} = 0 \tag{10.38}$$

If r is imaginary, equation (10.38) leads to the impossible result $\cosh(krH) = 0$. Therefore, r must be real and condition (10.38) gives

$$\cos(krH) = 0 \tag{10.39}$$

On substituting for r its value in (10.10), the condition leads to

$$kH(c^2/\alpha^2 - 1)^{1/2} = (n + \tfrac{1}{2})\pi, \qquad n = 0, 1, 2, 3, \ldots \tag{10.40}$$

Since r is real, that is $c > \alpha$, waves in the x_1 direction with velocity c are formed by reflections in the interior of the liquid layer. Waves propagate parallel to both limits of the layer and are called guided or channeled waves. Contrary to the previous case (10.25), in equation (10.38) the wave number k appeared. Thus, equation (10.40) implies that the velocity of propagation is a function of the wave number, $c(k)$, or of the frequency, $c(\omega)$. This means that waves are dispersed and equation (10.40) that relates the velocity to the frequency is called the dispersion equation. Wave dispersion is a consequence of introducing a finite dimension (the layer's thickness) into the problem, since the velocity depends on the relation between the wave length and the layer's thickness.

Another important result is the presence in the dispersion equation (10.40) of the integer n that takes an infinite number of discrete values. For each value of n, the relation between c and k is different; that is, we have a different type of wave propagation. Each of these forms of wave propagation is called a mode, just like in the problem of vibrations of finite elastic bodies (Chapter 4). Here the presence of modes is due to the finite dimension of the layer's thickness. Just like in Chapter 4, the lowest value of n corresponds to the fundamental mode and the other values to higher modes or harmonics.

10.2.1   Constructive interference

We have seen that the solution for guided waves corresponds to real values of r; that is, $c > \alpha$. Therefore we can consider that these waves are generated by constructive interference of body (acoustic) waves propagated in the interior of the liquid layer and reflected from its two surfaces. Hence we can also find the dispersion equation (10.40) by using the ray-theory approach for plane waves and the condition of constructive interference. This condition implies that waves that coincide on the same wave front must be in phase so that their amplitudes are summed. For a ray that is reflected on both surfaces (Fig. 10.5), the same wave front AB corresponds to rays that pass through A and through B. For constructive interference on this wave front, the two waves must be in phase; that is, the distance along the ray from A to B (AP + PQ + QB) must be an integer multiple of the wave length, taking into account the possible phase shifts at the points of reflection. The phase shift at the rigid surface is zero whereas that at the free surface is π (section 5.2). The condition of constructive interference can, then, be written as

$$\frac{2\pi}{\lambda_\alpha}(AP + PQ + QB) - \pi = 2\pi n \tag{10.41}$$

According to the ray geometry (Fig. 10.6)

$$AP + PQ + QB = 2H \cos i \tag{10.42}$$

Since $\sin i = \alpha/c$, $\cos i = (\alpha/c)(c^2/\alpha^2 - 1)^{1/2}$, and $k = k_\alpha \sin i$, by substitution into (10.41) we obtain the dispersion equation (10.40). Guided waves that appear on the

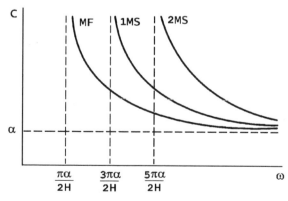

Fig. 10.6. Dispersion curves for guided waves in a liquid layer over a rigid medium. The fundamental mode and the two first higher modes with their cut-off frequencies are shown.

liquid layer are, thus, formed by constructive interference of acoustic waves reflected from its two surfaces.

10.2.2 The dispersion equation and curves

If we write (10.40) in terms of the frequency ω and solve for c, we find the explicit form of the dispersion equation:

$$c(\omega) = \frac{1}{\left\{ \dfrac{1}{\alpha^2} - \left[\left(n + \dfrac{1}{2} \right) \pi \right]^2 \dfrac{1}{H^2 \omega^2} \right\}^{1/2}} \tag{10.43}$$

The fundamental mode corresponds to $n = 0$; thus, according to (10.40), $kHr = \pi/2$. Higher modes correspond to $n \geq 1$. For real values of $c(\omega)$, the expression in the denominator must be larger than zero. The frequency ω_c corresponding to the zero value of the denominator is called the cut-off frequency, since there are no values of $c(\omega)$ for $\omega < \omega_c$. For any mode of order n, the cut-off frequency is

$$\omega_c = \frac{\pi (n + \frac{1}{2}) \alpha}{H} \tag{10.44}$$

The lower cut-off frequency, $\omega_c = \alpha \pi / (2H)$, corresponds to the fundamental mode. In all modes, the velocity c becomes infinite for the cut-off frequency $\omega = \omega_c$. With increasing ω, the velocity c decreases and, in the limit, when ω tend to infinity, c tends to α. The $c(\omega)$ curves are called dispersion curves (Fig. 10.6). For each mode there is a dispersion curve with values of the velocity for frequencies from the cut-off frequency to infinity. We can see that the lowest value of the velocity corresponds to the fundamental mode for a given frequency and increases with the order of the mode. Also, a given velocity corresponds to the lowest frequency for the fundamental mode and to higher frequencies with increasing order of the modes (Fig. 10.6).

10.2.3 Displacements

Using equations (10.33) and (10.37), we can determine the displacements of guided waves, by taking derivatives of the potential ϕ, according to (10.1) and (10.3):

$$u_1 = -2Ak \cos \left(krH \frac{x_3}{H} \right) \sin[k(x_1 - ct)] \tag{10.45}$$

$$u_3 = -2Akr \sin \left(krH \frac{x_3}{H} \right) \cos[k(x_1 - ct)] \tag{10.46}$$

The components u_1 and u_3 are shifted in phase by $\pi/2$, and the resulting motion in the vertical plane is elliptical and retrograde. For the fundamental mode, $kHr = \pi/2$, we obtain

$$u_1 = -2Ak \cos \left(\frac{\pi x_3}{2H} \right) \sin[k(x_1 - ct)] \tag{10.47}$$

$$u_3 = -\frac{\pi A}{H} \sin \left(\frac{\pi x_3}{2H} \right) \cos[k(x_1 - ct)] \tag{10.48}$$

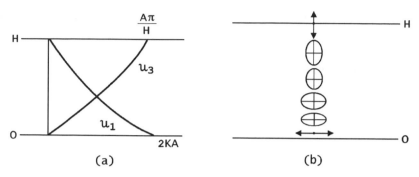

Fig. 10.7. (a) The amplitude of the displacement components u_1 and u_3 of guided waves in a liquid layer over a rigid medium for the fundamental mode. (b) A diagram showing a particle's motion on the vertical plane.

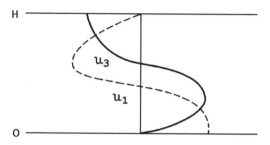

Fig. 10.8. Amplitudes with depth for the displacement components u_1 and u_3 for the first higher mode of guided waves in a liquid layer over a rigid medium.

For the free surface ($x_3 = H$), $u_1 = 0$ and the motion is vertical, whereas at the base of the layer ($x_3 = 0$), $u_3 = 0$ and the motion is horizontal. Inside the layer, the motion is elliptical; the major axis changes with depth from vertical to horizontal (the motion is circular for $x_3 = (2H/\pi) \tan^{-1}[\pi/(2hk)]$) (Fig. 10.7).

For the fundamental mode there is no depth inside the layer at which the amplitude either of u_1 or of u_3 is zero (Fig. 10.7(a)). For higher modes, there are values of x_3 inside the layer, equal to the order number of the mode, at which either u_1 or u_3 is zero (nodes), but they do not coincide. For example, for the first higher mode, $krH = 3\pi/2$, u_1 is zero for $x_3 = H/3$ and u_3 is zero for $x_3 = 2H/3$ (Fig. 10.8). This characteristic of guided waves is similar to the vibrations of an elastic rod of finite length (Chapter 4).

In conclusion, guided waves in a liquid layer over a rigid half-space present the following characteristics that are found in all cases of the propagation of waves in layers of finite thickness: (a) dispersion; that is, the velocity depends on the frequency; (b) an infinite number of modes corresponding to values of n, namely fundamental ($n = 0$) and higher ($n \geq 1$) modes; (c) for each mode there is a different dispersion curve with a cut-off frequency; and (d) displacements for higher modes have a number of amplitude nodes equal to their order number.

10.3 An elastic layer over a half-space. Love waves

The characteristics of Rayleigh waves on the free surface of a half-space do not agree with those of surface waves observed on the Earth. Ever since the installation of the first seismographs, it has been observed that surface waves are formed by dispersed trains and have transverse components of motion. Therefore, the theory deduced in section 10.1 is not sufficient to explain observations and we must introduce the presence of layers that affect the propagation of surface waves. The first to consider this problem was Love in 1911. He found that, in an elastic layer over a half-space, dispersed surface waves are produced with transverse components, which are now called Love waves.

In this problem, we consider only waves with a transverse component of displacement, u_2, that propagate in the direction x_1 with a velocity c, in a medium consisting of a layer of thickness H, density ρ' and shear velocity β', over a half-space of density ρ and velocity β (Fig. 10.9). We can write the displacements in the form of (10.15), but allowing for waves travelling in positive and negative directions of x_3 inside the layer leads to

$$u_2' = (A'\, e^{iks'x_3} + B'\, e^{-iks'x_3})\, e^{ik(x_1 - ct)} \tag{10.49}$$

$$u_2 = B\, e^{-iksx_3 + ik(x_1 - ct)} \tag{10.50}$$

where s is given by (10.11) and s' has a similar form obtained by replacing β' by β. In (10.50), according to the condition for surface waves, amplitudes must decrease with depth $(-x_3)$; therefore, s must be positive and imaginary, $s = i\bar{s}\,(c < \beta)$. We can not impose any condition on s', since x_3 inside the layer is limited to values between 0 and H. The boundary conditions are now as follows. (a) At the free surface $(x_3 = H)$, the component of stress τ_{32}' is null. (b) At the surface of contact between the layer and the half-space $(x_3 = 0)$, there is continuity of the stress $(\tau_{32}' = \tau_{32})$ and displacement $(u_2' = u_2)$. The second condition implies that the layer and half-space are welded together. The three boundary conditions in terms of the displacements are

$$x_3 = H: \qquad \mu' u_{2,3}' = 0 \tag{10.51}$$

$$x_3 = 0: \qquad \mu' u_{2,3}' = \mu u_{2,3} \tag{10.52}$$

$$u_2' = u_2 \tag{10.53}$$

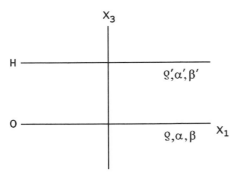

Fig. 10.9. An elastic layer of thickness H over an elastic half-space.

By substitution into (10.49) and (10.50), we find the following three equations:

$$A' e^{iks'H} - B' e^{-iks'H} = 0 \tag{10.54}$$

$$A' \mu' s' - B' \mu' s' + Bs\mu = 0 \tag{10.55}$$

$$A' + B' - B = 0 \tag{10.56}$$

A solution of this system of equations, apart from the trivial one $A' = B' = B = 0$, requires that the determinant be zero:

$$\begin{vmatrix} e^{iks'H} & -e^{-iks'H} & 0 \\ \mu's' & -\mu's' & \mu s \\ 1 & 1 & -1 \end{vmatrix} = 0 \tag{10.57}$$

By expanding the determinant we obtain

$$\frac{\mu s}{\mu' s'} = \frac{e^{iks'H} - e^{-iks'H}}{e^{iks'H} + e^{-iks'H}} \tag{10.58}$$

If s' is imaginary ($c < \beta'$), then, on making the substitution $s' = i\bar{s}'$, we obtain

$$\frac{\mu \bar{s}}{\mu' \bar{s}'} = - \tanh(k\bar{s}'H) \tag{10.59}$$

Since the left-hand side is always positive and the hyperbolic tangent is, too, this result is not possible. Therefore s' must be real ($\beta > c > \beta'$), so the velocity of the layer must be less than that of the half-space, and the velocities of Love waves have values between the two. Since s is imaginary and s' is real, from equation (10.58) we obtain

$$\frac{\mu(1 - c^2/\beta^2)^{1/2}}{\mu'(c^2/\beta'^2 - 1)^{1/2}} = \tan[kH(c^2/\beta'^2 - 1)^{1/2}] \tag{10.60}$$

On making the substitution $k = \omega/c$, we obtain

$$\frac{\mu(1/c^2 - 1/\beta^2)^{1/2}}{\mu'(1/\beta'^2 - 1/c^2)^{1/2}} = \tan[\omega H(1/\beta'^2 - 1/c^2)^{1/2}] \tag{10.61}$$

Just like in the case of guided waves in a liquid layer, the velocity of Love waves is a function of the frequency, $c(k)$, or $c(\omega)$ and equations (10.60) and (10.61) are dispersion equations. The tangent has positive values between zero and infinity for various intervals of its argument kHs', the first between 0 and $\pi/2$, the second between π and $3\pi/2$, etc. Each of them corresponds to a mode of propagation; the fundamental mode corresponds to the first interval ($0 \leq khs' \leq \pi/2$) and higher modes to the rest.

10.3.1 Constructive interference

In a similar manner to what we did in the case of guided waves in a liquid layer, we can also deduce the dispersion equation of Love waves (10.60), using the principle of constructive interference for SH rays reflected inside the layer corresponding to angles of incidence greater than the critical one. The condition of constructive interference requires that waves on the same wave front must be in phase. In Fig. 10.10, at points

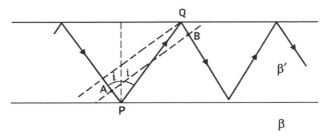

Fig. 10.10. Constructive interference on the wave front AB of SH waves with angles of incidence greater than the critical one inside an elastic layer over a half-space.

A and B waves must have the same phase; that is, the length $AP + PQ + QB$ must be a multiple of the wave length, taking into account changes in phase at points P and Q. The change of phase between incident and reflected SH waves with supercritical incidence at P is given by (5.74)

$$\theta = -2\tan^{-1}\left(\frac{\mu\bar{s}}{\mu's'}\right)$$

At the free surface (Q) there is no change in phase (section 5.4). Using (10.41) and (10.42), the condition for constructive interference is

$$2k_{\beta'}H\cos i - 2\tan^{-1}\left(\frac{\mu\bar{s}}{\mu's'}\right) = 2\pi n, \qquad n = 0, 1, 2, \dots \qquad (10.62)$$

where $k_{\beta'}$ is the wave number of SH waves in the layer. Its relation to k, the wave number corresponding to the velocity c, is $k_{\beta'} = k/\sin i$. Then, $\sin i = \beta'/c$ and $\cos i = s'\beta'/c$. On replacing these values and those of s and s' into (10.62) we obtain the same dispersion equation (10.60).

Since s' is real ($c > \beta'$), there exist in the layer SH waves that propagate upward and downward, reflecting from its base and free surface. In the half-space there are no transmitted waves (angles of incidence are greater than the critical one), only inhomogeneous waves represented by (10.50), where s is imaginary ($c < \beta$). We can conclude that Love waves are formed by the constructive interference of SH waves with supercritical incidence in the layer and inhomogeneous waves in the half-space.

10.3.2 Dispersion curves

The dispersion equation of Love waves (10.60) or (10.61) represents the dependence of the velocity on the frequency. Dispersion curves are represented by $c(k)$, $c(\omega)$, or $c(T)$, where T is the period ($T = 2\pi/\omega$). The forms of these curves depend on the parameters of the model β, β', and H. Owing to the periodicity of the tangent function, for each model there is an infinite number of modes of propagation and consequently an infinite number of dispersion curves. Modes depend on the range of the values of the argument of the tangent, the first ($n = 0$) corresponding to the fundamental mode, the rest ($n \geq 1$) to the higher modes:

$$n\pi \leq kH(c^2/\beta'^2 - 1)^{1/2} \leq \frac{2n+1}{2}\pi, \qquad n = 0, 1, 2, 3, \dots$$

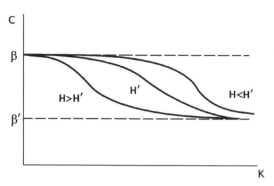

Fig. 10.11. Dispersion curves for the fundamental mode of Love waves for various thicknesses (H) of the layer.

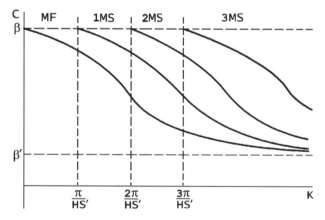

Fig. 10.12. Dispersion curves of Love waves in a layer over a half-space for the fundamental mode and the three first higher modes with their cut-off frequencies.

For the fundamental mode, according to (10.60), we have

$$0 \le kH(c^2/\beta'^2 - 1)^{1/2} \le \pi/2$$

For $kHs' = 0$, $k = 0$ and $c = \beta$. For $kHs' = \pi/2$, $k = \infty$ and $c = \beta'$.

In the fundamental mode all frequencies are present ($0 \le k \le \infty$) and the velocity of Love waves varies between the values of the S wave velocity in the half-space and the layer ($\beta' \le c \le \beta$). For the same velocities, the form of the curve depends on the layer thickness H (Fig. 10.11). The influence of the layer thickness (H) is conditioned by its relation to the wave length (λ). For $\lambda < H$, propagation is mainly influenced by the layer and $c \simeq \beta'$. For $\lambda > H$, propagation is conditioned by the half-space and $c \simeq \beta$. Therefore, for small thicknesses, the Love wave velocity approaches that of the half-space, $c \simeq \beta$; for relatively high frequencies ($\omega > 2\pi c/H$) and for large thicknesses this is true for lower frequencies (Fig. 10.11).

For higher modes, since the argument of the tangent does not start at zero, there is a limit for low frequencies, called the cut-off frequency (section 10.2). For each higher mode, the cut-off frequency has a different value. For a mode of order n, the cut-off wave number is

$$k_n = \frac{n\pi}{Hs'} \tag{10.63}$$

For the first higher mode we have

$$\pi \le kH(c^2/\beta'^2 - 1)^{1/2} \le 3\pi/2$$

For $kHs' = \pi$, $k = \pi/(Hs')$ and $c = \beta$. For $kHs' = 3\pi/2$, $k = \infty$ and $c = \beta'$. The cut-off frequency is $k_1 = \pi/(Hs')$ and the form of the dispersion curve is similar to that of the fundamental mode (Fig. 10.12).

In general, dispersion curves for higher modes start at the cut-off frequency with the value $c = \beta$, and, for very high frequencies, this tends toward $c = \beta'$. For a given frequency, the lowest velocity corresponds to the fundamental mode and the highest to the mode of highest order. A given velocity corresponds to lower frequencies in the fundamental mode and to higher frequencies in higher modes (Fig. 10.12). For a given distance, waves of lower frequencies of the fundamental mode arrive at the same time as those of higher frequencies of higher modes.

10.3.3 Displacements

Displacements of Love waves can be deduced from expressions (10.49) and (10.50). From (10.54), we have

$$B' = A' \, e^{i2ks'H} \tag{10.64}$$

On substituting this into (10.49) and multiplying and dividing by $\exp(ikHs')$, we obtain for the real part of the displacement in the layer

$$u'_2 = 2A' \cos\left[ks'H\left(1 - \frac{x_3}{H}\right)\right] \cos[k(s'H - x_1 - ct)] \tag{10.65}$$

Displacements in the half-space can be derived from (10.50). From (10.56) and (10.64) we have

$$B = 2A' \cos(ks'H) \, e^{iks'H} \tag{10.66}$$

By substitution into (10.50), we obtain for the real part

$$u_2 = 2A' \cos(ks'H) \, e^{k\bar{s}x_3} \cos[k(s'H + x_1 - ct)] \tag{10.67}$$

In the half-space, displacements decrease exponentially with depth $(-x_3)$. Inside the layer, displacements vary with depth depending on the value of $ks'H$. Therefore, the variation will be different for each mode.

For the fundamental mode, $0 \le kHs' \le \pi/2$:

$$k = 0, \qquad |u'_2| = 2A' \text{ for all values of } x_3$$

$$k = \infty, \qquad |u'_2| = \begin{cases} 2A' & \text{for } x_3 = H \\ 0 & \text{for } x_3 = 0 \end{cases}$$

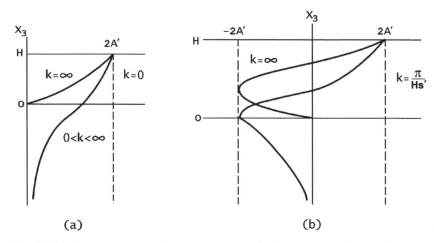

Fig. 10.13. The amplitude versus depth of Love waves in a layer over a half-space, for (a) the fundamental mode and (b) the first higher mode.

On the free surface $(x_3 = H)$, $|u_2'| = 2A'$ for all values of k. At the base of the layer $(x_3 = 0)$, the amplitude varies between $2A'$ and zero $(2A' > |u_2'| > 0)$, for wave numbers between zero and infinity $(0 < k < \infty)$. There are no nodes or values of x_3 inside the layer for which the amplitude is null (Fig. 10.13(a)). In the half-space, amplitudes decrease exponentially with depth, starting with the value at the base of the layer.

For the first higher mode, $\pi \leq kHs' \leq 3\pi/2$:

$$k = \pi/Hs', \qquad |u_2'| = \begin{cases} 2A' & \text{for } x_3 = H \\ 0 & \text{for } x_3 = H/2 \\ -2A' & \text{for } x_3 = 0 \end{cases}$$

$$k = \infty, \qquad |u_2'| = \begin{cases} 2A' & \text{for } x_3 = H \\ 0 & \text{for } x_3 = 2H/3 \\ -2A' & \text{for } x_3 = H/3 \\ 0 & \text{for } x_3 = 0 \end{cases}$$

At the free surface, amplitude is always $|u_2'| = 2A'$, for all frequencies and modes. Depending on the value of k, there is a node where amplitudes are zero, for x_3 between $H/2$ and $3H/2$ (Fig. 10.13(b)). At the base of the layer, amplitudes have values between 0 and $-2A'$ for k between π/Hs' and infinity.

In a similar way, we can find the positions of nodes for any higher mode. In each mode, the number of nodes is equal to the order number of the mode. The dispersion and existence of modes of propagation in Love waves, just like in the case of guided waves in a liquid layer (section 10.2), are consequences of the finite dimension of the layer thickness. The characteristics of propagation depend on the relation between the wave length and the layer thickness. As has been mentioned the situation is similar to that of vibrations of an elastic rod of finite length (section 4.3).

10.4   An elastic layer over a half-space. Rayleigh waves

In the previous section, we considered only the transverse component of displacements (u_2); now we must consider the longitudinal and vertical components (u_1 and u_3). The resulting surface waves are similar to the Rayleigh waves found for an elastic half-space (section 10.1). To determine the existence and properties of these waves we proceed in the same way as for the problem of Love waves. The displacement components u_1 and u_3 are derived from the scalar potentials ϕ and ψ according to (10.1) and (10.3). For surface waves that propagate in the x_1 direction with a velocity c, in a layer of thickness H and velocities α' and β', over a half-space of velocities α and β (Fig. 10.9), taking into account that amplitudes must decrease with depth in the half-space, the potentials are given by

$$\phi' = A' \exp[ikr'x_3 - ik(x_1 - ct)] + B' \exp[-ikr'x_3 - ik(x_1 - ct)] \tag{10.68}$$

$$\psi' = C' \exp[iks'x_3 - ik(x_1 - ct)] + D' \exp[-iks'x_3 - ik(x_1 - ct)] \tag{10.69}$$

$$\phi = A \exp[-ikrx_3 - ik(x_1 - ct)] \tag{10.70}$$

$$\psi = C \exp[-iksx_3 - ik(x_1 - ct)] \tag{10.71}$$

where r and s are given by (10.10) and (10.11), and similarly r' and s' (on replacing α by α' and β by β'). Since amplitudes must decrease with depth ($-x_3$), r and s must be imaginary and positive ($r = i\bar{r}$ and $s = i\bar{s}$), and consequently $c < \beta < \alpha$.

Just like in the case of Love waves, the boundary conditions are $x_3 = H$ for the free surface, the components of stress are null, $\tau_{33} = \tau_{31}' = 0$, the contact surface is $x_3 = 0$, and one has continuity of the stress and displacement components $\tau_{33}' = \tau_{33}$, $\tau_{31}' = \tau_{31}$, $u_1' = u_1$, and $u_3' = u_3$. As functions of the displacements, the boundary conditions result in the following equations:

$$x_3 = H: \qquad u_{3,1}' + u_{1,3}' = 0 \tag{10.72}$$

$$\lambda' u_{1,1}' + (\lambda' + 2\mu') u_{3,3}' = 0 \tag{10.73}$$

$$x_3 = 0: \qquad u_1' = u_1 \tag{10.74}$$

$$u_3' = u_3 \tag{10.75}$$

$$\mu'(u_{3,1}' + u_{1,3}') = \mu(u_{3,1} + u_{1,3}) \tag{10.76}$$

$$\lambda' u_{1,1}' + (\lambda' + 2\mu') u_{3,3}' = \lambda u_{1,1} + (\lambda + \mu) u_{3,3} \tag{10.77}$$

On substituting for the displacements in terms of the potentials ϕ, ϕ', ψ, and ψ', according to equations (10.68)–(10.71), we obtain

$$2r'(A' - B') + (1 - s'^2)(C' + D') = 0 \tag{10.78}$$

$$[\lambda'(1 + r'^2) + 2\mu' r'^2](A' + B') + 2\mu' s'(D' - C') = 0 \tag{10.79}$$

$$A' e^{ikr'H} + B' e^{-ikr'H} - s'C' e^{iks'H} + D' s' e^{-iks'H} = A e^{-ikrH} + sC e^{-iksH} \tag{10.80}$$

$$r'A' e^{ikr'H} - r'B' e^{-ikr'H} + C' e^{iks'H} - D' e^{-iks'H} = -rA e^{-ikrH} - C e^{-iksH} \tag{10.81}$$

$$\mu'[2r'(A'\,e^{ikr'H} + B'\,e^{-ikr'H}) + (1 - s'^2)(C'\,e^{iks'H} - D'\,e^{-iks'H})]$$
$$= \mu[2rA\,e^{-ikrH} + (1 - s^2)\,e^{-iksH}] \tag{10.82}$$

$$[\lambda'(1 + r'^2) + 2\mu'r'^2](A'\,e^{ikr'H} + B'\,e^{-ikr'H}) - 2\mu's'(D'\,e^{-iks'H} - C'\,e^{iks'H})$$
$$= [\lambda(1 + r^2) + 2\mu r^2]A\,e^{-ikrH} + 2\mu sC\,e^{-iksH} \tag{10.83}$$

Equations (10.78)–(10.83) form a system of six equations for six unknowns A', B', C', D', A, and C, the amplitudes of the potentials ϕ and ψ in the layer and half-space. Again, the condition for a solution is that the determinant of the system is null. Making the determinant equal to zero, we find an equation for c, the velocity of Rayleigh waves. Since this equation implies that the velocity $c(k)$ is a function of the frequency, Rayleigh waves in a layer over a half-space are dispersed.

For the particular case when Poisson's ratio in the layer and half-space is $\sigma = \frac{1}{4}$ ($\lambda = \mu$ and $\lambda' = \mu'$), the expressions are simplified. We can verify that

$$\lambda(1 + r^2) + 2\mu r^2 = \mu(1 + 3r^2) = \mu(1 + s^2) \tag{10.84}$$

If in the system of equations (10.78)–(10.83), we use as unknowns $A'\,e^{id'}$, $B'\,e^{-id'}$, $C'\,e^{ib'}$, $D'\,e^{-ib'}$, $A\,e^{-ia}$, and $C\,e^{-ib}$, where $d' = kr'H$, $b' = ks'H$, $a = krH$, and $b = ksH$, the determinant of the system can be written in the form

$$\begin{vmatrix} 2r'\,e^{-id'} & -2r'\,e^{id'} & (1 - s'^2)\,e^{-ib'} & (1 - s'^2)\,e^{ib'} & 0 & 0 \\ -(1 - s'^2)\,e^{-id'} & -(1 - s'^2)\,e^{id'} & 2s'\,e^{-ib'} & 2s'\,e^{ib'} & 0 & 0 \\ 1 & 1 & -s' & s' & -1 & -s \\ r' & -r' & 1 & -1 & r & 1 \\ 2\mu'r' & 2\mu'r' & \mu'(1 - s'^2) & -\mu'(1 - s'^2) & -2\mu r & -\mu(1 - s^2) \\ -\mu'(1 - s'^2) & -\mu'(1 - s'^2) & 2\mu's' & -2\mu's' & \mu(1 - s)^2 & -2\mu s \end{vmatrix}$$

The dispersion equation is obtained by expanding the determinant and putting it equal to zero. There are several ways of expanding this determinant. The one proposed by Love in 1911 results in the equation

$$\xi\eta' - \xi'\eta = 0 \tag{10.85}$$

where, substituting the variables used here in place of those used by Love,

$$\xi = (1 - s'^2)[X\cos d' + (Yr/r')\sin d'] + 2s'[Wr\sin b' - Z/s'\sin b'] \tag{10.86}$$

$$\xi' = (1 - s'^2)[Ws\cos d' + Zr'\sin d'] + 2s'[X\sin d' - (Ys/s')\cos b'] \tag{10.87}$$

$$\eta = (1 - s'^2)[Wr\cos b' + (Z/s')\sin b'] + 2r'[X\sin d' - (Yr/r')\cos d'] \tag{10.88}$$

$$\eta' = (1 - s'^2)[X\cos b' + (Ys/s')\sin b'] + 2r'[Ws\sin d' - (Z/r')\cos d'] \tag{10.89}$$

and, using $g = \mu/\mu'$,

$$X = gc^2/\beta^2 - 2(g - 1) \tag{10.90}$$

$$Y = c^2/\beta'^2 + 2(g - 1) \tag{10.91}$$

$$Z = gc^2/\beta'^2 - c^2/\beta^2 - 2(g - 1) \tag{10.92}$$

$$W = 2(g - 1) \tag{10.93}$$

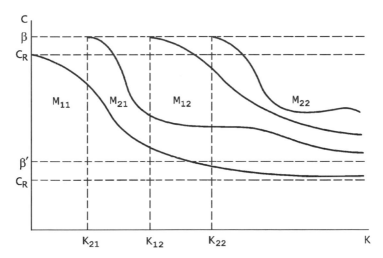

Fig. 10.14. Dispersion curves of Rayleigh waves in an elastic layer over a half-space for the fundamental mode (M_{11}) and the three first higher modes (symmetric M_{12}, and antisymmetric M_{21} and M_{22}).

For a solution that implies the existence of Rayleigh waves that decrease exponentially with depth in the half-space r and s are imaginary, while s' and r' are real. In some cases, however, r' can also be imaginary. These conditions imply that the velocity c of Rayleigh waves satisfies the condition $\alpha > \beta > c > \beta'$, while α' is, generally, less than c, but not necessarily. In a similar way to Love waves, Rayleigh waves in a layer over a half-space are formed by constructive interference of both P and SV waves reflected supercritically from the contact surface.

Solutions of the dispersion equation (10.85) give the velocity of Rayleigh waves as a function of the frequency $c(k)$ or $c(\omega)$. There are an infinite number of solutions corresponding to different modes. Modes in this case are separated into two types, symmetric (M_1) and antisymmetric (M_2), in analogy with the vibrations of an elastic layer. For symmetric modes, vertical displacements at the free surface and contact surface have opposite signs, whereas for antisymmetric modes they have the same sign. The particle motion is elliptical with a vertical major axis. At the free surface, the particle motion is retrograde for symmetric modes and prograde for antisymmetric ones.

The characteristics of dispersion curves are different for each mode and depend on the values of the model parameters H, β, β', α, and α'. The first two symmetric modes are M_{11} and M_{12} and the first two antisymmetric ones are M_{21} and M_{22} (Fig. 10.14). M_{11} represents the fundamental mode which is a symmetric mode and has all values of frequency ($0 \leq k \leq \infty$). All other modes, symmetric and antisymmetric, have a cut-off frequency (k_{1n}, k_{2n}). For M_{11} in the limit of low frequencies ($k = 0$) the velocity tends toward the Rayleigh wave velocity in the half-space ($c_R = 0.92\beta$), whereas for all the other modes the velocity for cut-off frequencies tends to that of S waves in the half-space (β). In the limit of high frequencies ($k = \infty$) for M_{11}, the velocity tends to that of Rayleigh waves in the layer ($c'_R = 0.92\beta'$) whereas for the other modes the limit is the S wave velocity in the layer (β').

10.5   Stoneley waves

If instead of a layer over a half-space we have two elastic half-spaces of different characteristics in contact, we find waves similar to surface waves related to the contact surface (Fig. 10.15). This case can be considered as the limit of the one-layer problem when its thickness tends to infinity. These waves are generated by constructive interference, their existence was shown by Stoneley (1924), and they are known by his name. The amplitudes of these waves decrease exponentially with the distance from the contact surface in both half-spaces. Thus, the potentials ϕ and ψ and displacements u_2 for waves propagating in the x_1 positive direction are given by

$$\phi = A \exp[-ikrx_3 + ik(x_1 - ct)] \tag{10.94}$$

$$\psi = B \exp[-iksx_3 + ik(x_1 - ct)] \tag{10.95}$$

$$u_2 = C \exp[-iksx_3 + ik(x_1 - ct)] \tag{10.96}$$

$$\phi' = A' \exp[ikr'x_3 + ik(x_1 - ct)] \tag{10.97}$$

$$\psi' = B' \exp[iks'x_3 + ik(x_1 - ct)] \tag{10.98}$$

$$u_2' = C' \exp[iks'x_3 + ik(x_1 - ct)] \tag{10.99}$$

The boundary conditions are the continuity of stress and displacement components across the contact surface ($x_3 = 0$):

$$\tau_{31} = \tau_{31}'; \qquad \tau_{32} = \tau_{32}'; \qquad \tau_{33} = \tau_{33}'$$

$$u_1 = u_1'; \qquad u_2 = u_2'; \qquad u_3 = u_3'$$

We proceed just like in the previous section, writing stresses as functions of displacements and these in terms of potentials. The conditions for τ_{32} and u_2 give the result $C = C' = 0$. Stoneley waves have no transverse component of displacement. From the

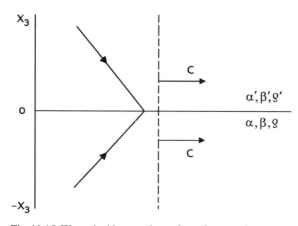

Fig. 10.15. Waves incident on the surface of contact between two elastic half-spaces. Stoneley waves are shown.

other four conditions we obtain

$$A + Bs = A' - B's' \tag{10.100}$$

$$-rA + B = r'A' + B' \tag{10.101}$$

$$\beta^2 \rho[2rA - (1 - s^2)B] = \beta'^2 \rho[2r'A - (1 - s'^2)B'] \tag{10.102}$$

$$\rho[\alpha^2(r^2 + 1) - 2\beta^2]A - 2\beta^2 \rho sB = \rho'[\alpha'^2(r'^2 + 1) - 2\beta'^2]A' - 2\beta'^2 s'\rho'B' \tag{10.103}$$

The condition for the existence of a solution is that the determinant of the system is null. For the particular case in which $\sigma = \frac{1}{4}$ ($\lambda = \mu$) in both media, the determinant is given by

$$\begin{vmatrix} 1 & s & -1 & s' \\ -r & 1 & -r' & -1 \\ 2r\mu & -(1 - s^2)\mu & -2r'\mu' & (1 - s'^2)\mu' \\ \mu(3r^2 - 1) & -2\mu s & -\mu'(3r'^2 - 1) & 2\mu's' \end{vmatrix} = 0$$

By expanding the determinant we obtain a fourth-order equation for c. The four roots are real only for certain values of the ratios ρ/ρ' and β/β'. If β' is less than β, the velocity of Stoneley waves c_S is in the range $\beta' < c_S < c_R$, where c_R is the Rayleigh wave velocity for the half-space of velocity β. Stoneley waves have characteristics similar to Rayleigh waves with generally prograde elliptical particle motion and a vertical major axis. The frequency does not appear in equations (10.100)–(10.103); therefore, waves are not dispersed and their velocity is constant, just like for Rayleigh waves in a half-space.

10.6   Surface waves in a spherical medium

In a flat medium, surface waves propagate along the surface and have cylindrical symmetry. Using the solution of the wave equation in cylindrical coordinates (3.113), for symmetry with respect to ϕ, we can write the displacements as

$$u(R, t) = \left(\frac{2}{\pi k R} \right)^{1/2} A \exp\left[i\left(kR - \omega t - \frac{\pi}{4} \right) \right] \tag{10.104}$$

The amplitude depends on the square root of the distance R along the surface. Thus the energy is constant along a circle of radius $2\pi R$.

In a spherical medium of radius a, the displacements of surface waves have the form of the solutions of the wave equation in spherical coordinates (3.137). For symmetry with respect to ϕ, they can be written as

$$u(\theta, t) = A_n \left(\frac{2}{\pi k} \right)^{1/2} \frac{1}{a} P_n(\cos\theta) \exp\left[i\left(\omega t - (n + 1)\frac{\pi}{2} \right) \right] \tag{10.105}$$

For wave lengths that are small relative to the radius, they correspond to high values of n and we can use the asymptotic expansion of $P_n(\cos\theta)$ which, for values of θ away from the poles ($\theta = 0$ and π), is given by

$$P_n(\cos\theta) \simeq \left(\frac{2}{2n\sin\theta}\right)^{1/2} \cos\left[\left(n+\frac{1}{2}\right)\theta - \frac{\pi}{4}\right] \tag{10.106}$$

Then, the displacements of surface waves may be written in a simplified way in terms of the angular distance Δ (Ben Menahem and Singh, 1981):

$$u_s \simeq A\left(\frac{\pi}{\sin\Delta}\right)^{1/2} \exp\left[i\left(ka\Delta - \omega t - \frac{3\pi}{4}\right)\right] \tag{10.107}$$

Similarly to the flat case, the amplitudes depend on the inverse of the square root of the distance ($\sin\Delta$) and the energy is constant on a circle of radius $2\pi\sin\Delta$. Surface waves in a sphere can be approximated by those of a flat medium if their wave length is small compared with the radius. The wave number corresponding to a spherical medium k_s can be obtained from that of the flat medium k_f, by introducing the correction (Ben Menahen and Singh, 1981)

$$k_s = \left(k_f^2 + \frac{9}{4a^2}\right) \tag{10.108}$$

From this expression we can find the relations between the phase velocities ($c = \omega/k$)

$$c_s = \omega c_f\left(\omega^2 + \frac{9c_f^2}{4a^2}\right)^{-1/2} \tag{10.109}$$

and group velocities ($U = d\omega/dk$)

$$U_s = \left(1 + \frac{9}{4k_f^2a^2}\right)U_f \tag{10.110}$$

An important phenomena in the propagation of surface waves in a sphere is the polar phase shift of $\lambda/4$ when waves cross the poles. This is due to the fact that harmonic waves are not exactly sinusoidal or cosinusoidal in the vicinity of the poles. This does not affect the determinations of group velocities but affects phase velocities. Thus, for determinations of phase velocities along paths that include the epicenter or its antipole, a phase shift of $\pi/2$ must be added for each pole or antipole crossing. This effect must be taken into account when determining phase velocities using surface waves that have circled the Earth (Brune *et al.*, 1961).

11 WAVE PROPAGATION IN LAYERED MEDIA

11.1 The equation for the displacement–stress vector

Many problems in seismology can be solved by representing the Earth as a stratified or layered medium, that is, one formed by layers of certain thicknesses and mechanical properties. For certain problems we can use a flat approximation of parallel horizontal layers and they are reduced to two dimensions. Layers of constant properties may be considered as an approximation for media whose elastic coefficients vary in a continuous form with depth. In layered or stratified media, problems are presented in discrete form and may be treated using matrix formulations. Solutions of problems of wave propagation in layered media using matrix formulation were introduced by Thomson (1950) and Haskell (1953), receiving the name of the Thomson–Haskell method. A similar formulation was proposed by Knopoff (1964). Gilbert and Backus (1966) introduced the concept of the propagator matrix that allows a more generalized formulation of the problem (Kennett, 1983).

In a half-space, we use Cartesian coordinates x horizontal and z vertical (positive downward). For monochromatic plane waves of frequency ω that propagate in the plane (x, z), according to (6.38) and (6.39), one component of the displacement of S waves can be expressed in the following forms:

$$u = [A \, e^{iksz} + B e^{-iksz}] \, e^{ik(x - ct)} \tag{11.1}$$

$$u = [A \, e^{i\omega qz} + B e^{-i\omega qz}] \, e^{i\omega(px - t)} \tag{11.2}$$

where c, k, p, s, and q were defined in section 6.3. In terms of the velocities, c and β, and angles of incidence i and f, s and q are given by (equations (6.40) and (6.41) upon replacing α by β) (Fig. 11.1):

$$s = \left(\frac{c^2}{\beta^2} - 1\right)^{1/2} = \cot i = \tan f \tag{11.3}$$

$$q = \left(\frac{1}{\beta^2} - \frac{1}{c^2}\right)^{1/2} = \frac{\cos i}{\beta} = \frac{\sin f}{\beta} \tag{11.4}$$

The relation between s and q is $ks = \omega q$.

In the general problem of an elastic medium, we have both P and S waves. Using expressions in the form of equations (11.1) and (11.2), in the exponential function we have s and q_β for S waves and r and q_α for P waves (r and q_α are defined as in (11.3) and (11.4) by replacing β by α).

Components of the stress tensor across surfaces normal to z ($z = x_3$ and $x = x_1$) are τ_{3j}. Since rays propagate in the (x, z) plane, the problem can be separated into two

205

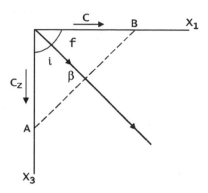

Fig. 11.1. The ray trajectory, and velocity components c in the x direction and c_z in the z direction. AB is the wave front.

parts: first, SH waves with displacements u_2 and stress component τ_{32}; and second, P and SV waves with displacements u_1 and u_3 and stress components τ_{31} and τ_{33}. According to (11.1) and (11.2), the displacement and stress can be expressed in the forms:

$$u_i(z, x) = [u_1(z), u_2(z), u_3(z)] \, e^{ik(x - ct)} \tag{11.5}$$

$$\tau_{3i}(z, x) = [\tau_{31}(z), \tau_{32}(z), \tau_{33}(z)] \, e^{ik(x - ct)} \tag{11.6}$$

Since the term for propagation in the horizontal direction is the same in both equations, we can consider the displacement and stress as functions of z only. The displacement and stress can be put together, defining the displacement–stress vector. For SH motion, it is given by

$$[u_2(z), \tau_{32}(z)] \tag{11.7}$$

and for P–SV motion it is given by

$$[u_1(z), u_3(z), \tau_{31}(z), \tau_{33}(z)] \tag{11.8}$$

For SH motion in an isotropic homogeneous elastic medium for waves propagating in the (x_1, x_3) plane, the relation between the stress and the displacement is $\tau_{32} = \mu u_{2,3}$ (2.17), which can now be written as

$$\frac{du_2}{dz} = \frac{1}{\mu} \tau_{32} \tag{11.9}$$

As we have already seen (in section 3.1), the transverse displacement $u_2(x, z, t)$ is a solution of the wave equation

$$\frac{\partial^2 u_2}{\partial x^2} + \frac{\partial^2 u_2}{\partial z^2} = \frac{\rho}{\mu} \frac{\partial^2 u_2}{\partial t^2} \tag{11.10}$$

By substituting (11.1) into (11.10) and considering that $\tau_{32,3} = \mu u_{2,33}$, we obtain the relation

$$\frac{d\tau_{32}}{dz} = (\mu k^2 - \rho \omega^2) u_2 = k^2 \rho (\beta^2 - c^2) u_2 \tag{11.11}$$

Equations (11.9) and (11.11) can be expressed as a single matrix equation for the displacement–stress vector of SH motion (11.7):

$$\frac{d}{dz}\begin{pmatrix} u_2 \\ \tau_{32} \end{pmatrix} = \begin{pmatrix} 0 & 1/\mu \\ k^2\rho(\beta^2 - c^2) & 0 \end{pmatrix}\begin{pmatrix} u_2 \\ \tau_{32} \end{pmatrix} \tag{11.12}$$

For P–SV motion, the derivation of an equation similar to (11.12) is more complicated. The displacement components u_1 and u_3 are not solutions of the wave equation. These must be expressed in terms of scalar potentials ϕ and ψ according to (3.91) and (3.92). The components of the stress τ_{31} and τ_{33} also must be written in terms of ϕ and ψ, just like in section 5.3. Finally, the matrix equation for the displacement–stress vector is (Kennett, 1983)

$$\frac{d}{dz}\begin{pmatrix} u_1 \\ u_3 \\ \tau_{31} \\ \tau_{33} \end{pmatrix} = \begin{pmatrix} 0 & k^2 & (\rho\beta^2)^{-1} & 0 \\ 1 - 2\beta^2/\alpha^2 & 0 & 0 & (\rho\alpha^2)^{-1} \\ \omega^2\rho h & 0 & 0 & k^2(1 - 2\beta^2/\alpha^2) \\ 0 & -\rho\omega^2 & -1 & 0 \end{pmatrix}\begin{pmatrix} u_1 \\ u_3 \\ \tau_{31} \\ \tau_{33} \end{pmatrix} \tag{11.13}$$

$$h = 4\beta^2 \frac{1}{c^2}\left(1 - \frac{\beta^2}{\alpha^2}\right) - 1$$

The matrix equations (11.12) and (11.13) for SH and P–SV motion show that the derivatives with respect to z of the components of the displacement–stress vector are linear functions of the same components. The matrix of proportionality coefficients contains the elastic parameters of the medium (α, β, and ρ), the frequency ω, and the horizontal velocity c. Equations (11.12) and (11.13) are fundamental for the development of matrix methods for the resolution of wave propagation in layered media.

11.2 The propagator matrix

Equations (11.12) and (11.13) can be written in a general matrix form as

$$\frac{d\boldsymbol{b}(z)}{dz} = \mathbf{A}(c, z)\boldsymbol{b}(z) \tag{11.14}$$

where $\boldsymbol{b}$ is the displacement–stress vector defined in (11.7) and (11.8) and $\mathbf{A}$ is a square matrix whose elements are the elastic parameters and density of the medium as functions of the depth ($\alpha(z)$, $\beta(z)$, and $\rho(z)$), frequency ω, and horizontal velocity c. In the case of layers of constant parameters, in each layer, the only variables are c and ω.

For equations of the type of (11.14), a matrix $\mathbf{B}$ that is a solution of the same equation is called a fundamental matrix of the equation (Gilbert and Backus, 1966). Thus we can write

$$\frac{d\mathbf{B}}{dz} = \mathbf{A}\mathbf{B} \tag{11.15}$$

For a fixed value of $z = z_0$, a reference depth, the vector $\boldsymbol{b}(z_0)$ is a solution of (11.14). For the same depth, there exists also a solution of (11.15) that is the constant matrix

$\mathbf{B}(z_0)$. We can normalize the problem and make $\mathbf{B}(z_0) = \mathbf{I}$, the unit matrix. We define now a new matrix $\mathbf{P}(z, z_0)$ that is a solution of equation (11.15) and thus a fundamental matrix:

$$\mathbf{P}(z, z_0) = \mathbf{B}^{-1}(z_0)\mathbf{B}(z) \tag{11.16}$$

Then, it follows that, for $z = z_0$, $\mathbf{P}(z_0, z_0) = \mathbf{I}$. If z is an arbitrary level, for an intermediate level z_1 between z_0 and z, according to (11.16), we can write

$$\mathbf{P}(z, z_0) = \mathbf{P}(z_1, z_0)\mathbf{P}(z, z_1) \tag{11.17}$$

If z_n is the nth level starting from z_0, we can write the matrix $\mathbf{P}(z_n, z_0)$, using (11.17), which relates the nth level to the zeroth level, as a product of the matrices that relate all the intermediate levels step by step:

$$\mathbf{P}(z_n, z_0) = \prod_{i=1}^{n} \mathbf{P}(z_i, z_{i-1}) \tag{11.18}$$

Considering the properties of the matrix $\mathbf{P}(z, z_0)$, we have that, for the zeroth level, $\boldsymbol{b}(z_0) = \mathbf{P}(z_0, z_0)\boldsymbol{b}(z_0)$, since $\mathbf{P}(z_0, z_0)$ is a unit matrix. For an arbitrary value of z, we can, then, write

$$\boldsymbol{b}(z) = \mathbf{P}(z, z_0)\boldsymbol{b}(z_0) \tag{11.19}$$

This follows from the fact that $\boldsymbol{b}(z)$ and $\mathbf{P}(z, z_0)$ are both solutions of the same equation according to (11.14) and (11.15). Equation (11.19) shows that we can determine the vector $\boldsymbol{b}(z)$ at an arbitrary level z from its value $\boldsymbol{b}(z_0)$ at a reference level z_0, by means of the matrix $\mathbf{P}(z, z_0)$. For this reason $\mathbf{P}(z, z_0)$ is called the propagator matrix. According to (11.18) and (11.19), if we have n levels, we can relate the displacement–stress vector for level n, $\boldsymbol{b}(z_n)$, to that of the zeroth level $\boldsymbol{b}(z_0)$:

$$\boldsymbol{b}(z_n) = \prod_{i=1}^{n} \mathbf{P}(z_i, z_{i-1})\boldsymbol{b}(z_0) \tag{11.20}$$

By changing the order of levels we can also write

$$\boldsymbol{b}(z_0) = \prod_{i=1}^{n} \mathbf{P}(z_i, z_{i-1})\boldsymbol{b}(z_n) \tag{11.21}$$

Equations (11.19)–(11.21) show the meaning of the propagator matrix $\mathbf{P}(z, z_0)$ which allows the determination of the displacement–stress vector $\boldsymbol{b}(z)$ at an arbitrary level from its value at a reference level, $\boldsymbol{b}(z_0)$. The form of equations (11.20) and (11.21) is specially useful when we have layered media with constant elastic parameters.

11.3   A layered medium with constant parameters

Wave propagation in a layered medium formed by n layers with constant parameters can be conveniently studied using the formulation in terms of the displacement–stress vector and the propagator matrix. According to equations (11.12) and (11.13), for each layer matrix $\mathbf{A}$ is constant. Let us start by considering the problem with only one

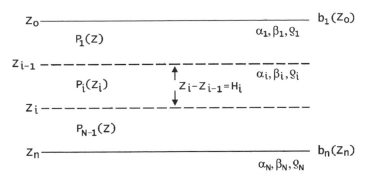

Fig. 11.2. A layered medium with constant parameters for each layer.

dimension. Equation (11.14) is reduced to

$$\frac{\mathrm{d}y(z)}{\mathrm{d}z} = ay(z) \tag{11.22}$$

where a is a constant. A solution of this equation is

$$y(z) = \mathrm{e}^{az} \tag{11.23}$$

If we know the value of the function for a reference level z_0, $y(z_0) = \mathrm{e}^{az_0}$, we can write

$$y(z) = \mathrm{e}^{a(z-z_0)}y(z_0) \tag{11.24}$$

This equation has the same form as (11.19), so we can define the propagator matrix, in this case with only one element, in the form

$$P(z, z_0) = \mathrm{e}^{a(z-z_0)} \tag{11.25}$$

For the general case when $b(z)$ is a vector of m elements and $\mathbf{A}$ is a square matrix of $m \times m$ constant elements, the solution of equation (11.14) is

$$b(z) = \mathrm{e}^{\mathbf{A}(z-z_0)}b(z_0) \tag{11.26}$$

The propagator matrix has the form

$$\mathbf{P}(z, z_0) = \mathrm{e}^{\mathbf{A}(z-z_0)} \tag{11.27}$$

If we have n levels corresponding to n layers, each with constant parameters, $\mathbf{A}_i = \mathbf{A}(\alpha_i, \beta_i, \rho_i, c, \omega)$ (Fig. 11.2), then, according to (11.18), the propagator matrix is given by

$$\mathbf{P}(z_n, z_0) = \prod_{i=1}^{n} \mathrm{e}^{\mathbf{A}_i(z_i - z_{i-1})} \tag{11.28}$$

The relation between the displacement–stress vector b for the nth level and that for the reference (zeroth) level (11.20) for a layered medium of constant parameters is

$$b(z_n) = \prod_{i=1}^{n} \mathrm{e}^{\mathbf{A}_i(z_i - z_{i-1})}b(z_0) \tag{11.29}$$

Thus, the propagator matrix is given by the product of the exponential functions of the matrix $\mathbf{A}$ for each layer.

11.3.1   Eigenvalues and eigenvectors

The solution of equation (11.29) can be simplified by using the eigenvalues and eigenvectors analysis of matrix $\mathbf{A}$ (Lanczos, 1957). If $\mathbf{A}$ has eigenvalues λ_k and eigenvectors u^k, these satisfy the equation

$$(A_{ij} - \lambda_k \delta_{ij}) u_i^k = 0 \tag{11.30}$$

The eigenvalues are solutions of the equation

$$|A_{ij} - \lambda_k \delta_{ij}| = 0 \tag{11.31}$$

The transpose matrix $\mathbf{A}^T$ has the same eigenvalues and eigenvectors v^k. If we form the matrix $\mathbf{U}$ with eigenvectors u^k, the matrix $\mathbf{V}$ with v^k and a diagonal matrix Λ with eigenvalues λ_k, according to the theorem of Eckart and Young, the matrix $\mathbf{A}$ and its inverse $\mathbf{A}^{-1}$ are given by

$$\mathbf{A} = \mathbf{U}\Lambda\mathbf{V}^T \tag{11.32}$$

$$\mathbf{A}^{-1} = \mathbf{U}\Lambda^{-1}\mathbf{V}^T \tag{11.33}$$

The matrices $\mathbf{V}$ and $\mathbf{U}$ satisfy the property that the product of one by the transpose of the other is the unit matrix, $\mathbf{U}\mathbf{V}^T = \mathbf{U}^T\mathbf{V} = \mathbf{V}\mathbf{U}^T = \mathbf{V}^T\mathbf{U} = \mathbf{I}$. Using this property and relation (11.32) and multiplying by $\mathbf{V}^T$ from the left-hand side in equation (11.14), we obtain

$$\frac{\mathrm{d}}{\mathrm{d}z}(\mathbf{V}^T b) = \Lambda(\mathbf{V}^T b) \tag{11.34}$$

Then, according to (11.26),

$$\mathbf{V}^T b(z) = \mathrm{e}^{\Lambda(z-z_0)} \mathbf{V}^T b(z_0) \tag{11.35}$$

If we multiply by $\mathbf{U}$ from the left-hand side, since $\mathbf{U}\mathbf{V}^T = \mathbf{I}$, we obtain

$$b(z) = \mathbf{U}\,\mathrm{e}^{\Lambda(z-z_0)}\mathbf{V}^T b(z_0) \tag{11.36}$$

The propagator matrix (10.27) is now given by

$$\mathbf{P}(z, z_0) = \mathbf{U}\,\mathrm{e}^{\Lambda(z-z_0)}\mathbf{V}^T \tag{11.37}$$

If we now multiply (11.36) by $\mathbf{V}^T$ from the left-hand side and substitute $w = \mathbf{V}^T b$, we obtain an equation similar to (10.26) for the new vector w:

$$w(z) = \mathrm{e}^{\Lambda(z-z_0)} w(z_0) \tag{11.38}$$

For the vector w the propagator matrix is

$$\mathbf{Q}(z, z_0) - \mathrm{e}^{\Lambda(z-z_0)} \tag{11.39}$$

The vector w is called the wave vector and its propagator matrix $\mathbf{Q}(z, z_0)$ is called the wave propagator.

Equations (11.36) and (11.38) relate values of the vectors b and w from one level to another. They are fundamental in the solution of problems of wave propagation in stratified media. By using the generalized Lanczos inverse matrix we avoid the calculation of the inverse of the matrix $\mathbf{A}$. This procedure is very practical since there are standard fast methods for calculating the eigenvalues and eigenvectors of a matrix.

11.3.2 The propagator matrix for SH motion

As an example, we present the determination of the propagator matrix of SH motion. According to equation (11.12), the matrix $\mathbf{A}$ for SH motion is given by

$$\mathbf{A} = \begin{pmatrix} 0 & (\rho\beta^2)^{-1} \\ \rho k^2(\beta^2 - c^2) & 0 \end{pmatrix} \tag{11.40}$$

Its transpose is

$$\mathbf{A}^{\mathrm{T}} = \begin{pmatrix} 0 & \rho k^2(\beta^2 - c^2) \\ (\rho\beta^2)^{-1} & 0 \end{pmatrix} \tag{11.41}$$

The eigenvalues of $\mathbf{A}$ are found by solving the corresponding equation (11.31); we obtain $\lambda_1 = iks$ and $\lambda_2 = -iks$. For $\mathbf{A}^{\mathrm{T}}$ the eigenvalues are the same. The matrix $\mathbf{\Lambda}$ is given by

$$\mathbf{\Lambda} = \begin{pmatrix} iks & 0 \\ 0 & -iks \end{pmatrix} \tag{11.42}$$

By solving equation (11.30) for $\mathbf{A}$ and $\mathbf{A}^{\mathrm{T}}$, we obtain the following eigenvectors, corresponding to each eigenvalue:

$$\boldsymbol{u}^1 = [1, iks\rho\beta^2], \qquad \boldsymbol{u}^2 = [1, -iks\rho\beta^2]$$
$$\boldsymbol{v}^1 = [1, (iks\rho\beta^2)^{-1}], \qquad \boldsymbol{v}^2 = [1, -(iks\rho\beta^2)^{-1}]$$

The matrices $\mathbf{U}$ and $\mathbf{V}$ are

$$\mathbf{U} = \begin{pmatrix} 1 & 1 \\ iks\rho\beta^2 & -iks\rho\beta^2 \end{pmatrix} \tag{11.43}$$

$$\mathbf{V} = \begin{pmatrix} 1 & 1 \\ (iks\rho\beta^2)^{-1} & -(iks\rho\beta^2)^{-1} \end{pmatrix} \tag{11.44}$$

On putting $ks\rho\beta^2 = a$ and $ks(z - z_0) = d$, according to (11.36), the propagator matrix is given by

$$\mathbf{P}(z, z_0) = \begin{pmatrix} 1 & 1 \\ ia & -ia \end{pmatrix} \begin{pmatrix} e^{id} & 0 \\ 0 & e^{-id} \end{pmatrix} \begin{pmatrix} 1 & 1/(ia) \\ 1 & -1/(ia) \end{pmatrix} \tag{11.45}$$

Taking the product and substituting for exponential functions sines and cosines, for real values of d, we obtain

$$\mathbf{P}(z, z_0) = 2 \begin{pmatrix} \cos d & (i/a)\sin d \\ ia\sin d & \cos d \end{pmatrix} \tag{11.46}$$

In (11.46) we have assumed that d is real, or in consequence that s is real. This means that $c > \beta$ and, therefore, that there are SH waves that propagate in the positive and negative directions of z, that is, waves going downward and upward. If s is imaginary, $c < \beta$, and we will have hyperbolic sines and cosines in (11.46). In this case, no SH waves are propagated in the z direction; waves are propagated in the x direction only. Furthermore, imposing the condition that their amplitudes decrease with depth, the solutions represent inhomogeneous waves.

11.4   SH motion in an elastic layer over a half-space

Let us now apply the propagator matrix to SH motion in a single elastic layer over a half-space (Cisternas, 1982). This problem has already been studied in section 10.3 for the propagation of Love waves. We use the notation $z = x_3$, $u(z) = u_2(z)$, and $\tau(z) = \tau_{32}(z) = \mu \, du/dz$. Subindexes 1 and 2 refer now to values of variables in the layer and the half-space, respectively (Fig. 11.3). The z coordinate is positive downward with $z = 0$ at the free surface and $z = H$ at the surface of contact between the layer and the half-space. For an arbitrary value of z, the displacement–stress vector $\boldsymbol{b}(z)$ (11.1) is given by

$$\begin{pmatrix} u(z) \\ \tau(z) \end{pmatrix} = \begin{pmatrix} A\,\mathrm{e}^{iksz} + B\,\mathrm{e}^{-iksz} \\ i\mu ksA\,\mathrm{e}^{iksz} - i\mu ksB\,\mathrm{e}^{-iksz} \end{pmatrix} \tag{11.47}$$

This vector can be expressed in terms of the product of a matrix and a vector formed by the displacement amplitudes A and B:

$$\begin{pmatrix} u(z) \\ \tau(z) \end{pmatrix} = \begin{pmatrix} \mathrm{e}^{iksz} & \mathrm{e}^{-iksz} \\ i\mu ks\,\mathrm{e}^{iksz} & -i\mu ks\,\mathrm{e}^{-iksz} \end{pmatrix} \begin{pmatrix} A \\ B \end{pmatrix} \tag{11.48}$$

The boundary conditions at the free surface (null stress) and at the contact surface (continuity of displacement and stress) are given by

$$\tau_1(0) = 0; \qquad u_1(H) = u_2(H); \qquad \tau_1(H) = \tau_2(H)$$

By substitution into (11.47), we can write the vectors in the layer, at the free surface $\boldsymbol{b}_1(0)$, at the contact surface $\boldsymbol{b}_1(H)$, and in the half-space at the same contact surface $\boldsymbol{b}_2(H)$ in the forms

$$\boldsymbol{b}_1(0) = \begin{pmatrix} A_1 + B_2 \\ 0 \end{pmatrix} \tag{11.49}$$

$$\boldsymbol{b}_1(H) = \begin{pmatrix} \mathrm{e}^{id_1} & \mathrm{e}^{-id_1} \\ ia_1\,\mathrm{e}^{id_1} & -ia_1\,\mathrm{e}^{-id_1} \end{pmatrix} \begin{pmatrix} A_1 \\ B_1 \end{pmatrix} \tag{11.50}$$

$$\boldsymbol{b}_2(H) = \begin{pmatrix} \mathrm{e}^{id_2} & \mathrm{e}^{-id_2} \\ ia_2\,\mathrm{e}^{id_2} & -ia_2\,\mathrm{e}^{-id_2} \end{pmatrix} \begin{pmatrix} A_2 \\ B_2 \end{pmatrix} \tag{11.51}$$

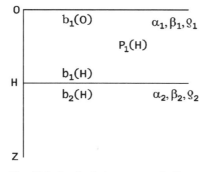

Fig. 11.3. An elastic layer over a half-space. Values of $\boldsymbol{b}(z)$ and $\boldsymbol{P}(z)$ are shown.

where

$$a_1 = \mu_1 s_1 k, \qquad d_1 = s_1 kH$$
$$a_2 = \mu_2 s_2 k, \qquad d_2 = s_2 kH$$

According to the definition of the propagator matrix (11.21), since $z_0 = 0$, $z_1 = H$ and, from the boundary condition on the contact surface between the layer and the half-space, $\boldsymbol{b}_1(H) = \boldsymbol{b}_2(H)$, we have

$$\boldsymbol{b}_1(0) = \mathbf{P}_1(H)\boldsymbol{b}_2(H) \tag{11.52}$$

From (11.46), the propagator matrix $\mathbf{P}_1(H)$ can be written in the form

$$\mathbf{P}_1(H) = 2 \begin{pmatrix} \cos d_1 & (1/a_1)\sin d_1 \\ a_1 \sin d_1 & \cos d_1 \end{pmatrix} \tag{11.53}$$

where we have considered that s_1 is real ($c > \beta_1$) and that SH waves propagate upward and downward inside the layer. By substitution of expressions (11.49), (11.53) and (11.51) into equation (11.52), we obtain

$$\begin{pmatrix} A_1 + B_1 \\ 0 \end{pmatrix} = 2 \begin{pmatrix} \cos d_1 & (1/a_1)\sin d_1 \\ a_1 \sin d_1 & \cos d_1 \end{pmatrix} \begin{pmatrix} e^{id_2} & e^{-id_2} \\ ia_2\, e^{id_2} & -ia_2\, e^{-id_2} \end{pmatrix} \begin{pmatrix} A_2 \\ B_2 \end{pmatrix} \tag{11.54}$$

This equation relates the amplitudes in the layer, A_1 and B_1, to those in the half-space, A_2 and B_2.

From equation (11.54) we can derive the dispersion equation of Love waves. For surface waves we impose the condition that there are no SH waves propagating in the z direction in the half-space; that is, s_2 is imaginary ($c < \beta_2$). Since amplitudes must decrease with z positive, it follows also that $B_2 = 0$. With these conditions, the equation resulting from the zeroth component of the vector $\boldsymbol{b}_1(0)$ in (11.54) is

$$A_2(a_1 \sin d_1\, e^{id_2} + ia_2 \cos d_1\, e^{id_2}) = 0 \tag{11.55}$$

The existence of Love waves requires that A_2 is not zero, since Love waves propagate in the x direction and attenuate in the z direction in the half-space. The expression inside the brackets must be zero and we obtain

$$\frac{\sin d_1}{\cos d_1} = -\frac{ia_2}{a_1} \tag{11.56}$$

If we substitute the values for a_1, d_1, a_2, and d_2 and put $s_2 = i\bar{s}_2$, we obtain

$$\tan(kHs_1) = \frac{\mu_2 \bar{s}_2}{\mu_1 s_1}$$

This is the same dispersion equation (10.60) we derived for Love waves in Chapter 10. As we saw in section 10.3, there is no solution for s_1 imaginary. Since s_1 is real, SH waves propagate inside the layer in both z directions (positive and negative). These waves are totally reflected from the contact surface so that there are only inhomogeneous waves (s_2 imaginary) in the half-space.

11.5   Waves in layered media

The general problem of wave propagation in layered media corresponds to $N - 1$ elastic layers over an elastic half-space. Each layer is characterized by its thickness H_i and elastic parameters β_i, α_i, and ρ_i. Some layers may be liquid ($\beta_j = 0$). Wave propagation in the z direction is specified by the values of s_i for SH imotion and by r_i and s_i for P–SV motion (Fig. 11.4). The boundary conditions are that the stress components across the free surface ($z = 0$) are null ($\tau_{3i} = 0$). At each surface of contact between any two layers and that between the last layer and the half-space, there is continuity of the displacement and stress ($\boldsymbol{b}_i(z_i) = \boldsymbol{b}_{i+1}(z_i), i = 1, \ldots, N - 1$). In terms of the propagator matrix (11.21), the relation between the displacement–stress vector at the free surface ($z = 0$) and that at the half-space beneath the last layer ($z = z_N$) is given by

$$\boldsymbol{b}_1(0) = \prod_{i=1}^{N-1} \mathbf{P}_i(H_i)\boldsymbol{b}_N(z_N) \tag{11.57}$$

where the subindex i refers to the value at each layer from 1 to $N - 1$ and N refers to the half-space. The matrices $\mathbf{P}_i$ and vectors $\boldsymbol{b}_i$ for each layer depend on the parameters α, β, and ρ, frequency ω, and velocity c. According to (11.37), the propagator matrix for each layer is given by

$$\mathbf{P}_i(H_i) = \mathbf{U}_i\, e^{\Lambda_i H_i}\mathbf{V}_i^{\mathrm{T}} \tag{11.58}$$

where Λ_i is the diagonal matrix formed by the eigenvalues of the matrix $\mathbf{A}_i$, and $\mathbf{U}_i$ and $\mathbf{V}_i$ are the matrices formed by the eigenvectors of the same matrix and its transpose.

11.5.1   SH motion

For SH motion, the matrix $\mathbf{A}$ is given by (11.40) and its transpose by (11.41). The diagonal matrix of the eigenvalues (11.42) in each layer is

$$\Lambda_i = \mathrm{diagonal}(iks_i, -iks_i) \tag{11.59}$$

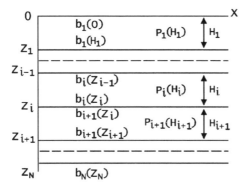

Fig. 11.4. A stratified medium with N layers of constant parameters for each layer. Values of $\boldsymbol{b}(z)$ and $\boldsymbol{P}(z)$ are shown.

According to (11.46), the propagator matrix in each layer is given by

$$\mathbf{P}_i(H_i) = 2 \begin{pmatrix} \cos(ks_iH_i) & (i/a_i)\sin(ks_iH_i) \\ ia_i\sin(ks_iH_i) & \cos(ks_iH_i) \end{pmatrix} \tag{11.60}$$

where $a_i = ks_i\rho_i\beta_i^2$ and $k = \omega/c$. In this expression we assume that s_i is real; that is, $\beta_i < c$, and that in each layer SH waves are propagated upward and downward. If, in a particular layer, k, there is total reflection, no waves are propagated to the following layer $(k+1)$; thus, s_{k+1} is imaginary and $\beta_{k+1} > c$.

If all layers have the same thickness $H_i = H$, the problem is somewhat simplified. Just like in the case of a single layer, the dispersion equation of Love waves for N layers is found by imposing the condition of transmission of SH waves in the layers and a decrease in amplitude in the half-space (s_N is imaginary and $B_N = 0$). The dispersion equation $(f(\omega, c) = 0)$ corresponds to the zeroth value of the amplitude in the half-space.

11.5.2 P–SV motion

The problem of P–SV motion is more complicated since the matrix $\mathbf{A}$ has 4×4 elements (11.13). For each layer the matrix $\mathbf{A}$ is given by

$$\mathbf{A}_i = \begin{pmatrix} 0 & k^2 & (\rho_i\beta_i^2)^{-1} & 0 \\ 1 - 2\beta_i^2/\alpha_i^2 & 0 & 0 & (\rho\alpha_i^2)^{-1} \\ \omega\rho_i h_i & 0 & 0 & k^2(1 - 2\beta_i^2/\alpha_i^2) \\ 0 & -\rho_i\omega^2 & -1 & 0 \end{pmatrix}$$

$$h_i = \frac{4\beta_i^2}{c^2}\left(1 - \frac{\beta_i^2}{\alpha_i^2}\right) - 1 \tag{11.61}$$

The eigenvalues are $\pm iks_i$ and $\pm ikr_i$. The diagonal matrix $\mathbf{\Lambda}$ is given by

$$\mathbf{\Lambda} = \text{diagonal}(ikr_iH_i, iks_iH_i, -ikr_iH_i, -iks_iH_i) \tag{11.62}$$

If r_i and s_i are real $(c > \alpha > \beta)$, exponential functions of $\mathbf{\Lambda}$ according to (11.58) represent P and SV waves that propagate in the positive and negative directions of z. If r_i or s_i have imaginary values for a layer, then we have total reflection from the layer above and inhomogeneous P or SV waves that propagate in the x direction, and their amplitudes decrease with depth.

The problem is solved by forming the matrices $\mathbf{A}_i$ and their transposes $\mathbf{A}_i^T$ for each layer and determining the matrices of eigenvalues $\mathbf{\Lambda}_i$ and of eigenvectors $\mathbf{U}_i$ and $\mathbf{V}_i$. From these matrices we find the propagator matrices $\mathbf{P}_i(H_i)$ (10.58). The product of these matrices according to (11.57) relates the displacement–stress vector at the surface to that of the half-space. Just like in the case of Love waves, imposing the conditions that waves propagate only in the x direction in the half-space and their amplitudes decrease with z (r_N and s_N have imaginary values and $B_N = D_N = 0$), we obtain the dispersion equation $(f(c, \omega) = 0)$ for Rayleigh waves.

The matrix formulation that has been presented in the context of the problem of surface waves is also applied to the determination of the problem of reflected and transmitted SH and P–SV waves in a layered medium. This is, in general, a more complex problem that varies with the depth of the focus. If the focus is at the surface, in the half-space there are only waves travelling downward, whereas in the layers there are waves going down and up (transmitted and reflected waves). In general, the matrix formulation is very convenient for the solution of problems of wave propagation in layered media with constant parameters. In our presentation we have given only the fundamental ideas of the method; a more complete discussion can be found in Aki and Richards (1980) and Kennett (1983).

12 WAVE DISPERSION. PHASE
AND GROUP VELOCITIES

12.1   Phase and group velocities

We have seen that surface waves in layered media are dispersed; that is, their velocity is a function of the frequency (or period). Thus, for an impulsive time function at the source, surface waves at some distance are formed by trains of waves, different frequencies arriving at different times. Arrival times, amplitudes, and phases for each frequency depend, then, on the dispersion equation. In section 3.4 we saw that, if the phase velocity is a function of the frequency, then the velocity of energy transport is not the same, but equal to the group velocity, or the velocity of propagation of wave groups. We will consider now wave dispersion and the relation between phase and group velocities.

The displacement of a sinusoidal wave of angular frequency ω and wave number k that propagates in the x direction is given by

$$u(x, t) = A \sin[(kx - \omega t) + \phi]$$ (12.1)

where the phase velocity, or the velocity of propagation of each value of the phase, is

$$c = \omega/k$$ (12.2)

For monochromatic waves in a homogeneous medium, c is constant and for each value of ω there is a single value of k. In this case, the velocity of energy transport or the group velocity is equal to the phase velocity (section 3.4). If the phase velocity is a function of the frequency $c(\omega)$, then we can also write $k(\omega)$ and $\omega(k)$, and we can use as the independent variable either k or ω. In the first case we are looking at the wave phenomenon from the point of view of its dependence on space and in the second, in terms of its dependence on time. As we saw in section 3.4, the group velocity is given by

$$U = d\omega/dk$$ (12.3)

If we take derivatives with respect to k in (12.2) we obtain

$$U = c + k\frac{dc}{dk}$$ (12.4)

For the dependence on ω, taking the derivative with respect to ω, we obtain

$$U = \frac{c}{1 - \frac{\omega}{c}\frac{dc}{d\omega}}$$ (12.5)

The simplest case of a wave with more than one frequency is that formed by the sum of two waves with the same amplitude and two similar frequencies, namely, $\omega_1 = \omega_0 - \Delta\omega$

217

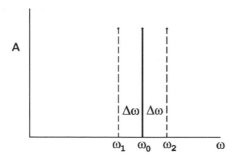

Fig. 12.1. Two frequencies at an interval $\Delta\omega$ from a central ω_0.

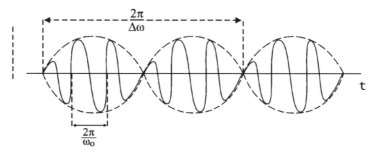

Fig. 12.2. The sum of two waves of frequencies at an interval $\Delta\omega$ from a central ω_0.

and $\omega_2 = \omega_0 + \Delta\omega$ (Fig. 12.1). The displacement for null initial phase, $\phi = 0$, is

$$u = A\sin(k_1 x - \omega_1 t) + A\sin(k_2 x - \omega_2 t)$$

This equation can be written as

$$u = 2A\sin\left(\frac{k_1 + k_2}{2}x - \frac{\omega_1 + \omega_2}{2}t\right)\cos\left(\frac{k_2 - k_1}{2}x - \frac{\omega_2 - \omega_1}{2}t\right) \tag{12.6}$$

and, on substituting for k_1, k_2, ω_1, and ω_2 in terms of ω_0 and $\Delta\omega$,

$$u = 2A\sin(k_0 x - \omega_0 t)\cos(\Delta k\, x - \Delta\omega\, t) \tag{12.7}$$

The displacement has the form of a sine wave of frequency ω_0 modulated by a cosine wave of frequency $\Delta\omega$. It is, then, formed by groups or packets of waves of frequency ω_0 with duration $\pi/\Delta\omega$ (Fig. 12.2). According to (12.2), the phase velocity of sine waves is $c = \omega_0/k_0$, whereas the maxima of the cosine function propagate with the velocity $U = \Delta\omega/\Delta k$. Since groups of waves correspond to maxima of the cosine function, this is the group velocity. Energy is associated with amplitude maxima and is propagated by the group velocity U. Even for a wave formed by two frequencies we can distinguish between the phase and group velocities.

A wave with a discrete content of N frequencies of values ω_k can be represented by the sum of the displacements of each of these frequencies:

$$u(x, t) = \sum_{k=1}^{N} A_k \cos\left[\omega_k\left(\frac{x}{c_k} - t\right) + \phi_k\right] \tag{12.8}$$

where c_k is the phase velocity corresponding to each frequency ω_k, A_k is the amplitude, and ϕ_k is the initial phase. If the content of frequencies is continuous between zero and infinity, then the displacement is expressed by the integral:

$$u(x, t) = \int_0^\infty A(\omega) \cos \left[\omega \left(\frac{x}{c(\omega)} - t \right) + \phi(\omega) \right] \tag{12.9}$$

$A(\omega)$ is the amplitude corresponding to each frequency and $\phi(\omega)$ is the initial phase. The wave displacement $u(x, t)$ is the sum of the contributions from all frequencies. The displacement observed at a particular distance x is a function of time $u(t)$ (a seismogram) and by means of Fourier transformation can be expressed as a function of frequency $u(\omega)$ (a complex spectrum) (section 3.2; Appendix 4).

12.2 Groups of waves

Let us consider now a wave formed by a continuous distribution of frequencies limited to a narrow band centered at a frequency ω_0 and a band width $2\Delta\omega$ (Fig. 12.3(a)). Using as a variable the wave number k, the displacements from equation (12.9), with zero initial phase for all frequencies and constant amplitude A_0, are given by

$$u(x, t) = A_0 \int_{k_0 - \Delta k}^{k_0 + \Delta k} \cos(kx - \omega t) \, dk \tag{12.10}$$

On performing a Taylor-series expansion of the phase around k_0 and taking only the term with the first derivative, we have

$$kx - \omega t = k_0 x - \omega_0 t + (k - k_0) \frac{d}{dk} (kx - \omega t)_{k_0} + \ldots \tag{12.11}$$

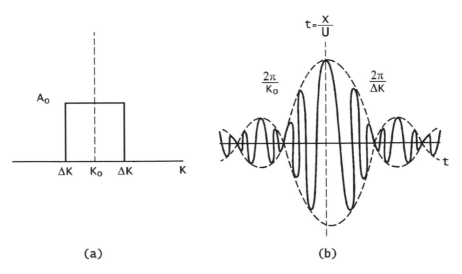

(a)   (b)

Fig. 12.3. A wave formed by a narrow band of frequencies centered at k_0 and of width $2 \, \Delta k$. (a) The band of frequencies. (b) The wave as a function of time, the product of $\cos(\omega_0 t)$ and $\sin X / X$.

On substituting into (12.10), we have

$$u = A_0 \int_{k_0 - \Delta k}^{k_0 + \Delta k} \cos\left[kx_0 - \omega t_0 - \left(x - \frac{d\omega}{dk}t\right)k_0 + \left(x - \frac{d\omega}{dk}t\right)k\right] dk \qquad (12.12)$$

The derivatives $d\omega/dk$ are evaluated at $k = k_0$. This integral can be written as

$$u = A_0 \int_{k_0 - \Delta k}^{k_0 + \Delta k} \frac{1}{b} \cos(a + bk)b \, dk \qquad (12.13)$$

$$a = k_0 \frac{d\omega}{dk}t - \omega_0 t = k_0 U t - \omega_0$$

$$b = x - \frac{d\omega}{dk}t = x - Ut$$

where U is the group velocity. On solving the integral,

$$u = \frac{A_0}{b}\{\sin[a + b(k_0 + \Delta k)] - \sin[a + b(k_0 - \Delta k)]\} \qquad (12.14)$$

Since $\sin(x + y) - \sin(x - y) = 2\sin x \cos y$, we obtain

$$u = \frac{2A_0}{b}\sin(\Delta k b)\cos(a + bk_0) \qquad (12.15)$$

On replacing the values of a and b and multiplying and dividing by Δk, we have

$$u(x, t) = A_0 \Delta k \frac{\sin X}{X} \cos(k_0 x - \omega_0 t) \qquad (12.16)$$

$$X = \left(x - \frac{d\omega}{dk}t\right)\Delta k = (x - Ut)\Delta k \qquad (12.17)$$

The displacement $u(x, t)$ is a cosine wave of wave number k_0, modulated by the function $\sin X/X$. This function has a maximum value equal to unity for $X = 0$, and zeros for $X = \pm\pi$, $\pm 2\pi$, etc. The main pulse is centered at $X = 0$, and the amplitudes of the other pulses decrease with $1/X$. As a function of time for a fixed distance, the displacement is formed by a group or packet, of width $2\pi/\Delta\omega$, of waves of frequency ω_0. The maximum of the wave group arrives at $t = x/U$ (Fig. 12.3(b)). The group velocity U is, then, the velocity of propagation of the wave packet formed by the envelope of the function $\sin X/X$. The phase velocity of waves contained in the packet with frequency ω_0 is $c = \omega_0/k_0$.

12.3   The principle of a stationary phase

The principle of a stationary phase is used in order to evaluate displacements of dispersed waves with a continuous distribution of frequencies with a broad spectrum (12.9). As a function of the wave number k and assuming all initial phase to be zero, equation (12.9) can be written as

$$u(x, t) = \int_{-\infty}^{\infty} A(k)\cos[\Phi(k)] \, dk \qquad (12.18)$$

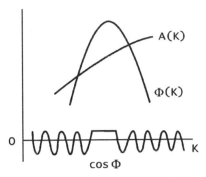

Fig. 12.4. The shapes of the functions $A(k)$, $\Phi(k)$, and $\cos\Phi$, showing the contribution of the stationary phase.

where

$$\Phi(k) = kx - \omega t \qquad (12.19)$$

If the amplitude $A(k)$ is a slowly varying function of k compared with the variation of the phase $\Phi(k)$, the integral (12.18) has significant values only for frequencies for which $\Phi(k)$ is stationary; that is, when $d\Phi/dk = 0$ (Fig. 12.4). For rapid variations of $\Phi(k)$, the cosine function changes sign and its integral becomes null, except for those values of $k = k_0$, for which $\Phi(k)$ does not vary ($d\Phi/dk = 0$). We call k_0 and ω_0 the wave number and frequency for which the phase is stationary. The integral (12.18) becomes, in a similar way to that in (12.10),

$$u(x, t) = A(k_0) \int_{k_0 - \Delta k}^{k_0 + \Delta k} \cos[\Phi(k)] \, dk \qquad (12.20)$$

This integral has values distinct from zero only for values of k near k_0.

If we take the derivative with respect to k of Φ and put it equal to zero,

$$\frac{d\Phi}{dk} = \frac{d}{dk}(kx - \omega t) = 0 \qquad (12.21)$$

then, since this is zero for $k = k_0$, and $d\omega/dk = U$, we obtain

$$x/t = U(k_0) \qquad (12.22)$$

where $U(k_0)$ is the group velocity corresponding to the wave number k_0. Then, at a distance x and time t, energy is contained in waves with wave number k_0 or frequency ω_0 that correspond to stationary values of the phase. According to the principle of a stationary phase, for that distance and time, the other frequencies do not contribute to the amplitude. For each value of x and t, k_0 has a different value. This result is similar to that of the previous section, in which k_0 was the central value of the wave-number band. Here we have all wave numbers but, for each pair of values of x and t, k_0 is the only value that makes the phase stationary.

An equivalent expression to (12.21) is obtained by taking derivatives of the phase Φ with respect to the frequency ω:

$$\frac{d\Phi}{d\omega} = \frac{d}{d\omega}(kx - \omega t) = \frac{x}{U} - t \qquad (12.23)$$

For the stationary phase, $d\Phi/d\omega = 0$, and the corresponding frequency ω_0, we obtain the same result as that in (12.22). The group velocity is, then, the velocity corresponding to the frequency that makes the phase stationary for a given distance and time. If we write Φ in terms of the phase velocity, the condition of the stationary phase results in

$$\frac{d\Phi}{d\omega} = \frac{d}{d\omega}\left[\omega\left(\frac{x}{c} - t\right)\right] = 0 \tag{12.24}$$

and, since for the stationary phase $x/t = U$, from (12.24) we obtain the relation between the group and phase velocities (12.5):

$$U = \left(\frac{1}{c} - \frac{\omega}{c^2}\frac{dc}{d\omega}\right)^{-1} \tag{12.25}$$

In order to determine the amplitudes corresponding to the stationary phase, we start with the equation

$$u(x, t) = \int_{-\infty}^{\infty} A(k)\, e^{i(kx - \omega t)}\, dk \tag{12.26}$$

where, for simplicity, we have assumed the initial phases to be zero. According to the principle of the stationary phase, this integral is null outside the values near the frequency that makes the phase stationary (12.20). On performing a Taylor expansion of the phase about the wave number k_0, including the second derivatives, we have

$$kx - \omega t = k_0 x - \omega_0 t + (k - k_0)\frac{d}{dk}[kx - \omega t]_{k_0 = k}$$

$$+ \frac{1}{2}(k - k_0)^2 \frac{d^2}{dk^2}[kx - \omega t]_{k = k_0} \tag{12.27}$$

As we have seen, for the stationary phase, the term of the first derivative is null. On substituting (12.27) into (12.26), we obtain

$$u(x, t) = A(k_0)\, e^{i(k_0 x - \omega_0 t)} \int_{-\infty}^{\infty} \exp\left(i\frac{1}{2}(k - k_0)^2 \frac{dU}{dk}t\right) dk \tag{12.28}$$

where we have used that

$$\frac{d}{dk}(kx - \omega t) = x - Ut \tag{12.29}$$

On making the change of variable

$$\sigma^2 = \frac{1}{2}(k - k_0)^2 \frac{dU}{dk}t \tag{12.30}$$

equation (12.28) takes the form

$$u(x, t) = A(k_0)\, e^{i(k_0 x - \omega_0 t)} \left[\frac{t}{2}\frac{dU}{dk}\right]_{k_0}^{-1/2} \int_{-\infty}^{\infty} e^{-i\sigma^2}\, d\sigma \tag{12.31}$$

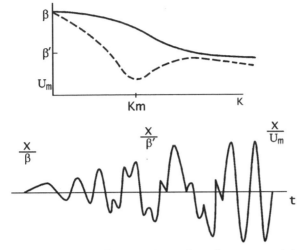

Fig. 12.5. Phase- and group-velocity dispersion curves and the corresponding train of dispersed waves at a distance x as functions of time.

The definite integral has the value $(i\pi)^{1/2}$. On substituting into (12.31), since $\sqrt{i} = \pm \exp(\pi/4)$, we obtain for the real part of (12.31)

$$u(x, t) = A(k_0) \left(\frac{2\pi}{\dfrac{x}{U} \dfrac{dU}{dk}} \right)^{1/2} \cos \left(k_0 x - \omega_0 t \pm \frac{\pi}{4} \right) \tag{12.32}$$

Then, for a given time and distance, energy is contained in a cosine wave of wave number k_0 and frequency ω_0, values corresponding to the stationary phase. Since dU/dk is in the denominator, the largest amplitudes correspond to $dU/dk = 0$, or Airy's phase. In this case, in the series expansion (12.27), the second term is now zero and we must take the third term. The amplitudes of Airy's phase are given by

$$u(x, t) = A(k_0) \left(\frac{2\pi}{\dfrac{x}{U} \dfrac{d^2 U}{dk^2}} \right)^{1/3} \cos \left(k_0 x - \omega_0 t \pm \frac{\pi}{4} \right) \tag{12.33}$$

An example of the distribution of amplitudes as a function of time in a dispersed train of surface waves is shown in Fig. 12.5. The phase velocity curve $c(k)$ is that of Love waves for a layer over a half-space (section 10.3). The group velocity curve $U(k)$ is derived from the phase velocity according to (12.4) (Fig. 12.5(a)). With this curve and equation (12.32), amplitudes observed at a distance x as a function of time are shown in Fig. 12.5(b). The waves observed are those corresponding to the frequencies that make the phase stationary. In consequence, they travel with the group velocity. Waves corresponding to the lowest frequencies arrive first, starting at time $t = x/\beta$. At time $t = x/\beta'$, waves with the highest frequencies that are superimposed on those of lower frequencies arrive. The last arrival, at $t = x/U_m$, is that of waves that travel with the minimum group velocity. At this time, waves with similar frequencies near k_m arrive together and are summed up to form Airy's phase. According to (12.32), on a seismogram recorded at a given distance, for each value of time, we observe only the wave corresponding to the frequency that makes the phase stationary.

12.4   Characteristics of dispersed waves

At a given distance, dispersed waves have the form of trains of waves with different frequencies arriving at different times, according to the frequency dependence of the velocity (the dispersion curve). Since the frequencies observed are those corresponding to stationary values of the phase, their times of arrival depend on the group velocity rather than on the phase velocity. This is an important aspect of dispersion, which is not always well understood, that can be further explained by considering the derivatives of the phase with respect to distance and time.

Taking derivatives of the phase Φ with respect to distance for a fixed time, we obtain

$$\frac{\partial}{\partial x}(kx - \omega t) = k + \frac{\partial k}{\partial x}(x - Ut) \tag{12.34}$$

For a wave number k_0 that corresponds to a stationary value of the phase, $U(k_0) = x/t$ and, consequently,

$$\frac{\partial \Phi}{\partial x} = k_0 \tag{12.35}$$

This means that, for a fixed time, at a distance x, energy arrives in a wave of wave number k_0, called the local wave number, which is different for every distance. Its wave length is $\lambda_0 = 2\pi/k_0$.

If we take the time derivative of the phase Φ, for a fixed distance, we obtain

$$\frac{\partial}{\partial t}(kx - \omega t) = -\omega + \frac{\partial k}{\partial t}(x - Ut) \tag{12.36}$$

Just like in the preceding case for the frequency ω_0 corresponding to the stationary phase, we have

$$\frac{\partial \Phi}{\partial t} = -\omega_0 \tag{12.37}$$

For a fixed distance, at each time, energy arrives in a wave of frequency ω_0 called the instantaneous frequency, which is different for every time. Its period is $T_0 = 2\pi/\omega_0$, and, for a fixed distance, in the train of dispersed waves, at each time, we observe only this period (Fig. 12.6). In conclusion, waves observed at a given distance and time correspond to a local wave number k_0 or instantaneous frequency ω_0, and propagate with the group velocity.

Let us consider now an increment in phase $\delta\Phi$, related to changes in distance and time and also in frequency and wave number:

$$\delta(kx - \omega t) = k\,\delta x + x\,\delta k - t\,\delta\omega - \omega\,\delta t \tag{12.38}$$

For frequencies and wave numbers corresponding to stationary values of the phase $x\,\delta k - t\,\delta\omega = 0$, and we obtain

$$\delta\Phi = k_0\,\delta x - \omega_0\,\delta t \tag{12.39}$$

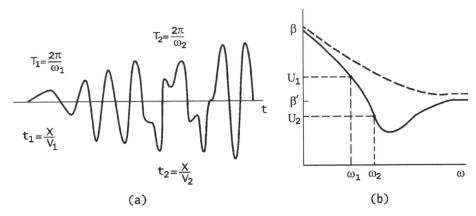

Fig. 12.6. (a) Arrivals of waves of instantaneous frequencies ω_1 and ω_2 in a dispersed wave train. (b) Phase- and group-velocity dispersion curves.

If we observe the same phase in a wave of the same frequency ω_0, at two points separated by a distance δx, with a time interval δt, then the phase increment is null ($\delta\Phi = 0$) and we can find that

$$\frac{\delta x}{\delta t} = \frac{\omega_0}{k_0} = c \tag{12.40}$$

where c is the phase velocity of waves with frequency and wave number ω_0 and k_0. This expression should be compared with equation (12.22) for the group velocity.

Using equations (12.22) and (12.40) we can determine both group and phase velocities from observations of dispersed wave trains. Consider a wave observed at distance x and time t, with instantaneous frequency ω_0; we obtain the group velocity $U(\omega_0) = x/t$. Consider now a wave observed at two nearby points separated by a distance δx, if the same phase for the same instantaneous frequency ω_0 arrives with a time difference δt, we can calculate the phase velocity $c(\omega_0) = \delta x/\delta t$. These ideas are at the base of methods used for determination of phase and group velocities from observations of dispersed wave trains.

12.5 The determination of group and phase velocities. Instantaneous frequencies

The determination of group and phase velocities from observations of dispersed surface waves is based on the ideas discussed previously. A seismogram is the recording of waves as a function of time for a given distance. The amplitudes of dispersed surface waves, according to (12.32), are given by

$$u(x, t) = A(\omega_0) \left(\frac{2\pi}{\dfrac{x \, dU}{U \, dK}} \right)^{1/2} \cos\left(k_0 x - \omega_0 t + \phi + \phi_I \pm \frac{\pi}{4} \right) \tag{12.41}$$

where we have introduced the initial phase ϕ and the phase shift ϕ_I produced by the instrument. For a given distance x (the epicentral distance), a wave of instantaneous frequency ω_0 arrives at time t. As has already been explained, these are the frequencies that correspond to stationary values of the phase. Methods that use directly recorded waves are, then, limited to these frequencies.

12.5.1   The group velocity

The group velocity can be determined from the record of surface waves at a single station. If we correct for the instrumental phase shift ϕ_I, the phase is given by

$$\Phi = kx - \omega t + \phi \pm \frac{\pi}{4} \tag{12.42}$$

On taking the derivative with respect to ω and, since $d\omega/dk = U$, solving for U, we obtain

$$U = \frac{x}{\dfrac{d\Phi}{d\omega} + \dfrac{d\phi}{d\omega} + t} \tag{12.43}$$

For a stationary phase $d\Phi/d\omega = 0$ and $\omega = \omega_0$. Assuming that the initial phase does not depend on frequency, we obtain for each instantaneous frequency ω_0

$$U(\omega_0) = \frac{x}{t(\omega_0)} \tag{12.44}$$

The method for determination of the group velocity consists in measuring times of arrival of peaks and troughs of waves in a dispersed train (these are phases $\Phi = 0$ and π) (Fig. 12.7(a)). These values are represented in a plot with respect to the order number (Fig. 12.7(b)). On doubling the intervals between pairs of values (or from peak to peak) we obtain the periods corresponding to instantaneous frequencies and, from the ordinates, we obtain their arrival times. On dividing the epicentral distance by each arrival time (12.44), we obtain the group velocity $U(T_0)$ for each period (Fig. 12.7(c)). This velocity corresponds to a mean value of the structure along the trajectory from the epicenter to the station.

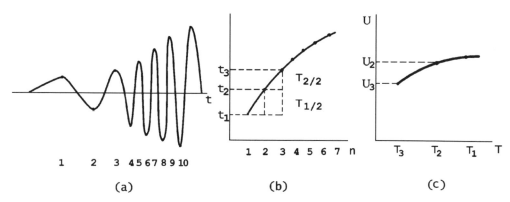

Fig. 12.7. Determination of the group velocity from the instantaneous frequencies at one station. (a) Identification of peaks and troughs. (b) Traveling times of peaks and troughs and determination of their periods. (c) The group velocities corresponding to each period.

12.5.2 The phase velocity

On a seismogram of dispersed surface waves, peaks correspond to values of the phase equal to multiples of 2π. The waves recorded are those corresponding to instantaneous frequencies (12.41), so that we can write

$$k_0 x - \omega_0 t + \phi + \phi_\mathrm{I} \pm \frac{\pi}{4} = 2N\pi \tag{12.45}$$

After correcting for the instrumental phase ϕ_I, on dividing (12.45) by k_0, we obtain for the phase velocity $(c = \omega/k)$

$$c(T_0) = \frac{x}{t - (\phi + N \pm \frac{1}{8})T_0} \tag{12.46}$$

If we want to determine c, we need to know the value of the initial phase ϕ corresponding to the azimuth to the station for each instantaneous frequency. This is possible only if we know the focal mechanism of the earthquake. Only in this case can we use equation (12.46) to determine the phase velocity. The method consists in selecting the peaks (or peaks and troughs) in the seismogram (phases $\Phi = 2N\pi$) and measuring their traveling times t_i and periods T_i. For the same periods we determine from the focal mechanism the initial phases ϕ and substitute them into equation (11.46). Values of N are found by trial and error, by successive substitutions 0, 1, 2, etc. and selecting those which give a reasonable value of the velocity. The value of N corresponds to the number of complete cycles that must be added for each period so that they correspond to the observed phase.

The necessity of knowing the focal mechanism if one is to calculate the phase velocity from a single station is the reason why this method is rarely used. This difficulty is avoided by using two stations. In this method two stations that are lined up with the epicenter are selected (that is, both stations and the epicenter lie on the same great circle). For the same instantaneous frequency ω_0, the phases corresponding to peaks at each station are given by

$$k_0 x_1 - \omega_0 t_1 + \phi + \phi_\mathrm{I} \pm \frac{\pi}{4} = 2L\pi \tag{12.47}$$

$$k_0 x_2 - \omega_0 t_2 + \phi + \phi_\mathrm{I} \pm \frac{\pi}{4} = 2M\pi \tag{12.48}$$

We subtract (12.47) from (12.48) and solve for $c = \omega/k$, putting $\Delta x = x_2 - x_1$, $\Delta t = t_2 - t_1$, and $N = M - L$, and obtain

$$c(T_0) = \frac{\Delta x}{\Delta t - NT_0} \tag{12.49}$$

N is now the number of complete cycles separating the phases of the two stations. In the two records we identify peaks of waves that correspond to the same period T_1, and subtract their traveling times from each other to get Δt_1 (Fig. 12.8(a)). The integer N is selected, just like in the previous case, by trial and error so that the dispersion curve is continuous and velocities are reasonable. In this way we obtain the phase velocity for periods present on the two seismograms (Fig. 12.8(b)). The phase velocities obtained correspond to a mean value of the structure along the trajectory between the two stations.

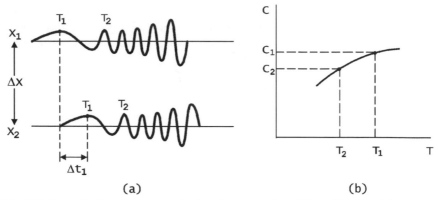

Fig. 12.8. Determination of the phase velocity using two stations. (a) Identification of the same phase for the same period at both stations and determination of the time interval. (b) The phase velocities corresponding to each period.

12.6   The determination of phase and group velocities. Fourier analysis

12.6.1   *Fourier analysis of seismograms*

The recording of a dispersed train of surface waves at a station located at a given distance from the epicenter is a real function of time $u(t)$. Its Fourier transform is a complex function of frequency $F(\omega)$ (Appendix 4):

$$F(\omega) = \int_0^\infty u(t)\,e^{-i\omega t}\,dt \tag{12.50}$$

$F(\omega)$ is the complex spectrum $F(\omega) = R(\omega) + iI(\omega)$ and can be expressed also as

$$F(\omega) = A(\omega)\,e^{i\Phi(\omega)} \tag{12.51}$$

where

$$A(\omega) = [R(\omega)^2 + I(\omega)^2]^{1/2} \tag{12.52}$$

$$\Phi(\omega) = \tan^{-1}[I(\omega)/R(\omega)] \tag{12.53}$$

$A(\omega)$ is the amplitude spectrum and $\Phi(\omega)$ is the phase spectrum. By using the inverse transform we pass from $F(\omega)$ to $u(t)$:

$$u(t) = \frac{1}{2\pi}\int_{-\infty}^\infty F(\omega)\,e^{i\omega t}\,d\omega \tag{12.54}$$

Seismic waves can be studied in the time domain $u(t)$ or in the frequency domain $F(\omega)$. Spectra show the contribution of each frequency to the waves observed. The record at a given distance of a dispersed train of surface waves, as we have already mentioned, shows at each time only the presence of waves corresponding to instantaneous frequencies. However, all frequencies are really present with more or less energy and their amplitudes and phases can be obtained by using Fourier transformation. In this way, we can calculate phase and group velocities for all frequencies, without being limited to instantaneous values like in the previous section.

In equations (12.50) and (12.54), $u(t)$ is a continuous real function of time defined from zero to infinity. In practice, it is only defined for a finite length and sampled at certain time intervals; that is, it is formed by a finite number of discrete values, u_i, for $i = 1$ to N at time intervals δt. Its Fourier transform is also discrete, F_k, for $k = 1$ to $N/2$, corresponding to frequencies between 0 and $\pi/\delta t$ at frequency intervals of $\delta\omega = 2\pi/(N\delta t)$. The highest frequency $\omega_N = \pi/\delta t$ is the Nyquist frequency and depends only on the sampling interval (Appendix 4).

Since early work by Satô (1955), many methods for determination of phase and group velocities of surface waves on the basis of Fourier analysis have been developed, especially since 1960 with the availability of digital computers and fast methods of computation of Fourier transform (Dziewonski and Hales, 1972). We show the basic ideas of two of these methods as examples.

12.6.2 The phase velocity

Fourier analysis can be easily applied to determination of the phase velocity from records of surface waves at two stations. The method is similar to that of instantaneous frequencies, but now we use all of the frequencies obtained from Fourier spectra. At each station the amplitude and phase spectrum, starting at times t_1 and t_2, is determined. For a distance Δx between stations and an interval $\Delta t = t_2 - t_1$ between the times of starting analysis at the stations, the difference between the phases is

$$\Phi_2(\omega) - \Phi_1(\omega) = \delta\Phi(\omega) + 2\pi N = k\,\Delta x - \omega\,\Delta t \tag{12.55}$$

The term $2\pi N$, as before, is added to compensate for the number of complete cycles separating the two phases. On dividing by k and solving for the phase velocity ($c = \omega/k$), we find an equation similar to (12.49):

$$c(\omega) = \frac{\Delta x}{(1/\omega)[\delta\Phi(\omega) + 2\pi N] + \Delta t} \tag{12.56}$$

If, for the two seismograms, we take the same starting time for calculating the Fourier transform, then $\Delta t = 0$. On expressing equation (12.56) as a function of the period, we obtain

$$c(T) = \frac{\Delta x}{T[\delta\Phi(T) + N]} \tag{12.57}$$

Equations (12.56) and (12.57) give the phase velocity for all frequencies obtained in the Fourier transform. In practice, since the seismogram is a sampled function of limited length, discrete Fourier transformation provides only a limited number of frequencies (Appendix 4). These limitations are present in all methods based on Fourier analysis of sampled functions of finite length.

12.6.3 The group velocity

Most methods used to obtain the group velocity using Fourier analysis are based on the application of a band-pass filter called a multiple filter that is centered on a frequency that takes various values (Landisman *et al.*, 1969). We will present only the basic ideas of this method. We define a Gaussian filter in the frequency domain centered on a certain frequency ω_n and with a width related to the constant α, in the form

$$H(\omega_n, \omega) = e^{-\alpha(\omega - \omega_n)^2/\omega_n^2} \tag{12.58}$$

The Gaussian function has the advantage that its transform is also a Gaussian function. The frequency ω_n takes successive values for which we want to calculate the group velocity. The Gaussian filter (12.58) is applied to the Fourier transform of the dispersed surface waves $u(t)$. The result, after eliminating the effects of initial and instrumental phases, can be written as

$$h(\omega_n, t) = \int_{-\infty}^{\infty} A(\omega) e^{-\alpha(\omega - \omega_n)^2/\omega_n^2} \cos(kx - \omega t) \, d\omega \tag{12.59}$$

where $h(\omega_n, t)$ is the inverse transform of the filtered function. Because of the shape of the filter, $A(\omega)$ and $k(\omega)$ can be expressed using a Taylor expansion about the central frequency ω_n:

$$A(\omega) = A(\omega_n) + (\omega - \omega_n)\frac{dA}{d\omega} = A_n + (\omega - \omega_n)A_n' \tag{12.60}$$

$$k(\omega) = k(\omega_n) + (\omega - \omega_n)\frac{dk}{d\omega} = k_n + (\omega - \omega_n)k_n' \tag{12.61}$$

On substituting these into (12.59), we obtain

$$h(\omega_n, t) = g(\omega_n, t) \cos(k_n x - \omega_n t + \varepsilon_n) \exp[-\omega_n^2(k_n' x - t)^2/(4\alpha)] \tag{12.62}$$

where

$$g(\omega_n, t) = (\pi/2)^{1/2}\omega_n\{A_n^2 + [A_n'\omega_n(k_n' x - t)]^2/(4\alpha^2)\}^2 \tag{12.63}$$

$$\varepsilon_n(\omega_n, t) = \tan^{-1}[A_n'\omega_n(k_n' x - t)/(2\alpha A_n)] \tag{12.64}$$

We can define the instantaneous amplitude $a(t)$ and phase $\Phi(t)$ for an arbitrary function $f(t)$ (Dziewonski and Hales, 1972) in the form

$$a(t) = [f(t)^2 + q(t)^2]^{1/2} \tag{12.65}$$

$$\Phi(t) = \tan^{-1}[q(t)/f(t)] \tag{12.66}$$

where $q(t)$ is the inverse transform of the Fourier transform of $f(t)$ after introducing a phase shift of $\pi/2$, or equivalently after interchanging its real and imaginary parts. The function $a(t)$ can be considered as the envelope of $f(t)$. On applying this operation to $h(\omega_n, t)$, we obtain

$$a_n(t) = g(\omega_n, t) \exp[-\omega^2(k_n' x - t)^2/(4\alpha)] \tag{12.67}$$

$$\Phi_n(t) = \omega_n t - k_n x - \varepsilon_n \tag{12.68}$$

From the principle of a stationary phase it follows that $d\Phi/dt = -\omega$ (12.37), where ω is the instantaneous frequency corresponding to each time. Maxima of $a(t)$ correspond to times for which $da/dt = 0$. Taking the derivative in (12.67) and putting it equal to zero, this is satisfied for $t_n = k'_n x$. However, $k'_n = dk_n/d\omega = U_n$ is precisely the group velocity associated with the maxima of $a_n(t)$ which is given by

$$U_n = x/t_n \tag{12.69}$$

where x is the distance from the station to the epicenter and t_n is the time corresponding to the arrival of the maximum of $a_n(t)$. Since this function is the envelope of $h(\omega_n, t)$, that is, of the original function $u(t)$ filtered by the Gaussian function centered at the frequency ω_n, this maximum represents the energy contained in $u(t)$ corresponding to the frequency ω_n. In consequence, t_n is the group traveling time associated with the frequency ω_n. By giving various values to ω_n, we find their corresponding group traveling times t_n and, according to (12.69), we find the values of their group velocities. The method is actually equivalent to the method of measuring the times of peaks and troughs on a seismogram, but it is not restricted to instantaneous frequencies. If, for the same frequency ω_n, there is more than one maximum of $a_n(t)$, they correspond to different modes of surface waves. The greatest traveling times (lowest velocities) correspond to the fundamental mode.

Since, in practice, we always use a sampled function u_m ($m = 1, \ldots, M$), the function a_{mn} has a matrix form and is called the energy diagram, where the subindex m refers to times and n refers to frequencies used in the filter. In this matrix, for each value of the frequency ω_n, we select the time t_n that corresponds to the maximum of a_{mn}. This time is the group traveling time associated with the maximum of energy corresponding to the frequency ω_n. By dividing the epicentral distance by t_n we find the group velocity U_n (12.69) corresponding to each frequency ω_n or period T_n (Fig. 12.9). This type of analysis can be used also to filter the signal with a selected distribution of group velocities. In this form, we can separate the energies propagated by the various modes, for example, that of the fundamental mode from those of higher modes. Filters of this type are called group-velocity filters.

12.7 Dispersion curves and the Earth's structure

12.7.1 Observations

Surface waves in the Earth are observed on seismograms of distant surface earthquakes as long trains of dispersed waves with large amplitudes. Dispersion is easily detected, first arrivals corresponding to waves of longer periods. As has been mentioned, the periods present on seismograms correspond to instantaneous frequencies. Love waves are registered only in the horizontal components whereas Rayleigh waves, which are polarized in the vertical plane, are registered both in horizontal and in vertical components. If we rotate the two horizontal components to make them coincide with the radial and transverse directions with respect to the orientation from the station to the epicenter, Love waves (LQ) are recorded only in the transversal component and Rayleigh waves (LR) are recorded only in the radial one (Fig. 12.10). Love waves of long periods (60–300 s) are also called G waves (in honor of B. Gutenberg). For these periods, the dispersion curve is practically flat with a velocity of about $4.4 \, \text{km s}^{-1}$ and waves have an almost impulsive form.

Energy Diagram

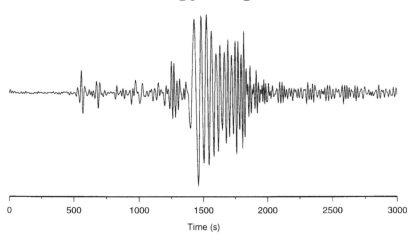

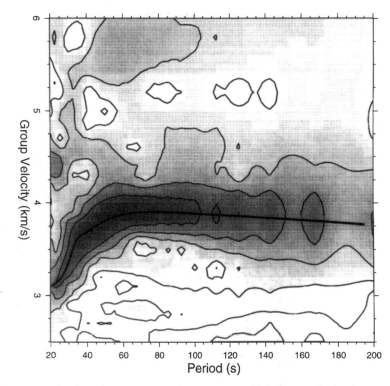

Fig. 12.9. Determination of the group velocity using the multiple-filter method. Dispersed surface waves and the energy diagram with group velocities and periods are shown for the Alaska earthquake of 20 November 1993 recorded at the HRV, $\Delta = 50°$ (courtesy of L. A. Rivera).

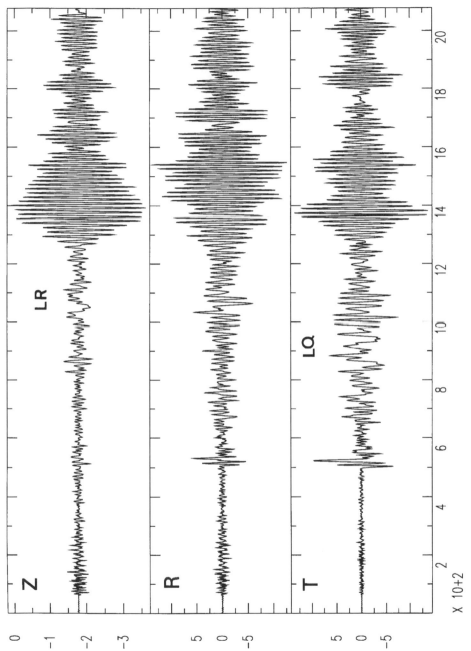

Fig. 12.10. Vertical, radial, and transverse components of the record of the broad-band station ANMO of a shallow earthquake in Peru (21 February 1996; $m_b = 6.7$.

233

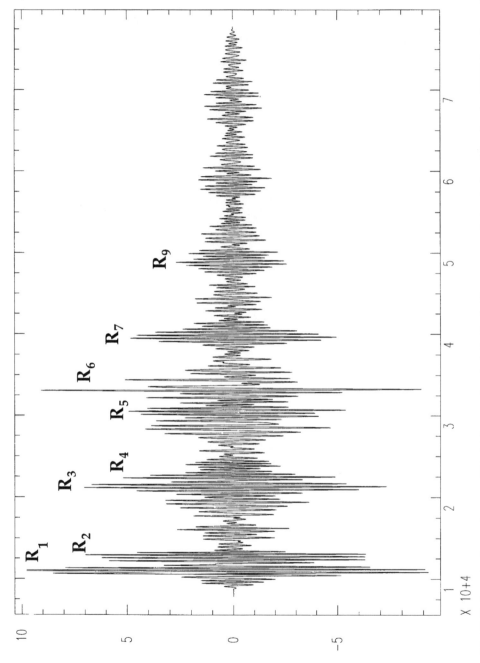

Fig. 12.11. Records of several groups of Rayleigh waves that circle the Earth. They were recorded at the broad-band station SFUC from the earthquake of the Balleny Islands (25 March 1998); $M = 8$.

Love and Rayleigh waves of short periods (8–12 s) in continental trajectories are channeled in the upper crust and are known as Lg and Rg waves. For periods between 60 and 300 s Love and Rayleigh waves travel mainly through the mantle and are called mantle waves. For large earthquakes, surface waves that travel around the Earth more than once are observed. These waves are designated with a subindex, G_1, G_2, G_3, etc. for Love waves, and R_1, R_2, R_3, etc. for Rayleigh waves (Fig. 12.11). Groups G_1 and R_1 are direct waves from the epicenter to the station and G_2 and R_2 are waves that arrive at the station traveling in the opposite direction. For higher subindexes, waves with an odd subindex circle the Earth, leaving the epicenter in the direction of the station; and those with an even subindex do it leaving in the opposite direction.

12.7.2 Interpretation

The forms of dispersion curves of phase and group velocities of Love and Rayleigh waves depend on the characteristics of the medium (section 10.3). Thus, their study allows the determination of the structure along their trajectories. Since surface waves travel along the surface of the Earth, their penetration into its interior depends on their wave lengths. For periods less than 60 s they are affected by the crust and for periods between 60 and 300 s they are affected by the mantle. They provide average values of the structure along their trajectory and do not give details of lateral variations. However, the analysis of surface waves along many trajectories permits the separation of effects of the various structures they have crossed. With this type of analysis, called regionalization, we can separate structures for different regions, such as oceanic regions with various ages of the sea floor, shields, orogenic regions, rifts, etc. Owing to the difficulty of interpreting surface waves of short periods (less than 10 s) because of the presence of large lateral heterogeneities, in general, regionalization does not give good results for small regions of the crust.

The first studies of the structure of the Earth's crust and mantle by means of the analysis of Rayleigh and Love waves were those by Ewing, Roehrbach and Carder in the 1930s. Later, during the 1950s, we have the work of Sato, Press, Oliver, and Wilson (Press, 1956; Oliver, 1962) among others. The rapid development both of long-period seismographic instrumentation and of digital computers during the 1960s gave a great impulse to these studies. During the 1960s and 1970s many studies of crust and mantle structures along a large number of trajectories, crossing all types of structures, oceanic, continental, shields, rifts, etc., were completed. Computers also made possible the calculation of theoretical dispersion curves for stratified media with many layers that are needed for interpretation of observations. Observed dispersion curves were at first inverted by comparison of observed and theoretical curves for various types of models. More recently, observed data have been inverted directly, including compensation for errors in observations and resolution of models.

One of the first results from the analysis of dispersion curves of surface waves was the difference between crust and upper mantle structures under oceans and those under continents. Purely oceanic trajectories show that group velocities of Rayleigh waves with periods 20–100 s have a very constant value of about $4\,\mathrm{km\,s^{-1}}$ and that those with shorter periods, 10–20 s, have a sharp fall due to the water layer. For continental

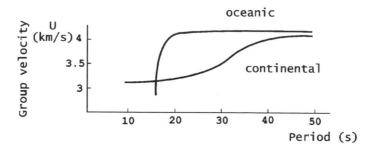

Fig. 12.12. Theoretical group-velocity dispersion curves for continental and oceanic structures. Continental crust, $H = 30$ km, $\beta = 3.5$ km s^{-1}, and upper mantle, $\beta = 4.6$ km s^{-1}; and oceanic crust $H = 10$ km, $\beta = 3.2$ km s^{-1}, and upper mantle, $\beta = 4.3$ km s^{-1}.

trajectories for periods shorter than 60 s, the group velocity decreases from 4 km s^{-1}, and for 10 s it reaches a minimum of about 3 km s^{-1}. This is due to the thickness of the crust being greater and to the presence of sediments with low velocities. For periods greater than 150 s, dispersion curves for continents and oceans are similar due to the influence of deeper parts of the mantle. These general characteristics can be explained by invoking the very simple model of oceanic and continental crusts and the upper mantle shown in Fig. 12.12. A summary of observed dispersion curves for group velocities of Rayleigh and Love waves for a large range of periods is shown in Fig. 12.13. Love waves have higher velocities than Rayleigh waves. For oceanic paths Love waves have a practically constant velocity of about 4.4 km s^{-1} for periods in the range 20–400 s. For periods shorter than about 80 s Love and Rayleigh waves for continental paths have lower values than they do for oceanic paths. There is a sharp drop in velocity both for Rayleigh and for Love waves of about 10–20 s period for oceanic paths. For Rayleigh waves there is minimum for a period of about 200 s that is produced by the influence of the low-velocity layer in the mantle. Characteristics of the crust and mantle found from the analysis of surface waves agree with those found from traveling times of body waves (sections 8.3 and 8.4).

A comparison among dispersion curves for various types of paths reveals that their structures are different (Knopoff, 1972). For Rayleigh waves and periods in the range 60–140 s, phase velocities are highest for continental shields and lowest for rift zones (Fig. 12.14). These curves reveal the effect of the differing thicknesses of the lithosphere and the influence of the astenosphere. The thickness of the lithosphere is greater for continental shields and smaller in rift zones with a shallower astenosphere. The mean curves shown in Fig. 12.14 show only the differences among very broad types of crust.

Since 1980, with the installation of global networks of seismographic stations with digital broad-band instrumentation (Chapter 21), it has been possible to analyze surface waves for various types of trajectories using not only dispersion curves but also amplitudes. These studies allow a more detailed regionalization of the structure of the crust and mantle, revealing lateral inhomogeneities through the whole mantle. Generally, these studies concern S wave velocity distributions obtained using spherical harmonics (Appendix 3). The best results have been found for harmonics of orders between four

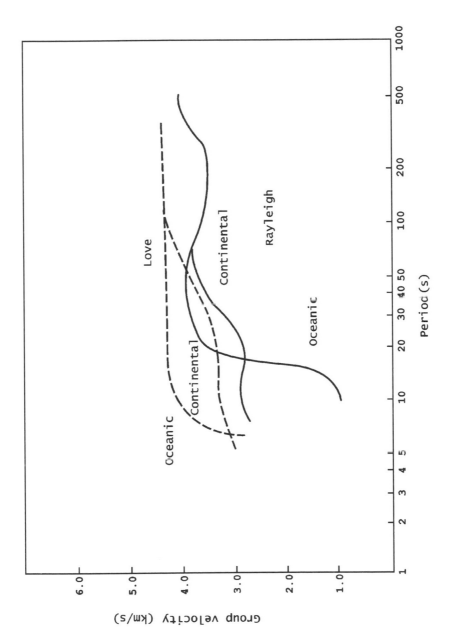

Fig. 12.13. Group-velocity dispersion curves for Rayleigh and Love waves (modified from Bullen and Bolt (1985)).

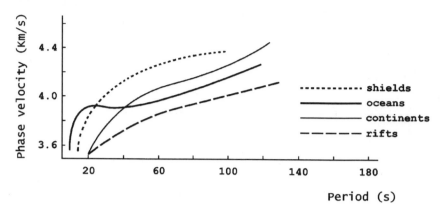

Fig. 12.14. Average phase-velocity dispersion curves of Rayleigh waves for regions of shields, oceans, continents, and rift regions (modified from Knopoff (1972)) (with permission from Elsevier Science).

and ten, applied to the resolution of mantle structure 300–1000 km deep (Dziewonski and Woodhouse, 1987). In this way, global three-dimensional models of the Earth's mantle have been obtained by the application of tomographic techniques to observations of surface waves.

13 FREE OSCILLATIONS OF THE EARTH

13.1 Wave propagation and modes of vibration

In Chapter 4, we considered the problem of free vibrations of an elastic body of finite dimensions, considering the vibrations of a string and a rod. Since the Earth has a finite radius and is bounded by a free surface, we must consider its free oscillations. Up to this point, we have treated wave propagation in the Earth without considering its finite dimensions. We find body waves traveling in its interior and the presence of its free surface generates surface waves. Body waves have relatively high frequencies (larger than 0.07 Hz) or short periods (less than 15 s). For an average velocity of $10 \, \mathrm{km \, s^{-1}}$ they correspond to wave lengths of about 150 km, which is small compared with the Earth's radius (6370 km). Hence, body waves' trajectories, traveling times and energy transport can be studied using the wave propagation approach and ray theory approximation in a flat or spherical medium, depending on distances. Surface waves extend to very low frequencies or large periods and their wave lengths reach values of the order of the Earth's dimensions. For example, waves of 400 s period, for a velocity of $4.5 \, \mathrm{km \, s^{-1}}$, have a wave length of 1800 km, about a third of the Earth's radius. For wave lengths of this order, the problem must be treated in the form of free oscillations or vibrations.

From the point of view of free oscillations, the Earth reacts to an earthquake by vibrating as a whole, in the same way as does a bell when it is hit. As we have seen in Chapter 4, the vibration of a finite elastic body is the sum of an infinite number of modes (harmonics) which correspond to frequencies with values that are multiples of the inverse of the body's dimensions. For an elastic body of finite dimensions, wave propagation and free vibrations are two different approaches to studying the same phenomenon. For wave lengths that are small compared with the dimensions of the body we can use wave propagation, but we can not do this if they are of the same order. Free-vibration or normal mode theory includes the complete phenomenon, but for small wave lengths we need a sum of many modes, which is not very practical. Thus, wave propagation is used for high frequencies and normal mode theory is used for low ones. The study of the Earth's free oscillations is based on the theory of vibrations of an elastic sphere. The problem increases in complexity as we proceed from an isotropic homogeneous sphere to models with radial distributions of elastic parameters and density, three-dimensional heterogeneities, and effects of gravity, rotation, and ellipticity.

13.2 Free oscillations of a homogeneous liquid sphere

As an introduction to the problem, let us first consider the free oscillations of a liquid sphere with constant density and bulk modulus. The components of displacements

239

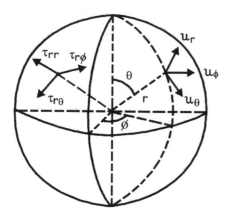

Fig. 13.1. Spherical coordinates and components of the displacements and stresses.

in spherical coordinates (Appendix 2) u_r, u_θ, and u_ϕ can be derived from a single scalar potential $\Phi(r, \theta, \phi, t)$ (section 3.10) (Fig. 13.1):

$$u_r = \frac{\partial \Phi}{\partial r} \tag{13.1}$$

$$u_\theta = \frac{1}{r} \frac{\partial \Phi}{\partial \theta} \tag{13.2}$$

$$u_\phi = \frac{1}{r \sin \theta} \frac{\partial \Phi}{\partial \phi} \tag{13.3}$$

According to (3.3), in the absence of body forces, the potential Φ satisfies the wave equation

$$\nabla^2 \Phi = \frac{1}{\alpha^2} \frac{\partial^2 \Phi}{\partial t^2} \tag{13.4}$$

where $\alpha^2 = K/\rho$ (K is the bulk modulus and ρ is the density) is the velocity of longitudinal (acoustic) waves. If Φ has a harmonic time dependence,

$$\Phi(r, \theta, \phi, t) = \Phi(r, \theta, \phi) \, e^{-i\omega t} \tag{13.5}$$

then equation (12.12) becomes the Helmholz equation (3.10),

$$(\nabla^2 + k^2)\Phi = 0 \tag{13.6}$$

where $k = \omega/\alpha$ is the wave number. By expressing the Laplacian in spherical coordinates (A2.30) and applying the method of separation of variables, $\Phi(r, \theta, \phi) = R(r)N(\theta)L(\phi)$, we obtain three equations for R, N, and L, just like (3.139), (3.140), and (3.141), where the constants of separation of variables are l and m (in section 3.10, we used n and m). These three equations, upon introducing into the second equation the change of variable $y = \cos \theta$, may be written in the following form:

$$\frac{\mathrm{d}}{\mathrm{d}r}\left(r^2 \frac{\mathrm{d}R}{\mathrm{d}r}\right) + [r^2 k^2 - l(l+1)]R = 0 \tag{13.7}$$

$$\frac{\mathrm{d}}{\mathrm{d}y}\left((1-y^2)\frac{\mathrm{d}N}{\mathrm{d}y}\right) + \left(l(l+1) - \frac{m^2}{1-y^2}\right)N = 0 \tag{13.8}$$

$$\frac{\mathrm{d}^2 L}{\mathrm{d}\phi^2} + m^2 L = 0 \tag{13.9}$$

These three equations have the form of the Sturm–Liouville equation (4.35). In equation (13.7), $p(r) = r^2$, $q(r) = r^2 k^2$, and $ns(r) = l(l+1)$; in (13.8), $p(y) = 1 - y^2$, $q(y) = -m^2/(1-y^2)$, and $ns(y) = l(l+1)$; and in (13.9), $p(\phi) = 1$, $q(\phi) = m^2$, and $ns(\phi) = 0$.

As we have seen in section 3.10, the solutions for $R(r)$ are given by spherical Bessel functions $j_1(kr)$, those of $N(\theta)$ by associate Legendre functions $P_l^m(\cos\theta)$, and those for $L(\phi)$ by harmonic functions $e^{\pm im\phi}$ (Appendix 3). In consequence, we obtain for the potential Φ a solution as a sum of normal modes (Chapter 4):

$$\Phi(r,\theta,\phi) = \sum_{l=0}^{\infty} j_1(kr) \sum_{m=-1}^{l} Y_l^m(\theta,\phi) \tag{13.10}$$

$$Y_l^m = (-1)^m \left(\frac{2l+1}{4\pi}\frac{(1-m)!}{(1+m)!}\right) P_l^m(\cos\theta)[C_l^m \cos(m\phi) + S_l^m \sin(m\phi)] \tag{13.11}$$

We impose the boundary condition that, for each point of the surface of the sphere, the normal component of the stress or pressure is null. According to (2.19) and (2.69),

$$P = -K\nabla^2\Phi = 0 \tag{13.12}$$

for $r = a$. By substitution into (13.6), putting k in terms of ω, the boundary condition becomes

$$\frac{K\omega^2}{\alpha^2}\Phi(a) = 0 \tag{13.13}$$

Since in the solution (13.10), the only part that depends on r is given by spherical Bessel functions, condition (13.13) implies that

$$\frac{K\omega^2}{\alpha^2} j_l\left(\frac{\omega a}{\alpha}\right) = 0 \tag{13.14}$$

This condition depends on the roots of $j_l(x)$ (Appendix 3). Using Rayleigh's formula to express spherical Bessel functions by sines and cosines (A3.14), for the first values of $l = 0$, 1, and 2, these are

$$j_0(x) = \frac{\sin x}{x} \tag{13.15}$$

$$j_1(x) = \frac{\sin x}{x^2} - \frac{\cos x}{x} \tag{13.16}$$

$$j_2(x) = \left(\frac{3}{x^3} - \frac{1}{x}\right)\sin x - \frac{3}{x^2}\cos x \tag{13.17}$$

Each of these functions has an infinite number of roots, which, for the first values of l, are

$$l = 0, \qquad x = 1.0\pi, \quad 2.0\pi, \quad 3.0\pi, \ldots$$
$$l = 1, \qquad x = 1.4303\pi, \quad 2.4590\pi, \quad 3.4709\pi, \ldots$$
$$l = 2, \qquad x = 1.8346\pi, \quad 2.8950\pi, \quad 3.9226\pi, \ldots$$

For each value of l, we use the subindex n for each of these roots. For example, according to equation (13.14), the roots for $l = 0$ correspond to

$$\frac{\omega a}{\alpha} = (n+1)\pi, \qquad n = 0, 1, 2, 3, \ldots \tag{13.18}$$

Thus, normal modes impose conditions on the possible values of the frequency. Since these values depend on both subindexes, l and n, possible frequencies are designated by ${}_n\omega_l$. For $l = 0$, according to (13.18), the frequencies are

$$_n\omega_0 = \frac{\alpha\pi}{a}(n+1) \tag{13.19}$$

and the periods are

$$_nT_0 = \frac{2a}{\alpha}(n+1) \tag{13.20}$$

Just like for an elastic string and a rod (Chapter 4), a liquid sphere vibrates at certain fixed frequencies that depend on its properties (K and ρ, or α) and its radius (a). There is an infinite number of modes of vibration, each with its own frequency that depends on two subindexes, n and l. For each value of l, $n = 0$ corresponds to the fundamental mode and $n \geq 1$ correspond to higher modes, harmonics, or overtones. The potential for $l = 0$, according to (13.10) and (13.11), is given by

$$_n\Phi_0 = C_0 \frac{\alpha}{_n\omega_0 r} \sin\left(\frac{_n\omega_0 r}{\alpha}\right) \tag{13.21}$$

In this case, the subindex m has only the value zero, and the potential does not depend on θ and ϕ. Then, according to (13.1), (13.2), and (12.3), displacements have only radial components (u_r), and these modes are called radial modes. According to (13.19) and (13.20), the frequencies and periods corresponding to the first three modes are

$$_0\omega_0 = \pi\alpha/a \qquad _0T_0 = 2a/\alpha$$
$$_1\omega_0 = 2\pi\alpha/a \qquad _1T_0 = a/\alpha$$
$$_2\omega_0 = 3\pi\alpha/a \qquad _2T_0 = 2a/(3\alpha)$$

The period of the fundamental mode ($n = 0$) is the longest possible and corresponds to the time that a wave takes to travel with a velocity α along the diameter of the sphere. For a sphere of the size of the Earth ($a = 6370\,\mathrm{km}$) with velocity $\alpha = 9\,\mathrm{km\,s^{-1}}$, $_0T_0$ equals 23.6 min.

For $l = 1$, we have three values of m, -1, 0, and 1, and there are also n roots of $j_1(\omega a/\alpha)$. Since $P_n^{-m} = (-1)^m P_n^m$, the potential Φ is given by

$$_n\Phi_1 = j_1(_n\omega_1 r/\alpha)[C_1^0 \cos\theta + (C_1^1 - C_1^{-1})\sin\theta\cos\phi + (S_1^1 + S_1^{-1})\sin\theta\sin\phi] \tag{13.22}$$

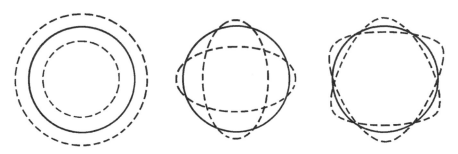

Fig. 13.2. Spheroidal modes for values of $l = 0$, 2, and 3.

According to the roots of $j_1 \left({}_n\omega_1 a/\alpha \right)$, we have the following frequencies for the fundamental and two first higher modes:

$$
{}_0\omega_1 = 1.4303\pi\alpha/a
$$

$$
{}_1\omega_1 = 2.4590\pi\alpha/a \tag{13.23}
$$

$$
{}_2\omega_1 = 3.4709\pi\alpha/a
$$

These frequencies are higher than those corresponding to modes for $l = 0$. For $l = 1$, the fundamental mode ($n = 0$) does not exist, since it implies a change in the position of the center of mass of the sphere that is not allowed for free vibrations. In general, the higher the mode order the higher the corresponding frequencies. Since, for $l \geq 1$, the potentials (13.10) and (13.11) depend on the three coordinates r, θ, and ϕ, the displacements have three components, u_r, u_θ, and u_ϕ. For each value of l, m takes $2l + 1$ values and there are $2l + 1$ potential functions, all corresponding to the same frequency. For example, for $l = 2$, m takes the values $m = -2, -1, 0, 1$, and 2 and there are five potentials for the same frequency ${}_n\omega_2$. This means that there are several eigenfunctions for one eigenvalue, which is known as a degeneracy problem (Chapter 4). For general, nondegenerate cases, the frequencies depend on the three indexes n, l, and m $\left({}_n\omega_l^m \right)$; modes depending on n are called radial modes, those depending on l are angular modes, and those depending on m are azimuthal modes.

If there is symmetry with respect to ϕ, that is for $m = 0$, then the displacements depend on θ according to the Legendre polynomials $P_l(\cos\theta)$. For $l = 0$ ($P_0 = 1$), the displacements are purely radial. For $l = 2$, the body takes an ellipsoidal form, alternatively elongated and flattened at the poles ($\theta = 0$). For higher values of l, the forms are symmetric with respect to the equatorial plane ($\theta = \pi/2$) if l is even and asymmetric if l is odd (Fig. 13.2).

13.3 Free oscillations of an elastic sphere

A first approximation to the problem of free oscillations of the Earth is that of a homogeneous elastic sphere. This problem was treated by Poisson in 1829 and in a more complete form by Lamb in 1882. In 1911 Love calculated the fundamental period of the vibrations of a sphere with the dimensions of the Earth and found a value near to 1 h.

The problem is similar to that of the oscillations of a liquid sphere. Displacements in spherical coordinates are derived from a scalar potential Φ and a vector potential   as we saw in section 3.10. The vector potential   can be separated into two potentials that are products of the unit vector in the radial direction and a scalar potential:

$$= \nabla \times S\mathbf{r} + T\mathbf{r} \tag{13.24}$$

$$\mathbf{u} = \nabla\Phi + \nabla \times \nabla \times S\mathbf{r} + \nabla \times T\mathbf{r} \tag{13.25}$$

In equation (13.25), the displacements are separated into three parts, the first term representing P wave motion, the second SV wave motion, and the third SH wave motion. In relation to surface waves, Φ and S represent Rayleigh wave motion and T represents Love wave motion. S is called the spheroidal potential and T is called the toroidal potential. When there is symmetry with respect to ϕ, the relations between these potentials and the scalar potentials Φ, ψ, and Λ used in section 3.10 are

$$\Phi = \Phi \tag{13.26}$$

$$\psi = \frac{1}{r^2}\frac{\partial S}{\partial \theta} \tag{13.27}$$

$$\Lambda = T/r \tag{13.28}$$

The components of displacements in the directions of the coordinates r, θ, and ϕ in terms of the potentials Φ, S, and T are given by

$$u_r = \frac{\partial \Phi}{\partial r} + \frac{1}{r\sin\theta}\left[\frac{\partial}{\partial \theta}\left(\frac{\sin\theta}{r}\frac{\partial S}{\partial \theta}\right) - \frac{1}{\sin\theta}\frac{\partial^2 S}{\partial \phi^2}\right] \tag{13.29}$$

$$u_\theta = \frac{1}{r}\frac{\partial \Phi}{\partial \theta} - \frac{1}{r}\frac{\partial}{\partial r}\left(\frac{\partial S}{\partial \theta}\right) + \frac{1}{\sin\theta}\frac{\partial T}{\partial \phi} \tag{13.30}$$

$$u_\phi = \frac{1}{r\sin\theta}\frac{\partial \Phi}{\partial \phi} + \frac{1}{r}\frac{\partial}{\partial r}\left(\frac{r}{\sin\theta}\frac{\partial S}{\partial \phi}\right) - \frac{1}{r}\frac{\partial T}{\partial \theta} \tag{13.31}$$

Spheroidal motion depending on the potentials Φ and S is a combination of P and SV displacements with u_r, u_θ, and u_ϕ components, whereas toroidal motion depending on the potential T is of SH type with u_θ and u_ϕ components only.

The boundary conditions on the free surface are that the components of the stress across it are null:

$$\tau_{rr} = \tau_{r\theta} = \tau_{r\phi} = 0 \tag{13.32}$$

For an isotropic medium, the components of the stress as functions of the displacements are given by

$$\tau_{rr} = (\lambda + 2\mu)\frac{\partial u_r}{\partial r} + \lambda\left[\frac{2u_r}{r} + \frac{1}{r\sin\theta}\left(\frac{\partial}{\partial \theta}(u_\theta \sin\theta) + \frac{\partial u_\phi}{\partial \phi}\right)\right] \tag{13.33}$$

$$\tau_{r\theta} = 2\mu\left(\frac{\partial u_\theta}{\partial r} - \frac{u_\theta}{r} + \frac{1}{r}\frac{\partial u_r}{\partial \theta}\right) \tag{13.34}$$

$$\tau_{r\phi} = 2\mu\left(\frac{1}{r\sin\theta}\frac{\partial u_r}{\partial \phi} + \frac{\partial u_\phi}{\partial r} - \frac{u_\phi}{r}\right) \tag{13.35}$$

where λ and μ are Lamé's coefficients.

In the absence of body forces and without considering rotation and gravity, Navier's equation (2.67) is

$$\alpha^2 \nabla(\nabla \cdot \boldsymbol{u}) - \beta^2 \nabla \times \nabla \times \boldsymbol{u} = \ddot{\boldsymbol{u}} \tag{13.36}$$

where α and β are the P and S waves' velocities. The problem is solved by expressing $\boldsymbol{u}$ in terms of the potentials Φ, S, and T according to equation (13.25) and applying the surface boundary conditions (13.32). The complete problem is rather complex and we will give here only the most basic parts.

13.4 Toroidal modes

The part of the problem that corresponds to the toroidal potential $T(r, \theta, \phi, t)$ can be treated separately, as was done with SH waves (Chapter 3) and Love-wave propagation (Chapter 10). If we assume a harmonic dependence on time, we can write the Helmholtz equation for T and the problem is similar to that with a liquid sphere:

$$(\nabla^2 + k_\beta^2) T = 0 \tag{13.37}$$

where $k_\beta = \omega/\beta$ is the wave number of S wave motion. Just like in (13.10), the solution can be written in terms of spherical Bessel, associated Legendre and harmonic functions:

$$T = \sum_{l=0}^{\infty} j_l(k_\beta r) \sum_{m=-l}^{l} P_l^m(\cos\theta)[C_l^m \cos(m\phi) + S_l^m \sin(m\phi)] \tag{13.38}$$

As mentioned before, the subindex l is the angular number and m is the azimuthal number. For each value of l there are $2l + 1$ values of m. The solution is given by an infinite sum of modes T_l^m that are called toroidal modes. Since the potential T represents transverse motion, these modes are similar to the torsional modes of vibration of an elastic rod (section 4.3). From (13.30) and (13.31), the toroidal displacements are given by

$$u_\theta = -\frac{1}{\sin\theta} \frac{\partial T}{\partial \phi} \tag{13.39}$$

$$u_\phi = -\frac{1}{r} \frac{\partial T}{\partial \theta} \tag{13.40}$$

The boundary condition at the surface of the sphere ($r = a$) is $\tau_{r\phi} = 0$. From (13.35), since for toroidal motion $u_r = 0$, we obtain

$$\frac{\partial u_\phi}{\partial r} - \frac{u_\phi}{r} = 0, \qquad r = a \tag{13.41}$$

On putting u_ϕ as a function of T according to (13.40), this condition gives

$$\frac{1}{r} \frac{\partial}{\partial r} \frac{\partial T}{\partial \theta} - \frac{1}{r^2} \frac{\partial T}{\partial \theta} = \frac{1}{r} \frac{\partial}{\partial \theta}\left(\frac{\partial T}{\partial r} - \frac{T}{r}\right) = 0, \qquad r = a \tag{13.42}$$

According to (13.38), T is a product of three parts, each depending on one only of the coordinates, $T(r, \theta, \phi) = T_r(r) T_\theta(\theta) T_\phi(\phi)$. Equation (13.42) imposes only a condition in the form

$$r\frac{\partial T_r}{\partial r} - T_r = 0, \qquad r = a \tag{13.43}$$

on $T_r(r)$. Since, according to (13.38), $T_r = j_l(k_\beta r)$, on putting $x = k_\beta r$, condition (13.43) gives

$$xj_l'(x) = j_l(x), \qquad x = k_\beta a \tag{13.44}$$

Using the relation $j_l' = [1/(l-1)]j_{l+1}(x)$, this condition may be written as

$$xj_{l+1}(x) = (l-1)j_l(x), \qquad x = k_\beta a \tag{13.45}$$

This equation has an infinite number of solutions for which we use the subindex n. For each value of l, the first value of n ($n = 0$) corresponds to the fundamental mode and the rest to higher modes, harmonics, or overtones.

For $l = 0$, $P_0(\cos\theta) = 1$ and $T_0 = j_0(k_\beta r)$. Then, $\partial T/\partial\theta = \partial T/\partial\phi = 0$ and $u_\theta = u_\phi = 0$; that is, there are no toroidal vibrations of order zero.

For $l = 1$ ($P_1^0 = \cos\theta$ and $P_1^1 = \sin\theta$), condition (13.45) corresponds to the roots of $j_2(x)$ which, according to (13.16), gives

$$\tan x = \frac{3x}{3 - x^2}, \qquad x = k_\beta a \tag{13.46}$$

The first three roots, corresponding to $n = 0$, 1, and 2, are $x = 1.8346\pi$, 2.8950π, and 3.9226π. Since $k_\beta = \omega/\beta$, the eigenfrequencies for $l = 1$ and $n = 0$, 1, and 2 are

$$_0\omega_1 = \frac{1.8346\pi\beta}{a}; \qquad _1\omega_1 = \frac{2.8950\pi\beta}{a}; \qquad _2\omega_1 = \frac{3.9226\pi\beta}{a}$$

For toroidal modes, a solution for $n = 0$ is not physically possible for free vibrations (in the absence of external torques) since this corresponds to a rigid oscillation of the whole sphere. For $n = 1$ and 2, if we use approximate values of a and β for the Earth ($a = 6370$ km and $\beta = 6$ km s^{-1}), then the corresponding periods in minutes are $_1T_1 = 12.22$ and $_2T_1 = 9.02$.

The potential function $T(r, \theta, \phi)$ for $l = 1$ and $m = -1$, 0, and 1 is, according to (13.38),

$$_nT_1 = j_1(_n\omega_1 r/\alpha)\{C_1^0\cos\theta + \sin\theta[(C_1^1 - C_1^{-1})\cos\phi + (S_1^1 + S_1^{-1})\sin\phi]\} \tag{13.47}$$

For each value of n, this can be considered as a sum of three eigenfunctions, $_nT_1^0$, $_nT_1^1$, and $_nT_1^{-1}$, that correspond to the same eigenvalue, $_n\omega_1$. Therefore, the problem is a degenerate one.

For $l = 2$, condition (13.44), by substitution of (13.17), gives

$$\tan x = \frac{(12 - x^2)x}{12 - 5x^2} \tag{13.48}$$

The first three roots of this equation, corresponding to $n = 0$, 1, and 2, are $x = 0.796\pi$, 2.271π, and 3.346π. The corresponding periods, for the given values of a and β, are $_0T_2 = 44.46$ min, $_1T_2 = 15.58$ min, and $_2T_2 = 10.58$ min. In this case, we have a fundamental mode ($_0T_2$) which has the longest period (lowest frequency) of all toroidal

Table 13.1. *Periods (in minutes) of toroidal modes*

l	n			
	0	1	2	3
1	–	13.45	7.63	5.22
2	43.82	12.60	7.49	5.17
3	28.35	11.56	7.28	5.08

Fig. 13.3. Toroidal modes for values of $l = 1, 2,$ and 3.

modes. The period of $_0T_2$ (44.46 min) is greater than the time (35.39 min) which an SH wave takes to travel along the diameter of the sphere.

For toroidal modes the index l refers to the order of associated Legendre functions and in consequence to the distribution of displacements on spherical surfaces depending on the angle θ. For $l = 1$, the fundamental mode implies that whole-body oscillations around the origin of the θ axis, that are not possible as free vibrations, occur. They are possible for higher modes since internal parts oscillate in different senses and there is no change in total angular momentum. For $l = 2$, the two hemispheres oscillate in opposite senses. For higher values ($l \geq 3$), there are as many zones oscillating in opposite senses as the order number of the mode (Fig. 13.3). The index m refers to the dependences of displacements on the azimuthal angle ϕ and takes values from -1 to 1. The index n refers to the number of roots in the r dependences of displacements for each configuration of l and m (Fig. 13.4). Since, for a homogeneous sphere, the eigenfrequencies do not depend on the index m, displacements that are different according to their values of m correspond to the same frequencies; that is, the problem is a degenerate one. The same happens if elastic properties depend on the radius only, but not for the complete heterogeneous problem. For each value of l and m, $n = 0$ corresponds to the fundamental mode and $n \geq 1$ correspond to the higher modes or overtones.

The eigenperiods in minutes for toroidal modes of low order corresponding to a model of the Earth with radial symmetry are given in Table 13.1 (Dziewonski and Gilbert, 1972). Periods decrease with increasing mode-order number. The longest period, 43.82 min, corresponds to the fundamental mode of the second-order mode $_0T_2$. The values in Table 13.1 are similar to those found for a homogeneous sphere with the assumed value of S wave velocity (6 km s^{-1}).

13.5   Spheroidal modes

The problem of spheroidal modes is more complicated since it implies solutions for two potentials, Φ and S, and the displacements correspond to P–SV motion. Let us consider briefly a homogeneous sphere of velocities α and β, and density ρ. The boundary conditions at the free surface are that the stresses are null:

$$\tau_{rr} = \tau_{r\theta} = 0, \qquad r = a \tag{13.49}$$

Wave equations in spherical coordinates have solutions for potentials Φ and S of the same type as those found for T (13.38). Different spheroidal modes are represented by $_nS_l$.

For the lowest order $l = 0$, according to (13.38), the potentials Φ and S are functions of r only. Then, according to (13.29)–(13.31), there are only displacements u_r, and $u_\theta = u_\phi = 0$. These modes are called radial modes. The boundary condition $\tau_{rr} = 0$ gives the equation (Ben Menahem and Singh, 1981)

$$\cot x = \frac{1}{x} - \frac{1}{4}\left(\frac{\alpha}{\beta}\right)^2 x \tag{13.50}$$

where $x = k_\alpha a$. For $\alpha = \sqrt{3}\beta$ ($\lambda = \mu$), the first roots are $x = 0.816\pi$, 1.929π, 2.936π, and 3.966π, corresponding to harmonics of orders $n = 0, 1, 2$, and 3. If, for an approximation to the Earth, we substitute $a = 6370$ km and $\alpha = 9$ km s^{-1}, the corresponding periods in minutes are 28.91, 12.23, 8.03, and 5.95. The fundamental mode $_0S_0$ corresponds to an expansion and contraction of the sphere without its form changing (Fig. 13.2). For higher values of n, there are as many nodal surfaces inside the sphere as the order number for which the motion is null and changes sign.

$\ell=2$

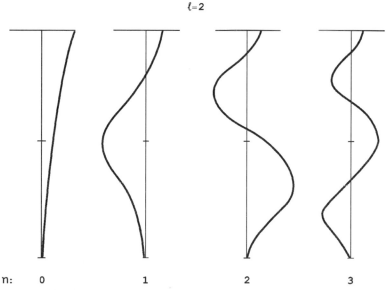

n: 0   1   2   3

Fig. 13.4. The distribution of amplitude with radius for the fundamental ($n = 0$) and higher order harmonics ($n = 1, 2$, and 3) for the toroidal mode of order $l = 2$.

Table 13.2. *Periods (in minutes) of spheroidal modes*

	n			
l	0	1	2	3
0	20.45	10.25	6.62	5.10
1	–	41.30	17.66	11.79
2	53.80	24.5	15.08	8.48
3	35.53	17.72	13.43	8.14

For $l = 1$, the fundamental mode $_0S_1$ does not exist, since it corresponds to a displacement of the center of mass, which is not possible in free vibrations, as we saw for the liquid sphere (Lapwood and Usami, 1981). We have values for the higher modes $_1S_1$, $_2S_1$, $_3S_1$, etc.

For $l = 2$, displacements are symmetric with respect to the plane normal to the origin of the axis θ. The surface takes an ellipsoidal form that is alternately flattened and elongated at the poles (Fig. 13.2). The fundamental mode $_0S_2$ has the lowest frequency of all modes and no nodes of displacements in the interior of the sphere. For higher modes ($n \geq 1$), just like for toroidal modes, there are in the interior of the sphere as many nodal surfaces as the order of the mode.

Eigenperiods in minutes for spheroidal modes of lowest order for a model of the Earth with radial symmetry are given in Table 13.2 (Dziewonski and Gilbert, 1972). The longest period corresponds to the mode $_0S_2$ and is greater than the period of the corresponding toroidal mode $_0T_2$. Thus, this is the longest period for free oscillations of the Earth. For the same order, periods of spheroidal modes are longer than those of toroidal modes.

13.6   Effects on free oscillations

We have considered the properties of the free oscillations of an isotropic homogeneous elastic sphere. If the elastic properties vary with the radius the solutions are similar. With the Earth there are circumstances that deviate from this simple model. First of all, we have the effect of gravity that affects mainly spheroidal modes. Secondly, the presence of a liquid core results in the existence of toroidal modes in the mantle only. This influence is greater for low-order modes. For example, spheroidal modes have periods significantly different than those in models without a liquid core. These two effects do not change the spherical symmetry of the problem.

Three additional effects derive from the ellipticity of the shape of the Earth, its lateral heterogeneity (the dependences of θ and ϕ on the elastic properties), and its rotation. These three factors separate the problem from that of spherical symmetry. The problem is no longer degenerate and the values of the eigenfrequencies are modified. For each value of l and n, there are now several frequencies, depending on the index m with a maximum of $2l + 1$. These frequencies are called multiplets (Fig. 13.5).

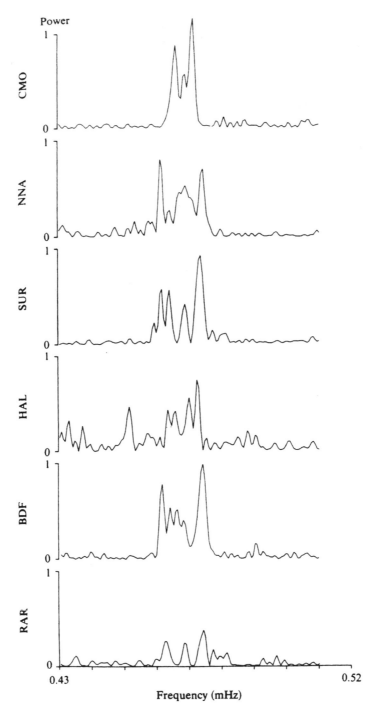

Fig. 13.5. The splitting of the spectral peak of mode $_0S_3$ from observations at six stations of the IDA network of the Indonesia earthquake of 19 August 1977 (Buland *et al.*, 1979) (with permission from Macmillan Magazine Ltd).

The most important of these effects is due to the Earth's rotation. Centrifugal and Coriolis forces have axial symmetry, so the problem no longer has spherical symmetry. This effect produces splitting of the eigenfrequencies (Dahlen, 1980). If the angle ϕ is measured on a plane normal to the axis of the Earth's rotation, a perturbation, due principally to the Coriolis force, displaces the eigenfrequencies associated with modes of angular order l by a quantity related to the index m, $\delta\omega = mb$, where b is a function of the index l and the quotient $\Omega/_n\omega_l$ (Ω is the Earth's frequency of angular rotation and $_n\omega_l$ are the unperturbed eigenfrequencies). Since this quotient is small for high eigenfrequencies, this effect is observed only for low frequencies.

Two other effects are due to the Earth's ellipticity and deviations of elastic properties and density from radial symmetry. These effects are small and also produce splittings of eigenfrequencies. If the lack of radial symmetry in the Earth's composition were very large, then, besides splitting of eigenfrequencies we could have coupling of toroidal and spheroidal modes. Although their effects are small, heterogeneities in the Earth's composition can be detected in the analysis of free oscillations.

13.7 Observations

Free oscillations of the Earth can be observed in the analysis of seismograms of large earthquakes ($M > 7$). In large shocks enough energy is released to generate low-frequency free oscillations that can be observed by means of long-period seismographs and gravity meters. However, the eigenfrequencies of the various modes are not observed directly on seismograms, but must be obtained from the peaks of their power spectra (Appendix 4). The eigenfrequencies of the various modes are found from observations taken from spectral analysis of long-period records using long time windows in order to obtain sufficient precision for peaks of the spectra (Fig. 13.6). To increase the signal-to-noise ratio, methods involving the stacking of several records for the same earthquake are used to correct for phase differences.

The first observations of free oscillations of the Earth were obtained from long-period records of the two large ($M > 8$) earthquakes in Kamchatka in 1952 and Mongolia in 1957, but the earthquakes that provided better data were those in Chile in 1960, the Kurile Islands in 1964, and Alaska in 1964 (Alsop *et al.*, 1961; Benioff *et al.*, 1961). Splittings of spectral peaks due to the rotation of the Earth, especially for low-order modes, were observed even in the earliest studies. Observed values of periods in minutes for low-order toroidal and spheroidal fundamental modes are given in Table 13.3 (Derr, 1969).

Comparison of observed values of eigenperiods of free oscillation and those calculated from theoretical models is used to obtain global models of the Earth's interior. If we compare corresponding values of Table 13.3 with those of Tables 13.1 and 13.2, we find that the model with radial symmetry is not a bad approximation for very-low-order modes. Since the greatest energy for each mode corresponds to a particular depth, they provide information about the structure at such a depth. Thus, mode data may contribute to determination of the structures at depths for which the structure is not well known from body-wave data. For example, mode analysis gives a value of S

Table 13.3. *Observed periods (in minutes) of low-order modes*

l	$_0T_l$	$_0S_l$
2	44.011	53.883
3	28.463	35.559
4	21.739	25.786
5	17.938	19.821
6	15.422	16.065

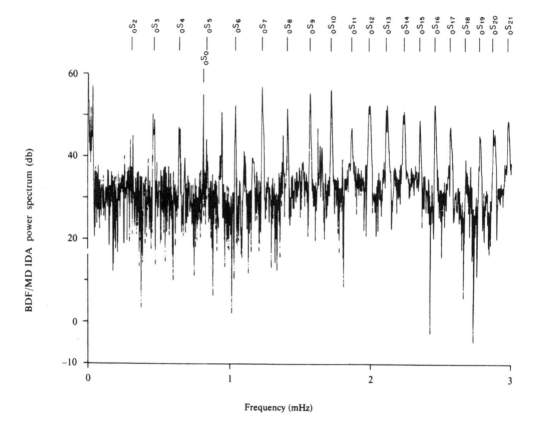

Fig. 13.6. Spectral peaks of spheroidal modes observed at the BDF station of the IDA network (Buland *et al.*, 1979) (with permission from Macmillan Magazine Ltd).

wave velocity of $3.5\,\mathrm{km\,s^{-1}}$ for the rigidity of the inner core. The splitting of spectral peaks due to lateral heterogeneities is also used in mode studies of the Earth's interior. Modern models of the Earth's interior integrate data from body waves, surface waves, and free oscillations covering a wide range of frequencies.

14 ANELASTICITY AND ANISOTROPY

14.1 Anelasticity and damping

In all previous chapters we have considered the mechanical behavior of perfectly elastic bodies. We know, however, that the Earth can not behave in this way. According to the second law of thermodynamics, in all physical deformable media, there is dissipation of energy in the form of heat through mechanisms known as internal friction. As we have seen, in a perfectly elastic medium, the amplitudes of waves decrease with distance due to geometric spreading, which depends on factors of $1/r$ for spherical waves (body waves) and $1/\sqrt{r}$ for cylindrical waves (surface waves) (sections 3.9 and 3.10). The lack of perfect elasticity adds a further decrease in wave amplitude due to anelastic attenuation.

Loss of energy in the Earth due to internal friction is responsible for the attenuation of seismic waves with distance and time. Concrete mechanisms of internal friction are complex and depend on the atomic and molecular structures of crystals in minerals, the presence of small cracks and fractures, and the inclusion of liquids in rocks. Since these mechanisms depend on the nature of the materials through which waves propagate, their effect is called intrinsic attenuation. From the very first studies (Jeffreys, 1957; Lomnitz, 1957) anelastic attenuation of seismic waves in the Earth has been an important subject of seismology (Jackson and Anderson, 1970; Minster, 1980).

14.1.1 Anelasticity

According to Hooke's law, the deformation (strain) of a perfectly elastic body is proportional to the applied stress (2.14). Once stresses are removed, the body instantaneously recovers its initial form. In one dimension, an elastic body may be represented by a spring and the strain–stress relation is given by (Fig. 14.1(a))

$$\sigma = \mu e \tag{14.1}$$

where σ represents the stress, e is the elongation or strain, and μ is the coefficient of elasticity of the spring.

A different type of mechanical behavior corresponds to a viscous body for which, according to Stokes' law, the applied stress is proportional to the time derivative of the strain. In one dimension, a viscous body can be represented by a dashpot with viscosity coefficient η, and the relation between the stress and the strain (Fig. 14.1(b)) is

$$\sigma = \eta \frac{\mathrm{d}e}{\mathrm{d}t} = \eta \dot{e} \tag{14.2}$$

253

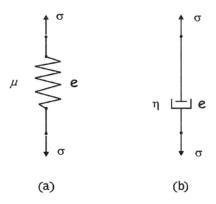

Fig. 14.1. Mechanical models of elastic and viscous bodies: (a) elastic (a spring) and (b) viscous (a dashpot).

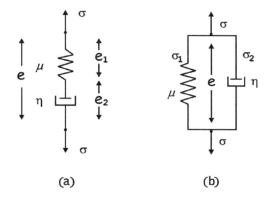

Fig. 14.2. A combination of a spring and dashpots to represent viscoelastic bodies (a) in series – a Maxwell body; and (b) in parallel – a Kelvin–Voigt body.

Imperfectly elastic bodies can be considered as having properties intermediate between those of elastic and viscous bodies, and are called viscoelastic bodies. Their behavior can be understood in terms of one-dimensional mechanical models consisting in combinations of springs and dashpots. A Maxwellian body consists in an elastic (spring) and a viscous (dashpot) element in series (Fig. 14.2(a)). If a stress σ acts on both elements, the deformations in each of them are different and given by

$$\sigma = \mu e_1 \tag{14.3}$$

$$\sigma = \eta \dot{e}_2 \tag{14.4}$$

The total deformation of the system is $e = e_1 + e_2$, and its rate is given by

$$\dot{e} = \frac{\dot{\sigma}}{\mu} + \frac{\sigma}{\eta} \tag{14.5}$$

A Kelvin–Voigt body is represented by an elastic (spring) and a viscous (dashpot) element in parallel (Fig. 14.2(b)). The deformation in the spring and dashpot is the same e, but the stresses acting on each element are different. The total stress is the sum $\sigma = \sigma_1 + \sigma_2$, namely

$$\sigma = \mu e + \eta \dot{e} \tag{14.6}$$

In both models, when a stress is applied, the spring is deformed immediately but the dashpot takes some time to respond. From their combined action it results that the system presents a time delay of the elastic deformation with respect to the applied stress, an inelastic deformation, and a relaxation time for the system to recover its initial state when the stress is removed. The anelastic behavior of the Earth's materials does not agree with either of the two simple models and more complex systems with more than two elements have been proposed. One of them is the standard linear solid consisting in a combination of three elements, two springs and a dashpot.

14.1.2 Harmonic excitation of a Maxwellian body

Certain properties of the response of imperfect elastic bodies may be understood by studying the excitation of a Maxwellian body by a harmonic stress:

$$\sigma = \sigma_0 \sin(\omega t) \tag{14.7}$$

On substituting into equation (14.5) and integrating, we obtain for the deformation at a time t $(0 < t < \pi/\omega)$

$$e(t) = \frac{\omega \sigma_0}{\mu} \int_0^t \cos(\omega \tau) \, d\tau + \frac{\sigma_0}{\eta} \int_0^t \sin(\omega \tau) \, d\tau \tag{14.8}$$

Solving the integrals gives

$$e(t) = \frac{\sigma_0}{\mu} \left\{ \left[1 + \left(\frac{\mu}{\eta \omega} \right)^2 \right]^{1/2} \sin(\omega t + \phi) + \frac{\mu}{\eta \omega} \right\} \tag{14.9}$$

where $\phi = \tan^{-1}[\mu/(\eta \omega)]$. A perfect elastic response (for one spring only) is

$$e(t) = \frac{\sigma_0}{\mu} \sin(\omega t) \tag{14.10}$$

On comparing (14.9) and (14.10) we see the difference between the responses of Maxwellian and pure elastic bodies. This difference can be expressed by a parameter Q defined as

$$\frac{1}{Q} = \frac{\mu}{\eta \omega} \tag{14.11}$$

Equation (14.9) becomes

$$e(t) = \frac{\sigma_0}{\mu} \left[\left(1 + \frac{1}{Q^2} \right)^{1/2} \sin(\omega t + \phi) + \frac{1}{Q} \right] \tag{14.12}$$

where $\phi = \tan^{-1}(1/Q)$. The factor $1/Q$ represents how much the response of a Maxwellian body differs from that of a perfectly elastic one. If it is zero (Q is infinite), the problem reduces to one of perfect elasticity. The presence of the element of viscosity modifies the amplitude of the deformation and introduces a phase shift between the stress and the strain (Fig. 14.3).

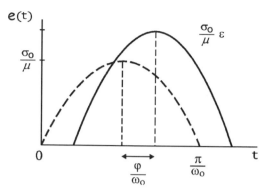

Fig. 14.3. The response of a Maxwell body to a harmonic excitation of frequency ω_0. The dashed line is for an elastic body.

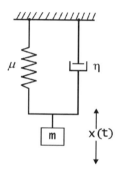

Fig. 14.4. A mechanical model of a system with damped harmonic motion.

14.1.3   *Damped harmonic motion. The Q coefficient*

The effect of anelastic deformation on wave propagation can be understood by considering damped harmonic motion. Let us consider a mass m, suspended by a spring of elasticity coefficient μ, and a dashpot of viscosity η mounted in parallel, in a configuration similar to that of a Kelvin–Voigt body (Fig. 14.4). The equation of motion of the mass is given by

$$\ddot{x} + \frac{\eta}{m}\dot{x} + \frac{\mu}{m}x = 0 \tag{14.13}$$

In a system without damping (one with no dashpot), the motion is harmonic with a frequency $\omega_0 = (\mu/m)^{1/2}$. The damping effect of the dashpot can be represented by the coefficient Q, which now is defined by

$$\frac{1}{Q} = \frac{\eta}{m\omega_0} \tag{14.14}$$

On replacing the values of ω_0 and Q, we rewrite equation (14.13) as

$$\ddot{x} + \frac{\omega_0}{Q}\dot{x} + \omega_0^2 x = 0 \tag{14.15}$$

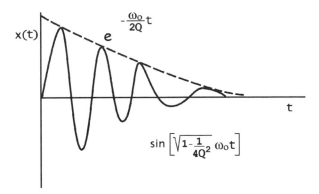

Fig. 14.5. The decrease of amplitude with time in damped harmonic motion.

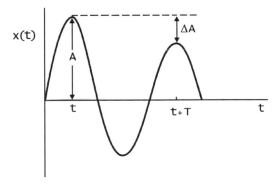

Fig. 14.6. The decrease of amplitude during one period in damped harmonic motion.

The solution of this equation is

$$x(t) = A \exp\left(-\frac{\omega_0}{2Q}t\right) \sin(\omega t + \phi) \tag{14.16}$$

$$\omega = \left(1 - \frac{1}{4Q^2}\right)^{1/2} \omega_0$$

Equation (14.16) represents damped harmonic motion of frequency ω that has been modified with respect to ω_0, the frequency of the undamped system (Fig. 14.5). For $Q = \frac{1}{2}$, the motion is exponentially decreasing with no harmonic component and the system is said to have critical damping. For larger values ($Q > \frac{1}{2}$) the motion is a damped harmonic motion. For smaller values ($Q < \frac{1}{2}$) ω is imaginary and the motion is exponentially decreasing.

In damped harmonic motion, amplitudes decrease exponentially with time. For two times separated by a period $T = 2\pi/\omega$, the ratio of amplitudes according to (14.16) (Fig. 14.6) is given by

$$\frac{x(t)}{x(t+T)} = \exp\left(\frac{\pi}{Q\left(1 - \frac{1}{4Q^2}\right)^{1/2}}\right) \tag{14.17}$$

The logarithm of this quotient is called δ, the logarithmic decrement,

$$\delta = \frac{\pi}{Q\left(1 - \frac{1}{4Q^2}\right)^{1/2}} \simeq \frac{\pi}{Q}, \qquad Q \gg 1 \tag{14.18}$$

If we consider the decrease in amplitude during one period,

$$\Delta A = x(t) - x(t + T) = A(1 - e^{-\delta}) \tag{14.19}$$

then, for large values of Q, according to (14.18), on taking a series expansion of the exponential function, we obtain

$$\frac{1}{\pi}\frac{\Delta A}{A} = \frac{1}{Q} \tag{14.20}$$

This expression defines the coefficient Q from the point of view of damped harmonic motion. The factor $1/Q$ represents the ratio of the decrease in amplitude during one period and the initial amplitude. This definition is valid for values of Q that are sufficiently large with respect to unity (the approximation of (14.18)).

In conclusion, we have seen that the inelastic behavior of a material may be represented by the coefficient Q, which is called the quality factor and can be defined in various ways. For damped harmonic oscillations, $1/Q$ represents the ratio of the decrease in amplitude during one cycle and its initial value. This definition of Q can be used to characterize any type of process in media in which the amplitude of motion decreases due to there being a lack of perfect elasticity.

14.2   Wave attenuation. The quality factor Q

Wave propagation, as we saw in Chapter 3, implies a variation of motion in space and time. Thus, attenuation of wave motion can be considered in time or in space. For a given location wave motion is attenuated with time and, for a given time, it is attenuated with distance. For wave motion, similarly to equation (14.20), we can define the quality factor $Q(\omega)$ as a function of frequency in the form

$$\frac{1}{Q(\omega)} = \frac{1}{2\pi}\frac{\Delta E}{E} \tag{14.21}$$

In this definition, $1/Q$ represents the ratio of the elastic energy ΔE dissipated during one cycle of harmonic motion of frequency ω and the maximum or the mean energy E accumulated during the same cycle.

If we consider a harmonic wave of amplitude A that is attenuated so that, after one period or one wave length, its amplitude is $A\exp(-\pi/Q)$, then, since the energy is proportional to the square of the amplitude (section 3.4), the energy dissipated in one cycle is

$$\Delta E = A^2\left[1 - \exp\left(-\frac{2\pi}{Q}\right)\right] \tag{14.22}$$

On taking the ratio $\Delta E/E$, we obtain equation (14.21) in the same way as we derived equation (14.20). On taking the amplitude ratio $\Delta A/A$, we obtain equation (14.20) and define for wave propagation

$$\frac{1}{Q} = \frac{1}{\pi} \frac{\Delta A}{A} \tag{14.23}$$

Since wave phenomena can be considered as variations in time or in space, the energy dissipated during one cycle can be considered as the dissipation during one period or in one wave length. In this form we define temporal (Q_t) and spatial (Q_s) quality factors. According to (14.21) and (14.23), Q_t represents the wave attenuation with time during one period for a fixed point of space and Q_s represents the attenuation at a given time along a wave-length distance.

Wave attenuation can also be considered by assigning complex values to the frequency and wave number. For a harmonic elastic wave,

$$u(x, t) = A \exp[i(k'x - \omega't)] \tag{14.24}$$

where the wave number and frequency are now complex quantities:

$$k' = k + ik^* \tag{14.25}$$

$$\omega' = \omega - i\omega^* \tag{14.26}$$

Equation (14.24) becomes

$$u(x, t) = A \exp[i(kx - \omega t) - (k^*x + \omega^* t)] \tag{14.27}$$

At a fixed time, amplitudes attenuate with distance; and, at a given distance, amplitudes attenuate with time:

$$u(x) = A\,e^{-k^*x} \cos(kx - \omega t) \tag{14.28}$$

$$u(t) = A\,e^{-\omega^* t} \cos(kx - \omega t) \tag{14.29}$$

According to definitions of Q_t and Q_e we can easily deduce that

$$\frac{1}{Q_t} = \frac{2\omega^*}{\omega} \tag{14.30}$$

$$\frac{1}{Q_e} = \frac{2k^*}{k} \tag{14.31}$$

Since the phase velocity is $c' = \omega'/k'$, it has also a complex value $c' = c + ic^*$. If $\omega^* \ll \omega$ and $k^* \ll k$, the imaginary part of the phase velocity is

$$c^* = c\left(\frac{\omega^*}{\omega} + \frac{k^*}{k}\right) \tag{14.32}$$

In conclusion, Q_t is related to the imaginary part of the frequency and Q_e is related to that of the wave number. A monochromatic wave of frequency ω and phase velocity c travels a distance x in time t and $k^*x = \omega^* t$. Since $c = x/t$, by using (14.30) and (14.31) we obtain that $Q_t = Q_e$. For nondispersed waves the spatial and temporal attenuations are equal.

14.3   The attenuation of body and surface waves

14.3.1   *Body waves*

The attenuation of body waves can be expressed by taking complex values for the velocities of P and S waves, namely $\alpha' = \alpha + i\alpha^*$ and $\beta' = \beta + i\beta^*$. Since the attenuation of body waves is measured from amplitudes at various distances, the imaginary parts of velocities are related to the spatial quality factor Q_e. For P and S waves we can define quality factors in a similar way to (14.31):

$$\frac{1}{Q_\alpha} = \frac{2\alpha^*}{\alpha} \tag{14.33}$$

$$\frac{1}{Q_\beta} = \frac{2\beta^*}{\beta} \tag{14.34}$$

The complex velocities can now be expressed in terms of the corresponding Q factors:

$$\alpha' = \alpha\left(1 + \frac{i}{2Q_\alpha}\right) \tag{14.35}$$

$$\beta' = \beta\left(1 + \frac{i}{2Q_\beta}\right) \tag{14.36}$$

We can also consider complex values for the elasticity coefficients μ and K (the rigidity and bulk modulus), $\mu' = \mu + i\mu^*$ and $K' = K + iK^*$, and define the quality factors Q_μ and Q_K:

$$\frac{1}{Q_\mu} = 2\left(\frac{\mu^*}{\mu}\right)^{1/2} \tag{14.37}$$

$$\frac{1}{Q_K} = 2\left(\frac{K^*}{K}\right)^{1/2} \tag{14.38}$$

According to (2.64) and (2.65), the relations between Q_μ and Q_K and Q_α and Q_β (for values of Q_α and Q_β larger than unity) are

$$\frac{1}{Q_\beta} = \frac{1}{Q_\mu} \tag{14.39}$$

$$\frac{1}{Q_\alpha} = \frac{4}{3}\left(\frac{\beta}{\alpha}\right)^2 \frac{1}{Q_\mu} + \left[1 - \frac{4}{3}\left(\frac{\beta}{\alpha}\right)^2\right]\frac{1}{Q_K} \tag{14.40}$$

In most seismologic problems, it is assumed that there is no dissipation of energy in purely compressive or dilational processes and therefore $Q_K = \infty$. Under this hypothesis, from (14.39) and (14.40), we obtain

$$\frac{1}{Q_\alpha} = \frac{4}{3}\left(\frac{\beta}{\alpha}\right)^2 \frac{1}{Q_\beta} \tag{14.41}$$

If $\sigma = 0.25$, $\alpha = \sqrt{3}\beta$ and the relation gives $Q_\alpha = \frac{9}{4}Q_\beta$.

In the ray theory approximation for body-wave propagation (Chapters 6–8), we are interested in the attenuation along a ray from the focus to the observation point. The

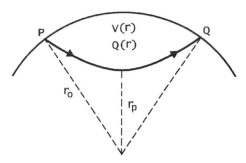

Fig. 14.7. The trajectory of P or S waves inside the Earth with the radial dependences of the velocity $v(r)$ and attenuation $Q(r)$.

attenuation of the amplitude of a monochromatic P wave in the Earth's interior, according to (14.28) and (14.31), is given by

$$A = A_0 \exp\left(-\frac{\omega s}{2\alpha Q_\alpha}\right) = A_0\, e^{-\omega t^*} \tag{14.42}$$

where A and A_0 are the amplitudes at the observation point and the focus, and s is the distance traveled along the ray. For a homogeneous medium $t^* = t/(2Q_\alpha)$, where $t = s/\alpha$ is the traveling time of P waves. A similar relation can be written for S waves using β and Q_β. For a spherical Earth of radial symmetry, $v(r)$ $(\eta(r) = r/v(r))$, the quality factor $Q(r)$ depends also on the radius. Using (8.20) for a ray with a surface focus and a ray parameter p, we obtain (Fig. 14.7)

$$t^* = \int_{r_p}^{r_0} \frac{\eta^2(r)\,dr}{rQ(r)[\eta^2(r) - p^2]^{1/2}} \tag{14.43}$$

If $\bar{Q}$ is the mean value of $Q(r)$ along the ray and t is the traveling time, then an approximation of (14.43) is $t^* = t/(2\bar{Q})$. For the Earth, for surface foci and epicentral distances between 30 and 90 degrees, t^* is practically constant with values of 1 s for P waves and 5 s for S waves. This means that S waves attenuate faster than do P waves.

Since we do not know the amplitude at the focus, the attenuation is usually determined from amplitude ratios of waves observed at different distances. For body waves, we need observations along similar ray paths, so that the attenuation of the amplitude is referred to a certain distance Δs along the ray inside the Earth. In an approximate form, the attenuation with epicentral distance between two stations may be found from

$$\ln\left(\frac{A_2(\omega)}{A_1(\omega)}\right) = \ln C - \gamma(\omega)\,\Delta x \tag{14.44}$$

where $\gamma(\omega)$ is the overall attenuation of amplitude with horizontal distance and C depends on the geometric spreading. According to (14.42), $\gamma = \omega\,\Delta x/(2\bar{\alpha}\bar{Q})$, and $\bar{\alpha}$ and $\bar{Q}$ are average values corresponding to the increment in epicentral distance. Equation (14.44) illustrates the difference between the influences of geometric spreading and anelastic attenuation on the decrease in amplitude with distance.

14.3.2   Surface waves

The anelastic attenuation of surface waves with distance and time can be expressed by using the coefficients γ_e and γ_t:

$$u(x) \simeq A \exp \left[\gamma_e x + i\omega \left(\frac{x}{c} - t \right) \right] \tag{14.45}$$

$$u(t) \simeq A \exp \left[\gamma_t t + i\omega \left(\frac{x}{c} - t \right) \right] \tag{14.46}$$

According to equations (14.28)–(14.31), the attenuation coefficients for each frequency, in terms of the quality factors, are given by

$$\gamma_e = \frac{\omega}{2cQ_e} \tag{14.47}$$

$$\gamma_t = \frac{\omega}{2Q_t} \tag{14.48}$$

For a dispersed wave, energy propagates with the group velocity U. For amplitudes corresponding to instantaneous frequencies, the phase velocity must be replaced by the group velocity (section 12.1). If ω_0 is the instantaneous frequency, then, for given values of x and t, the time that the corresponding wave takes to travel through the medium is $t = x/U$. Then its attenuation in time and space is given by

$$A \exp \left(-\frac{\omega_0 t}{2Q_t} \right) = A \exp \left(-\frac{\omega_0 x}{2UQ_t} \right) \tag{14.49}$$

Since, for a distance x, the amplitude is attenuated by $\gamma_e x$, by substitution of (14.47) into (14.49) we deduce for dispersed waves the relation

$$\frac{1}{Q_t} = \frac{U}{c} \frac{1}{Q_e} \tag{14.50}$$

For dispersed waves, the temporal and spatial quality factors are different. In a homogeneous half-space Rayleigh waves are not dispersed and we can define for them a quality factor Q_R that is related to those of P and S waves:

$$\frac{1}{Q_R} = m \frac{1}{Q_\alpha} + (1 - m) \frac{1}{Q_\beta} \tag{14.51}$$

$$m = \frac{(2 - b)(1 - b)}{(2 - b)(1 - b) - b/[a(1 - a)(2 - 3b)]} \tag{14.52}$$

$$a = (c/\alpha)^2, \qquad b = (c/\beta)^2$$

where c is the velocity of Rayleigh waves. In a layered medium or one with a depth-dependent velocity distribution, Q values for Rayleigh and Love waves depend on the distributions of $Q_\alpha(r)$, $Q_\beta(r)$, $\alpha(r)$, and $\beta(r)$. Because the depth to which surface waves penetrate depends on their wave length, their Q values are also functions of the frequency, $Q_R(\omega)$ and $Q_L(\omega)$.

The attenuation of surface waves can be determined from spectral amplitude ratios of waves of the same frequency at two stations at different distances along the same great circle path relative to the epicenter. For a spherical Earth, according to (10.107),

$$\ln \left(\frac{A_2(\omega)}{A_1(\omega)} \right) = \frac{1}{2} \ln \left(\frac{\sin \Delta_2}{\sin \Delta_1} \right) - \gamma(\omega)\, \Delta x \tag{14.53}$$

where the Δ are angular distances from the epicenter to the stations and Δx is the distance between the stations. The coefficient $\gamma(\omega)$ represents the anelastic attenuation of surface waves along the distance between the two stations for each frequency. From $\gamma(\omega)$ we can find values of Q by using (14.47).

14.4 The attenuation of free oscillations

In Chapter 13, we discussed free oscillations of the Earth considered as a perfectly elastic sphere. If the elasticity is not perfect, the amplitude of each mode of the free oscillations decreases with time due to anelastic attenuation. According to (14.29) and (14.30), the attenuation with time depends on the quality factor Q_t. For a homogeneous Earth, the attenuation of toroidal modes depends on Q_β and that for spheroidal modes depends on a combination of Q_β and Q_α. For an Earth with radial symmetry, Q_α and Q_β are functions of the radius and Q corresponding to free oscillations varies according to the eigenfrequencies corresponding to each mode $Q(_n\omega_1)$. The maximum amplitude for each mode corresponds to a certain depth and thus it is influenced by the Q value corresponding to that depth.

We have seen that free oscillations are detected in peaks of the power spectrum corresponding to the frequencies of each mode. In a perfectly elastic medium, the displacements of each mode are harmonic functions of time and their power spectra are delta functions of eigenfrequencies $\delta(\omega - \omega_n)$ (Fig. 14.8(a)). In the presence of anelasticity, the displacements of each mode are represented by damped harmonic motion:

$$u_n(t) = A_n\, e^{-\gamma_n t + i\omega_n t} \tag{14.54}$$

According to (14.30), $\gamma_n = \omega_n/(2Q)$. The power spectrum of (14.54) is

$$|U(\omega)|^2 = \frac{A_n^2}{\gamma_n^2 + (\omega - \omega_n)^2} \tag{14.55}$$

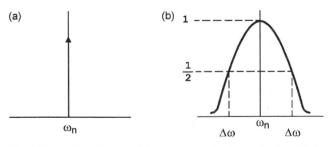

Fig. 14.8. The broadening of the spectral peak of a mode of oscillation due to attenuation.

The spectrum now has a maximum amplitude A_n/γ_n centered at a frequency ω_n and a certain width $\Delta\omega$ (Fig. 14.8(b)). If $\Delta\omega$ is the width of the spectrum when its amplitude is half its maximum value, we can obtain from (14.55) that

$$\frac{1}{Q_t} = \frac{\Delta\omega}{\omega_n} \tag{14.56}$$

Hence Q_t can be found from the width of the spectrum. Since the attenuation increases with $1/Q$, the larger the attenuation the wider the peaks of the spectrum. Broadening of spectral peaks is the main effect of anelasticity on free oscillations. This is a very general effect that applies also to observations in the time domain of propagating pulses in an imperfectly elastic medium. In all cases, the larger the attenuation the wider the pulses become.

14.5   The attenuation of coda waves

Waves observed in the last part of a seismogram are called coda waves. This name seems to have been given by Jeffreys in 1929 to waves arriving after surface waves. Today this name is applied to waves of near earthquakes that arrive after Lg waves with amplitudes decreasing exponentially with time (Aki, 1969) (Fig. 14.9). The attenuation of coda waves has become an important area of research in seismology (Aki and Chouet, 1975; Herraiz and Espinosa, 1987). The attenuation of coda waves is caused by two different effects, anelasticity and scattering of waves due to their inter-action with obstacles or heterogeneities in the medium. In the first case, as we have seen, energy is lost by conversion into heat through internal friction (intrinsic attenuation). In the second, energy is distributed through space so that part of it does not arrive at the observation point. Both factors contribute to the attenuation of amplitudes observed at a given distance relatively near to the focus. For such short distances, waves propagate mainly through the crust and are affected by its anelasticity and heterogeneity. The total attenuation is given by the coda Q factor, Q_c, which includes both effects:

$$\frac{1}{Q_c} = \frac{1}{Q_i} + \frac{1}{Q_s} \tag{14.57}$$

Q_i represents the intrinsic attenuation and is approximately equal to Q_β, indicating that coda waves are principally transverse waves, and Q_s accounts for the attenuation due to scattering phenomena. A formula proposed by Dainty (1981) for Q_s is

$$\frac{1}{Q_s} = \frac{gv}{\omega} \tag{14.58}$$

where v and ω are the velocity and frequency, and $g = \Delta I/(IL)$ is the dispersion coefficient given by the wave's energy I and the fraction of it ΔI that is lost when it crosses a layer of thickness L where heterogeneities are present. Many other models are used to explain coda waves' attenuation. They range from pure diffusion, for which $Q_c = Q_i$, to simple and multiple scattering with complex interaction of waves with obstacles and heterogeneities. In principle, the attenuation of coda waves allows the determination both of the anelasticity (Q_i) and of the heterogeneity (Q_s) of crustal material. However, it is not easy to separate the two effects that contribute to Q_c. The amplitudes of coda

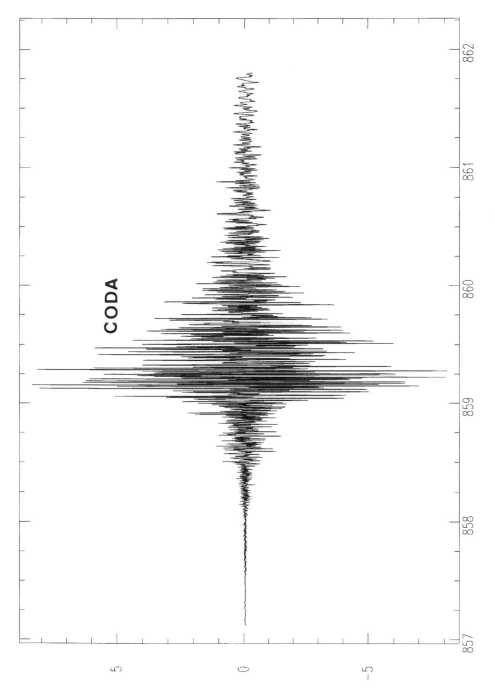

Fig. 14.9. A seismogram of a near earthquake, for which coda waves are shown (the SFUC BB station, for the earthquake of 21 May 1997, $\Delta = 324\,\mathrm{km}$).

waves attenuate with time in the form

$$A(\omega, t) = A_0 \exp\left(-\frac{\omega t}{2Q_c}\right) \tag{14.59}$$

Experimentally it has been found that Q_c varies with frequency in the form

$$Q_c(\omega) = Q_0\left(\frac{\omega}{\omega_0}\right)^n \tag{14.60}$$

where Q_0 is the value corresponding to the reference frequency ω_0. The exponent n varies in the range 0.2–0.4 for high values of Q_0 and is close to unity for low values.

14.6   Attenuation in the Earth

Studies of seismic waves' attenuation lead to the determination of the Q distribution in the interior of the Earth. In the first models with radial symmetry, $Q(r)$, the velocity distribution was kept constant and independent from the attenuation. In the formulation of complex velocities, (14.35) and (14.36), this implies that the real and imaginary parts are considered independently. More recent determinations use simultaneous inversion of attenuation and velocities (Anderson and Archambeau, 1964; Lee and Solomon, 1978). The observations used for determination of the attenuation inside the Earth are those of the amplitudes of body, P, S, PcP, ScS, surface, Rayleigh, and Love waves, and free oscillations.

A widely used model of the distribution of Q inside the Earth is that known as SL8 (Anderson and Hart, 1978) (Fig. 14.10). Its most important features are a lithosphere (0–80 km) with moderate to high values of Q_β in the range 200–500, an upper mantle (80–500 km) with low values of about 110, and a lower mantle (500–2880 km) with a gradual increase from 150 to 500 of Q_β with depth. Q_β is null in the outer core and varies in the range 400–800 in the inner core. Q_K is infinite in the mantle and equal to Q_μ in the inner core. This model is consistent with observations of practically constant values of t^* for distances between 30 and 90 degrees of 1 s for P waves and 5 s for S waves. The attenuation in the Earth's interior is principally due to dissipation of energy in shear motion. The absence of energy loss in compressive processes leads to an infinite value of Q_K in the mantle. However, the attenuation of radial modes of free oscillations seems to indicate that it has a large but finite value. Most seismic data can be explained by invoking relatively simple models of the Q distribution with radial symmetry and independence from frequency. For high frequencies, however, there seems to be such a dependence.

For the lithosphere and upper mantle (0–300 km) there are more detailed models of the distribution of Q with depth. In the crust, values are relatively low; $Q \simeq 160$. Under the crust (50–100 km) values are higher; $Q \simeq 500$. In the astenosphere or low-velocity layer (100–200 km), values are low; $Q \simeq 125$. The values of Q_c are related to the conditions in the upper crust. In the shallowest layers, there are strong variations in Q_c for different regions with values in the range 120–600. Since Q_c depends on the presence of heterogeneities, low values indicate that the crust is very heterogeneous, which can be associated with seismically active zones, whereas high values can be

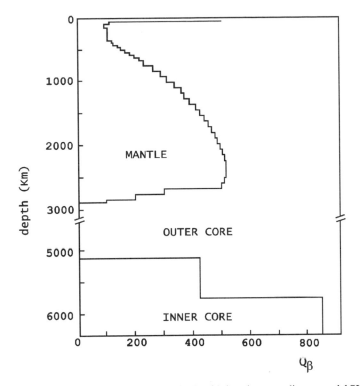

Fig. 14.10. The distribution of Q in the Earth's interior according to model SL8 (Anderson and Hart, 1978) (with permission from the American Geophysical Union).

ascribed to the more homogeneous crust of stable regions. Thus values of Q_c can be correlated to seismicity.

14.7 Anisotropy

In previous chapters we have always assumed isotropy for the material of the Earth. We have seen that this material is not perfectly elastic; we will see now that it is not perfectly isotropic. Consideration of the lack of isotropy or anisotropy in the Earth is becoming an important subject in seismology and it is related to geodynamic processes. Although deviations from the conditions of isotropy are small, their effect on the propagation of seismic waves can be observed and provides important information (Crampin, 1977; Babuska and Cara, 1991). Only the most fundamental ideas about wave propagation in anisotropic media are presented.

For an elastic body the relation between stress and strain is given by Hooke's law (2.14):

$$\tau_{ij} = C_{ijkl}e_{kl} \tag{14.61}$$

As we saw in section 2.2, C_{ijkl}, the tensor of elasticity coefficients, has 21 independent components. For an isotropic body, these are reduced to two (the Lamé coefficients λ

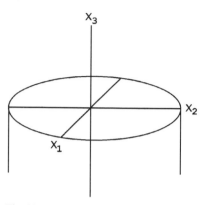

Fig. 14.11. The system of axes in a medium with hexagonal symmetry and the principal axis in the x_3 direction.

and μ) and the tensor is given by (2.16)

$$C_{ijkl} = \lambda\delta_{ij}\delta_{kl} + \mu(\delta_{ik}\delta_{jl} + \delta_{il}\delta_{jk}) \tag{14.62}$$

Thus, it has the following values:

$$C_{1111} = C_{2222} = C_{3333} = \lambda + 2\mu$$

$$C_{1122} = C_{1133} = C_{2233} = \lambda$$

$$C_{1212} = C_{1313} = C_{2323} = \mu$$

Apart from these components and those related to them by symmetry, the components are zero. According to (2.41), the strain energy is given by

$$W = \tfrac{1}{2}\lambda(e_{11} + e_{22} + e_{33})^2 + \mu e_{ij}e_{ij} \tag{14.63}$$

For complete anisotropy without any kind of symmetry, the elastic tensor has 21 independent components. If there is some kind of symmetry this number is reduced. For example, there are nine for orthorhombic symmetry, five for hexagonal symmetry, and three for cubic symmetry.

Hexagonal or cylindrical symmetry is often assumed in seismologic problems. This symmetry has a principal axis and is called transverse symmetry since any direction normal to this axis has the same properties (Fig. 14.11). According to Love (1945), the five independent elastic coefficients for a medium with hexagonal symmetry and its principal axis in the x_3 direction are

$$C_{1111} = C_{2222} = A$$

$$C_{3333} = C$$

$$C_{3311} = C_{3322} = F$$

$$C_{2323} = C_{1313} = L$$

$$C_{1212} = N$$

$$C_{1122} = C_{2211} = A - 2N$$

The strain energy is given by

$$W = \tfrac{1}{2}A(e_{11}^2 + e_{22}^2) + \tfrac{1}{2}Ce_{33}^2 + F(e_{11} + e_{22})e_{33} + (A - 2N)e_{11}e_{22}$$
$$+ \tfrac{1}{2}L(e_{13}^2 + e_{31}^2) + Ne_{12}^2 \tag{14.64}$$

On comparing (14.64) with (14.63) we can see that, even for this relatively simple case of anisotropy, the expression becomes more complicated. With the Earth, there are several situations that may be treated using this type of anisotropy. For example, in finely stratified media with elastic properties alternating from layer to layer, the material behaves as a whole like an anisotropic medium with hexagonal symmetry with its principal axis normal to the layers. Since the properties of the medium normal to that axis are the same, this is also called a transverse isotropic medium. Another case with the same characteristics is that of a material with cracks aligned in a particular direction that constitutes the principal axis of symmetry.

14.8   Wave propagation in anisotropic media

14.8.1   Body waves

Some properties of wave propagation in anisotropic media can be understood by studying hexagonal symmetry or transverse isotropy. Let us consider the case of the principal axis being in the direction of x_3. Monochromatic plane waves propagating in the x_3 and x_1 directions are given by

$$u_i = A_i \sin\left[\omega\left(\frac{x_3}{c} - t\right)\right] \tag{14.65}$$

$$u_i = B_i \sin\left[\omega\left(\frac{x_1}{c} - t\right)\right] \tag{14.66}$$

Using the same procedure as that used in section 3.6 for an isotropic medium, for propagation in the x_3 direction, we substitute (14.65) into the homogeneous equation of motion (equation (2.59) with $\mathbf{F} = 0$) and obtain

$$\left(\frac{1}{\rho}C_{i3k3} - c^2\delta_{ik}\right)A_i = 0 \tag{14.67}$$

For hexagonal symmetry, on substituting for the components of the elastic tensor in terms of the five independent coefficients A, C, F, L, and N, equation (14.67) is given in matrix form by

$$\begin{bmatrix} L/\rho - c^2 & 0 & 0 \\ 0 & L/\rho - c^2 & 0 \\ 0 & 0 & C/\rho - c^2 \end{bmatrix} \begin{bmatrix} A_1 \\ A_2 \\ A_3 \end{bmatrix} = 0 \tag{14.68}$$

Then, there are two velocities, given by $(C/\rho)^{1/2}$ and $(L/\rho)^{1/2}$. The first corresponds to waves with only the A_3 component, and, since this is the direction in which waves propagate, it corresponds to P waves with the velocity $\alpha = (C/\rho)^{1/2}$. The second corresponds to waves with components A_1 and A_2, that is, S waves, and their velocity is $\beta = (L/\rho)^{1/2}$. The displacements are given by

$$u_i^P = (0, 0, A_3) \sin\left[\omega\left(\frac{x_3}{\alpha} - t\right)\right]; \qquad \alpha = (C/\rho)^{1/2} \tag{14.69}$$

$$u_i^S = (A_1, A_2, 0) \sin\left[\omega\left(\frac{x_3}{\beta} - t\right)\right]; \qquad \beta = (L/\rho)^{1/2} \tag{14.70}$$

Wave propagation in this case is similar to that of an isotropic medium.

For propagation in the x_1 direction, on substituting (14.66) into the equation of motion, just like in the previous case, we obtain

$$\left(\frac{1}{\rho} C_{i1k1} - c^2 \delta_{ik}\right) B_i = 0 \tag{14.71}$$

Proceeding as before, we have

$$\begin{bmatrix} A/\rho - c^2 & 0 & 0 \\ 0 & N/\rho - c^2 & 0 \\ 0 & 0 & L/\rho - c^2 \end{bmatrix} \begin{bmatrix} B_1 \\ B_2 \\ B_3 \end{bmatrix} = 0 \tag{14.72}$$

In this case, however, we have three velocities and therefore three different waves. Since waves propagate in the x_1 direction, B_1 corresponds to P waves and their velocity is $\alpha = (A/\rho)^{1/2}$. The displacements with amplitudes B_2 and B_3 are perpendicular to the direction of propagation and correspond to S waves. There are now two different S waves with different velocities, one with a displacement in the x_2 direction and the velocity $\beta_1 = (N/\rho)^{1/2}$ and another with a displacement in the x_3 direction and the velocity $\beta_2 = (L/\rho)^{1/2}$. The displacements of the three waves are

$$u_i^P = (B_1, 0, 0) \sin\left[\omega\left(\frac{x_1}{\alpha} - t\right)\right]; \qquad \alpha = (A/\rho)^{1/2} \tag{14.73}$$

$$u_i^{S1} = (0, B_2, 0) \sin\left[\omega\left(\frac{x_1}{\beta_1} - t\right)\right]; \qquad \beta_1 = (N/\rho)^{1/2} \tag{14.74}$$

$$u_i^{S2} = (0, 0, B_3) \sin\left[\omega\left(\frac{x_1}{\beta_2} - t\right)\right]; \qquad \beta_2 = (L/\rho)^{1/2} \tag{14.75}$$

By comparison of equations (14.69) and (14.70) with (14.73)–(14.75) we can draw the following conclusions. In an anisotropic medium with hexagonal symmetry, P waves propagate with different velocities along the principal axis of symmetry (x_3) and along a direction normal to it (x_1). In the first case there is only one type of S wave and in the second there are two. If x_3 is the vertical direction, then, for waves traveling in the x_1 direction, S1 corresponds to SH and S2 to SV. In this way, SH and SV components propagate with different velocities and are two different waves. This phenomenon is known as S wave splitting, since the SV and SH components arrive with a time delay between them (Fig. 14.12). Owing to the type of symmetry, this phenomenon takes place for any orientation of propagation in the plane normal to the principal axis. Thus, for example, in the Earth S wave splitting occurs during horizontal propagation in media with fine layering of layers with alternating high and low rigidities. In the

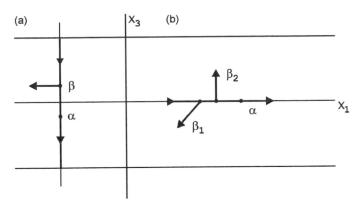

Fig. 14.12. The propagation of P (α) and S (β) waves in a medium with hexagonal symmetry and the principal axis in the x_3 direction: (a) waves in the x_3 direction; and (b) waves in the x_1 direction, SH waves have velocity β_1 and SV waves have velocity β_2.

same medium, P waves propagate with different velocities in the vertical and horizontal directions.

In general, for any type of anisotropy, there are always three types of waves propagating with three different velocities. Choosing the three components of displacement adequately, they are called quasi-P, quasi-SH, and quasi-SV waves. The velocities for these three types of waves change according to the type of symmetry present in the medium. Because of these properties, anisotropy is detected by observations of changes in P wave velocity along two perpendicular directions and by observations of S wave splitting. For both effects it is not necessary that the whole medium be anisotropic; only a certain part of it need be. The time delay between SH and SV components produced in the anisotropic part is preserved while waves travel along the isotropic part.

14.8.2 Surface waves

The main effect of anisotropy on surface waves is that they can not always be separated into Rayleigh and Love waves, unlike in isotropic media. The radial and transverse components are coupled, forming generalized dispersed surface waves that are a combination of Rayleigh and Love waves. To summarize, we can distinguish three effects of anisotropy on the propagation of surface waves. First, there is a discrepancy in the relation between the phase velocities of Rayleigh and Love waves with respect to that of isotropic media. This effect is the reason behind the difficulty in finding a unique model of layered isotropic media that agrees with observations of both types of data. Second, there are discrepancies in phase velocities found for trajectories along different azimuths in the same region. Third, there is a departure of the plane of polarization of Rayleigh waves from vertical orientation.

Because of the relatively long wave lengths of surface waves, the type of anisotropy that affects them is produced by oriented heterogeneities in some preferential directions. If heterogeneities are randomly distributed, the medium as a whole behaves isotropically, whereas if they are consistently oriented, they produce an effect of anisotropy. The distinction between heterogeneity and anisotropy is not always an easy one to make and depends on the relation between the dimensions of heterogeneities and the wave lengths. For anisotropy produced by heterogeneities with some preferential orientation, if the resulting symmetry axis is in a particular horizontal direction, then, along that direction, Rayleigh waves propagate with higher velocities and Love waves with lower ones than in the isotropic case. For a perpendicular trajectory, the effect is opposite, with lower velocities for Rayleigh waves and higher ones for Love waves.

14.9   Anisotropy in the Earth

Observations of the propagation of seismic waves have revealed anisotropy in the material of the Earth's interior. Essentially two types of anisotropy have been observed. The first type has a symmetry with the principal axis in the vertical direction (transverse isotropy) and is due principally to stratifications or horizontal alignments of structural or mineralogic nature. This results in splitting of S waves with delays in traveling times of SV and SH components and different phase velocities for Rayleigh and Love waves depending on their trajectories. The second type is due to preferential alignments of crystals, cracks, or heterogeneities along a particular azimuth (azimuthal anisotropy). This produces effects on the velocities of propagation of waves along trajectories in a particular azimuth in comparison with those perpendicular to them. Some authors distinguish between anisotropy due to stratification and that due to fractures and cracks.

In the shallow part of the crust, anisotropy due to sediment stratification has been observed. A similar effect has also been observed in the lower crust, which has a laminated structure. Another type of anisotropy observed in the crust is due to the presence of cracks. For a determined stress regime, cracks are oriented in the direction of compression and normal to tensional stresses. Since the crust is subject to tectonic stresses, this may be a very general situation (Crampin, 1978). In subcortical oceanic lithosphere, sea floor spreading may also be responsible for anisotropy. The flow of material from oceanic ridges produces a preferential orientation of olivine crystals along flow lines that is preserved through the aging process, resulting in azimuthal anisotropy.

In the upper mantle, the astenosphere under the lithosphere is a region of strong anisotropy related to plastic flow of material that follows the motion of lithospheric plates. Two symmetries may be present, one with a vertical principal axis produced by horizontal flow that results in higher SH than SV velocities and another with azimuthal anisotropy along flow lines with higher seismic velocities. For regions with predominantly vertical flow, the opposite effect is observed, with SV velocities higher than SH velocities. These two effects have been observed in surface wave propagation along different trajectories in the Pacific Ocean. Rayleigh waves propagating along trajectories that coincide with plate-motion directions have higher velocities than do those with perpendicular trajectories. This anisotropy may extend down to 300 km depth and is associated with the orientation of olivine crystals along flow lines. Variations in velocities for different trajectories may be of magnitude 3–10% (Leveque and Cara, 1985; Nishimura and Forsyth, 1989).

For depths below 400 km, observations reveal no appreciable anisotropy and the lower mantle can be considered isotropic. The region of transition between the mantle and the core (the CMB) is considered anisotropic by some authors, but data are not yet sufficient for a definitive conclusion to be drawn. The solid material of the inner core, according to increasing evidence from observations of PKIKP waves and free oscillation modes of low order, is thought to be strongly anisotropic with hexagonal symmetry and a principal axis of symmetry in the direction of the axis of the Earth's rotation (Shearer *et al.*, 1988). The origin of this anisotropy is assigned to the preferential alignment of iron crystals parallel to the axis of rotation.

15 FOCAL PARAMETERS OF EARTHQUAKES

15.1 Earthquakes and faults

The causes of earthquakes have interested man since antiquity. As was mentioned in section 1.1, various ideas have been proposed from the time of the ancient Greek natural philosophers to our days. During the 19th century systematic field studies after earthquakes were started and the first attempts to relate them to tectonic processes were made by Mallet (Naples, Italy, 1857), Koto (Neo, Japan, 1891), and Oldham (Assan, India, 1897) among others. With the increase in number of field observations and in precision of localization of epicenters, the correlation between earthquakes and faults became clearer. Authors such as Suess, Koto, Montessus de Ballore, and Sieberg assigned the cause of earthquakes to stresses accumulated in the Earth's crust by tectonic processes and their release by its fracture. The first mechanical model was presented by Reid (1911) in order to explain the origin of the San Francisco earthquake of 1906. His theory, known as elastic rebound, proposes that earthquakes take place by fracturing of the Earth's crust with the total or partial release of the elastic strain accumulated in a region owing to tectonic stress. According to plate tectonics, which was developed in 1960, tectonic stresses are ultimately related to the relative motion of lithospheric plates.

An earthquake can be considered to be produced by rupturing of part of the Earth's crust with a relative displacement of its two sides and the release of the accumulated elastic strain that had been produced by tectonic processes. The place where earthquakes originate is called the focal region or focus. The parameters that define the focus are those that describe the motion of a fracture or fault. These are the following (Fig. 15.1): The azimuth, ϕ, is the angle between the trace of the fault (the intersection of the fault plane with the horizontal) and North ($0° \leq \phi \leq 360°$); the angle is measured so that the fault plane dips to the right-hand side. The dip, δ, is the angle between the fault plane and the horizontal at a right angle to the trace ($0° \leq \delta \leq 90°$). The slip or rake, λ, is the angle between the direction of relative displacement or slip and the horizontal measured on the fault plane ($-180° \leq \lambda \leq 180°$); λ is negative for normal faults and positive for reverse faults. According to the values of δ and λ, we have different types of faults, namely, for $\delta = 90°$ and $\lambda = 0°$, strike slip faults; for $\delta = 90°$ and $\lambda = 90°$, vertical faults; for $\delta > 0°$, inclined faults for which, $\lambda = 0°$, $180°$ or $-180°$, the motion is horizontal, for $\lambda = 90°$ or $-90°$, the motion is vertical, and, for other values of λ, the motion has vertical and horizontal components, of reverse or normal type according to its sign. The slip or displacement, Δu, is the distance traveled in the relative motion of a point on one side with respect to one on the other. If Δu varies along the fault plane, its mean value is $\Delta \bar{u}$. The area of the fault is S (for a rectangular fault $S = LD$, where L is its length and D is its width; for a circular fault $S = \pi a^2$, where

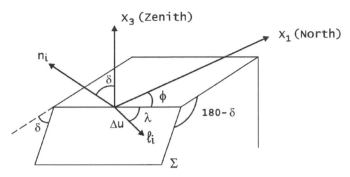

Fig. 15.1. Parameters of the motion on a fault.

a is its radius). Thus, the orientation of the motion of the fault is given by the three angles ϕ, δ, and λ, and its dimensions are given by its area S and mean displacement Δu.

The location of the focus is given by its geographic coordinates, its depth, and the time of occurrence or origin time. Owing to the fault's dimensions, the focal coordinates refer to a certain point, for example, where rupture starts, and the origin time refers to the initiation of faulting. The reduction of the focus to a point is called the point-source approximation. The definitions of the parameters of the focal location used this approximation.

15.2   The location of an earthquake's focus

The concept of an earthquake's point focus from which waves propagate in all directions was introduced by Mallet and is fundamental to the determination of its location and time of origin. The focus at a certain depth is called the hypocenter and its horizontal projection is called the epicenter. The hypocentral parameters are the geographic latitude and longitude of the epicenter (ϕ and λ), the focal depth (h) and the origin time (t_0). The desirability of being able to represent of earthquakes' foci on maps and the difficulty, sometimes, of determining their depth justifies the use of the concept of an epicenter. Since the process of rupturing in earthquakes has dimensions and a duration, the point focus is a simplification. We will see that the point focus has different meanings depending on how it is determined.

Lists and catalogs of earthquakes have been being published since the 17th century. The first epicenter determinations were based on damage and were started in the second half of the 19th century; today they are known as macroseismic determinations. One of the first modern catalogs was published by Mallet in 1858. The first instrumental determinations were started in 1904, when the Central Bureau of the International Association of Seismology started their publication in Strasbourg. In 1911, the Seismological Committee of the British Association for the Advancement of Science started the publication of an earthquake catalog that in 1918 became the International Seismological Summary (ISS). The catalogues of the International Seismological Center (ISC) have been being published since 1963. Other agencies such as the United States Coast and Geodetic Survey (USCGS), later the National Earthquake Information

Center (NEIC) of the United States Geological Survey (USGS), started to publish hypocenter determinations. In many countries there are institutions that perform hypocenter determinations and publish catalogs of local earthquakes.

15.2.1   *Macroseismic determination of epicenter locations*

The first determinations of the locations of foci of earthquakes were based on information regarding damage to buildings, fractures and cracks in the ground, and other effects. Michell in 1760 proposed, for the first time, methods based on this type of observation. However, the first to perform a true determination of an earthquake's focus was Mallet in 1862, for the Naples earthquake of 1857. By using the orientation of the lines along which he thought waves had propagated, on the basis of the orientations of cracks and other evidence, he determined in his own words 'the real position of the place upon the earth's surface vertically above that one beneath, whence the shock emanated' (Davison, 1927). This is, I think, the first clear reference to the concepts of the epicenter and hypocenter. Early work in determination of foci was done by Milne, Omori, and Mercalli, among many others.

Macroseismic epicenter determination is based on field observation of the effects of an earthquake on the ground and on buildings and other structures. From these observations intensities are assigned to each location and an intensity or isoseismal map is drawn, as we will see later (section 15.3). The macroseismic epicenter is located at the central point of maximum intensity (Fig. 15.2). The depth of the focus can also be determined from intensity maps, as we will see later, from the distribution of the zones of various intensities. This method was used until the installation of seismographic instruments at the beginning of this century. The method is still used today for

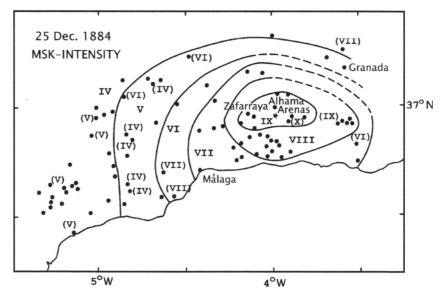

Fig. 15.2. An intensity map of the Andalucian earthquake of 25 December 1884 and the location of the macroseismic epicenter (courtesy of D. Muñoz).

determination of the epicenters of historical earthquakes, that is, those that occurred before the instrumental period. The study of historical earthquakes is an important part of the evaluation of the seismicity of a region. Large earthquakes may be separated by long periods of time, so it is necessary to extend the study as far into the past as possible. These studies require a search for historical documents and a careful evaluation of the information about damage produced by earthquakes in order to draw reliable intensity maps and estimate their locations and sizes.

The concept of the macroseismic epicenter is not equivalent to instrumental determination based on the times of arrival of seismic waves. An instrumentally determined hypocenter corresponds to the point and time of initiation of the fracture process whereas a macroseismic epicenter is situated at the central point of the zone of maximum damage. This point may represent, in an approximate way, the projection onto the surface of the central point on the fracture plane. In this sense, it may approximate the horizontal projection of the centroid of the fault area. This concept was introduced in some methods for the determination of the focal mechanism (section 19.4).

15.2.2 Graphical methods

Instrumental hypocenter determinations are based on the arrival times of seismic waves, mainly P and S waves, recorded on seismograms at stations distributed around the epicenter. The first methods were graphical ones using a map for short distances and a globe for larger or teleseismic distances. Graphical methods determine the location of the epicenter, its origination time, and an estimation of its focal depth.

For short distances ($\Delta < 1000\,\mathrm{km}$), the problem is solved on a map on which we have situated the positions of seismographic stations. The method consists in an iterative graphical process of successive approximations. Let us consider N stations ($N > 4$) where we have measured the arrival times of P and S waves of a local earthquake, namely, t_i^P and t_i^S, $i = 1, \ldots, N$. It is necessary to have travel times tables or curves for these waves. We start by determining the time interval between arrivals of P and S waves at one or several stations ($\delta t = t^S - t^P$). From the travel times curve, this interval (δt) at one station gives us the distance Δ and traveling time of P waves (t'^P) corresponding to that station (Fig. 15.3(a)). The origin time is determined by subtracting the travel time from the arrival time, $t_0 = t^P - t'^P$. From this value of the origin time, we obtain the travel times for all of the stations $t_i'^P = t_i^P - t_0$ and, from them, using travel times curves or tables, we determine the distances Δ_i for all of the stations. Using these distances (Δ_i) as radii, we draw circles with their centers at the positions of all of the stations. In the absence of errors, these circles will cross at a point that corresponds to the epicenter whose coordinates (ϕ_0 and λ_0) are found from the map (Fig. 15.3(b)). If the circles do not cross at a point, the origin time t_0 is changed a little and the process is repeated until all or most circles nearly cross at one point.

Figure 15.3 shows an ideal example using five stations. In general, all the arcs of circles can not be made to cross at a single point, but their intersection defines a small area that reflects errors present in determinations of arrival times and travel times curves. The epicenter is located at the center point of this area. If the focus is at a certain depth and we have used travel times for a shallow focus, the circles will not cross at a point. In this

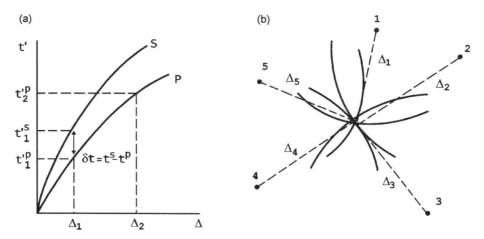

Fig. 15.3. Graphical determination of the origin time and epicenter. (a) Determination of epicentral distance from the S − P time interval $(\delta t = t^S − t^P)$ and of the distance Δ_2 from the P wave travel time $t_2'^P$. (b) Location of the epicenter from the intersection of the epicentral distances at five stations.

case, we have to repeat the process with travel times of different depths (e.g. 20, 40, 80 km, etc.) until a good fit is found. In this way, the depth of focus is also determined. We need a minimum of three stations in order to determine the epicenter and four for adjusting its depth. For large distances, the method is similar but a terrestrial globe is used for the stations' locations. In this case, the depth of the focus can be estimated first using the time interval between the arrivals of the pP and P rays (section 9.5). Determination of focal depths by graphical methods are not very exact and, in many cases, they indicate only whether the focus is shallow, intermediate, or deep. In general, if the stations do not surround the epicenter, determinations are not very precise.

15.2.3 *Numerical methods*

The theoretical basis of numerical methods for hypocenter determination was developed relatively early with the work of Geiger (1910) and Inglada (1928). Their application, however, was not generalized until the development of digital computers around 1960 made fast resolution of the problem for a large number of observations possible. The basis of all methods consists in the linearization of equations representing arrival times and their solution using least squares methods. Various methods have been developed by many authors; we will present only the fundamental ideas common to many of them. The first algorithms and computer programs for hypocenter determination were proposed around 1960 (Bolt, 1960). Among those developed for regional distances, a widely used one is that known by the name of HYPO, of which there are several updated versions (Lee and Lahr, 1971).

The arrival times of seismic waves t_i (generally P and S waves, though other phases may also be used) are recorded at N stations $(N > 4)$ with geographic coordinates ϕ_i and λ_i. The arrival times can be considered as nonlinear functions of the coordinates of stations (ϕ_i and λ_i), focal parameters (coordinates, depth, and origination time) (ϕ_0, λ_0, h, and t_0), and the distribution of velocities of seismic waves in the Earth's interior. The problem can be linearized by using a Taylor expansion about a set of approximate initial values of the focal parameters ($\phi_0', \lambda_0', h_0'$, and t_0') thought to be sufficiently near the real ones that we can write

$$t_i = t_i' + \delta t + \frac{\partial t_i}{\partial \phi} \delta\phi + \frac{\partial t_i}{\partial \lambda} \delta\lambda + \frac{\partial t_i}{\partial h} \delta h; \qquad i = 1, \ldots, N \tag{15.1}$$

where t_i' are the arrival times at each station calculated from the initial solutions ($\phi_0', \lambda_0', h_0'$, and t_0') and partial derivatives are evaluated for that solution. We define residuals as the differences between observed and calculated arrival times for each station:

$$r_i = t_i - t_i'; \qquad i = 1, \ldots, N \tag{15.2}$$

By substituting (15.2) into (15.1) and expressing the N equations in matrix form, we obtain

$$r_i = A_{ij}\,\delta x_j; \qquad i = 1, \ldots, N; \; j = 1, \ldots, 4$$
$$r = \mathbf{A}\,\delta x \tag{15.3}$$

The matrix $\mathbf{A}$ ($4 \times N$) is formed by taking the partial derivatives of traveling times for each station with respect to the coordinates of the epicenter, depth, and origin time, which are calculated from travel times curves or tables. The four components of the vector δx formed by the increments of the four focal parameters are the four unknowns of the problem. The method is iterative. From the first initial values of focal parameters, by solving equation (15.3), we obtain the first set of increments δx that we add to the initial values to give us new values of the parameters. These new values are used again as initial values and the process is achieved when the increments become small, that is, of the order of errors in observations, or when the overall error has a minimum value. Since the problem is overdetermined, the system (15.3) consists in N equations with four unknowns. For its solution we use a least squares method that minimizes the sum of the squares of the residuals:

$$\varepsilon^2 = \frac{1}{N} \sum_{i=1}^{N} r_i^2$$

Several methods may be used for the solution of equation (15.3). Solutions give minimum-error estimates of the parameters and of their standard deviations. A solution can be obtained by multiplying (15.3) from the left-hand side by the transpose of $\mathbf{A}$ and then finding the inverse of the (4×4) matrix $\mathbf{A}^T\mathbf{A}$:

$$\delta x = (\mathbf{A}^T\mathbf{A})^{-1}\mathbf{A}^T r \tag{15.4}$$

Another solution can be found by using the generalized inverse matrix (Lanczos, 1957). As we saw in section 11.3, the matrix $\mathbf{A}$ can be factored as

$$\mathbf{A} = \mathbf{U}\mathbf{\Lambda}\mathbf{V}^T \tag{15.5}$$

where Λ is a diagonal matrix formed by the square roots of the eigenvalues of $A^T A$, V is a matrix formed by the eigenvectors of $A^T A$, and U is a matrix formed by the eigenvectors of AA^T. The generalized inverse of A is given by

$$A^{-1} = U^T \Lambda^{-1} V \tag{15.6}$$

Thus the solution of (15.3) is

$$\delta x = U^T \Lambda^{-1} V r \tag{15.7}$$

From the matrices U and V, we can form the covariance matrix $C = V\Lambda^{-2}V^T$, whose diagonal elements are the variances of the parameters, the resolution matrix $R = VV^T$, whose elements indicate the relative resolution of each parameter, and the information density matrix $D = UU^T$, whose elements indicate which observations contribute the most information to the problem.

If the matrix A is nearly singular, the problem becomes unstable. One way to avoid this is to introduce an attenuation factor by replacing the matrix $A^T A$ in (15.4) by $A^T A + kI$, where k has a small value, before determining the inverse (Marquardt, 1963). In this form we eliminate the occurrence on the diagonal of elements with values near to zero. In the case of the generalized inverse (15.7), the problem manifests itself in the presence of very small or zero eigenvalues. The matrix Λ is replaced by $\Lambda + kI$, so that eigenvalues near to zero become finite.

Several methods are also used in order to speed up the convergence, such as the introduction of a weighting procedure for observations, use of a scaling factor, and centering. A scaling factor is introduced so that the columns of the matrix A have norm unity. Centering consists in replacing the origin time by the weighted mean of the other terms, so that the matrix A satisfies the condition of being centered. These procedures are commonly used in statistical regression problems.

15.2.4 *Joint hypocenter determination*

Often the problem of hypocenter determination is not well conditioned, even though the number of equations is greater than the number of unknowns. Some of the equations need not be really independent and in many cases the origin time and depth of focus are coupled together. Also, travel times are deduced from average tables or simplified models of the Earth's velocity distribution and do not correspond to real values. For regional distances, heterogeneities of the crust and upper mantle may affect hypocenter determinations. One solution is to introduce station corrections into the equations, but this is not easy insofar as they depend on ray paths and must be known *a priori*.

A solution for this problem is given by the joint hypocenter determination (JHD), that is, the determination of groups of hypocenters using the same group of stations. The method, proposed by Douglas (1967) and developed by Dewey (1972), basically consists in joint determination of hypocenters of M earthquakes using observations from the same set of N stations, introducing as new unknowns N station corrections, which are the same for all events. For simplicity, let us consider that at each station we read only the arrival times of P waves for all shocks. In this case we have NM equations and $4M$ parameters of the M hypocenters. If we add N station corrections, the number of unknowns is $4M + N$. The method uses one earthquake, usually the largest and best-recorded one, as a calibration event for which the solution is supposed to be known previously and the number of unknowns is $4(M - 1) + N$. The linearized equations (15.3) are

$$r_{jk} = A_{skj}\,\delta x_{sk} + \delta g_j, \qquad j = 1, \ldots, N, \quad k = 1, \ldots, M, \quad s = 1, \ldots, 4 \qquad (15.8)$$

where the subindex j refers to stations, k refers to events and s refers to hypocenter parameters. The δg_j terms are station corrections or anomalies in traveling times for each station. The unknowns are four increments δx_{sk} of the parameters of each event (a total of $4M$) and N corrections δg_j for each station. The method is used for earthquakes that are relatively near to each other, for example, aftershocks or swarms of earthquakes, so that the same station corrections can be used for all shocks. Sometimes, this condition may render the problem unstable, since time corrections and traveling times may become linearly dependent. Since the number of equations is greater than the number of unknowns, the problem is solved by a least squares method, just like the normal hypocenter determination.

15.3 Seismic intensity

The first way to describe the size of an earthquake is in terms of its intensity on the basis of observations on the Earth's surface of damage to buildings and other structures and ground effects such as fractures, cracks and landslides. Traditionally intensity is represented by degrees given by Roman numbers, using scales on which each degree is defined in a descriptive way. Although intensity applies directly to the degree to which an earthquake is felt at a particular location, it can be used also to designate the size of an earthquake. For this purpose the maximum intensity I_{max} or epicentral intensity I_0 is used. These two concepts are generally considered equivalent. This need not always be correct. For offshore earthquakes the maximum damage is on the coast, so I_{max} does not correspond to I_0. Instrumental epicenters, in some cases, do not correspond to the region of maximum damage or maximum felt intensity and hence these two values are different.

The first attempt to classify the damage caused by an earthquake was that by D. Pignataro in Italy in 1783. The first scales of intensity were developed by de Rossi and Forel, in Italy and Switzerland, who together proposed in 1883 the Rossi–Forel scale divided into ten degrees represented by the Roman numbers I–X. A modification of this scale was proposed by Mercalli in 1902, first with ten degrees and later, after a proposal of Cancani, with twelve (I–XII). This scale has served as the basis for later scales. In North America, a scale named the Modified Mercalli (MM) scale, was proposed by Wood and Newman in 1931 and revised by Richter in 1956. In Europe, the most commonly used scale is the MSK scale published by Medvedev, Sponheuer, and Karnik in 1967. The two last scales are practically equivalent. The MSK scale has been updated to the European Macroseismic Scale 1992 (Grünthal, 1993) (Table 15.1).

The assignment of degrees of intensity from field observations after an earthquake is not free from a certain amount of subjectivity. Although descriptions of degrees of

Table 15.1. *The European Macroseismic Scale 1992 (an updated version of the MSK scale)* *(abridged from Grünthal (1993))*

Type of structure: masonry (five classes from rubble stone to reinforced brick), reinforced concrete (RC) (four classes from RC without antiseismic design (ASD) to RC with a high level of ASD), and wood

Vulnerability (classes A–F): A, rubble stone, adobe; B, stone, unreinforced brick; C, brick with RC floors, RC without ASD; D, reinforced brick, RC with minimum ASD; E, RC with moderate ASD; and F, RC with high ASD (ranges of vulnerability are given for each type of structure)

Classification of damage: (grades 1–5): 1, negligible to slight (no structural damage); 2, moderate (slight structural, moderate nonstructural); 3, substantial to heavy damage (moderate structural, heavy nonstructural); 4, very heavy (heavy structural, very heavy nonstructural); and 5, destruction (very heavy structural, near or total collapse)

Effects: (a) on humans, (b) on objects and nature, and (c) on buildings

Degrees of intensity

I   Not felt. (a) Not felt.

II   Scarcely felt. (a) Felt by very few.

III   Weak. (a) Felt indoors by a few. (b) Hanging objects swing.

IV   Largely observed. (a) Felt indoors by many, outdoors by a few. (b) Doors and glasses rattle, furniture shakes.

V   Strong. (a) Felt indoors by most, outdoors by a few, strong shaking, people awake. (b) Objects swing, some fall down, doors open or shut, window panes break. (c) Grade 1 damage to a few buildings.

VI   Slightly damaging. (a) Felt by most indoors and many outdoors, many people frightened. (b) Small objects fall, furniture shifts, glassware breaks, animals frightened. (c) Grade 1 damage to many buildings, grade 2 damage to a few.

VII   Damaging. (a) Most people frightened, find it difficult to stand. (b) Furniture shifted and overturned, objects fall, water splashes. (c) Many buildings of class B and a few of C suffer grade 2 damage, many buildings of Class A suffer grade 4 damage, especially to their upper parts.

VIII   Heavily damaging. (a) Many find it difficult to stand. (b) Furniture overturned, objects fall, tombstones displaced or overturned, waves seen on soft ground. (c) Many class C buildings suffer grade 2 damage, many class B and a few class C buildings suffer grade 3 damage, many class A and a few class B buildings suffer grade 4 damage, a few class A buildings suffer grade 5 damage, a few class D buildings suffer grade 2 damage.

IX   Destructive. (a) General panic, people thrown to the ground. (b) Many monuments and columns fall or are twisted. Waves seen on soft ground. (c) Many class C buildings suffer grade 3 damage, many class B and a few class C buildings suffer grade 4 damage, many class A and a few class B buildings suffer grade 5 damage, many class D buildings suffer grade 2 damage and a few suffer grade 3 damage, a few class E buildings suffer grade 2 damage.

X   Very destructive. (c) Many class C buildings suffer grade 4 damage, most class A, many class B and a few class C buildings suffer grade 5 damage, many class D buildings suffer grade 3 damage and a few suffer grade 4 damage, many class E buildings suffer grade 2 damage and a few suffer grade 3 damage, a few class F buildings suffer grade 2 damage.

Table 15.1. (*cont.*)

XI   Devastating. (c) Most class C buildings suffer grade 4 damage, most class B and many
     class C buildings suffer grade 5 damage, many class D buildings suffer grade 4 damage
     and a few suffer grade 5 damage, many class E buildings suffer grade 3 damage and a few
     suffer grade 4 damage, many class F buildings suffer grade 2 damage and a few suffer
     grade 3 damage.
XII Completely devastating. (c) Practically all structures above and below ground are
     destroyed.

intensity for the scales are well defined, the same situation may be accorded different
degrees by different observers.

15.3.1   *Isoseismal or intensity maps*

Isoseismal or intensity maps are drawn from observed intensities with lines
separating regions of different degrees on a map (Fig. 15.2). The first such map seems
to have been drawn by P. N. C. Egen for the earthquake of 1828 in the Netherlands,
using his own scale of six degrees (Davison, 1927). Intensity maps, despite their lack
of precision, are a very important means of establishing distributions of ground vibration levels due to earthquakes. These maps have a great deal of information regarding
the extent and intensity of shaking of the ground, and the response of buildings and
other structures. From the point of view of intensity, the size of an earthquake depends
not only on its maximum value but also on the extents of the areas with various degrees
of intensity.

The distribution of intensity on the Earth's surface shown on isoseismal maps depends
not only on the size of an earthquake but also on its focal depth and the attenuation of
the shaking of the ground with distance. If the epicentral intensity is I_0, the intensity I, at
a certain distance Δ, can be expressed by writing

$$I = I_0 - a \log \left(\frac{1}{h} (\Delta^2 + h^2)^{1/2} \right) - b \left((\Delta^2 + h^2)^{1/2} - h \right) \tag{15.9}$$

where h is the focal depth, a is a coefficient related to geometric spreading (section 7.9),
and b is related to the anelastic attenuation (section 14.3). Thus, from the intensity map
of an earthquake we can estimate its size given by I_{max} or I_0, the macroseismic epicenter,
depth of focus, and values of the coefficients a and b, which give information on how
intensities are attenuated in the region near the epicenter (the near field) (Fig. 15.4).

Despite the lack of precision, the information provided by isoseismal maps is very
important and complementary to that provided by analysis of instrumentally recorded
seismic waves. For historical earthquakes, this is the only information available. Even
for recent earthquakes, this information is very important, especially from the point
of view of engineering. The study of earthquakes does not end with the analysis of the
shaking of the ground but rather extends also to consideration of damage to buildings
and the responses of persons affected by them.

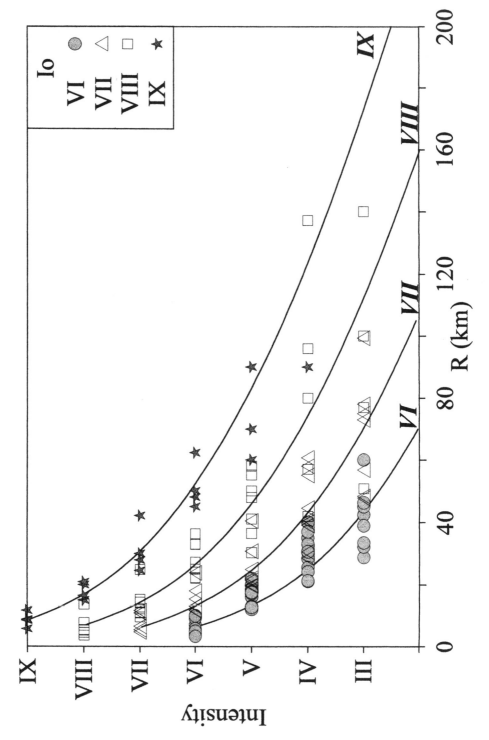

Fig. 15.4. The attenuation of the intensity with distance for earthquakes in the Iberian peninsula (courtesy of C. Lopez Casado).

15.4 Magnitude

Intensity, by its very definition, is an indirect measure of the size of an earthquake. A very shallow earthquake can produce very high intensities in a limited region although its size need not be large. For this reason, the maximum intensity is not always a good indication of the size of an earthquake. Measurement of the size of an earthquake must be done in terms of the energy released at its focus, independently of the damage caused. Since earthquakes are caused by fracturing of crustal material, quantification of their size must represent in some way the energy released by such a phenomenon.

15.4.1 *Scales of magnitude*

The idea of measuring the size of an earthquake by means of an instrumental estimation of the energy released at the focus led Richter (1935) to the creation of the first scale of magnitude. The concept of magnitude is based on the fact that amplitudes of seismic waves depend on the energy released at the focus after it has been corrected for their attenuation during their propagation. Using observations of the amplitudes of waves of earthquakes in California at seismographic stations for regional distances ($\Delta < 600 \, \text{km}$), Richter defined the magnitude M in the form

$$M = \log A - \log A_0 \tag{15.10}$$

where A is the maximum amplitude of waves measured on a seismogram in mm (usually Lg waves) and A_0 is a calibration function that depends on distance. Richter defined his scale using records of a particular instrument, the Wood–Anderson torsion seismograph (with amplification 2800 and period 0.85 s). A_0 corresponds to the amplitude that would be recorded at a given distance for an earthquake of magnitude $M = 0$ (Table 15.2). Calibration of the scale was achieved by assigning the value $M = 3$ to an earthquake that, at a distance of 100 km, is recorded by a Wood–Anderson seismograph with a maximum amplitude of $A = 1 \, \text{mm}$ ($\log A_0 = -3$, for $\Delta = 100 \, \text{km}$). This definition is applicable only to surface earthquakes at regional distances and in its original or modified form it is today known as the local magnitude M_L.

Richter's magnitude for local earthquakes (15.10) can be reformulated in terms of ground motion measured by any type of seismograph with a period near to 1 s by

Table 15.2. *The calibration term in Richter's magnitude*

Δ (km)	$-\log A_0$	Δ (km)	$-\log A_0$
10	1.5	150	3.3
20	1.7	200	3.5
30	2.1	300	4.0
40	2.4	400	4.5
50	2.6	500	4.7
100	3.0	600	4.9

Table 15.3. *The calibration term of the magnitude* m_b *(P_Z) for a shallow shock*

Δ (degrees)	σ	Δ (degrees)	σ
20	6.00	50	6.85
25	6.45	60	6.90
30	6.65	70	6.95
35	6.70	80	6.90
40	6.70	90	7.00
45	6.80	100	7.40

using the formula

$$M_L = \log A + 2.56 \log \Delta - 1.67 \tag{15.11}$$

where A is the maximum amplitude of ground motion (usually corresponding to Lg waves) in micrometers, that is, the amplitude corrected for the instrumental amplification and Δ is the distance in kilometers ($\Delta < 600$ km).

The extension of the definition of magnitude to earthquakes at large distances (teleseisms) ($\Delta > 600$ km) was done by Gutenberg and Richter between 1936 and 1956. Two scales were defined in terms of ground motion recorded at a distance Δ, one for body waves (generally P waves) and the other for surface waves (Rayleigh waves) (Gutenberg and Richter, 1942, 1956). The first is given by

$$m_b = \log(A/T) + \sigma(\Delta, h) \tag{15.12}$$

where A is the amplitude of the ground motion of body waves (corrected for the instrument's response), T is the period, and $\sigma(\Delta, h)$ is a calibration term that depends on the distance and focal depth (Table 15.3). Remember that the energy propagated in a wave is proportional to the square of A/T (section 3.4). Generally, A is measured as the maximum amplitude of the P wave group on seismograms of short periods ($T \approx 1$ s), namely vertical component instruments.

For surface waves, the magnitude scale is valid only for surface earthquakes at distances greater than 15°. The formula has the general form

$$M_S = \log(A/T) + \alpha \log \Delta + \beta \tag{15.13}$$

where A is the maximum amplitude in micrometers of ground motion of Rayleigh waves, T is the period (approximately 20 s), Δ is the distance from the epicenter in degrees, and α and β are two calibration constants. The original values given to these constants by Gutenberg (1945) are $\alpha = 1.656$ and $\beta = 1.818$. In 1964, the IASPEI adopted the values $\alpha = 1.66$ and $\beta = 3.3$ (Vaněk *et al.*, 1962).

For earthquakes for which there are no instrumental records, but only macroseismic information concerning the damage they caused (historical earthquakes), magnitudes can be estimated from the epicentral intensity I_0. A relation proposed by Sponheuer (1960) is

$$M = 0.661 I_0 + 1.7 \log h - 1.4 \tag{15.14}$$

There are other magnitude scales based on amplitudes of certain waves. One introduced by Nuttli (1974), which is valid for regional distances, uses the maximum amplitude of Lg waves (section 12.6). These waves are the maximum recorded amplitudes on short-period seismograms for continental paths at regional distances:

$$M_{\text{Lg}} = \log A + 0.83 \log \Delta + \gamma(\Delta - 0.09)\log e + 3.81 \tag{15.15}$$

where A is the amplitude of the ground motion of Lg waves in micrometers, Δ is the distance in degrees, and γ is an attenuation coefficient that is different for each region (for the central USA, $\gamma = 0.07$; for California, $\gamma = 0.53$).

For near earthquakes ($\Delta < 200$ km), owing to the large magnification of modern seismographs, even relatively small earthquakes saturate records and maximum amplitudes can not be measured. This situation has led to a scale of magnitude based on the duration of a seismic signal instead of its amplitude. The first attempt to use the duration to determine magnitudes was by Bisztricsany in 1958. The magnitudes for local earthquakes based on durations have formulas such as

$$M_\tau = a \log \tau - b + c\Delta \tag{15.16}$$

where τ is the duration of the earthquake signal in seconds and the constants a, b, and c are adjusted so that values of M_τ correspond to those of M_{L}. For California these constants are $a = 2.2$, $b = 0.87$, and $c = 0.0035$ (Lee *et al.*, 1972).

A different type of magnitude scale was introduced by Kanamori (1977) in order to avoid the problem of saturation that afflicts all other scales (Hanks and Kanamori, 1979). This scale is based on the determination of the seismic moment and is called the moment magnitude scale:

$$M_{\text{W}} = \tfrac{2}{3}\log M_0 - 10.7 \tag{15.17}$$

where M_0 is the scalar seismic moment that will be defined later, which is determined from amplitude spectra at low frequencies or observations of fault areas and slippage.

15.4.2 The saturation of magnitude scales

Most magnitude scales depend on the frequency of the waves used for their determination. For this reason it is not possible to define a single scale that is valid for the whole range of observed magnitudes (approximately from -1 to 9). Since the definition of the two teleseismic scales m_b and M_{S}, it has been observed that they coincide only for values of about 6.5. For smaller magnitudes m_b is larger and for greater ones M_{S} is larger (Fig. 15.5). The relation between the two magnitudes established by Gutenberg and Richter (1956) is

$$m_b = 0.63M_{\text{S}} + 2.5 \tag{15.18}$$

This indicates that the size of small earthquakes ($M < 6.5$) is better measured by m_b and that of large ones ($M > 6.5$) by M_{S}. This is an example of the saturation of the scales. The scale for m_b becomes saturated at about 6.5 and larger earthquakes do not give greater values. The scale for M_{S} that underestimates the size of small earthquakes ($M < 6.5$) behaves well for those in the range 6.5–8, but saturates above that value.

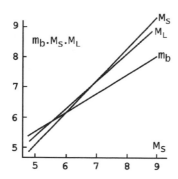

Fig. 15.5. The relation among the magnitudes M_S, m_b, and M_L.

This phenomenon is due to the fact that the amplitude spectrum is displaced toward low frequencies with increasing size of earthquakes.

As we have mentioned, the M_S scale is saturated at about $M = 8$ and sizes of very large earthquakes are not measured well. These earthquakes produce fractures hundreds of kilometers long with displacements of several meters and waves of 20 s are not representative of the energy radiated. This problem is solved with Kanamori's moment magnitude M_W. This scale does not depend on frequency and can be used for the whole range of sizes of earthquakes from very small to very large up to values of about 9.5. However, its determination is not so simple as are those of other scales (direct measurement of an amplitude or duration), since it requires the determination of the seismic moment, from spectra of seismic waves or other methods.

15.5   Seismic energy

The first reference to the energy produced by an earthquake was made by Bassani in 1895 in the study of the Florence earthquake of the same year. Later Reid, Galitzin and Navarro-Neumann between 1911 and 1916 estimated the energies of some large earthquakes. The energy propagated by seismic waves is proportional to the square of their amplitudes (section 3.4) and, thus, the magnitude is proportional to the logarithm of the energy. Gutenberg and Richter (1942, 1956) established the first empirical relations between the magnitude and the energy:

$$\log E_S = 2.4 m_b - 1.3 \tag{15.19}$$
$$\log E_S = 1.5 M_S + 4.2 \tag{15.20}$$

where E_S in joules is the energy propagated in seismic waves, which is often called the seismic energy. According to (15.20), an earthquake of $M_S = 8$ has a seismic energy of 10^{18} J (10^{25} erg). For comparison, a nuclear explosion of 5 megatons (Amchitka, Alaska 1971) has an energy of 10^{16} J, and is equivalent to an earthquake of magnitude 6.7. If we calculate the approximate energy of all earthquakes that happen in 1 year we obtain a value in the range 10^{18}–10^{19} J. About 90% of this energy corresponds to earthquakes with magnitudes equal to and larger than 7. This energy is approximately equal to the global energy consumption in 1 year.

The total energy E released by an earthquake is the sum of the seismic energy E_S and the energy E_R dissipated by inelastic phenomena and in the form of heat at the focus:

$$E = E_S + E_R \tag{15.21}$$

Since the only part of E we can measure is the seismic energy, we can consider this as a fraction of the total energy by using the seismic efficiency coefficient η,

$$E_S = \eta E \tag{15.22}$$

This coefficient has a value less than unity that is difficult to estimate since we can not measure the total energy released by earthquakes. For nuclear explosions whose yields are known, values obtained for η vary from case to case according to the conditions of the medium in which they are produced. The mechanism of explosions is very different than that of earthquakes, so the results are not comparable.

15.6 The seismic moment, stress drop, and average stress

The magnitude of an earthquake is related to the energy released and is independent from the mechanism of its generation. Another measure of the size of an earthquake is the seismic moment M_0, which was introduced by Aki (1966). It is based on the idea that earthquakes are caused by shear fractures in the Earth's crust and defined as

$$M_0 - \mu \, \overline{\Delta u} \, S \tag{15.23}$$

where μ is the shear or rigidity modulus, $\overline{\Delta u}$ is the mean value of the slip or displacement on the fault plane, and S is the area of the fault plane. In cgs units the seismic moment is given in dyn cm and in SI units it is in N m. The seismic moment includes the area of the fault, slip, and strength of the material, and thus constitutes a good physical measure of the size of an earthquake.

In a simplified model of fracture, the relative slip Δu of the two sides of a fault is due to the shear stress acting, which, at a given moment, exceeds the strength of the material or the friction that maintains the fault locked. If the shear stresses acting on the fault plane before and after an earthquake are σ_0 and σ_1, we can define two new parameters, namely, the average stress $\bar{\sigma}$ (the mean value of the stresses acting before and after the earthquake) and the stress drop $\Delta\sigma$ (the difference between them) (Fig. 15.6):

$$\bar{\sigma} = \tfrac{1}{2}(\sigma_0 + \sigma_1) \tag{15.24}$$

$$\Delta\sigma = \sigma_0 - \sigma_1 \tag{15.25}$$

The drop in stress represents the part of the stress acting that is employed in producing the slip of the fault. If $\sigma_1 = 0$, the stress drop is total and $\Delta\sigma = 2\bar{\sigma}$. Owing to the friction between the two sides of a fault there is always some residual stress σ_1 after the fracturing has finished. Only if there is no friction would the stress drop be total. The initial stress σ_0 is the tectonic stress responsible for the strain in the focal region.

In a simplified form, the total release of energy during fracturing can be expressed by

$$E = \bar{\sigma} \, \overline{\Delta u} \, S \tag{15.26}$$

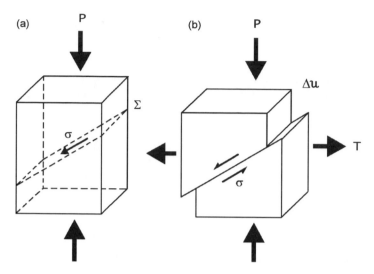

Fig. 15.6. The stress acting before (a) and after (b) the occurrence of a shear fracture of slip Δu and drop in stress $\Delta\sigma = \sigma_0 - \sigma_1$.

($\bar{\sigma}S$ represents a force). On substituting into this the seismic moment (15.23), we obtain

$$E = \frac{\bar{\sigma}}{\mu} M_0 \tag{15.27}$$

If the stress drop is total, equation (15.26) gives

$$E = \frac{\Delta\sigma}{2\mu} M_0 \tag{15.28}$$

This expression relates the total energy released by an earthquake to the seismic moment and the stress drop. For a shear fracture, the stress drop is proportional to the deformation of the fault, $\Delta\sigma = \Delta u/L'$, where L' is a length dimension of the fault plane (for example, for a circular fault $L' = a$, the radius, whereas for rectangular faults $L' = D$, the width). The stress drop is then given by

$$\Delta\sigma = C\mu\frac{\overline{\Delta u}}{L'} \tag{15.29}$$

where C is an adimensional factor that depends on the shape of the fracture (e.g. $C = 7\pi/16$ for a circular fault). For a circular fault, by substituting (15.29) into (15.23), we obtain the relation between the seismic moment and the stress drop:

$$M_0 = \frac{16}{7}a^3\,\Delta\sigma \tag{15.30}$$

Then, if we know the seismic moment and the dimensions of the fracture, we can determine the stress drop:

$$\Delta\sigma = \frac{7}{16a^3} M_0 \tag{15.31}$$

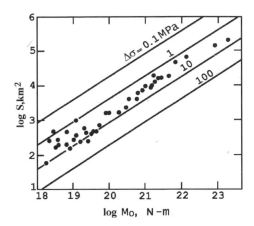

Fig. 15.7. The relation between the fault area S and the seismic moment M_0 with lines of equal drop in stress $\Delta\sigma$ (modified from Kanamori and Anderson (1975)) (with permission from the Seismological Society of America).

Since the fault radius is present in this equation raised to the power of three, small errors will produce large errors in determinations of the stress drop. If we substitute the fault area ($S = \pi a^2$) into (15.31), we obtain

$$M_0 - \frac{16\,\Delta\sigma}{7\pi^{3/2}}S^{3/2} \tag{15.32}$$

and, on taking logarithms,

$$\log M_0 = \frac{3}{2}\log S + \log\left(\frac{16\,\Delta\sigma}{7\pi^{3/2}}\right) \tag{15.33}$$

From this equation it follows that, if the stress drop is constant for all earthquakes, then $\log S$ is proportional to $\frac{2}{3}\log M_0$. That this hypothesis is valid for a large range of magnitudes has been shown empirically (Kanamori and Anderson, 1975). For earthquakes of moderate and large magnitudes ($m > 5$), $\Delta\sigma$ has values in the range 1–10 MPa (10 and 100 bars) and a mean value of 6 MPa (60 bars) (Fig. 15.7). Kanamori and Anderson (1975) suggested that earthquakes that take place at plate boundaries (interplate shocks) have lower stress drops (of 3 MPa (30 bars)) than do those in plate interiors (intraplate shocks), for which stress drops are about 10 MPa (100 bars). The mean stress drop (6 MPa) is of the same order of magnitude as the value suggested by Tsuboi (1956) for the critical strain of the Earth's crust.

Constancy of the stress drop is required in the definition of Kanamori's moment magnitude M_W. The formula for the moment magnitude (15.17) is obtained by substitution of (15.28) into (15.20), assuming a constant value for $\Delta\sigma/\mu = 10^{-4}$ and solving for M_S, which is now renamed M_W. The moment magnitude is, then, the magnitude derived from the seismic moment, under the hypothesis of a constant stress drop that satisfies the relation of Gutenberg and Richter between the surface waves magnitude and the energy.

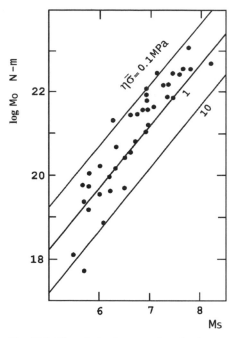

Fig. 15.8. The relation between the seismic moment M_0 and the surface wave magnitude M_S with lines of equal apparent average stress $\eta\bar{\sigma}$ (modified from Kanamori and Anderson (1975)) (with permission from the Seismological Society of America).

By substituting for the energy in terms of the seismic energy (15.22) we can define from (15.27), the apparent average stress

$$\eta\bar{\sigma} = \mu \frac{E_S}{M_0} \tag{15.34}$$

A simple way to relate the magnitude M_S to the seismic moment M_0 is by using equations (15.20) and (15.34), resulting in

$$\log M_0 = \frac{3}{2} M_S + 11.8 - \log\left(\frac{\eta\bar{\sigma}}{\mu}\right) \tag{15.35}$$

If $\eta\bar{\sigma}$ is constant, we have a linear relation between $\log M_0$ and M_S with a slope equal to $\frac{3}{2}$. Observations agree with this hypothesis, although there is a certain dispersion in the data, especially for very large earthquakes, that may be due to saturation of the M_S scale (Kanamori and Anderson, 1975) (Fig. 15.8). In this case, there is also some evidence that mean stresses for intraplate earthquakes (5 MPa) are larger than those for interplate earthquakes (1.5 MPa).

If the drop in stress is total, we can make the approximation that $\eta\bar{\sigma} \approx \Delta\sigma/2$. According to Orowan's fracture model, the residual energy E_R corresponds to the energy lost by friction and, similarly to in (15.26), it is given (Orowan, 1960) by

$$E_R = \sigma_f \overline{\Delta u}\, S \tag{15.36}$$

where σ_f is the friction stress during fracturing. The situation for which $\sigma_1 = \sigma_f$, known as Orowan's condition, implies that the stress drop is total and thence $\eta\bar{\sigma} \approx \Delta\sigma/2$. This means that the stress that does not contribute to the slip is lost in friction as heat. When this condition is satisfied, the energy given by (14.28) is the minimum estimation of the total energy E. If the stress drop is not complete, the final stress can be greater or less than the friction stress and the average stress can be greater or less than half the stress drop.

In conclusion, the seismic moment M_0 can be considered the best measure of the size of an earthquake. This quantity assumes the mechanism of shear fracture. This model introduces also the concepts of the stress drop and average stress. Through the relations of these quantities to the energy released during fracturing, we have found useful relations among them.

16 THE SOURCE MECHANISM

16.1 The representation of the source. Kinematic and dynamic models

We saw in Chapter 15 that earthquakes are produced by fractures in the Earth's crust. In Reid's model of elastic rebound, faulting is caused by the sudden release of accumulated elastic strain when the strength of the material is overcome. In seismology the problem of the source mechanism consists in relating observed seismic waves to the parameters that describe the source. In the direct problem, theoretical seismic wave displacements are determined from source models, whereas in the inverse problem, the parameters of source models are derived from observed wave displacements. The first step in both problems is to define the seismic source in terms of a mechanical model that represents the physical fracture. These models or representations of the source are defined by parameters whose number depends on their complexity. Simple models are defined by a few parameters whereas more complex ones require a larger number of parameters (Madariaga, 1983; Udías, 1991; Koyama, 1997).

Fracturing can be approached in two different ways, kinematic and dynamic. Kinematic models of the source consider the slip of the fault without relating it to the stresses that cause it. Fracturing is described purely in terms of the slip vector as a function of the coordinates on the fault plane and time. From models of this type, it is relatively simple to determine the corresponding elastic displacement field. The second approach considers the complete fracture process relating the fault slip to the stress acting on the focal region. A complete dynamic description must be able to describe fracturing from the properties of the material of the focal region and the stress conditions. Dynamic models present greater difficulties and their solutions, in many cases, can be found only by numerical methods.

16.2 Equivalent forces. Point sources

The first mathematical formulation of the mechanism of earthquakes was presented by Nakano (1923) using the ideas already developed by Lamb (1904) and Love (1945). Nakano used the point-source approximation, which is valid if observation points are at a sufficiently large distance compared with source dimensions and wave lengths are also large. Thus he could represent the source by a system of body forces acting at a point. Since these forces must represent the fracture phenomenon they are called equivalent forces.

The problem may be stated as follows. Let us consider an elastic medium of volume V surrounded by a surface S. In its interior there is a small region of volume V_0,

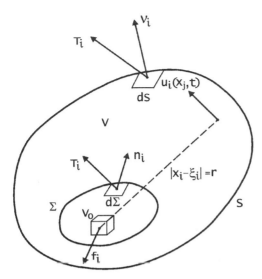

Fig. 16.1. The stress and displacement in a medium of volume V surrounded by a surface S. The body force f and stress T are acting in the focal region of volume V_0, surrounded by the surface Σ.

surrounded by a surface Σ, which we will call the focal region, where fracture takes place (Fig. 16.1). This process can be represented by a distribution of body forces $F(\xi_i, t)$ acting per unit volume inside V_0. If it is assumed that no other body forces (gravity, etc.) are present, the equation of motion (2.55) can be written as

$$\int_{V-V_0} [\rho\ddot{u}_i(x_i, t) - \tau_{ij,j}(x_i, t)]\,dV = \int_{V_0} F_i(\xi_i, t)\,dV \qquad (16.1)$$

where ξ_i are the coordinates inside of the focal region and x_i are those outside of it. Only elastic displacements and stresses outside of the focal region are considered. From the static case (2.57), the body forces F_i are formally related to stresses inside of V_0 by

$$F_i = -\tau_{ij,j} \qquad (16.2)$$

In the case of a point source, if the volume V is an infinite medium, the equation (16.1), according to (2.57), is given by

$$\rho\ddot{u}_i - \tau_{ij,j} = F_i \qquad (16.3)$$

where F_i are forces at a point that is selected as the origin of the x_i coordinates where the elastic displacements $u(x, t)$ are evaluated. These forces are the limit of the forces acting on V_0 as it tends to zero:

$$F_i(t) = \lim_{V_0 \Rightarrow 0} \int_{V_0} F_i(\xi_k, t)\,dV \qquad (16.4)$$

For a homogeneous medium, equation (16.3) can be expressed in terms of displacements, using (2.59), as

$$\rho\ddot{u}_i - C_{ijkl}u_{k,lj} = F_i \qquad (16.5)$$

This equation allows the determination of the elastic displacements $u(x, t)$ produced by a force or system of forces F acting at the origin of the coordinates. In the inverse problem, from the observed elastic displacements we can obtain certain characteristics of these forces.

16.2.1   The formulation using Green's function

A more convenient formulation of the problem can be obtained by using the representation theorem in terms of Green's function (section 2.8). According to (2.88), if body forces are limited to the focal region V_0 (Fig. 16.2) and on its surface Σ the stresses and displacements are null, we obtain for a volume V surrounded by a surface S that

$$u_i = \int_{-\infty}^{\infty} \mathrm{d}\tau \int_{V_0} F_k G_{ki} \, \mathrm{d}V + \int_{-\infty}^{\infty} \mathrm{d}\tau \int_{S} (G_{ji}T_j - u_j C_{jkln}G_{li,n}\nu_k) \, \mathrm{d}S \qquad (16.6)$$

where $T_i = \tau_{ij}\nu_j$ is the stress vector, ν_i is the normal to the surface element $\mathrm{d}S$, and G_{ki} is Green's function of the medium, defined by equation (2.76) in section 2.7. Green's function, a tensor, is continuous throughout the volume V and represents the effect of propagation in the medium. As we saw in section 2.7, Green's function is the solution of the equation of motion for an impulsive force and depends on the characeristics (C_{ijkl} and ρ) of the medium. If the medium is infinite, conditions on the surface S are homogeneous (the stress and displacement are null) and equation (16.6) becomes (Fig. 16.3)

$$u_i(x_s, t) = \int_{-\infty}^{\infty} \mathrm{d}\tau \int_{V_0} F_k(\xi_s, \tau)G_{ki}(x_s, t; \xi_s, \tau) \, \mathrm{d}V \qquad (16.7)$$

The function G_{ki} acts as a 'propagator' of the effects of the forces F_k from the points where they are acting (ξ_s inside V_0) to points x_s outside V_0 where the elastic displacements u_i are produced. For a point focus at the origin of coordinates we have

$$u_i(x_s, t) = \int_{-\infty}^{\infty} F_k(\tau)G_{ki}(x_s, t - \tau) \, \mathrm{d}\tau \qquad (16.8)$$

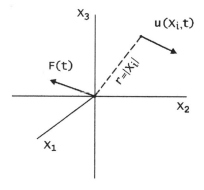

Fig. 16.2. A body force $F(t)$ acting at the origin of coordinates and elastic displacements $u(x, t)$ at a point at a distance r.

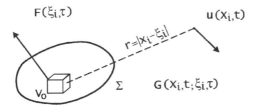

Fig. 16.3. A body force $F(\xi_i, t)$ acting on a source volume V_0 and elastic displacements $u(x, t)$ at a distance r in an infinite medium.

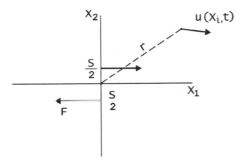

Fig. 16.4. A single couple of forces acting at the origin and elastic displacements $u(x, t)$ at a distance r.

The elastic displacements are now given by the time convolution of the forces acting at the focus with Green's function of the medium.

From the point of view of the representation of the seismic source in terms of equivalent forces, there are two ways to find the elastic displacements. The first consists in solving directly the equation of motion (16.1). This implies solving a second-order inhomogeneous differential equation for displacements or solving a homogeneous equation and introducing the forces as boundary conditions. In both cases, the problem is not easy. The second consists in using equation (16.6), (16.7), or (16.8). In this case we have to have prior knowledge of Green's function. Since Green's function is the solution of the equation of motion, this equation must be solved anyway. However, the advantage of the second approach is that, for a given medium, the equation of motion must be solved only once to find Green's function, whereas in the first, it must be solved for each system of forces. For example, in the point-source problem, the first approach requires for each system of forces the solution of equation (16.5) in the same medium. In the second approach, equation (2.77) is solved only once to find Green's function. Then, for each system of forces, we apply equation (16.8), which is a convolution of each system of forces with Green's function.

16.2.2 Single and double couples

Several systems of forces have been proposed to represent the source of an earthquake. For point sources, the most common are those of a couple of forces (SC, a single couple) and two couples perpendicular to each other without a resulting moment (DC, a double couple) (Figs. 16.4 and 16.5). The second system is also equivalent to two linear

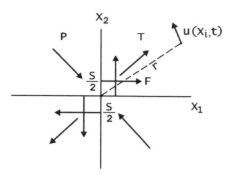

Fig. 16.5. A double couple of forces without a moment acting at the origin and the displacement $u(x, t)$ at a distance r. An equivalent system of two linear dipoles of pressure and tension forces P and T is shown.

dipoles of forces (with arms in the same direction as the forces) corresponding to pressure and tension acting at 45° to the couples. Both models were thought to represent shear fracture, but, as we will see, this is true only for the second. For models of extended sources, distributions of single or double couples on a plane surface were used.

For a point source, elastic displacements due to a couple of forces can be derived from those from a single force. If u_i^1 are the displacements due to a force acting at the origin in the x_1 direction, those of a couple of forces in the plane (x_1, x_2) with forces along the x_1 axis and the arm in the x_2 direction (Fig. 16.4) are derived by performing a Taylor expansion for each force of the couple, displaced by $s/2$ from the origin along the x_2 axis. For the force in the positive direction of x_1 and shifted by $s/2$ from the origin in the positive direction of x_2, the elastic displacement is

$$u_i^+ = u_i^1 + \frac{s}{2} u_{i,2}^1 \tag{16.9}$$

For the force in the negative direction of x_1 and shifted by $s/2$ in the negative direction of x_2, the displacement is

$$u_i^- = -u_i^1 + \frac{s}{2} u_{i,2}^1 \tag{16.10}$$

The displacement due to a single couple is the sum of the two:

$$u_i^{SC} = s u_{i,2}^1 \tag{16.11}$$

For a double couple in the (x_1, x_2) plane, with forces in the directions of x_1 and x_2 (Fig. 16.5), using (16.11), the elastic displacement is given by

$$u_i^{DC} = s(u_{i,2}^1 + u_{i,1}^2) \tag{16.12}$$

If we substitute (16.11) into (16.8), we obtain the displacement due to a SC, as defined above, in terms of Green's function:

$$u_i^{SC} = \int_{-\infty}^{\infty} M(\tau) G_{i1,2}(t - \tau) \, d\tau \tag{16.13}$$

where $M(t) = F(t)s$ is the moment of the couple. For a DC in the x_1 and x_2 directions, using (16.12), we obtain

$$u_i^{DC} = \int_{-\infty}^{\infty} M(\tau)[G_{i1,2}(t-\tau) + G_{i2,1}(t-\tau)]\,d\tau \tag{16.14}$$

For a SC in an arbitrary orientation, with forces in the direction of the unit vector l and the arm in that of n, where $n \cdot l = 0$, and a DC with the second couple with forces in the direction of n and its arm in that of l, general expressions are

$$u_i^{SC} = \int_{-\infty}^{\infty} M l_k n_l G_{ik,l}\,d\tau \tag{16.15}$$

$$u_i^{DC} = \int_{-\infty}^{\infty} M(l_k n_l + n_k l_l)G_{ik,l}\,d\tau \tag{16.16}$$

Let us consider now two perpendicular linear dipoles with opposite signs. The linear dipole with forces in the positive direction corresponds to tension, whereas that with forces in the negative direction corresponds to pressure. If the forces are in the directions of x_1' and x_2', in a similar form to that in (16.12), the elastic displacements are

$$u_i^{TP} = s(u_{i,1}^{l1} - u_{i,2}^{l2}) \tag{16.17}$$

If the system of coordinates (x_1', x_2') in (16.17) is rotated by $45°$ with respect to (x_1, x_2) of equation (16.12), the two expressions can be shown to be equivalent. If the tension and pressure forces are defined by the scalar moment M and unit vectors T and P, in a similar form to that in (16.16), the elastic displacement in terms of Green's function is

$$u_i^{TP} = \int_{-\infty}^{\infty} M(T_k T_k - P_l P_l)G_{ik,l}\,d\tau \tag{16.18}$$

For the two equivalent systems, relations between the unit vectors P and T and n and l are

$$P = \frac{1}{\sqrt{2}}(n - l) \tag{16.19}$$

$$T = \frac{1}{\sqrt{2}}(n + l) \tag{16.20}$$

$$B = n \times l = P \times T \tag{16.21}$$

where B is the unit vector normal to the plane of the forces. This vector is known as the null axis, since there is no component of forces in its direction. In this way, for a DC point source, we can define two orthogonal systems of axes in the directions of the unit vectors n, l, and B and P, T, and B to specify the orientation of the source. We will see that the second system corresponds to the principal axes of stress.

16.3 Fractures and dislocations

If an earthquake is produced by fracturing of the Earth's crust, a mechanical representation of its source in terms of fractures or dislocations in an elastic medium can be achieved. The theory of elastic dislocations was developed by Volterra in 1907

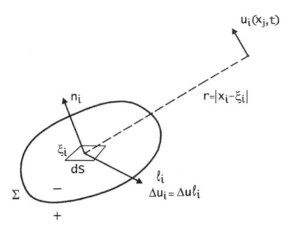

Fig. 16.6. A dislocation Δu on a surface Σ and the elastic displacement $u(x, t)$ at a distance r in an infinite medium.

and discussed by Love (1945). Its first applications to the problem of seismic sources were by Vvedenskaya (1956), Keylis-Borok (1956), Steketee (1958), Knopoff and Gilbert (1960) and Burridge and Knopoff (1964).

A dislocation consists in an internal surface inside an elastic medium across which there is a discontinuity of displacement or strain. Here we will consider only displacement dislocations, that is, those for which there is a discontinuity of displacement but the stress is continuous. The problem will be formulated using the representation theorem in terms of Green's function (2.88). The focal region consists in an internal surface Σ with two sides (positive and negative). This surface can be considered to be derived from the focal volume V_0 that is flattened to form a surface with both sides together without any volume. The coordinates on this surface are ξ_i and the normals at the points are $n_i(\xi_k)$. From one side to the other of this surface there is a discontinuity in displacement or slip (Fig. 16.6) so that

$$u_i^+(\xi_k, t) - u_i^-(\xi_k, t) = \Delta u_i(\xi_k, t) \tag{16.22}$$

where the plus and minus signs refer to the displacements at each side of the surface Σ. If there are no body forces ($F = 0$), the stresses are continuous through Σ (their integral is null) and the conditions on the external surface S are homogeneous (all integrals on S are null), then equation (2.88) results in

$$u_n(x_s, t) \int_{-\infty}^{\infty} d\tau \int_{\Sigma} \Delta u_i(\xi_s, \tau) C_{ijkl} n_j(\xi_s) G_{nk,l}(\xi_s, \tau; x_s, t) \, dS \tag{16.23}$$

In consequence, the seismic source is represented by a dislocation or discontinuity in displacement given by the slip vector Δu on the surface Σ, which corresponds to the relative displacement of the two sides of a fault. This is, then, an inelastic displacement that, once it has been produced, does not go back to the initial position. In the most general case, $\Delta u(\xi_i, \tau)$ can have a different direction for each point ξ_i of the surface Σ and, at each of these points, varies with time, starting from a zero value at $t = 0$, to a maximum value at a certain time. The normal to the surface Σ, given by the unit vector $n(\xi_i)$, can have different directions at points of the surface, but usually is

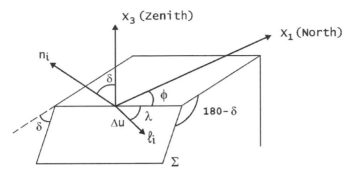

Fig. 16.7. Parameters of the orientation of a fault or shear fracture ϕ, δ, and λ, and unit vectors *n* and *l*.

considered to be constant; that is, Σ is a plane. Green's function *G* includes the effects of the medium on propagation from points ξ_i on the surface Σ to points x_i where elastic displacements u_i are evaluated. To solve the problem, according to equation (16.23) we must first know the derivatives of Green's function for the medium, which are also known as excitation functions.

Equation (16.23) corresponds to a kinematic model of the source; that is, a model in which the elastic displacements *u* are derived from the slip vector Δu, which represents the inelastic displacement of the two sides of a fault of surface Σ. The slip is assumed to be known rather than being derived from the stress conditions in the focal region as is done in dynamic models. In equation (16.23) the derivatives of Green's function include derivatives of the delta function. If we change the order of integration, integration over time of the product of Δu with the derivatives of the delta function results in time derivatives $\Delta \dot{u}$, that is, the slip velocity. Thus, elastic displacements depend not on the slip but rather on the slip velocity. This means that the source radiates elastic energy only while it is moving. When motion at the source stops it ceases to radiate energy.

As a particular case, let us consider an isotropic medium of coefficients λ and μ, a plane surface Σ (with *n* constant) and a constant slip Δu with the same direction defined by the unit vector *l*. The integrand of (16.23) becomes

$$\Delta u(t)[\lambda l_k n_k \delta_{ij} + \mu(l_i n_j + l_j n_i)]G_{ni,j} \tag{16.24}$$

The geometry of the source is now defined by the orientations of the two unit vectors *n* and *l*. These two vectors, referred to the geographic system of axes (North, East, and nadir), define the orientation of the source, namely, *n* is the orientation of the fault plane and *l* is that of the slip (Fig. 16.7). Since they are unit vectors, each has only two independent components. For $i = j$, expression (16.24) gives the component of the displacement normal to the fault plane which implies changes in volume. If *l* and *n* are perpendicular, the slip is along the fault plane, there are no changes in volume, and it represents a shear fracture. In this case ($n \cdot l = 0$), there are only three independent components of *n* and *l*.

In the kinematic model of a dislocation on a plane surface with a constant slip, the parameters of the source are the elastic coefficients of the focal region λ and μ, four independent components of *n* and *l* defining the orientation of the fault plane and slip, the magnitude of the slip Δu, and the area *S* of the fault. There are eight parameters that, when added to the four of the hypocenter (ϕ_0, λ_0, h, and t_0), sum up to 12. If the source is a shear fracture the number of parameters is only ten.

16.4   The Green function for an infinite medium

The problem of determining Green's function is not an easy one and depends on the characteristics of each medium. As we saw in section 2.7, Green's function is the solution of the equation of motion for an impulsive force in time and space ((2.76) and (2.77)). Let us consider Green's function for an infinite, homogeneous, isotropic elastic medium with force acting at the origin and $t = 0$. According to (2.61) and (2.77), Green's function is the solution of the equation

$$\rho \ddot{G}_{ij} - (\lambda + \mu)G_{ik,kj} - \mu G_{ij,kk} = \delta(x_s)\delta(t)\delta_{ij} \tag{16.25}$$

In terms of the velocities α and β (2.63), we also have

$$\ddot{G}_{ij} - \alpha^2 \nabla(\nabla \cdot \boldsymbol{G}_{ij}) + \beta^2 \nabla \times (\nabla \times \boldsymbol{G}_{ij}) = \frac{1}{\rho}\delta(x_s)\delta(t)\delta_{ij} \tag{16.26}$$

We will start with a simplified problem in only one dimension.

16.4.1   The radial force

In a problem with spherical symmetry and an impulsive radial force, elastic displacements have only radial components and the problem can be solved in one dimension (Fig. 16.8). Green's function has only one component, $G_{ij} = u(r, t)$. Since for this case $\nabla \times \boldsymbol{u} = 0$, equation (16.26) becomes

$$\ddot{u}(r, t) - \alpha^2 \nabla^2 u(r, t) = \frac{1}{\rho}\delta(r)\delta(t) \tag{16.27}$$

On substituting the Laplacian in spherical coordinates for a dependence on r alone (A2.30), we have

$$\frac{\partial^2 u}{\partial t^2} = \frac{\alpha^2}{r}\frac{\partial^2}{\partial r^2}(ru) + \frac{1}{\rho}\delta(r)\delta(t) \tag{16.28}$$

If there are no forces, equation (16.28) for a harmonic dependence results in Helmholtz's equation (3.122) and its solution (3.125) is

$$u(r, t) = \frac{A}{r}\exp\left[i\omega\left(t - \frac{r}{\alpha}\right)\right] \tag{16.29}$$

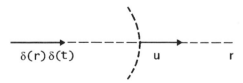

Fig. 16.8. An impulsive radial force at the origin and the displacement at a distance r.

Let us now consider Poisson's equation,

$$\nabla^2 u(r) = \frac{1}{\alpha^2 \rho} \delta(r) \qquad (16.30)$$

Its solution is

$$u(r) = \frac{\delta(r)}{4\pi\alpha^2 \rho r} \qquad (16.31)$$

We can consider that the solution of (16.28), in analogy with (16.29) and (16.31), can be written as

$$u(r, t) = \frac{\delta(t - r/\alpha)}{4\pi\alpha^2 \rho r} \qquad (16.32)$$

Then, for an impulsive radial force, the elastic radial displacements have the form of an impulse that depends on $(t - r/\alpha)$. This impulse travels with a velocity α (P wave), arrives at a distance r at time $t = r/\alpha$, and its amplitude decreases with distance as $1/r$. Equation (16.32) represents Green's function for a radial force with spherical symmetry in an infinite medium.

16.4.2 *An impulsive force in an arbitrary direction*

The complete problem of Green's function corresponds to an impulsive force in an arbitrary direction (Aki and Richards, 1980). Using Cartesian coordinates, we start with the particular case of a force in the x_1 direction applied at the origin of coordinates (Fig. 16.9). Equation (16.26), putting $G_{ij} = u_i$, is (2.63), namely

$$\ddot{u} = F/\rho + \alpha^2 \nabla(\nabla \cdot u) - \beta^2 \nabla \times (\nabla \times u) \qquad (16.33)$$

where F is given by

$$F = \delta(t)\delta(x_1, x_2, x_3)(1, 0, 0) \qquad (16.34)$$

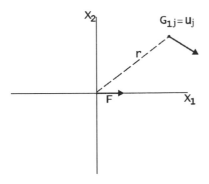

Fig. 16.9. An impulsive force F acting at the origin in the x_1 direction and the elastic displacement u equivalent to Green's function G_{1i}.

Just as in section 2.6, we introduce the potentials ϕ and ψ for displacements and Φ and Ψ for forces ((2.68) and (2.71)),

$$u = \nabla\phi + \nabla \times \psi \tag{16.35}$$

$$F = \nabla\Phi + \nabla \times \Psi \tag{16.36}$$

On replacing (16.35) and (16.36) into (16.33) we obtain (2.72) and (2.73),

$$\ddot{\phi} - \alpha\nabla^2\phi^2 = \Phi/\rho \tag{16.37}$$

$$\ddot{\psi} - \beta^2\nabla^2\psi = \Psi/\rho \tag{16.38}$$

We introduce now a new vectorial function W from which we can deduce the force potentials:

$$\Phi = \nabla \cdot W \tag{16.39}$$

$$\Psi = -\nabla \times W \tag{16.40}$$

By substitution into (16.36), we obtain

$$\nabla^2 W = F \tag{16.41}$$

This equation has the form of Poisson's equation (16.30) and its solution can be given as

$$W = \frac{F}{4\pi r} \tag{16.42}$$

where r is the distance from the origin, where the force is applied to a point of coordinates x_i where we evaluate the displacements u_i. If F is distributed in a volume V, wherein the coordinates of each point are ξ_i, the solution of (16.41) is

$$W(x_i) = \frac{1}{4\pi}\int_V \frac{F(\xi_i)}{r}\,dV \tag{16.43}$$

where $r = |x_i - \xi_i|$. In our case, F is given by (16.34) and W is

$$W = \frac{\delta(t)}{4\pi r}(1,0,0) \tag{16.44}$$

According to (16.39), (16.40), and (16.44), the force potentials Φ and Ψ are

$$\Phi = \frac{\delta(t)}{4\pi}\frac{\partial}{\partial x_1}\left(\frac{1}{r}\right) \tag{16.45}$$

$$\Psi = \frac{\delta(t)}{4\pi}\left[0, \frac{\partial}{\partial x_3}\left(\frac{1}{r}\right), -\frac{\partial}{\partial x_2}\left(\frac{1}{r}\right)\right] \tag{16.46}$$

Equations (16.37) and (16.38) have the same form as (16.27); therefore, their solutions can be written in a similar form to (16.32), where the right-hand terms are Φ/ρ and Ψ/ρ, respectively. Since the force potentials are defined in the whole volume V, the solutions must be written in integral form as in (16.43):

$$\phi = \frac{1}{4\pi\alpha^2\rho}\int_V \frac{\Phi(t - r/\alpha)}{r}\,dV \tag{16.47}$$

$$\psi = \frac{1}{4\pi\beta^2\rho}\int_V \frac{\Psi(t - r/\beta)}{r}\,dV \tag{16.48}$$

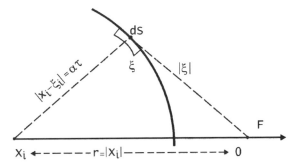

Fig. 16.10. A spherical surface with its center at x_i with radius $\alpha\tau$ for the evaluation of integral (16.49) (ξ is the variable of integration).

In our problem, the volume V represents the whole of the infinite space. We must notice that the potentials are defined over the whole volume under consideration. This may sound strange, especially for force potentials. Even when forces are acting at a point (the origin of coordinates), the potentials Φ and Ψ that represent them are defined over the whole volume of the problem, in our case the infinite space. By substituting expression (16.45) into (16.47), we obtain the solution of the displacement potential ϕ for an impulsive force in the x_1 direction:

$$\phi = \frac{1}{16\pi^2\rho\alpha^2} \int_V \frac{\delta(t - r/\alpha)}{r} \frac{\partial}{\partial x_1}\left(\frac{1}{r}\right) dV \tag{16.49}$$

where V represents the infinite medium. A similar expression is found for ψ by using (16.46) and (16.48). The evaluation of these integrals is performed in the following manner. The integral over the volume V (in our case the infinite medium) is separated into two parts, first an integral over a spherical surface S, with its center at x_i, the point where the displacements are evaluated, and a radius equal to $\alpha\tau$ for ϕ and to $\beta\tau$ for ψ (Fig. 16.10), and second an integral over the time τ from zero to infinity. In this form we cover the infinite medium. With these substitutions equation (16.49) becomes

$$\phi = \frac{1}{16\pi^2\rho\alpha^2} \int_0^\infty \frac{\delta(t - \tau)}{\tau} \int_S \frac{\partial}{\partial\xi_1} \frac{1}{|\xi_i|} dS \, d\tau \tag{16.50}$$

where $|\xi_i|$ is the distance from the origin to an element dS of the spherical surface S, $|x_i| = r$ is the distance from the origin to the point where u is evaluated, and $|x_i - \xi_i| = \alpha\tau$ is the distance from the point x_i to the spherical surface (Fig. 16.10). We change the orders of differentiation and integration in the integral over the surface S. Then, it can be shown that, when the spherical surface includes the origin ($\tau > r/\alpha$), the integral over S of $1/|\xi_i|$ is constant and equal to $4\pi\alpha\tau$ and consequently its derivative is zero. If the surface does not include the origin ($\tau < r/\alpha$), the integral has the value $4\pi^2\alpha^2\tau^2/r$ (Aki and Richards, 1980). Thus, we obtain

$$\int_S \frac{\partial}{\partial\xi_l} \frac{1}{|\xi_i|} dS = 4\pi^2\alpha^2\tau^2 \frac{\partial}{\partial x_1} \frac{1}{r}$$

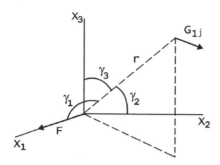

Fig. 16.11. The geometry of an impulsive force F in the x_1 direction and displacements (Green's function) G_{1j} at a distance r and direction cosines γ_i.

and, on substituting this into (16.50), we have

$$\phi = \frac{1}{4\pi\rho} \frac{\partial}{\partial x_1} \frac{1}{r} \int_0^{r/\alpha} \tau\delta(t - \tau)\,d\tau \tag{16.51}$$

The derivative with respect to ξ_1 now becomes one with respect to x_1 and the integral over τ extends only to $\tau = r/\alpha$ since, for $\tau > r/\alpha$, the integrand is zero. In a similar manner we obtain the expression for the vector potential ψ, using (16.48) and (16.46):

$$\psi = \frac{1}{4\pi\rho} \left(0, \frac{\partial}{\partial x_3} \frac{1}{r}, -\frac{\partial}{\partial x_2} \frac{1}{r}\right) \int_0^{r/\beta} \tau\delta(t - \tau)\,d\tau \tag{16.52}$$

By taking the corresponding derivatives according to (16.35) in (16.51) and (16.52), we obtain the displacement u, that is, Green's function for a force in the x_1 direction or G_{i1}. The derivatives of $1/r$ can be expressed in terms of the direction cosines γ_i of the vector r from the point ξ_i where the force is applied (in our case the origin) to the point x_i where Green's function is evaluated (Fig. 16.11). This is easily shown, since

$$r = [(x_1 - \xi_1)^2 + (x_2 - \xi_2)^2 + (x_3 - \xi_3)^2]^{1/2}$$

Its derivative with respect to x_i is

$$\frac{\partial r}{\partial x_1} = \frac{x_1 - \xi_1}{r} = \cos(r, x_1) = \gamma_1$$

and, in general,

$$\frac{\partial r}{\partial x_i} = \gamma_i \tag{16.53}$$

The derivatives with respect to ξ_i are

$$\frac{\partial r}{\partial \xi_i} = -\gamma_i$$

It is also easy to show that

$$\frac{\partial}{\partial x_i} \frac{1}{r} = -\frac{\gamma_i}{r^2} \tag{16.54}$$

$$\frac{\partial \gamma_i}{\partial x_j} = -\frac{1}{r}(\gamma_i\gamma_j - \delta_{ij}) \tag{16.55}$$

After taking the corresponding derivatives and substituting in the direction cosines, we obtain

$$G_{i1} = \frac{1}{4\pi\rho} \left[\frac{1}{r^3} (3\gamma_1\gamma_i - \delta_{1i}) \int_{r/\alpha}^{r/\beta} \tau\delta(t - \tau)\,d\tau \right.$$
$$\left. + \frac{1}{r\alpha^2} \gamma_1\gamma_i\delta\left(t - \frac{r}{\alpha}\right) - \frac{1}{r\beta^2}(\gamma_1\gamma_i - \delta_{1i})\delta\left(t - \frac{r}{\beta}\right) \right] \tag{16.56}$$

This equation corresponds to a force in the x_1 direction. We can generalize this result for Green's function corresponding to a force in an arbitrary direction given by the vector ν_j:

$$G_{ij} = \frac{1}{4\pi\rho} \left[\frac{1}{r^3} (3\gamma_i\gamma_j - \delta_{ij}) \int_{r/\alpha}^{r/\beta} \tau\delta(t - \tau)\,d\tau \right.$$
$$\left. + \frac{1}{r\alpha^2} \gamma_i\gamma_j\delta\left(t - \frac{r}{\alpha}\right) - \frac{1}{r\beta^2}(\gamma_i\gamma_j - \delta_{ij})\delta\left(t - \frac{r}{\beta}\right) \right] \tag{16.57}$$

This is the expression for Green's function for an infinite, homogeneous, isotropic elastic medium with velocities α and β. This is a very important result in elastodynamics, which gives the elastic displacement field for the most fundamental type of source. It constitutes the basic building block of seismic source studies.

A similar, fundamental problem is the static solution for a constant force acting at a point in the direction of the unit vector ν_j. For an infinite, homogeneous, isotropic elastic medium, the static displacements are solutions of the equation

$$\alpha^2 \nabla(\nabla \cdot \boldsymbol{u}) - \beta^2 \nabla \times (\nabla \times \boldsymbol{u}) = \boldsymbol{F}/\rho$$

The solution can be found in a similar way to that in the previous problem by expressing the displacement and force in terms of potentials. This leads to equations of the form of Poisson's equation (Lay and Wallace, 1995). The result for the displacement, in terms of the direction cosines γ_i, is

$$S_{ij} = \frac{F}{8\pi\rho r} \left[\left(\frac{1}{\beta^2} - \frac{1}{\alpha^2} \right) \gamma_i\gamma_j + \left(\frac{1}{\beta^2} + \frac{1}{\alpha^2} \right) \delta_{ij} \right] \tag{16.58}$$

The subindex j indicates the direction of the force and, just like in (16.57), the displacements are given by a tensor. This expression, known as Sommigliana's tensor, is the fundamental equation in elastostatics.

16.5 The separation of near and far fields

The first term of Green's function in equation (16.57) depends on the distance as r^{-3} and the other two have r^{-1} dependences. Thus, the displacement represented by the first term is attenuated more rapidly with distance and for this reason is called the near field. This term depends both on α and on β, and is a displacement of mixed P and S motion. The second and third terms constitute the far field where P and S waves are separated. In both cases, near and far fields, displacements have two parts, one, called the radiation pattern, depends on the direction cosines and expresses the spatial distribution of amplitudes, and another that depends on time or the wave form.

16.5.1   The near field

The time dependence of the near field (16.57) can be rewritten (Knopoff, 1967) as

$$\int_{r/\alpha}^{r/\beta} \tau \delta(t - \tau)\, d\tau = \int_{-\infty}^{\infty} \tau \delta(t - \tau) \left[H\left(\tau - \frac{r}{\alpha}\right) - H\left(\tau - \frac{r}{\beta}\right) \right] d\tau \qquad (16.59)$$

where $H(t)$ is the step or Heaviside function. According to the properties of step and delta functions, the integral on the right-hand side of (16.59) results in

$$\int_{-\infty}^{\infty} [\ \ ]\, d\tau = tH\left(t - \frac{r}{\alpha}\right) - tH\left(t - \frac{r}{\beta}\right) \qquad (16.60)$$

Each of these two terms can be written as

$$tH\left(t - \frac{r}{\alpha}\right) = \left(t - \frac{r}{\alpha}\right) H\left(t - \frac{r}{\alpha}\right) + \frac{r}{\alpha} H\left(t - \frac{r}{\alpha}\right) \qquad (16.61)$$

The term $(t - r/\alpha)H(t - r/\alpha)$ is a ramp function of slope unity. The complete expression for the near field can be written as

$$G_{ij}^{\mathrm{NF}} = \frac{1}{4\pi\rho} (3\gamma_i\gamma_j - \delta_{ij}) \left\{ \frac{1}{r^3} \left[\left(t - \frac{r}{\alpha}\right) H\left(t - \frac{r}{\alpha}\right) - \left(t - \frac{r}{\beta}\right) H\left(t - \frac{r}{\beta}\right) \right] \right.$$
$$\left. + \frac{1}{r^2} \left[\frac{1}{\alpha} H\left(t - \frac{r}{\alpha}\right) - \frac{1}{\beta} H\left(t - \frac{r}{\beta}\right) \right] \right\} \qquad (16.62)$$

The time dependence of the near field now has two parts that depend on distance as r^{-3} and r^{-2}, both depending on the velocities of P and S waves. The first part is the difference between two ramp functions and the second is the difference between two step functions of different amplitudes. The result is shown in Fig. 16.12. The part that depends on r^{-3} is formed by a ramp of unit slope starting at $t = r/\alpha$ and continuing until $t = r/\beta$. From this time onward the displacement has a constant amplitude of $1/\beta - 1/\alpha$. The part that depends on r^{-2} is a step function starting at $t = r/\alpha$ and amplitude $1/\alpha$ followed at $t = r/\beta$ by one of amplitude $1/\alpha - 1/\beta$. The displacement in the near field has a part that remains constant with time. The radiation pattern is common to the complete near-field displacement.

16.5.2   The far field

The far field (the part that depends on $1/r$) of Green's function is formed by separate P and S waves:

$$G_{ij}^{\mathrm{P}} = \frac{1}{4\pi\rho\alpha^2 r} \gamma_i\gamma_j \delta\left(t - \frac{r}{\alpha}\right) \qquad (16.63)$$

$$G_{ij}^{\mathrm{S}} = \frac{-1}{4\pi\rho\beta^2 r} (\gamma_i\gamma_j - \delta_{ij}) \delta\left(t - \frac{r}{\beta}\right) \qquad (16.64)$$

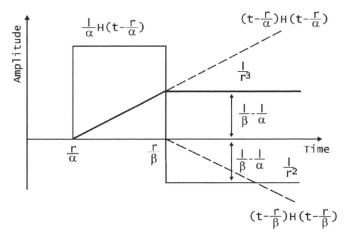

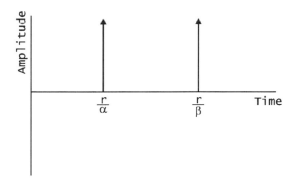

Fig. 16.12. The near-field displacement of Green's function for an infinite medium as a function of time at a distance r.

Fig. 16.13. The far-field displacement of Green's function for an infinite medium (P and S waves) at a distance r.

The time dependence is in both cases a delta function. The far field is, then, formed by two impulses that propagate with velocities α and β, that is, P and S waves of impulsive form (Fig. 16.13).

The radiation patterns for P and S waves are different ((16.63) and (16.64)). To represent the radiation patterns we take polar coordinates (r, θ) with their center at the focus and consider the distribution of normalized amplitudes. If the force is in the x_1 direction, the normalized components of the displacement of P waves in the (x_1, x_3) plane are given (Fig. 16.14) by

$$G_{11}^{P} = \gamma_1 \gamma_1 = \cos \theta \cos \theta \tag{16.65}$$

$$G_{31}^{P} = \gamma_3 \gamma_1 = \cos \theta \sin \theta \tag{16.66}$$

It can easily be seen that the displacement is in the radial direction, as expected for

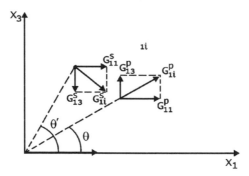

Fig. 16.14. Components of Green's function in the far field corresponding to P and S waves for a force in the x_i direction.

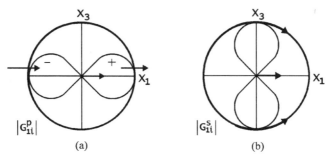

Fig. 16.15. The radiation pattern of Green's function on the (x_1, x_3) plane for a force in the x_1 direction: (a) P waves and (b) S waves.

P waves, and that its modulus is

$$|G_{i1}^P| = \cos\theta \tag{16.67}$$

For S waves, the normalized displacement components are

$$G_{11}^S = -(\gamma_1\gamma_1 - 1) = \sin\theta\sin\theta \tag{16.68}$$

$$G_{31}^S = -\gamma_1\gamma_3 = -\sin\theta\cos\theta \tag{16.69}$$

The resulting displacement is in the transverse direction and its modulus is

$$|G_{i1}^S| = \sin\theta \tag{16.70}$$

The displacements of P and S waves are in the radial and transverse directions correspondingly for each type of wave. By giving values from 0 to 360° to θ, we obtain radiation patterns. In both cases the pattern has two lobes (Fig. 16.15). For P waves, the displacements are radial, in the right lobe outward, that is, compressions and in the left inward, dilations, with maxima at $\theta = 0$ and π (Fig. 16.15(a)). For S waves, the displacements are transverse ones that converge toward the direction of the force in both lobes with maxima at $\theta = \pi/2$ and $3\pi/2$ (Fig. 16.15(b)). P waves have a nodal plane (x_2, x_3) normal to the force and S waves are in the plane (x_1, x_2) that contains the force.

16.6   A shear dislocation or fracture. The point source

Let us consider the seismic source represented by a shear dislocation fracture, with fault plane Σ of area S, normal $\boldsymbol{n}$, and slip $\Delta u(\xi, \tau)$ in the direction of the unit vector $\boldsymbol{l}$, contained on the plane so that $\boldsymbol{l}$ and $\boldsymbol{n}$ are perpendicular $(\boldsymbol{n} \cdot \boldsymbol{l} = 0)$. For an infinite, homogeneous isotropic medium, the displacement, according to (16.23) and (16.24), is

$$u_k = \int_{-\infty}^{\infty} d\tau \int_{\Sigma} \Delta u\, \mu (l_i n_j + l_j n_i) G_{ki,j}\, dS \tag{16.71}$$

If the distance from the observation point to the source is large in comparison with the source dimensions $(r \gg \Sigma)$ and the wave lengths are also large, the problem can be approximated by a point source and equation (16.71) takes the form

$$u_k(t) = \mu S (l_i n_j + l_j n_i) \int_{-\infty}^{\infty} \Delta u(\tau) G_{ki,j}(t - \tau)\, d\tau \tag{16.72}$$

The displacements are given by the time convolution of the slip with the derivatives of Green's function.

For the far field, Green's functions for P and S waves are given by (16.63) and (16.64). The derivatives for P waves are

$$G_{ki,j}^{\mathrm{P}} = \frac{1}{4\pi\rho\alpha^2} \frac{\partial}{\partial \xi_j} \left[\frac{1}{r} \gamma_i \gamma_k \delta \left(t - \frac{r}{\alpha} \right) \right] \tag{16.73}$$

If in the derivatives we keep only the terms that depend on the least negative power of r $(1/r)$, then we obtain

$$\frac{1}{r} \gamma_i \gamma_k \frac{\partial}{\partial \xi_j} \delta \left(t - \frac{r}{\alpha} \right) = -\frac{1}{r\alpha} \gamma_i \gamma_k \dot{\delta} \left(t - \frac{r}{\alpha} \right) \frac{\partial r}{\partial \xi_j}$$

On substituting in the direction cosine, we obtain for P waves

$$G_{ki,j}^{\mathrm{P}} = \frac{1}{4\pi\rho\alpha^3 r} \gamma_i \gamma_k \gamma_j \dot{\delta} \left(t - \frac{r}{\alpha} \right) \tag{16.74}$$

In a similar form for S waves,

$$G_{ki,j}^{\mathrm{S}} = \frac{-1}{4\pi\rho\beta^3 r} (\gamma_i \gamma_k - \delta_{ik}) \gamma_j \dot{\delta} \left(t - \frac{r}{\beta} \right) \tag{16.75}$$

This approximation is consistent with the far field. Now, we substitute (16.74) and (16.75) into (16.72) and take into account the property of the derivative of the delta function:

$$\int_{-\infty}^{\infty} \Delta u(\tau) \dot{\delta} \left(t - \frac{r}{\alpha} - \tau \right) d\tau = \Delta \dot{u} \left(t - \frac{r}{\alpha} \right) \tag{16.76}$$

The final result, after this substitution, gives for the displacements of the P and S waves in the far field

$$u_j^{\mathrm{P}} = \frac{\mu S}{4\pi\rho\alpha^3 r}(n_k l_i + n_i l_k)\gamma_i\gamma_k\gamma_j\,\Delta\dot{u}\left(t - \frac{r}{\alpha}\right) \tag{16.77}$$

$$u_j^{\mathrm{S}} = \frac{\mu S}{4\pi\rho\beta^3 r}(n_k l_i + n_i l_k)(\delta_{ij} - \gamma_i\gamma_j)\gamma_k\,\Delta\dot{u}\left(t - \frac{r}{\beta}\right) \tag{16.78}$$

It is important to notice, as we mentioned in section 16.3, that elastic displacements depend on the slip velocity or rate of slip. The source radiates elastic energy only while it is moving and ceases so to do when it stops. If the source time function is a step function $\Delta u(t) = \Delta u\, H(t)$, its derivative is the delta function, and we obtain from (16.77) and (16.78)

$$u_j^{\mathrm{P}} = \frac{M_0}{4\pi\rho\alpha^3 r}(n_k l_i + n_i l_k)\gamma_i\gamma_k\gamma_j\delta\left(t - \frac{r}{\alpha}\right) \tag{16.79}$$

$$u_j^{\mathrm{S}} = \frac{M_0}{4\pi\rho\beta^3 r}(n_k l_i + n_i l_k)(\delta_{ij} - \gamma_i\gamma_j)\gamma_k\delta\left(t - \frac{r}{\beta}\right) \tag{16.80}$$

where we have substituted in the seismic moment $M_0 = \mu\,\Delta u\, S$ (15.23). Therefore, for a step source time function, elastic displacements in the far field for P and S waves are impulses that arrive at a distance r at times, $t = r/\alpha$ and $t = r/\beta$.

16.6.1　*The radiation pattern*

The radiation pattern consists in the spatial distribution of amplitudes around the source. Let us consider a shear fracture on the plane (x_1, x_2), with slip in the x_1 direction, that is, $n = (0, 0, 1)$ and $l = (1, 0, 0)$ (Fig. 16.16). In a similar form to that for Green's function, the normalized displacements in the (x_1, x_3) plane in polar coordinates

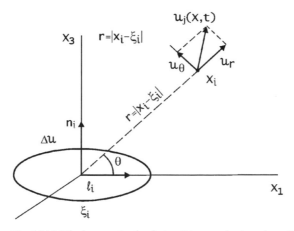

Fig. 16.16. Displacements $u(x, t)$ at a distance r due to a shear dislocation with slip in the x_1 direction on a plane normal to the x_3 axis.

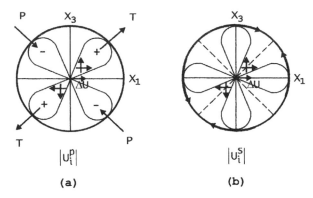

|u_i^p| |u_i^s|

(a)   **(b)**

Fig. 16.17. The radiation pattern in the plane (x_1, x_3) due to a point shear dislocation (Fig. 16.16) and the equivalent DC and PT system of forces: (a) P waves and (b) S waves.

(r, θ) for P waves are

$$u_1^P = 2\gamma_1^2\gamma_3 = \sin(2\theta)\cos\theta \tag{16.81}$$

$$u_3^P = 2\gamma_3^2\gamma_1 = \sin(2\theta)\sin\theta \tag{16.82}$$

and those for S waves are

$$u_1^S = (1 - 2\gamma_1^2)\gamma_3 = -\cos(2\theta)\sin\theta \tag{16.83}$$

$$u_3^S = (1 - 2\gamma_3^2)\gamma_1 = \cos(2\theta)\cos\theta \tag{16.84}$$

We can see that the displacement of P waves is in the radial direction and that of S waves is in the transverse direction. If we define the components u_r and u_θ in these two directions, we obtain

$$u_r^P = \sin(2\theta) \tag{16.85}$$

$$u_\theta^S = \cos(2\theta) \tag{16.86}$$

In both cases, the radiation pattern has four lobes or quadrants. For P waves the lobes have alternating directions of motion, outward or positive (compression) and inward or negative (dilation). There are two nodal planes, (x_1, x_2) and (x_3, x_2), the first corresponds to the fault plane and the second, corresponding to the normal to this and to the direction of Δu, is called the auxiliary plane. The maxima of the displacement are at 45° to the directions of l and n (Fig. 16.17(a)). In the four lobes of the radiation pattern of S waves, the motion changes direction. The maxima coincide with the directions of l and n and the nodal planes are at 45° to them (Fig. 16.17(b)). In both cases, the radiation pattern is symmetric and we can interchange n and l without changing the result. This is a consequence of the expression (16.71) being symmetric with respect to n and l. For this reason, the radiation patterns of P and S waves do not distinguish the fault plane from the auxiliary plane. This ambiguity is present in the methods used to determine the orientation of the fault plane from far-field displacements of P and S waves.

To study the radiation pattern in three dimensions we use spherical coordinates (r, θ, ϕ). The focus is located at the center of a sphere of unit radius (the focal sphere) and displacements are evaluated for points on its surface. A system of Cartesian

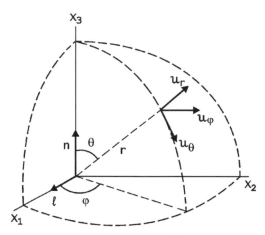

Fig. 16.18. Components of the displacement $u(r, \theta, \phi)$ in spherical coordinates for the point shear dislocation of Fig. 16.16.

coordinates (x_1, x_2, x_3) with its origin at the center of the sphere, called the source system, is defined so that (x_1, x_2) is the fault plane, n is in the x_3 direction and l is in that of x_1. For a point on the surface of the sphere, the direction cosines of r with respect to the three axes are

$$\gamma_1 = \sin \theta \cos \phi \qquad (16.87)$$

$$\gamma_2 = \sin \theta \sin \phi \qquad (16.88)$$

$$\gamma_3 = \cos \theta \qquad (16.89)$$

where θ is measured from x_3 and ϕ is measured from x_1. At each point of the spherical surface, we define a system of Cartesian coordinates with unit vectors (e_r, e_θ, e_ϕ) in the directions of the increments of r, θ, and ϕ. The components of displacements in these directions correspond to P waves and two components of S waves, respectively (Fig. 16.18). The normalized amplitudes are given by

$$\text{P:} \qquad u_r = \sin(2\theta) \cos \phi \qquad (16.90)$$

$$\text{S1:} \qquad u_\theta = \cos(2\theta) \cos \phi \qquad (16.91)$$

$$\text{S2:} \qquad u_\phi = \cos \theta \sin \phi \qquad (16.92)$$

Displacements of P waves have two nodal planes, $\theta = \pi/2$, (x_1, x_2), and $\phi = \pi/2$, (x_2, x_3). Displacements of S1 component have a nodal plane for $\phi = \pi/2$, (x_2, x_3), and are null also for points of intersection of the surface of the focal sphere and the solid angle $\theta = \pi/4$ and $3\pi/4$. Displacements of S2 have two nodal planes for $\theta = \pi/2$, (x_1, x_2), and $\phi = 0$, (x_1, x_3).

16.6.2   *The geometry of a shear fracture*

As we have seen, the orientation of a shear fracture is given by two orthogonal unit vectors n and l. These two vectors must be expressed in relation to the geographic reference system defined by the axes x_1, x_2, and x_3, positive in North, East, and nadir

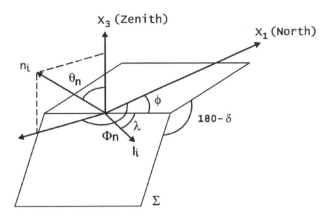

Fig. 16.19. Relations between the angles ϕ, δ, and λ, and θ_n and ϕ_n for a shear fault.

directions. With respect to these axes we define now the spherical coordinates θ measured from x_3 and ϕ measured from x_1 on the (x_1, x_2) (horizontal) plane. In reference to this system the vector $\boldsymbol{n}$ is given by

$$n_1 = \sin\theta_n \cos\phi_n \tag{16.93}$$

$$n_2 = \sin\theta_n \sin\phi_n \tag{16.94}$$

$$n_3 = \cos\theta_n \tag{16.95}$$

We do this in a similar way for the vector $\boldsymbol{l}$ (l_1, l_2, l_3). Since $\boldsymbol{l}$ and $\boldsymbol{n}$ are orthogonal unit vectors, only three components are independent. We saw in section 15.1 that a fault or shear fracture can also be defined by the angles ϕ, δ, and λ (Fig. 15.1). These angles can be expressed in terms of those defining the geographic orientations of the vectors $\boldsymbol{l}$ and $\boldsymbol{n}$ (Fig. 16.19):

$$\phi = \phi_n + \pi/2 \tag{16.96}$$

$$\delta = \theta_n \tag{16.97}$$

$$\lambda - \sin^{-1}\left(\frac{\cos\theta_l}{\sin\theta_n}\right) \tag{16.98}$$

The components of $\boldsymbol{n}$ and $\boldsymbol{l}$ referred to the geographic axes $(x_1, x_2,$ and $x_3)$ can be written in terms of ϕ, δ, and λ, in the form

$$n_1 = -\sin\delta \sin\phi \tag{16.99}$$

$$n_2 = \sin\delta \cos\phi \tag{16.100}$$

$$n_3 = -\cos\delta \tag{16.101}$$

$$l_1 = \cos\lambda \cos\phi + \cos\delta \sin\lambda \sin\phi \tag{16.102}$$

$$l_2 = \cos\lambda \sin\phi - \cos\delta \sin\lambda \cos\phi \tag{16.103}$$

$$l_3 = -\sin\lambda \sin\delta \tag{16.104}$$

In every case, the orientation of the source is given uniquely by three parameters, namely, ϕ, δ, and λ, or θ_n, ϕ_n, and θ_l. Since, as we have mentioned already, there is

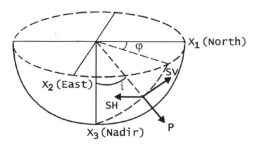

Fig. 16.20. Displacements of P waves and SV and SH components of S waves, in spherical geometry for a source at the center.

always an ambiguity with respect to the vectors n and l, the source orientation given by (θ_n, ϕ_n) and (θ_l, ϕ_l) does not assume a distinction between fault and auxiliary planes. We assumed that this is the orientation of the fault plane given in terms of ϕ, δ, and λ. If we do not know which one is the fault plane, we must give values of ϕ, δ, and λ for both nodal planes.

Displacement of P waves and of SV and SH components of S waves can be referred to the geographic coordinate axes through the direction of the seismic ray. In the focal sphere, this is a straight line from the center to the surface. If ϕ is the azimuth of the ray measured from North and i is the take-off angle of the ray measured from the downward vertical, then the direction cosines of the ray with respect to the geographic axes (North, East, and nadir) are

$$\gamma_1 = \sin i \cos \phi \tag{16.105}$$

$$\gamma_2 = \sin i \sin \phi \tag{16.106}$$

$$\gamma_3 = \cos i \tag{16.107}$$

The components of P, SV, and SH displacements along the geographic axes (Fig. 16.20) are

$$P_1 = P \sin i \cos \phi, \qquad SV_1 = SV \cos i \cos \phi, \qquad SH_1 = -SH \sin \phi$$

$$P_2 = P \sin i \sin \phi, \qquad SV_2 = SV \cos i \sin \phi, \qquad SH_2 = SH \cos \phi$$

$$P_3 = P \cos i, \qquad SV_3 = -SV \sin i, \qquad SH_3 = 0$$

where the P, SV, and SH wave amplitudes depend on the location of the observation point with respect to the source orientation. For points on the focal sphere, the observation point is given by (ϕ, i) and the orientation of the source is given by (ϕ, δ, λ) or $(\theta_n, \phi_n, \theta_l)$.

16.7   The source time function

The source time function (STF) $\Delta u(t)$ represents the slip's dependence on time and is an important characteristic of the focal mechanism. We have already considered the most simple STF, namely, the step function. There are other functions; some

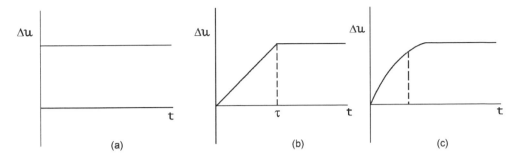

Fig. 16.21. The source time function for a slip $\Delta u(t)$: (a) a step function; (b) a ramp function with a rise time τ; and (c) an exponential function with a rise time τ.

commonly used ones, including the step function (Fig. 16.21), are

$$\Delta u(t) = \Delta u \, H(t) \tag{16.108}$$

$$\Delta u(t) = \begin{cases} \Delta u \dfrac{t}{\tau}, & 0 \le t \le \tau \\[2mm] \Delta u, & t > \tau \end{cases} \tag{16.109}$$

$$\Delta u(t) = \Delta u \, H(t)(1 - e^{-t/\tau}) \tag{16.110}$$

In all cases, the slip of the fault starts at $t = 0$, and, once it has reached its maximum value Δu, it stays constant. The fault does not return to its initial state. In the first case (16.108), $\Delta u(t)$ has the form of a step or Heaviside function such that the slip reaches its maximum value instantaneously at time $t = 0$. In the second case (16.109), $\Delta u(t)$ increases linearly from $t = 0$ to $t = \tau$, and at that time reaches its maximum value. This STF introduces a new parameter of the source, namely, τ, the time taken for the slip to reach its maximum value, or the rise time. In the third case (16.110), $\Delta u(t)$ is a continuous function for $t > 0$. The slip reaches its maximum value asymptotically with time. For the rise time, $\Delta u(\tau) = 0.63 \, \Delta u$.

We have seen in equations (16.77) and (16.78) that elastic displacements depend on the slip velocity $\Delta \dot{u}$. For this reason, the time dependence of the slip velocity is often also called the STF. For the first two models ((16.108) and (16.109)) we obtain

$$\Delta \dot{u}(t) = \Delta V \, \delta(t) \tag{16.111}$$

$$\Delta \dot{u}(t) = \Delta V \, [H(t) - H(t - \tau)] \tag{16.112}$$

In these two models, at $t = 0$, the slip velocity jumps instantaneously from 0 to its maximum value ΔV (Figs. 16.22(a) and (b)). In the first, the slip velocity is an impulse and in the second it has a duration τ with constant value. More realistic is to define a STF with a slip velocity that increases from zero to its maximum value and then decreases to zero after a time τ. A model that satisfies these conditions is a triangular

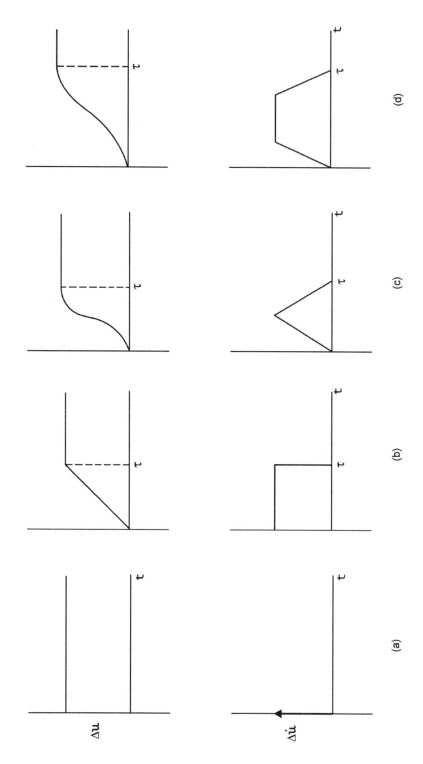

Fig. 16.22. The source time function for a slip velocity $\Delta\dot{u}(t)$ and its relation to that for a slip $\Delta u(t)$: (a) an impulsive function; (b) a rectangular function with duration τ; (c) a triangular function of duration τ; and (d) a trapezoidal function of duration τ.

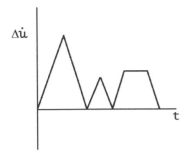

Fig. 16.23. The source time function for a slip velocity $\Delta\dot{u}(t)$ for a complex source with several events.

function (Fig. 16.22(c)):

$$\Delta\dot{u} = \begin{cases} 0, & t < 0 \\ \Delta V \dfrac{2t}{\tau}, & 0 \leq t \leq \dfrac{\tau}{2} \\ \Delta V \dfrac{2(\tau - t)}{\tau}, & \dfrac{\tau}{2} < t \leq \tau \\ 0, & t > \tau \end{cases} \tag{16.113}$$

The slip velocity increases linearly from zero at $t = 0$ to reach its maximum value (ΔV) at $t = \tau/2$ and then decreases to zero for $t = \tau$. During the first part of the process, the slip acceleration ($\Delta\ddot{u}$) is positive, whereas in the second it is negative. If we want to increase the duration of the source process we can use a STF of trapezoidal form (Fig. 16.22(d)). In this case, the slip velocity maintains its maximum value for a certain time before decreasing to zero at $t = \tau$.

The models of the STF we have mentioned represent simple sources consisting of a single event. A complex source can be represented by a STF consisting of several triangles or trapezoids of different heights (Fig. 16.23). In this way we represent with a point source a mechanism that has several accelerations ($\Delta\ddot{u} > 0$), decelerations ($\Delta\ddot{u} < 0$), and stops ($\Delta\dot{u} = 0$), during the total process of fracturing.

16.8 The equivalence between forces and dislocations

We have seen that the source of earthquakes can be represented by systems of forces (16.7) or by displacement dislocations (16.23). We will see now that the double-couple system of forces is equivalent to a shear dislocation or fracture. For a point source, the elastic displacement due to a single couple with forces in the x_1 direction and the couple's arm s in the x_3 direction ($\boldsymbol{l} = (1,0,0)$ and $\boldsymbol{n} = (0,0,1)$), according to (16.15), is

$$u_i^{\text{SC}} = \int_{-\infty}^{\infty} M G_{i1,3} \, \mathrm{d}\tau \tag{16.114}$$

Where $M = Fs$ is the moment of the couple. For a double couple with forces in the x_1 and x_3 directions ($\boldsymbol{l}_1 = (1,0,0)$, $\boldsymbol{n}_1 = (0,0,1)$, $\boldsymbol{l}_2 = (0,0,1)$, and $\boldsymbol{n}_2 = (1,0,0)$), by

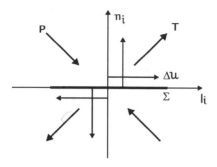

Fig. 16.24. The equivalence between a double couple (and PT system) and a point shear dislocation.

substitution into (16.16), we obtain

$$u_i^{DP} = \int_{-\infty}^{\infty} M[G_{i1,3} + G_{i3,1}] \, d\tau \qquad (16.115)$$

We substitute Green's function for the far-field P waves (16.63) into (16.114) and (16.115), and, putting $M(t) = MH(t)$, we obtain

$$u_i^{DC:P} = 2u_i^{SC:P} = \frac{M}{4\pi\rho\alpha^3 r} \gamma_1 \gamma_3 \gamma_i \delta\left(t - \frac{r}{\alpha}\right) \qquad (16.116)$$

For displacements of S waves in the far field, by substituting (16.64) into (16.114) and (16.115), we get

$$u_i^{SC:S} = \frac{-M}{4\pi\rho\beta^3 r}(\gamma_i \gamma_1 \gamma_3 - \delta_{i1}\gamma_3)\delta\left(t - \frac{r}{\beta}\right) \qquad (16.117)$$

$$u_i^{DC:S} = \frac{-M}{4\pi\rho\beta^3 r}(2\gamma_i \gamma_1 \gamma_3 - \delta_{i1}\gamma_3 - \delta_{i3}\gamma_1)\delta\left(t - \frac{r}{\beta}\right) \qquad (16.118)$$

The radiation pattern for P waves is the same as those for SC and DC models. The first studies of focal mechanisms were based on the polarity distribution of P waves, which is the same for both models. This explains the discussion about which model really corresponded to the seismic source. The radiation pattern of S waves, however, is different for each model. Analysis of S waves, especially after 1960, proved that the DC model was the one that is consistent with observations.

Let us consider now elastic displacements in the far field due to a point shear dislocation (16.72). For the slip orientation given by $\boldsymbol{n} = (0, 0, 1)$ and $\boldsymbol{l} = (1, 0, 0)$, that is, a fracture on the plane (x_1, x_2), with slip in the x_1 direction, equation (16.72) becomes

$$u_k = \mu S \int_{-\infty}^{\infty} \Delta u(G_{k1,3} + G_{k3,1}) \, d\tau \qquad (16.119)$$

This expression is equivalent to (16.115). In consequence, a double couple with one couple in the direction of the slip and the other perpendicular to the fault plane is equivalent to a shear fracture (Fig. 16.24). The same result is obtained when we compare expressions (16.79) and (16.80) for the far field of P and S waves due to a shear dislocation, substituting $\boldsymbol{l} = (1, 0, 0)$ and $\boldsymbol{n} = (0, 0, 1)$ with the expressions for a DC (16.116) and (16.118). In the comparison of the two sets of expressions, we find that the moment of the couple of forces $M = sF$ is equivalent to the seismic moment $M_0 = \mu \, \Delta u \, S$ of the shear fracture; in both cases the dimensions are force times distance.

Another more general manner to show this equivalence is to start with the equality (Aki and Richards, 1980)

$$\int_V F_k G_{ik} \, \mathrm{d}V = \int_\Sigma \Delta u \, \mu (n_k l_j + n_j l_k) G_{ik,j} \, \mathrm{d}S \tag{16.120}$$

In this equation, we search for the set of body forces F_k distributed inside of the volume V which corresponds to the shear dislocation on the surface Σ located inside of V. Inside of the volume V the coordinates are ξ_i and η_i on the surface Σ. We consider the same particular case for $\boldsymbol{n} = (0, 0, 1)$ and $\boldsymbol{l} = (1, 0, 0)$:

$$\int_\Sigma \mu \, \Delta u \, (G_{i1,3} + G_{i3,1}) \, \mathrm{d}S = \int_V F_k G_{ik} \, \mathrm{d}V \tag{16.121}$$

The integral on the left-hand side can be expressed in the form of an integral over the volume V by using the delta function $\delta(\xi_i - \eta_i)$. This function is zero for all points ξ_i in V that do not coincide with the points η_i on the surface Σ, that is, for all points of the volume that are not on the surface. The left-hand side of equation (16.121) becomes

$$\int_V \mu \, \Delta u \, (G_{i1,3} + G_{i3,1}) \delta(\xi_n - \eta_n) \, \mathrm{d}V \tag{16.122}$$

According to the properties of the derivatives of the delta function, the derivatives of G_{ij} can be written using those of $\delta(\xi_i - \eta_i)$ in the form

$$G_{ij,k} = \int_V \frac{\partial}{\partial \xi_k} [\delta(\xi_n - \eta_n)] G_{ij} \, \mathrm{d}V \tag{16.123}$$

On substituting (16.123) into (16.122), we obtain

$$\int_V \mu \, \Delta u \left(\frac{\partial \delta}{\partial \xi_3} G_{i1} + \frac{\partial \delta}{\partial \xi_1} G_{i3} \right) \delta(\xi_n - \eta_n) \, \mathrm{d}V = \int_V F_k G_{ik} \, \mathrm{d}V \tag{16.124}$$

where both integrals are defined over the same volume. If we shrink the volume to a point, for a point source, the part in brackets can be taken out of the integral and the integral becomes

$$\int_V \mu \, \Delta u \, \delta(\xi_n - \eta_n) \, \mathrm{d}V = \mu \, \Delta u \, S = M_0$$

Then, equation (16.124) becomes

$$F_k G_{ik} = M_0 \left(\frac{\partial \delta}{\partial \xi_3} G_{i1} + \frac{\partial \delta}{\partial \xi_1} G_{i3} \right) \tag{16.125}$$

In this equation we can identify the components of the force F_k that are equivalent to the shear dislocation:

$$F_1 = M_0 \frac{\partial \delta}{\partial \xi_3}, \qquad F_2 = 0, \qquad F_3 = M_0 \frac{\partial \delta}{\partial \xi_1} \tag{16.126}$$

Since the delta function represents an impulsive force, its derivatives with respect to a particular coordinate represent a dipole or a couple with its arm in that direction. For example, $\partial\delta/\partial x_3$ is a couple with its arm in the x_3 direction. Thus F_k in (16.126) is formed by two couples in the plane (x_1, x_3), in the directions of x_1 and x_3, or a double couple. The two couples have the same moment, equal to the seismic moment M_0. In consequence, a double couple on the plane (x_1, x_3) is the force system equivalent to a shear fracture on the plane (x_1, x_2), with slip in the x_1 direction. In general, the system of equivalent forces for a shear dislocation is a double couple with one couple in the direction of the slip and the other normal to the fault plane. Here we have shown this equivalence for a particular case and a point source. The equivalence can be shown in a more general form and also for an extended fracture. In the latter case, the equivalent forces are a distribution of double couples on the fault plane (Aki and Richards, 1980). Owing to this equivalence, the term double-couple source is often used as a synonym for a shear fracture.

17 THE SEISMIC MOMENT TENSOR

17.1   The definition of the moment tensor

In Chapter 15 we introduced the scalar seismic moment as a measure of the size of an earthquake. In the formulation of the theory of the source mechanism an important concept is that of the seismic moment tensor M_{ij}, and the moment tensor density per unit volume or unit surface m_{ij} (Jost and Herrmann, 1989). The relation between them is

$$M_{ij} = \int_V m_{ij} \, dV = \int_S m_{ij} \, dS \qquad (17.1)$$

The seismic moment tensor was first proposed by Gilbert (1970), who related it to the total drop in stress $\Delta\sigma$ of earthquakes. Backus and Mulcahy (1976) clarified that the moment tensor represents only that part of the internal drop in stress that is dissipated in inelastic deformations at the source.

If we consider an elastic medium in which only elastic processes occur, then, in the absence of body forces, the equation of motion (2.56) is

$$\rho \ddot{u}_i = \tau_{ij,j} \qquad (17.2)$$

Since in the real or physical situation there are besides elastic also inelastic processes occurring, the total stress is given by σ_{ij} and τ_{ij} correspond only to the pure elastic model. The equation of motion for the real situation is, then,

$$\rho \ddot{u}_i = \sigma_{ij,j} \qquad (17.3)$$

If we define the moment tensor density m_{ij} as the stress in excess of the purely elastic stress or the stress glut, it will be given by the difference

$$m_{ij} = \tau_{ij} - \sigma_{ij} \qquad (17.4)$$

If we substitute σ_{ij} from (17.4) into (17.3), we obtain

$$\rho \ddot{u}_i = \tau_{ij,j} - m_{ij,j} \qquad (17.5)$$

If we compare this equation with (16.3), in which the seismic source was represented by equivalent body forces F_i, we get

$$F_i = -m_{ij,j} \qquad (17.6)$$

Equivalent body forces can be derived from the moment tensor and both can be used to represent the seismic source. Equation (17.6) clarifies the meaning of (16.2), in which body forces were related to stresses in the focal region. Equivalent body forces correspond only to the stresses responsible for inelastic processes in the source region. The moment tensor according to (17.5) represents, precisely, those stresses that are directly related to inelastic displacements at the source of an earthquake. Just like for equivalent body forces, the seismic moment is defined only inside of the focal region, the region where the inelastic processes take place, and is zero outside of it.

If we substitute m_{ij}, according to (17.6), into (16.7), we can express elastic displacements outside the focal region u_i in terms of m_{ij} and the corresponding Green function:

$$u_i = \int_{-\infty}^{\infty} d\tau \int_{V_0} -m_{kj,j} G_{ik} \, dV \tag{17.7}$$

Integrating by parts with respect to the spatial coordinates gives

$$u_i = \int_{-\infty}^{\infty} m_{kj} G_{kj} \, d\tau + \int_{-\infty}^{\infty} d\tau \int_{V_0} m_{kj} G_{ik,j} \, dV$$

In the absence of external forces and torques, the sum of all internal forces and moments is null; then, by an appropriate choice of the origin of coordinates, $m_{kj} G_{kj} = 0$, and we obtain

$$u_i = \int_{-\infty}^{\infty} d\tau \int_{V_0} m_{kj} G_{ik,j} \, dV \tag{17.8}$$

If the moment tensor is defined only on a surface Σ, we use m_{ij}, the moment tensor density per unit surface, and we write (17.8) as a surface integral:

$$u_i = \int_{-\infty}^{\infty} d\tau \int_{\Sigma} m_{kj} G_{ik,j} \, dS \tag{17.9}$$

Equations (17.8) and (17.9) show that elastic displacements outside of the focal region can be derived from the seismic moment tensor and the derivatives of Green's function integrated over the focal region (V_0 or Σ). Since we have not specified its form, m_{ij} can represent a very general type of source. It corresponds to any system of internal body forces according to (17.6), provided that the nett effect of their sum and the sum of their moments are null. The moment tensor is, thus, a very convenient form in which to represent the source of an earthquake in a general way.

For a point source, equations (17.8) and (17.9) can be written in a compact form using an asterisk to express time convolution:

$$u_i = M_{kj} * G_{ik,j} \tag{17.10}$$

The physical meaning of the moment tensor can be understood in relation to the equivalent body forces. According to (16.7), the elastic displacements are given by

$$u_i(x_s, t) = \int_{-\infty}^{\infty} d\tau \int_{V_0} F_k(\xi_s, \tau) G_{ik}(x_s, t; \xi_s, \tau) \, dV \tag{17.11}$$

If we perform a Taylor expansion of G_{ik} around the origin, $\xi_k = 0$, the first three terms are

$$G_{ik}(\xi_s) = G_{ik}(0) + \xi_s \frac{\partial G_{ik}}{\partial \xi_s} + \frac{1}{2} \xi_n \xi_s \frac{\partial^2 G_{ik}}{\partial \xi_n \, \partial \xi_s} + \cdots \tag{17.12}$$

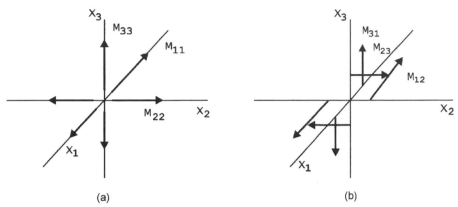

Fig. 17.1. Representations of the components of the seismic moment tensor: (a) for $i = j$ and (b) for $i \neq j$ (only three components are shown).

Taking only the first two terms and substituting into (17.11), the first term is zero by virtue of the condition that the sum of internal forces must be zero:

$$\int_{V_0} F_k(0) G_{ik}(0, x) \, dV = 0 \tag{17.13}$$

Therefore, we obtain

$$u_i = \int_{-\infty}^{\infty} d\tau \int_{V_0} F_k \xi_j G_{ik,j} \, dV \tag{17.14}$$

By comparison of (17.14) with (17.8), we find that

$$m_{jk} = \xi_j F_k \tag{17.15}$$

The moment tensor density m_{ij} corresponds to the first nonzero term in the Taylor expansion of (17.12), and thus it is called the first-order moment tensor. This term is associated with the first derivatives of Green functions. We can also derive moment tensors of higher order that are associated with higher derivatives of Green functions, for example, the second-order moment tensor, the third term in (17.12), which represents its variation with space.

According to (17.15), the components of m_{ij} correspond to force couples or dipoles. The components m_{11}, m_{22}, and m_{33} are linear dipoles without moments, that is, the arm is in the same direction as the forces. The other components have their arms perpendicular to the forces and are couples with moments (Fig. 17.1). The condition of zero nett moment implies that the tensor is symmetric, $m_{ij} = m_{ji}$; couples with opposite moment must be equal. We have seen that Green's function represents displacements due to impulsive forces, whereas its derivatives represent displacements due to couples or dipoles of impulsive forces. In consequence, according to equations (17.8) and (17.9), elastic displacements are given by the convolution of distributions of dipoles or couples of forces representing the source (moment tensor) with displacements due to couples of impulsive forces (derivatives of Green's function).

The components of the moment tensor are expressed in relation to a coordinate system of reference, usually, the geographic system, with its origin at the focus of the

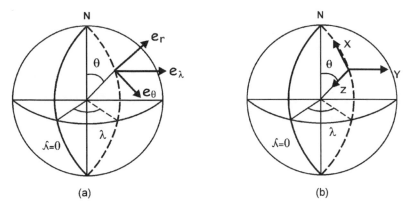

Fig. 17.2. Systems of coordinates at the focus referred to the geocentric geographic system of the Earth (r, θ, ϕ). (a) The system formed by unit vectors (e_r, e_θ, e_ϕ) (zenith, South, East). (b) The system in geographic directions (x, y, z) or (x_1, x_2, x_3) (North, East, nadir).

earthquake, for example, the Cartesian coordinate system (x_1, x_2, x_3) or (x, y, z), positive in the directions North, East, and nadir (Fig. 17.2(b)) or also North, West, and zenith. Another system that also is used is referred to as geocentric spherical coordinates of the focus (r, θ, ϕ), where r is in the radial direction, θ is the geocentric colatitude and ϕ is the geocentric longitude. At the focus a Cartesian coordinate system with unit vectors e_r, e_θ, and e_ϕ (in the directions of positive increments of r, θ, and ϕ) is formed. This system has positive axes in the directions zenith, South, and East (Fig. 17.2(a)). The correspondence among the six components of the moment tensor in the three systems is

$$M_{11} = M_{xx} = M_{\theta\theta}$$
$$M_{22} = M_{yy} = M_{\phi\phi}$$
$$M_{33} = M_{zz} = M_{rr}$$
$$M_{12} = M_{xy} = -M_{\theta\phi}$$
$$M_{13} = M_{xz} = M_{r\theta}$$
$$M_{23} = M_{yz} = -M_{r\phi}$$

17.2   The moment tensor and elastic dislocations

If we compare equations (17.9) and (16.23), we can define the moment tensor density corresponding to a dislocation with slip Δu on a surface Σ of normal n as

$$m_{ij} = C_{ijkl} \, \Delta u_k \, n_l \tag{17.16}$$

and that for an isotropic medium as

$$m_{ij} = \lambda n_k \, \Delta u_k \, \delta_{ij} + \mu(\Delta u_i \, n_j + \Delta u_j \, n_i) \tag{17.17}$$

If the slip direction is given by the unit vector l equation (17.17) becomes

$$m_{ij} = \Delta u \, [\lambda l_k n_k \delta_{ij} + \mu(l_i n_j + l_j n_i)] \tag{17.18}$$

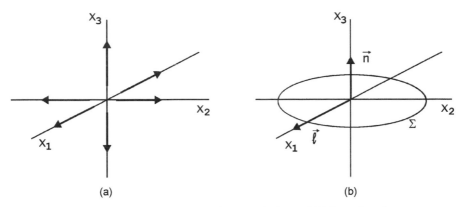

Fig. 17.3. Representations of the source: (a) for an explosion and (b) for a shear fracture.

From this expression we can find the moment tensor for various types of sources, by specifying the orientations of n_i and l_i.

17.2.1 An explosive source

An explosive source may be considered as an expansion along the three coordinate axes. This situation is represented by linear dipoles (l and n in the same direction) along each axis, that is, $(1, 0, 0)$, $(0, 1, 0)$, and $(0, 0, 1)$. The moment tensor is the sum of the three and, using (17.18), we obtain (Fig. 17.3(a))

$$m_{ij} = K \, \Delta u \begin{pmatrix} 1 & 0 & 0 \\ 0 & 1 & 0 \\ 0 & 0 & 1 \end{pmatrix} \tag{17.19}$$

where $K = \lambda + \frac{2}{3}\mu$ is the bulk modulus (2.22). The sum of elements of the principal diagonal gives the increase in volume per unit volume:

$$m_{11} + m_{22} + m_{33} = 3K \, \Delta u \tag{17.20}$$

17.2.2 Shear fracture

In a shear fracture slip, Δu is along the fault plane; that is, n and l are perpendicular. Using equation (17.18) and the definition of M_0 after integration over the source surface of area S, the moment tensor for a point source is

$$M_{ij} = M_0(l_i n_j + l_j n_i) \tag{17.21}$$

For a particular case in which the fault plane is the (x_1, x_2) plane, that is, $n = (0, 0, 1)$, and the slip is in the x_1 direction, $l = (1, 0, 0)$ (Fig. 17.3(b)), we obtain

$$M_{ij} = M_0 \begin{pmatrix} 0 & 0 & 1 \\ 0 & 0 & 0 \\ 1 & 0 & 0 \end{pmatrix} \tag{17.22}$$

The sum of the principal diagonal is null, indicating that there is no change in volume.

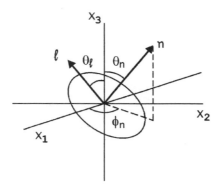

Fig. 17.4. Orientations of the unit vectors l and n, and definitions of the angles θ_n, ϕ_n, and θ_l.

The unit vectors n and l referred to the geographic reference system can be written in terms of the angles θ_n and ϕ_n, and θ_l and ϕ_l, respectively (Fig. 17.4). By substitution into (17.21) of the expressions for the components of n, according to (16.93)–(16.95), and similarly for l, we obtain for the six components of the normalized moment tensor

$$m_{11} = 2 \sin \theta_n \cos \phi_n \sin \theta_l \cos \phi_l$$

$$m_{22} = 2 \sin \theta_n \sin \phi_n \sin \theta_l \sin \phi_l$$

$$m_{33} = 2 \cos \theta_n \cos \theta_l$$

$$m_{12} = \sin \theta_l \cos \phi_l \sin \theta_n \sin \phi_n + \sin \theta_l \sin \phi_l \sin \theta_n \cos \phi_n$$

$$m_{13} = \sin \theta_l \cos \phi_l \cos \theta_n + \cos \theta_l \sin \theta_n \cos \phi_n$$

$$m_{23} = \sin \theta_l \sin \phi_l \cos \theta_n + \sin \phi_n \sin \theta_n \cos \theta_l \qquad (17.23)$$

As we have seen, a shear fracture can also be specified in terms of the angles ϕ, δ, and λ (section 16.6). Using the relations between ϕ, δ, and λ, and n_i and l_i, equations (16.99)–(16.104), from (17.21), the components of the moment tensor (Fig. 17.5) are

$$m_{11} = -\sin \delta \cos \lambda \sin(2\phi) - \sin(2\delta) \sin^2 \phi \sin \lambda$$

$$m_{22} = \sin \delta \cos \lambda \sin(2\phi) - \sin(2\delta) \cos^2 \phi \sin \lambda$$

$$m_{33} = \sin(2\delta) \sin \lambda$$

$$m_{12} = \sin \delta \cos \lambda \cos(2\phi) + \tfrac{1}{2} \sin(2\delta) \sin(2\phi) \sin \lambda$$

$$m_{13} = -\sin \lambda \sin \phi \cos(2\delta) - \cos \delta \cos \lambda \cos \phi$$

$$m_{23} = \cos \phi \sin \lambda \cos(2\delta) - \cos \delta \cos \lambda \sin \phi \qquad (17.24)$$

In (17.23), the expressions are symmetric with respect to n and l, and do not imply that one can select the fault plane from the two possible planes. In (17.24), the equations are related to the orientation of motion on the selected fault plane (Fig. 17.5). Naturally, the result is the same for values of ϕ, δ, and λ of the second plane.

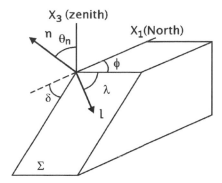

Fig. 17.5. Relation between unit vectors l and n, and the angles ϕ, δ, and λ that define the motion in a fault.

17.3   Eigenvalues and eigenvectors

As we saw in the discussion of stress and strain tensors (section 2.1), we can also perform an eigenvalue and eigenvector analysis of the moment tensor. Since this tensor is symmetric, its eigenvalues are real and its eigenvectors are mutually orthogonal. They satisfy the equation

$$(M_{ij} - \delta_{ij}\sigma)\nu_k = 0 \tag{17.25}$$

where the three eigenvalues σ_1, σ_2, and σ_3 are the roots of the cubic equation resulting from putting the determinant of (17.25) equal to zero. By substituting each eigenvalue into (17.25) we obtain the three eigenvectors ν_k^1, ν_k^2, and ν_k^3, which form the principal axes. In reference to these axes, the moment tensor has the form

$$M_{ij} = \begin{pmatrix} \sigma_1 & 0 & 0 \\ 0 & \sigma_2 & 0 \\ 0 & 0 & \sigma_3 \end{pmatrix} \tag{17.26}$$

In this system, the moment tensor is formed by three linear dipoles in the direction of the principal axes and thus represents the principal stresses. If we order the eigenvalues $\sigma_1 > \sigma_2 > \sigma_3$, then σ_1 corresponds to the greatest stress, σ_3 to the least stress, and σ_2 to the intermediate stress. The sum of the elements of the principal diagonal is the first invariant of the tensor and has the same value for any reference system:

$$M_{11} + M_{22} + M_{33} = \sigma_1 + \sigma_2 + \sigma_3 \tag{17.27}$$

This sum represents the change in volume, as we saw for the explosive source. Thus we can define the isotropic part of the moment tensor as

$$\sigma_0 = \tfrac{1}{3}(\sigma_1 + \sigma_2 + \sigma_3) \tag{17.28}$$

If we subtract this from the tensor M_{ij}, we obtain the deviatoric tensor M'_{ij}, the sum of whose diagonal elements is always zero and does not include changes in volume:

$$M'_{ij} = M_{ij} - \delta_{ij}\sigma_0 \tag{17.29}$$

According to (17.29), the moment tensor can be separated into two tensors, one isotropic, $M_{ij}^0 = \delta_{ij}\sigma_0$, and the other deviatoric, M_{ij}',

$$M_{ij} = M_{ij}^0 + M_{ij}' \tag{17.30}$$

Changes in volume are thus separated from other parts of the moment tensor. The moment tensor that represents an explosive source (17.19) is purely isotropic and that for a shear fracture (17.22) is purely deviatoric.

If we represent the moment tensor for an explosive source referred to its principal axis, we obtain the same result as that in (17.19). An explosive source is purely isotropic and any reference system is equivalent to the principal axes. For the shear fracture of (17.22) (the fault plane is normal to x_3 and slip is in the x_1 direction), the eigenvalues of the matrix (17.22) are 1, -1, and 0. The eigenvectors are found by substitution into (17.25), resulting in $(1/\sqrt{2}, 0, 1/\sqrt{2})$ for $\sigma = 1$, and $(1/\sqrt{2}, 0, -1/\sqrt{2})$ for $\sigma = -1$. The tensor referred to its principal axes is

$$M_{ij} = M_0 \begin{pmatrix} 1 & 0 & 0 \\ 0 & 0 & 0 \\ 0 & 0 & -1 \end{pmatrix} \tag{17.31}$$

This tensor represents two linear dipoles of positive and negative forces or tension and pressure forces along the principal axes, that is, in the (x_1, x_3) plane at 45° to the direction of slip. Thus, for a shear fracture, eigenvectors corresponding to the eigenvalues σ_1 and σ_3 define the principal axes of stress or pressure and tension axes, P and T. The third axis corresponding to the zero eigenvalue is the null axis B. In terms of the unit vectors P and T, the moment tensor is given by

$$M_{ij} = M_0(T_i T_i - P_j P_j) \tag{17.32}$$

This result is analogous to that found in section 16.2 regarding the equivalence of a double couple to pressure and tension forces at 45° to the couples.

17.4   Types of sources and separation of the moment tensor

We have already said that the moment tensor represents a very general type of source. The analysis of its eigenvalues indicates, in each case, the type of source. The most general case corresponds to three different eigenvalues, $\sigma_1 \neq \sigma_2 \neq \sigma_3$, whose sum is not zero, $\sigma_1 + \sigma_2 + \sigma_3 \neq 0$. Then, the source has changes in volume and, after separation of the isotropic part (17.29), the deviatoric part is of a general type, not necessarily a shear fracture or double couple.

If $\sigma_1 = \sigma_2 = \sigma_3$, as we have seen, the source is an isotropic expansion or contraction depending on the sign. In each case $\sigma_1 + \sigma_2 + \sigma_3$ represents the increase or decrease in volume. For positive sign the source represents an explosion.

For sources without nett volume changes, $\sigma_1 + \sigma_2 + \sigma_3 = 0$, the moment tensor is purely deviatoric. This condition is often imposed on earthquake sources. In this case, only two of the eigenvalues are independent, since $\sigma_2 = -\sigma_1 - \sigma_3$. For a shear fracture or double-couple source, the moment tensor is deviatoric and must satisfy the conditions $\sigma_3 = -\sigma_1$ and $\sigma_2 = 0$.

Earthquake sources are thought to be shear fractures or nearly so. However, this need not always be the case and even the possibility of the occurrence of changes in volume can not be completely ruled out. Methods of inversion of the moment tensor from observations (17.7) do not impose any condition and result, for some earthquake sources in the presence of certain amounts of anisotropic and nondouble-couple components. For this reason, it is convenient to separate the moment tensor into three parts, one isotropic, corresponding to changes in volume, one of pure shear fracture or a double couple (DC), and a third that may be of various kinds (Strelitz, 1989). This analysis is called partition or separation of the moment tensor and can be expressed by writing

$$\mathbf{M} = \mathbf{M}^0 + \mathbf{M}^{DC} + \mathbf{M}^R \tag{17.33}$$

The isotropic part (17.28) has already been defined. Partition of the deviatoric part $(\mathbf{M}^{DC} + \mathbf{M}^R)$ can be done in several ways. The simplest is to separate this part into two DCs, major and minor. To do this we take into account that, for a deviatoric tensor, $\sigma_2 = -\sigma_1 - \sigma_3$, and obtain

$$
\begin{pmatrix}
\sigma_1 & 0 & 0 \\
0 & \sigma_2 & 0 \\
0 & 0 & \sigma_3
\end{pmatrix}
=
\begin{pmatrix}
\sigma_1 & 0 & 0 \\
0 & -\sigma_1 & 0 \\
0 & 0 & 0
\end{pmatrix}
+
\begin{pmatrix}
0 & 0 & 0 \\
0 & -\sigma_3 & 0 \\
0 & 0 & \sigma_3
\end{pmatrix}
\tag{17.34}
$$

The two DCs have different orientations, the major DC with moment $M_0 = \sigma_1$ and the minor one with $M_0 = \sigma_3$.

A more efficient separation is that proposed by Knopoff and Randall (1970):

$$
\begin{pmatrix}
\sigma_1 & 0 & 0 \\
0 & \sigma_2 & 0 \\
0 & 0 & \sigma_3
\end{pmatrix}
=
\begin{pmatrix}
\frac{1}{2}(\sigma_1 - \sigma_3) & 0 & 0 \\
0 & 0 & 0 \\
0 & 0 & -\frac{1}{2}(\sigma_1 - \sigma_3)
\end{pmatrix}
+
\begin{pmatrix}
-\sigma_2/2 & 0 & 0 \\
0 & \sigma_2 & 0 \\
0 & 0 & -\sigma_2/2
\end{pmatrix}
\tag{17.35}
$$

As before $\sigma_2 = -\sigma_1 - \sigma_3$. The first term is a DC source. The second is called a compensated linear vector dipole (CLVD). Its physical meaning is a sudden change in the shear modulus in a direction normal to the fault plane, without changes in volume. The source represented by a DC plus a CLVD corresponds to a shear fracture in which, during the rupture process, the shear modulus in the focal region changes suddenly. This separation represents the best solution that maximizes the DC part of the source.

In conclusion, a seismic point source of general type can be represented by the moment tensor. This source may involve changes in volume, shear fracture and sudden changes in rigidity at the source, and thus can be separated in the form

$$\mathbf{M} = \mathbf{M}^0 + \mathbf{M}^{DC} + \mathbf{M}^{CLVD} \tag{17.36}$$

This partition separates shear fracture, considered the standard model for the source of earthquakes, from other effects that may also occur. The isotropic part is presupposed to be zero in many problems. A deviatoric source is formed by the DC plus CLVD sum. A deviation from a pure DC is sometimes represented by $\delta = |\sigma_3/\sigma_1|$, the ratio of the greatest and least eigenvalues. For a pure DC, $\delta = 1$.

When the moment tensor is obtained from observations, the presence of non-DC components may be due to errors in observations or to propagation effects that have not been taken into account, rather than to the source itself. There is always a certain amount of ambiguity in distinguishing between effects that are due to the source and those due to propagation. Perfect separation of these two effects is not always possible.

17.5   Displacements due to a point source

According to (17.10), displacements due to a point source can be expressed by a time convolution of the moment tensor with derivatives of the Green function:

$$u_i(x_n, t) = \int_{-\infty}^{\infty} M_{kj}(\tau) G_{ik,j}(t - \tau)\, d\tau \tag{17.37}$$

Derivatives of the Green function for P and S waves in the far field for an infinite, homogeneous isotropic medium are given by (16.74) and (16.75). For P waves, time convolution according to (16.76) is given by

$$\int_{-\infty}^{\infty} M_{ij}(\tau)\delta\left(t - \frac{r}{\alpha} - \tau\right) d\tau = \dot{M}_{ij}\left(t - \frac{r}{\alpha}\right) \tag{17.38}$$

If we separate the modulus and time dependence from the orientation in the form $M_{ij}(t) = M_0(t)m_{ij}$, according to (16.77) and (16.78), we obtain the elastic displacements of P and S waves in the far field:

$$u_k^{\mathrm{P}} = \frac{\dot{M}_0(t - r/\alpha)}{4\pi\rho\alpha^3 r}\gamma_i\gamma_j\gamma_k m_{ij} \tag{17.39}$$

$$u_k^{\mathrm{S}} = \frac{\dot{M}_0(t - r/\beta)}{4\pi\rho\beta^3 r}(\delta_{ik} - \gamma_i\gamma_k)\gamma_j m_{ij} \tag{17.40}$$

Elastic displacements depend on the time derivative of the moment or the moment rate. Displacements for P waves and for SV and SH components of S waves are (calling the factors in (17.39) and (17.40) A and B)

$$u_{\mathrm{P}} = A\gamma_i\gamma_j m_{ij} \tag{17.41}$$

$$u_{\mathrm{SV}} = -B\,\mathrm{SV}_i\,\gamma_j m_{ij} \tag{17.42}$$

$$u_{\mathrm{SH}} = B\,\mathrm{SH}_i\,\gamma_j m_{ij} \tag{17.43}$$

where SV_i and SH_i are unit vectors in the SV and SH directions and we have taken into account that $\mathrm{SV}_i\gamma_i = 0$ and $\mathrm{SH}_i\gamma_i = 0$. Since γ_i is a unit vector in the ray's direction, P displacements are in the same direction and those of S are perpendicular (Fig. 16.20). If the problem is referred to geographic axes (North, East, and nadir), for a homogeneous medium, γ_i, SV_i, and SH_i can be given in terms of the azimuth ϕ and take-off angle i of the ray:

$$\gamma_1 = \sin i \cos\phi$$

$$\gamma_2 = \sin i \sin\phi \tag{17.44}$$

$$\gamma_3 = \cos i$$

$$\mathrm{SV}_1 = \cos i \cos\phi$$

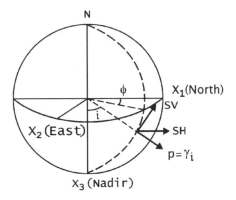

Fig. 17.6. Displacements of P, SH, and SV waves, and their relations to the angles i and ϕ of the ray at the focus.

$$SV_2 = \cos i \sin \phi \qquad (17.45)$$

$$SV_3 = -\sin i$$

$$SH_1 = -\sin \phi$$

$$SH_2 = \cos \phi \qquad (17.46)$$

$$SH_3 = 0$$

The displacements u_P, u_{SV}, and u_{SH} can finally be given in terms of i and ϕ (Fig. 17.6) and components of the moment tensor, by substituting equations (17.44)–(17.46) into equations (17.41)–(17.43):

$$u_P = A\{\sin^2 i\,[\cos^2 \phi\, m_{11} + \sin^2 \phi\, m_{22} + \sin(2\phi)\, m_{12}]$$

$$+ \cos^2 i\, m_{33} + \sin(2i)(\cos \phi\, m_{13} + \sin \phi\, m_{23})\} \qquad (17.47)$$

$$u_{SV} = B\{\tfrac{1}{2}\sin(2i)[\cos^2 \phi\, m_{11} + \sin^2 \phi\, m_{22} - m_{33} + \sin(2\phi)\, m_{12}]$$

$$+ \cos(2i)(\cos \phi\, m_{13} + \sin \phi\, m_{23})\} \qquad (17.48)$$

$$u_{SH} = B\{\sin i\,[\tfrac{1}{2}\sin(2\phi)\, m_{22} - \tfrac{1}{2}\sin(2\phi)\, m_{11} + \cos(2\phi)\, m_{12}]$$

$$+ \cos i\,(\cos \phi\, m_{23} - \sin \phi\, m_{13})\} \qquad (17.49)$$

These three equations are for a general form of the moment tensor without assuming any particular condition.

17.6 The temporal dependence

For a point-source moment tensor, if all of its components have the same time dependence, the source time function is given by $M_0(t)$. As we saw in (17.39) and (17.40), the displacements depend on the moment rate $\dot{M}(t)$ and its time dependence is also called the STF. This function represents the form according to which the moment rate changes with time and its integral is the scalar seismic moment M_0.

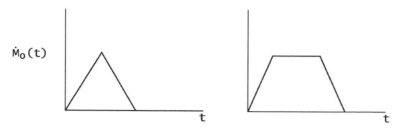

Fig. 17.7. The temporal dependence of $\dot{M}_0(t)$ for a simple source.

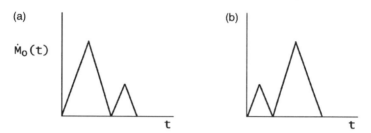

Fig. 17.8. The temporal dependence of $\dot{M}_0(t)$ for a complex source. (a) The main moment release occurs at the beginning. (b) The main moment release occurs at the end.

Just like for the time dependence of the slip rate, for a simple source a commonly used function for $\dot{M}_0(t)$ is a triangle or a trapezoid (Fig. 17.7). We have discussed the properties of this STF in section 16.7 (Fig. 16.22). Since the size of an earthquake is given by M_0, the same size will result in two different functions of the moment rate, one with a greater modulus and shorter duration and another a with smaller modulus and longer duration. We must remember that this is a point-source representation, whereby the duration of the moment rate function is related to an extended source with a specific duration of the fracturing process and thus to its dimensions. For a complex source, the moment rate may be represented by several triangles of different sizes (Fig. 17.8). The relative sizes of these sources show how the moment's radiation with time or moment release takes place. In some cases, the greatest part of the moment release occurs during the first part of the shock and a minor part follows later (Fig. 17.8(a)). Another possibility is that there is first a small moment release and then, later, the main part (Fig. 17.8(b)). The small event may be considered as an aftershock in the first case and as a foreshock in the second. Since these events are parts of the fracturing process, the total duration of the source includes the complete moment release.

17.7   Inversion of the moment tensor

According to equations (17.8) and (17.9), for a point source (17.10), the elastic displacements are linear functions of the components of the moment tensor and derivatives of Green functions. In an explicit form, this is given in equations (17.47)–(17.49) for P and S waves in the far field of an infinite, homogeneous isotropic medium. This linear dependence makes it possible to determine the six components of the moment tensor from observations of the elastic displacements by linear inversion.

For a point source, the elastic displacements in the far field can be expressed in the time domain as a convolution (17.37):

$$u_k(t) = M_{ij}(t) * G_{ki,j}(t) \tag{17.50}$$

Taking the Fourier transform, in the frequency domain, the relation is a product of their transforms:

$$\bar{u}_k(\omega) = \bar{M}_{ij}(\omega)\bar{G}_{ik,j}(\omega) \tag{17.51}$$

In both cases, inversion consists in the determination of the six components of M_{ij} from observed values of elastic waves u_i and components of appropriate functions for $G_{ki,j}$ that are assumed to be known. Depending on the waves used (P, S, LR, or LQ) and the characteristics of the problem, the medium can be approximated by an infinite homogeneous medium, a half-space, or a layer medium using a flat or spherical Earth. Naturally, the Green functions become more complicated with increasing complexity of the model. Derivatives of Green functions have in general 27 components that for some particular cases can be reduced in number to eight.

If we impose no conditions, there are six components of the moment tensor to be determined. If the source is assumed to be purely deviatoric, $M_{11} + M_{22} + M_{33} = 0$, that is, $M_{33} = -M_{11} - M_{22}$ and the number of unknowns is five. This is a linear condition and the problem remains linear. If we assumed the source to be a shear fracture or DC, the condition is that the determinant of M_{ij} is zero, and the problem ceases to be linear. For this reason this condition is not imposed for linear inversions.

For the linear problem, equations (17.50) and (17.51) may be written in matrix form as

$$\boldsymbol{u} = \mathbf{GM} \tag{17.52}$$

We need a number of observations larger than the number of unknowns (six or five), and solutions are obtained by a least squares procedure (15.4):

$$\mathbf{M} = (\mathbf{G}^{\mathrm{T}}\mathbf{G})^{-1}\mathbf{G}^{\mathrm{T}}\boldsymbol{u} \tag{17.53}$$

Or, using the generalized inverse (15.7),

$$\mathbf{M} = \mathbf{U}^{\mathrm{T}}\mathbf{\Lambda}^{-1}\mathbf{V}\boldsymbol{u} \tag{17.54}$$

where $\mathbf{U}$ and $\mathbf{V}$ are matrices formed by the eigenvectors of $\mathbf{G}^{\mathrm{T}}\mathbf{G}$ and $\mathbf{GG}^{\mathrm{T}}$, respectively, and $\mathbf{\Lambda}$ is a diagonal matrix formed by the eigenvalues of $\mathbf{G}^{\mathrm{T}}\mathbf{G}$. In practice, there are several methods for the solution of this problem; some are presented in Chapter 19. Some methods use linear inversion whereas others use iterative procedures minimizing the differences between theoretical and observed displacements.

The observed data are the amplitudes of seismic waves recorded by seismographs. The problem can be solved in the time (17.50) or frequency (17.51) domain. In the frequency domain, the problem is linear either for the real or for the imaginary part of the spectrum but not for the spectral amplitude. In all cases, observations at different locations and at different times or frequencies are used so that the problem is well conditioned. Generally, we obtain the six components of the moment tensor or five assuming that there is no change in volume. Since solutions do not necessarily correspond to a DC source, the moment tensors obtained are separated into a DC and a non-DC part, as we saw in section 17.4. The main advantage of using the moment tensor formalism is that the source problem can be solved by linear inversion. Also we can represent the source in a general form and investigate whether there are volume changes and non-DC components.

18 MODELS OF FRACTURE

18.1 Source dimensions. Kinematic models

In Chapters 16 and 17 we considered in some detail the characteristics and displacement field corresponding to point sources. A more complete representation of the seismic source must include its dimensions and consider its effects on wave radiation. The first considerations of the dimensions of the seismic focus proposed models consisting in spherical cavities of finite radii with uniform distributions of stresses on their surfaces (Jeffreys, 1931; Nishimura, 1937; Scholte, 1962).

The first models for extended sources of shear fracture were kinematic models consisting in slip that propagates with a constant velocity over a surface of finite area. Ben Menahem (1961, 1962) described extended sources in terms of distributions of single and double couples propagating with a certain velocity over a rectangular surface, and determined the corresponding displacements of body and surface waves. Berckhemer (1962) studied the effect of a circular fracture of finite radius that propagates from its center on the width of temporal pulses. Burridge and Knopoff (1964) treated shear dislocations that propagate over a certain area and showed their equivalence to propagating double couples. Haskell (1964, 1966) proposed a rectangular model of fracture, and Savage (1966) proposed an elliptical fault and studied the effects on the spectra of body and surface waves. Brune (1970) presented a model with shear stresses suddenly applied to a circular fault, and studied elastic displacements in near and far fields. More recent kinematic models include propagating shear fractures on finite faults with variable slip, rupture velocity, and rise time (Hartzell, 1989).

Let us consider first some general characteristics of kinematic models of extended sources represented by a surface Σ over which a shear dislocation $\Delta u(\xi_i)$ propagates with a constant velocity v in one direction, from the origin ($\xi_i = 0$) to a final point over a distance L (Fig. 18.1) (Aki and Richards, 1980). The velocity of the fracture's propagation is assumed to be constant and less than the velocity of wave propagation ($v < \beta < \alpha$); that is, we treat subsonic fractures (a common value is $v = 0.7\beta$). From equations (16.71) and (16.77), the displacements of P waves in the far field for an infinite, homogeneous, isotropic medium can be written as

$$u_i^P(x_j, t) = \frac{\mu}{4\pi\alpha^3\rho} \int_\Sigma \frac{R(n_k, l_k, \gamma_k)}{r} \Delta\dot{u}\left(\xi_i, t - \frac{r}{\alpha}\right) dS \tag{18.1}$$

where $r = |x_i - \xi_i|$ is the distance from the point of observation x to a point of the source ξ_i where the slip Δu is located at each moment and $R(n_k, l_k, \gamma_k)$ is the radiation pattern that depends on the orientation of the source (l, n) and the position of the observation point (γ_i). If we are interested only in the wave form as a function of time at a certain

337

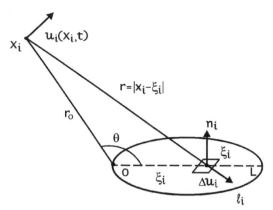

Fig. 18.1. An extended source of dimension L, with slip Δu at ξ_i and an elastic displacement u at x_i.

distance r_0 from the origin of ξ_i (Fig. 18.1), we need only consider the integral (section 16.3)

$$u(t) = \int_\Sigma \Delta \dot{u}\left(\xi_i, t - \frac{r}{\alpha}\right) \mathrm{d}S \tag{18.2}$$

Expressing r in terms of a Taylor expansion about r_0, we obtain

$$r = r_0 + \xi_i \frac{\partial r}{\partial \xi_i} + \frac{1}{2} \xi_i^2 \frac{\partial^2 r}{\partial \xi_i^2} + \cdots \tag{18.3}$$

and, since (16.53) applies,

$$\frac{\partial r}{\partial \xi_i} = -\gamma_i \tag{18.4}$$

If we keep only the first-order term of ξ_i and neglect those involving higher powers, we obtain

$$r = r_0 - \xi_i \gamma_i \tag{18.5}$$

Since the first neglected term is ξ_i^2/r_0, and the maximum value of $|\xi_i|$ is L, the approximation given by (18.5) is good for displacements of wave length λ greater than the neglected term, that is, for $\lambda r_0 \gg L^2$. For these wave lengths, elastic displacements of P waves are given by (18.2):

$$u(t) = \int_\Sigma \Delta \dot{u}\left(\xi_i, t - \frac{r_0 - \xi_i \gamma_i}{\alpha}\right) \mathrm{d}S \tag{18.6}$$

Its Fourier transform is

$$U(\omega) = \int_\Sigma \mathrm{d}S \int_{-\infty}^{\infty} \Delta \dot{u}(t) \exp\left[-\mathrm{i}\omega\left(t - \frac{r_0 - \xi_i \gamma_i}{\alpha}\right)\right] \mathrm{d}t \tag{18.7}$$

Since the transform of Δu is

$$\Delta U(\omega) = \int_{-\infty}^{\infty} \Delta u(t)\, \mathrm{e}^{-\mathrm{i}\omega t}\, \mathrm{d}t \tag{18.8}$$

we obtain for $U(\omega)$

$$U(\omega) = e^{i\omega r_0/\alpha} \int_{\Sigma} i\omega \, \Delta U(\omega, \xi_i) \, e^{-i\omega\gamma_i\xi_i/\alpha} \, ds \tag{18.9}$$

where we have used that, if the transform of $\Delta u(t)$ is $\Delta U(\omega)$, that of $\Delta \dot{u}(t)$ is $i\omega \, \Delta U(\omega)$. Thus, the transform of elastic displacements $U(\omega)$ has the form of a spatial transform over the fault plane of the transform of the slip $\Delta U(\omega)$. In the exponential, $\omega\gamma_i/\alpha = k\gamma_i$ is the projection of the wave number k onto the fault plane Σ.

If the slip has a step-function time dependence $\Delta u(t, \xi_i) = \Delta u(\xi_i) \, H(t)$, its transform is $\Delta u(\xi_i)/(i\omega)$. By substituting into (18.9), we obtain

$$U(\omega) = e^{i\omega r_0/\alpha} \int_{\Sigma} \Delta u(\xi_i) \, e^{-i\omega\xi_i\gamma_i/\alpha} \, dS \tag{18.10}$$

If we take the limit for low frequencies, then, when ω tends to zero, we obtain

$$U(0) \simeq \mu \int_{\Sigma} \Delta u(\xi_i) \, dS \simeq \mu \, \Delta\bar{u} \, S \simeq M_0 \tag{18.11}$$

For low frequencies, the spectral amplitudes are proportional to the seismic moment. The proportionality depends on the factors present in (18.1). For high frequencies, as they tend to infinity, if the slip does not change sign, the spectral amplitudes tend to zero. The form in which $U(\omega)$ tends to zero from the limit of $U(0)$ depends on the form of the slip function.

In general, the form of amplitude spectra corresponding to a source with finite dimensions is the following: $U(\omega)$ is constant for a range of low frequencies and starts to decrease from a certain frequency that, as will be shown later, is proportional to the inverse of the source dimensions. The envelope of the spectrum in the high frequencies has a frequency dependence of $\omega^{-\varepsilon}$, where ε has values between zero and three, and generally equals two (Aki, 1967). Thus, high-frequency spectra are limited to a certain range. For point sources and a step source time function, we saw that the far-field elastic displacements of P and S waves are impulses (delta functions) ((16.79) and (17.80)) and in consequence their spectra are constant for all frequencies. The source dimensions limit the high-frequency spectra to a certain maximum value. The larger the source dimension the lower the frequencies present in the spectrum. These results have been obtained without specifying the form of the source and are common characteristics for all sources of finite dimensions such that slip starts at zero and ends with a constant value.

18.2 Rectangular faults. Haskell's model

A simple kinematic model of finite dimensions, known as Haskell's model, is a rectangular fault of length L and width W, such that the slip Δu propagates only along the L direction with a constant velocity v (the slip moves instantaneously along W) (Haskell, 1964). The coordinate along L is ξ, with its origin at one end of the fault and Δu has only one component (Fig. 18.2). Fractures that propagate only in one sense (from 0 to L) are called unilateral fractures and those that propagate in both senses (from 0 to $L/2$ and from 0 to $-L/2$) are called bilateral fractures. For unilateral

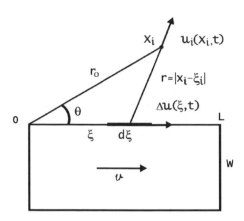

Fig. 18.2. The rectangular fault of Haskell's model.

fractures, according to (18.6), considering the radiation pattern, the dependence on distance, and the other factors of (18.1), the form of P waves in the far field is given by

$$u(x_i, t) = W \int_0^L \Delta \dot{u}\left(\xi, t - \frac{r_0 - \xi \cos \theta}{\alpha}\right) d\xi \tag{18.12}$$

If the slip moves in the positive direction of ξ with a constant fracture velocity v, then $\Delta u(\xi, t) = \Delta u(t - \xi/v)$ and we obtain

$$u(x_i, t) = W \int_0^L \Delta \dot{u}\left[t - \frac{r_0}{\alpha} - \frac{\xi}{\alpha}\left(\frac{\alpha}{v} - \cos \theta\right)\right] d\xi \tag{18.13}$$

If we make the substitution

$$d = \frac{r_0}{\alpha} + \frac{\xi}{\alpha}\left(\frac{\alpha}{v} - \cos \theta\right) \tag{18.14}$$

then the Fourier transform of $u(x_i, t)$ is

$$U(x_i, \omega) = W \int_0^L d\xi \int_{-\infty}^{\infty} \Delta \dot{u}(t - d) \, e^{-i\omega(t - d)} \, dt \tag{18.15}$$

However, we have that

$$\int_{-\infty}^{\infty} \Delta \dot{u}(t - d) \, e^{-i\omega(t - d)} \, dt = i\omega \, \Delta U(\omega) \, e^{-i\omega d} \tag{18.16}$$

where $\Delta U(\omega)$ is the transform of $\Delta u(t)$, the transform of $\Delta \dot{u}(t)$ is $i\omega \, \Delta U(\omega)$, and that of $\Delta u(t - d) = \Delta U(\omega) \exp(-i\omega d)$. Therefore, equation (18.15) becomes

$$U(x_i, \omega) = W i\omega \, \Delta U(\omega) \, e^{-i\omega r_0/\alpha} \int_0^L \exp\left[-i\frac{\xi\omega}{\alpha}\left(\frac{\alpha}{v} - \cos \theta\right)\right] d\xi \tag{18.17}$$

To evaluate the integral in (18.17), we make the substitution $b = -(\omega/\alpha)(\alpha/v - \cos \theta)$ and obtain

$$\int_0^L e^{ib\xi} \, d\xi = \frac{2}{b} \sin\left(\frac{bL}{2}\right) \exp\left(i\frac{bL}{2}\right) = L \frac{\sin X}{X} \, e^{iX} \tag{18.18}$$

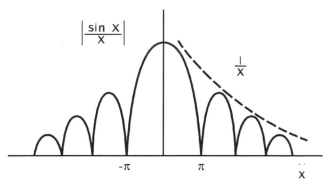

Fig. 18.3. The function $\sin X/X$.

where

$$X = \frac{bL}{2} = -\frac{\omega L}{2\alpha}\left(\frac{\alpha}{v} - \cos\theta\right) \tag{18.19}$$

The final form for the transform of elastic displacements of P waves $U(x_i, \omega)$, according to (18.17), is

$$U(x_i, \omega) = WL\omega\,\Delta U(\omega)\,\frac{\sin X}{X}\,\exp\left[-i\left(\frac{\omega r_0}{\alpha} - X - \frac{\pi}{2}\right)\right] \tag{18.20}$$

where we have replaced $i = e^{i\pi/2}$. The form of the amplitude spectrum depends on the factor $\sin X/X$. We have discussed the form of this function in section 12.2. It has the value unity for $X = 0$ and roots for X equal to integer multiples of π, and its envelope decreases as $1/X$ (Fig. 18.3). Since, for fixed values of θ and L, X depends on ω, in the limit when ω tends to zero (low frequencies), the factor equals unity and for high frequencies its envelope decreases with $1/\omega$.

The form of the amplitude spectrum depends also on the form of $\Delta U(\omega)$, the transform of the source time function (STF) (18.20). If $\Delta u(t) = \Delta u\,H(t)$ its transform is $\Delta U(\omega) = \Delta u/(i\omega)$. From (18.20) we obtain that $U(\omega)$ is proportional to the seismic moment $(M_0 = \mu L W\,\Delta u)$ for the limit of low frequencies and decreases as $1/\omega$ for high frequencies. If the STF has a rise time τ that Δu takes to attain its maximum value at each point of the fault plane (section 16.7), the spectrum depends on the transform of the STF. For example, the transforms of the STF given by (16.109) and (16.110) are

$$\Delta u(t) = \begin{cases} \Delta ut/\tau_0, & 0 < t < \tau \\ \Delta u, & t \geq t \end{cases}; \qquad \Delta U(\omega) = \frac{\Delta u(1 - e^{-i\omega\tau})}{\omega^2\tau} \tag{18.21}$$

$$\Delta u(t) = \Delta u\,H(t)(1 - e^{-t/\tau}); \qquad \Delta U(\omega) = \frac{\Delta u}{(1 + i\omega\tau)i\omega} \tag{18.22}$$

In both cases, the transforms depend on $1/\omega^2$. If we substitute these values of $\Delta U(\omega)$ into (18.20), the envelope of $U(\omega)$ decreases with the frequency as $1/\omega^2$. If we represent the spectrum with respect to the logarithm of the frequency, its form is a flat part for low frequencies and, from a certain frequency ω_c, called the corner frequency, its envelope is

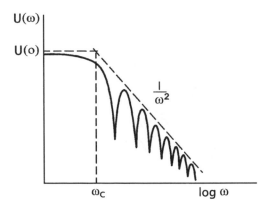

Fig. 18.4. The form of the amplitude spectrum of seismic waves for an extended fault with finite dimensions and rise time.

a straight line of slope -2 (Fig. 18.4). This form of the spectrum is due to the combined effect of the source dimensions and the rise time. If we consider the particular case in which $\theta = \pi/2$ and ω_c corresponds to $X = \pi/2$, we obtain $\omega_c = 2v/L$; that is, the corner frequency is proportional to the inverse of the source length. Observed spectra of seismic waves exhibit these characteristics, indicating the finite dimensions of the source and the existence of the rise time (Aki, 1967).

The influence of the source dimensions can be isolated by means of the directivity function $D(\omega)$ defined by Ben Menahem (1961) as the quotient of spectral amplitudes of waves that leave the source in opposite directions, that is, with angles θ and $\theta + \pi$. According to (18.19) and (18.20), this quotient is

$$D(\omega) = \frac{\sin\{[\omega L/(2c)](c/v - \cos\theta)]\}(c/v + \cos\theta)}{\sin\{[\omega L/(2c)](c/v + \cos\theta)]\}(c/v - \cos\theta)} \tag{18.23}$$

where c is the wave velocity. This function has a series of maxima and minima for frequencies that depend on L and v, and can be used to determine the source dimensions and velocity of fracture propagation. This is easier for surface waves, since θ represents the azimuth at the focus with respect to the trace of the fault.

Another effect of equation (18.20) is on the form of the radiation pattern (section 16.5, Fig. 16.17). If the wave length is much larger than the source dimensions ($\lambda \gg L$), X tends to zero and $\sin X/X$ is unity for all values of θ. Amplitudes are not affected and the radiation pattern corresponds to that of a point source. If the wave length is of the same order as the dimensions ($\lambda \approx L$), amplitudes are affected by the factor $\sin X/X$ that depends on θ and the radiation pattern is modified. According to (18.19), this factor is maximum for $\theta = 0$ and minimum for $\theta = \pi$; that is, amplitudes are larger in the same direction as that of fracture propagation ($\theta = 0$) and smaller in the opposite direction ($\theta = \pi$) (Fig. 18.5). This effect is called the focusing of energy in the direction of fracture propagation and is a phenomenon that occurs for all propagating sources.

The kinematic model of a rectangular unilateral fracture with a constant rupture velocity has shown us the effects of the source dimensions on the radiated displacement

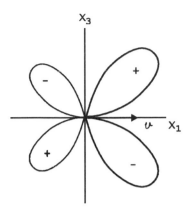

Fig. 18.5. The effect of fracture propagation on the radiation pattern of P waves.

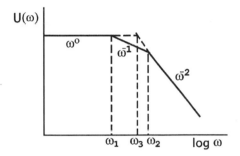

Fig. 18.6. The amplitude spectrum of seismic waves according to Savage's model.

field. Amplitude spectra of displacements have constant values proportional to the seismic moment at low frequencies and these values decrease with the frequency for high frequencies, starting from the corner frequency. If the STF includes a rise time, this decrease corresponds to $1/\omega^2$. The radiation pattern is also affected by dimensions, with more energy radiated in the azimuth corresponding to the direction of propagation of the rupture.

Haskell's model with bilateral fracture, a rupture velocity of $v = 0.9\beta$ and a STF given by (18.22) has two corner frequencies ω_1 and ω_2 instead of one (Savage, 1972). For frequencies between zero and ω_1, the spectrum is flat; between ω_1 and ω_2 it decreases as ω^{-1}; and for frequencies higher than ω_2, it decreases as ω^{-2} (Fig. 18.6). A third corner frequency ω_3 is defined by the intersection of the flat part and the decay as ω^{-2}. For P and S waves ω_1, ω_2, and ω_3 are given by

$$\text{P:} \quad \omega_1 = \frac{\alpha}{2L} \qquad\qquad \text{S:} \quad \omega_1 = \frac{3.6\beta}{L}$$

$$\omega_2 = \frac{2.4\alpha}{W} \qquad\qquad \omega_2 = \frac{4.1\beta}{W}$$

$$\omega_3^2 = \frac{2.9\alpha^2}{LW} \qquad\qquad \omega_3^2 = \frac{14.8\beta^2}{LW}$$

The corner frequencies of P waves are always lower than those of S waves. Usually, observed corner frequencies ω_c correspond to ω_3, and from this value we can obtain the source dimensions:

$$(LW)^{1/2} = \frac{1.7\alpha}{\omega_c^P} = \frac{3.8\beta}{\omega_c^S}$$ (18.24)

The difference between ω_1 and ω_2 depends on the relation between L and W. If $W \ll L$, that is, the fault is long and narrow, then the difference is large, whereas, if $L \approx W$, the three frequencies practically coincide.

18.3   Circular faults. Brune's model

Another fundamental model of an extended seismic source is that of a circular fault known as Brune's model (Brune, 1970). This model consists in a circular fault plane with finite radius on which a shear stress pulse is applied instantaneously (Fig. 18.7). Since this model specifies the stress on the fault, this is not exactly a kinematic model. Because the stress pulse is applied instantaneously on the whole fault area, there is no fracture propagation. The shear pulse generates a shear wave that propagates perpendicularly to the fault plane. Adapting Brune's notation to the one we have used, we call $\Delta\sigma$ Brune's effective shear stress and Δu the displacement on the fault plane (that is, for $x = 0$, where x is the distance normal to the fault plane). The stress pulse has a time dependence given by a step function and, for a distance x, is

$$\Delta\sigma(x, t) = \Delta\sigma \, H\left(t - \frac{x}{\beta}\right)$$ (18.25)

The shear displacement Δu, for $x = 0$, is obtained by integration of (18.25), since, in this case, $\sigma = \mu \, \partial u / \partial x$:

$$\Delta u(t) = H(t) \frac{\Delta\sigma}{\mu} \beta t$$ (18.26)

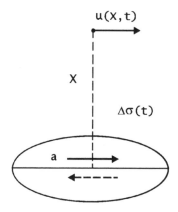

Fig. 18.7. Brune's model of a circular fault.

Its Fourier transform is

$$\Delta U(\omega) = -\frac{\Delta \sigma \, \beta}{\mu \omega^2} \tag{18.27}$$

The effective stress is the difference between the tectonic stress σ_0 acting on the fault plane and the friction stress σ_f ($\Delta \sigma = \sigma_0 - \sigma_f = \varepsilon \sigma_0$). This coincides with the drop in stress defined in (15.25) with $\sigma_1 = \sigma_f$. For a total drop in stress ($\Delta \sigma = \sigma_0$ and $\varepsilon = 1$) the displacement of S waves in the far field at distance r, not including the radiation pattern and the dependence on distance, is

$$u(t) = \frac{\Delta \sigma \, \beta}{\mu} \left(t - \frac{r}{\beta} \right) \exp \left[-b \left(t - \frac{r}{\beta} \right) \right] \tag{18.28}$$

Its spectrum is

$$U(\omega) = \frac{\Delta \sigma \, \beta}{\mu} \frac{1}{\omega^2 + b^2} \tag{18.29}$$

$$b = \frac{2.33\beta}{a} \tag{18.30}$$

where a is the radius of the fault. The spectrum (18.29) has a flat part at low frequencies as they tend to zero and decreases as ω^{-2} for high frequencies, starting at the corner frequency $\omega_c = b$. If the stress drop is not total, then, for small ε ($\varepsilon \simeq 0.01$), the spectrum decreases as ω^{-1}. The fault radius can be deduced from the corner frequency of S waves:

$$a = 2.33\beta/\omega_c \tag{18.31}$$

Brune's model is commonly used to obtain fault dimensions from spectra of S waves for earthquakes of small-to-moderate size ($M < 6$), for which the circular fault is a good approximation. We have mentioned (section 15.1) that earthquakes take place in the brittle part of the crust (about 20 km thickness) or the seismogenic layer. For dimensions less than 20 km ($M < 6$), fault planes are contained inside the seismogenic layer. Fractures start at a point and grow unhindered in all directions with near circular form ($L \simeq W$) and can be approximated by Brune's model. Larger earthquakes have larger dimensions, so, since their widths are limited to about 20 km, their lengths must be larger than their widths ($L > W$). In these cases, Haskell's rectangular model is a better approximation.

18.4   Nucleation, propagation, and arrest of a rupture

Haskell's model does not include the effect either of the beginning or nucleation of a rupture or of its cessation or arrest. The first kinematic model that included both effects was proposed by Savage (1966). Savage's model consists in an elliptical fault in which slip begins at one of the foci and stops when it reaches the border of the ellipse. The model can be simplified for a circular fault of radius a, for which the slip Δu (which is constant for all points) begins at the center, propagates radially with a constant rupture velocity v and circular rupture fronts and stops at the circular border. We use

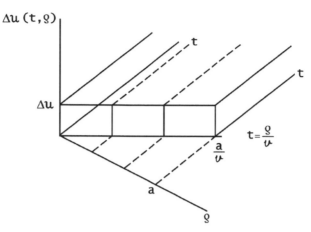

Fig. 18.8. The slip as a function of time and distance inside the fault for Savage's model of initiation and cessation of rupturing.

polar coordinates (ρ, ϕ) on the fault plane, ρ with its origin at the center of the fault and ϕ measured from x_1. Then, the time dependence of the slip is a step function and the slip is a function only of ρ given in the form

$$\Delta u(\rho, t) = \Delta u \, H\left(t - \frac{\rho}{v}\right)[1 - H(\rho - a)] \tag{18.32}$$

At the center of the fault ($\rho = 0$), the slip at $t = 0$ passes instantaneously from zero to Δu. For a value at a distance ρ ($0 < \rho < a$) from the center, the slip is 0 until $t = \rho/v$, when it becomes Δu. For $\rho = a$, rupture stops and, for $\rho \geq a$, the slip is zero for all values of t (Fig. 18.8).

For a point of observation on the x_3 axis, that is, over the center of the fault at a distance r_0 (Fig. 18.9(a)), the form of P waves, according to (18.2), is given by

$$u(r_0, t) = \int_0^{2\pi} \int_0^a \Delta \dot{u}\left(\rho, t - \frac{r}{\alpha}\right) \rho \, d\rho \, d\phi \tag{18.33}$$

Since rupture propagates in the ρ direction with a velocity v, as in (18.13), the slip can be written as $\Delta u(t - r/\alpha - \rho/v)$. Substituting into (18.33) and with the approximation that, for $r_0 \gg a$, $r = r_0$, after integration over ϕ, we have

$$u(r_0, t) = 2\pi \int_0^a \Delta \dot{u}\left(t - \frac{r_0}{\alpha} - \frac{\rho}{v}\right) \rho \, d\rho \tag{18.34}$$

Taking the time derivative in (18.32) and substituting it into (18.34), the displacement is given by

$$u(r_0, t) = 2\pi \Delta u \, H\left(t - \frac{r_0}{\alpha}\right) \int_0^a \delta\left(t - \frac{\rho}{v} - \frac{r_0}{\alpha}\right)[1 - H(\rho - a)] \rho \, d\rho \tag{18.35}$$

To evaluate this integral we use the relation

$$\int_{-\infty}^{\infty} f(x) \delta(ax - b) \, dx = \frac{1}{a} f\left(\frac{b}{a}\right) \tag{18.36}$$

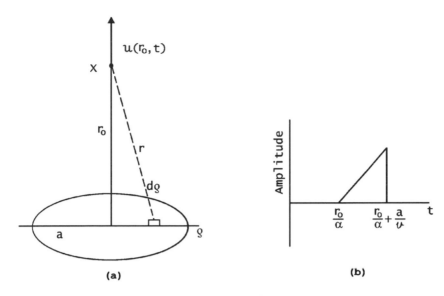

Fig. 18.9. Effects of initiation and cessation of a circular fault for points on a perpendicular axis at its center. (a) The fault and observation point. (b) The elastic displacement of P waves at a distance r_0 along the perpendicular axis.

The integral of (18.35) can be extended to the interval $(-\infty, \infty)$ since it is zero outside the interval $(0, a)$, and thus, by applying (18.36), we obtain

$$u(r_0, t) = 2\pi \Delta u\, v^2 H\left(t - \frac{r_0}{\alpha}\right)\left(t - \frac{r_0}{\alpha}\right)\left\{1 - H\left[v\left(t - \frac{r_0}{\alpha}\right) - a\right]\right\} \qquad (18.37)$$

According to this expression, for

$$t \le \frac{r_0}{\alpha} \qquad \text{and} \qquad t \ge \frac{r_0}{\alpha} + \frac{a}{v}, \qquad u(r_0, t) = 0 \qquad (18.38)$$

whereas for

$$\frac{r_0}{\alpha} < t < \frac{r_0}{\alpha} + \frac{a}{v}, \qquad u(r_0, t) = 2\pi \Delta u\, v^2\left(t - \frac{r_0}{\alpha}\right) \qquad (18.39)$$

Displacement starts at $t = r_0/\alpha$ and increases linearly with time until $t = a/v + r_0/\alpha$, when it drops to zero (Fig. 18.9(b)). According to the approximation used ($r = r_0$), the time of inception of the discontinuity corresponds to the arrival of the signal from the cessation of the fracture at the border ($\rho = a$), which is called the stopping phase. The displacement drops discontinuously to zero and the velocity and acceleration become infinite.

In Savage's model the slip passes instantaneously from zero to its maximum value at each point of the fault as the rupture propagates from the center outward. The slip velocity is a pulse that propagates in the same way until it reaches the border of the fault. Since elastic displacements depend on the slip velocity, other models specify this value directly (as was done for the STF in section 16.7) (Molnar *et al.*, 1973). For a

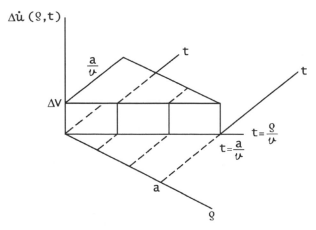

Fig. 18.10. The source model for the slip velocity in a fault with initiation and cessation.

circular fault the slip velocity can be expressed as

$$\Delta \dot{u}(\rho, t) = \Delta V \left[H\left(t - \frac{\rho}{v} \right) - H\left(t + \frac{\rho}{v} - \frac{a}{v} \right) \right] H(a - \rho) \tag{18.40}$$

In this model, the slip velocity takes a constant value $\Delta \dot{u} = \Delta V$ at each point of the fault as the rupture front arrives at $t = \rho/v$ that remains constant as long as $t < (a - \rho)/v$ and ceases for $t \geq (a - \rho)/v$. This means that the slip velocity persists at each point of the fault until fracture stops at the border $\rho = a$ (Fig. 18.10). It is not physically possible to stop motion in the fault, since this implies that information on the cessation of the fault at the border instantaneously (with infinite velocity) reaches all points of the fault. This can be solved by introducing a finite velocity, for example, the velocity of P waves, to bring the information on the cessation of rupture at the border to all points of the interior of the fault. This can be done by replacing the second term inside the square brackets of (18.40) by $H[t - a/v - (a - \rho)/\alpha]$. Now the slip velocity becomes zero at each point of the fault as a P wave arrives from the fault border, once the motion has stopped. This wave is called the healing front, since it heals the fault by stopping its motion. In kinematic models the healing of the rupture inside the fault must be introduced in some way.

Models represented by (18.32) and (18.40) have the slip and slip-velocity time dependences of a step function. Just like for point sources, we can introduce a rise time into these models. As the rupture front reaches each point of the interior of the fault, the slip velocity starts to increase from zero to a maximum value during a time τ. The slip velocity is brought down to zero at each point of the fault as a healing front arrives from the border of the fault where it has stopped. More realistic kinematic models can be established with various shapes, generally rectangular, in which the slip and slip velocities decrease gradually with rupture near to the border of the fault, and the fracture velocity, maximum slip, slip velocity, and rise time vary along the fault plane (Archuleta, 1984; Mendoza and Hartzell, 1989). In general, kinematic models of faulting can be made to correspond quite realistically to conditions on a fault, but they are not completely exempt from a certain arbitrariness. Some conditions must be imposed on the faulting process *a priori*, such as the velocity of rupture propagation, stopping at the border, and the healing process.

18.5 Dynamic models of fracture

The kinematic models we have considered up to this point are, naturally, simplifications of real fractures and include certain arbitrary factors in the definition of the slip and conditions at the fault border. The physical problem of fracture is a dynamic problem in which the slip has to be considered a consequence of stress conditions and the strength of material in the focal region. Dynamic models of the seismic source take these conditions into consideration and are based on the theory of the generation and propagation of fractures in stressed media. From this point of view, the mechanism of an earthquake is represented by a shear fracture produced by the drop in stress in the focal region. Fracture initiates at a point of the fault when the stress acting on the fault plane exceeds a critical value, propagates with a certain velocity, and finally stops when conditions impede its further propagation. A complete dynamic model must, then, include the whole of the fracturing process, its initiation or nucleation, propagation and arrest, derived from stress conditions and properties of the material in the focal region. Two determinant factors are tectonic stresses that are a consequence of lithospheric plate motion and mechanical properties of rocks in the fault region. Among the first studies of fracture dynamics applied to earthquakes were those of Keylis-Borok (1959), Kostrov (1964, 1966), Burridge (1969), Freund (1972, 1979), and Madariaga (1976). These studies were based on the work on fractures of crystals and metals published between 1920 and 1950 by Griffith, Starr, and Irwin. The dynamic problem is more complicated than the kinematic problem; therefore, we will present only the more basic principles in a simplified form.

Let us consider the energy produced by the fracture process given by (15.26). From the dynamic point of view, a fracture is produced by a drop in stress $\Delta\sigma = \sigma_0 - \sigma_f$, where σ_0 is the sum of the tectonic stresses acting before faulting (the shear component) and σ_f is the friction between the two sides of the fault. By substitution into (15.26), assuming that $\bar{\sigma} = \Delta\sigma/2$, we obtain

$$E = \tfrac{1}{2}\Delta\sigma\,\Delta u\,S + \sigma_f\,\Delta u\,S \tag{18.41}$$

where the first term represents the seismic energy due to the slip on the fault and the second represents the residual energy lost by friction. From the dynamic point of view, the stress drop is the parameter that determines the fracture and the slip is its consequence.

18.5.1 The static problem

Let us start with the static problem of a fracture free from stress. For a circular fault of radius a, the shear stress before faulting is σ_0 and, after faulting, the stress inside the fault ($\rho < a$) is null. Then, $\Delta\sigma_s = \sigma_0$ is the static drop in stress. In absence of body forces ($F = 0$), for the static case, the equation of motion (2.57) is $\tau_{ij,j} = 0$. For shear slip on the fault $u(\rho)$, in a homogeneous isotropic elastic medium, the

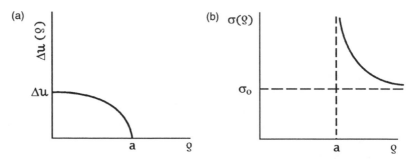

Fig. 18.11. The static model of fracture with a total drop in stress: (a) slip and (b) stress.

equation becomes

$$\mu \nabla^2 u = 0$$

The conditions of displacement and stress inside and outside of the fault are

$\rho = 0,$	$u(0) = \Delta u$
$\rho \geq a,$	$u(\rho) = 0$
$\rho < a,$	$\sigma(\rho) = 0$
$\rho = \infty,$	$\sigma(\rho) = \sigma_0$

Under these conditions, the solution for the displacement inside of the fault is

$$\Delta u(\rho) = \frac{\sigma_0}{\mu}(a^2 - \rho^2)^{1/2}, \qquad \rho < a \tag{18.42}$$

From this expression, we can derive the stress outside of the fault ($\rho > a$), since $\sigma(\rho) = \mu \, \partial u/\partial \rho$:

$$\sigma(\rho) = \frac{\sigma_0 \rho}{(\rho^2 - a^2)^{1/2}}, \qquad \rho > a \tag{18.43}$$

Equations (18.42) and (18.43) describe the static distribution of slip inside of the fault and stress outside of it (Fig. 18.11). Since the drop in stress is total, $\Delta\sigma_s = \sigma_0$, equation (18.42) establishes the relation between the slip and the stress drop. We must notice from (18.43) that a complete drop in stress inside of the fault implies that the stress becomes infinite outside of it at its border ($\rho = a$), which is not physically possible.

18.5.2   The dynamic problem

The dynamic problem requires the solution of the fracture problem as a function of time. The fracture front propagates with a certain velocity and, as it advances, material becomes fractured. Behind the front, the stress becomes zero for a total drop in stress or has a residual value that depends on the friction. Let us consider a simple case with a plane rupture front unlimited in the x_2 direction that advances in the x_1 direction with a constant velocity v.

The relation between the direction of the slip in the fracture plane and its direction of propagation defines three modes of fracture (Fig. 18.12). In mode I, tensional fracture,

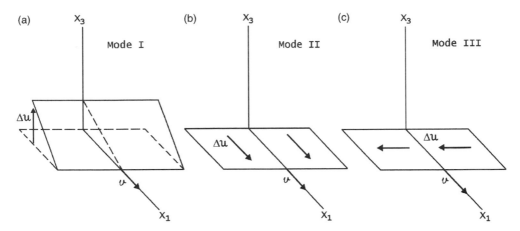

Fig. 18.12. Modes of fracture: (a) mode I, tensional fracture; (b) mode II, in-plane shear fracture; and (c) mode III, antiplane shear fracture.

Δu is normal to the fracture plane and to the direction of propagation. As the fracture propagates, the two sides of the fault separate and therefore the drop in stress is always total. In mode II, in-plane shear fracture, Δu is contained in the fracture plane and has the same direction as that of its propagation. In mode III, antiplane shear fracture, Δu is contained in the fracture plane with its direction normal to that of its propagation. From the point of view of seismology, mode I has little application, since earthquakes are assumed to be produced by shear fracture. In mode II, along the x_1 axis (defined as the direction of propagation of the fracture) we observe P and SV waves, whereas in mode III we observe only SH waves. The last case has a simpler solution.

The dynamic problem of the propagation of a fracture is centered on the energetic situation at the rupture front (Kostrov and Das, 1988). Let us consider a tensional fracture (mode I) whereby the stress drop is total. For the fracture front to advance, new fracture surface must be created and a certain amount of energy must be consumed. For this, an elastic energy flux from the part which has not been fractured to the fracture front is necessary. We call the energy necessary to create a unit of new fracture surface the specific effective surface energy or Griffith's energy γ. The value of γ is a characteristic of each material. When the flux of energy to the rupture front G equals the energy necessary to create new fracture surface ($G = \gamma$), then fracture progresses. This is called Griffith's fracture condition. To produce an element of new fracture surface dS, the energy necessary is $2\gamma\,dS$ (the factor of two is due to there being two sides of a fracture). This energy comes from the stress drop $\Delta\sigma$ behind the rupture front and is given by $G = \Delta\sigma\,\Delta u\,dS$. For a perfect brittle fracture in elastic material, the material ahead of the fracture front (which is not yet fractured) is continuous ($u^+ = u^-$) and that behind (which has already been fractured) is discontinuous ($u^+ - u^- = \Delta u$). If drop in stress is total, then, for each point behind the rupture front, the stress is zero ($\sigma = 0$). There is, then, a discontinuity at the rupture front and there is no transition between unfractured and fractured material.

The problem of the relation between the stress drop and the slip in the dynamic problem is not easy. In a simplified form, the problem consists in the solution of the

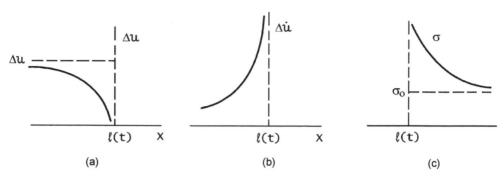

Fig. 18.13. The situation at the rupture front $(x = l(t))$ for dynamic fracture: (a) slip, (b) slip velocity, and (c) stress.

equation of motion (2.56) for the relative displacement (slip) of the two sides of the fault plane $\Delta u(x, t) = u^+ - u^-$, for $x < vt$, where x is the direction of propagation of the rupture with velocity v (for simple cases v is constant). The boundary conditions require that, as the fracture front advances on the fault plane, the stress drops by $\Delta\sigma = \sigma_0 - \sigma_f$. If the fracture front is planar and propagates in the direction of x, with a constant velocity v, its position at each moment is given by $l(t) = vt$. The slip inside the fault can be written as

$$\Delta u(x, t) = \frac{\Delta u}{a}(v^2 t^2 - x^2)^{1/2}, \qquad x < vt \tag{18.44}$$

and the slip velocity is

$$\Delta\dot{u}(x, t) = \frac{V}{(v^2 t^2 - x^2)^{1/2}}, \qquad x < vt \tag{18.45}$$

If the drop in stress is total ($\sigma_f = 0$), the stress inside of the fault is zero and that outside of it, in a similar form to (18.43), is

$$\sigma(x, t) = \frac{K}{(x^2 - v^2 t^2)^{1/2}}, \qquad x > vt \tag{18.46}$$

Here V is the velocity intensity dynamic factor and K is the stress intensity dynamic factor. As the rupture front advances (vt increases in the x direction), the slip (18.44) increases from zero ahead of the rupture front to a constant value inside of it (Fig. 18.13(a)). The slip velocity (18.45) becomes infinite at the rupture front (Fig. 18.13(b)). The stress (18.46) becomes infinite when it approaches the rupture front from outside (Fig. 18.13(c)) and has a constant value or zero inside. The dynamic factors K and V are related to the flux of energy from unfractured material to the rupture front

$$G = \frac{\pi}{2v}KV \tag{18.47}$$

For mode III (antiplane), K and V are related by

$$K = V\frac{\mu}{2v}(1 - v^2/\beta^2)^{1/2} \tag{18.48}$$

For a circular fracture that grows from its center, according to Madariaga (1976), the relation between Δu and $\Delta \sigma$ is

$$\Delta u(\rho, t) = \frac{\Delta \sigma}{\mu} C(v) v (t^2 - \rho^2 / v^2)^{1/2}, \qquad t > \rho / v \qquad (18.49)$$

where $v < \beta$, and $C(v)$ is a factor with a value near unity. The growth of the fracture is assured by the constant flux of energy from the material that has not yet been fractured. If the medium is homogeneous, then rupture, once it has started, can not stop and grows indefinitely. This is due to the constant conditions of the material ahead of the rupture front.

Since, in an homogeneous medium, rupture, once it has started, can not be stopped by itself, stopping must be introduced as a condition, for example, for a circular fault by imposing that the limit is at a predetermined value of the radius $\rho = a$. The beginning of the fracture must also be introduced as an added condition; for example, fracture starts when the applied stress exceeds a certain critical value. This value represents the maximum stress the material can support without breaking. Dynamic models of fracture allow determination of the slip and its propagation over the fault plane from a specified drop in stress. Thus, we can solve for the elastic displacement field from the dynamic conditions in the source region. The solution of dynamic problems, even for simple cases, is difficult and in many cases they must be solved numerically.

18.6 The complexity of a fracture

Under homogeneous conditions, the dynamic problem of the propagation of a rupture implies certain unrealistic border conditions concerning the stress and slip velocity and their nucleation and arrest. To solve some of these problems we must introduce inhomogeneities into the medium and complexities into the fracture process.

18.6.1 The cohesive zone

According to (18.46), there is a discontinuity of stress at the rupture front. As we approach it from the outside, the stress becomes infinite and drops to zero or a constant value inside. Another discontinuity occurs for the slip velocity (18.45), which becomes infinite immediately behind the rupture front. This is a situation that is not physically possible, since no material of finite strength can sustain an infinite stress or move with an infinite velocity. These two inconsistencies of the homogeneous model follow from the fact that material is either purely elastic (unfractured) ahead of the front or fractured behind it. To avoid this situation we must consider the existence of a transition zone immediately ahead of the fracture front where material behaves in an inelastic way.

The first fracture model with a transition zone was proposed by Barenblatt (1959). The transition zone is called the cohesive zone. In this zone cohesive forces act to oppose the advance of the fracture and hold the stress immediately ahead of the rupture front finite, eliminating the stress singularity. In the cohesive zone of width d, the stress has a finite mean value σ_c that is larger than the applied tectonic stress σ_0 and reduces to the friction stress behind the rupture front (Fig. 18.14(a)). The value of σ_c is related to

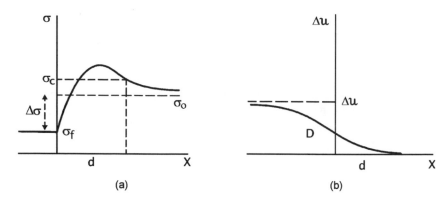

Fig. 18.14. The cohesive zone ahead of the rupture front: (a) cohesive stress and (b) the critical slip and length of the cohesive zone.

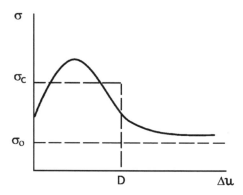

Fig. 18.15. The dependence of the stress on the slip in the slip-weakening model.

Griffith's energy in the form

$$\gamma = \frac{2\sigma_c^2 d}{\mu \pi C(v)} \tag{18.50}$$

where $C(v)$ is a factor that depends on the velocity of fracture and has a value near unity for subsonic fractures. The slip does not become zero in the cohesive zone and there is no singularity in the slip velocity at the rupture front (Fig. 18.14(b)).

In the model with weakening slip, the cohesive stress is taken to be dependent on the slip (Ida, 1972; Palmer and Rice, 1973). This model assumes that the stress inside of the fracture is a function of the slip $\sigma(\Delta u)$ in such a way that it has a finite value for $\Delta u = 0$, and decreases with increasing slip to a final value equal to the friction stress σ_f for Δu larger than a certain critical value $\Delta u = D$ (Fig. 18.15). The average stress, between $\Delta u = 0$ and $\Delta u = D$, equals the cohesive stress. The critical value of the slip D is related to Griffith's energy and the cohesive stress in the form

$$\gamma = \tfrac{1}{2}\sigma_c D \tag{18.51}$$

This relation shows that the energy dissipated in the creation of a unit fracture surface equals the product of the cohesive stress and the critical slip. From (18.50) and (18.51), we can derive a relation between d and D:

$$d = \frac{\mu \pi C(v)}{4\sigma_c} D \tag{18.52}$$

For earthquakes, values of d and D are small relative to the total dimensions of the fracture. For a fracture several kilometers long, d is only some meters and D some centimeters.

18.6.2 *Barriers and asperities*

We have seen that homogeneous models of fracture with uniform slip and constant rupture velocity are not very realistic. The simple fact that rupture must stop at the border of the fault indicates that the conditions can not be homogeneous. In the Earth, faults cross rocks of various strengths, change direction often, and present jumps, joints, and bends. Analysis of observed wave forms from earthquakes also reveals greater complexity than would be expected from homogeneous fractures. This is especially so for the complex form of high-frequency waves in the near field. Another item of evidence for the complexity of the source is the observation of practically constant values of stress drops (in the range 1–10 MPa) for earthquakes of magnitudes larger than five. Since it has been shown in laboratory experiments that rocks can support larger stresses without breaking, the observed drops in stress are really average values for the whole of the fracture process. All these observations show that earthquake sources are complex fracture processes. Two models have been proposed to explain this complexity, namely, models with barriers and asperities.

The barrier model (Das and Aki, 1977; Aki, 1979) assumes that fracture takes place under uniform conditions of stress on the fault fracture, but with different strengths in the material (Fig. 18.16). Zones in the fault surface with high strengths form barriers that make fracture propagation difficult or impede it. When the rupture front reaches a zone of barriers it stops and, if the barrier is sufficiently strong, it will not break.

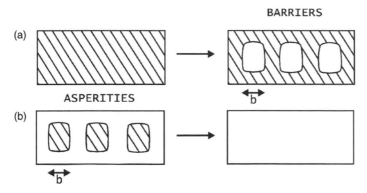

Fig. 18.16. Models of complex sources: (a) barriers and (b) asperities.

Fracture may stop there or continue behind the barrier, leaving an unruptured zone on the fault plane. Once the fracture process has finished on the whole fault plane, stress is released in fractured zones and accumulated in the unbroken barriers. After the earthquake, the distribution of stress on the fault plane becomes heterogeneous. For a large earthquake there may be several barriers and the fracturing process is the sum of several ruptures separated by the barriers. Unbroken barriers may break later, giving rise to aftershocks (section 20.2). Papageorgiou and Aki (1983) have proposed a barrier model in which a rectangular fault is formed by several elementary circular fractures separated by barriers. In this model we can distinguish between the global stress drop, which is the mean value over the whole fault area, and the local stress drop due to the breakage of each individual elementary fracture. The latter is generally larger than the former. In this way we can explain the observed low values of drops in stress and the high energies in high-frequency waves of strong motion instruments in the near field.

The model of asperities (Kanamori and Stewart, 1978; Madariaga, 1979), consists in a fault with a heterogeneous distribution of stress on its surface, with zones of high and low values (Fig. 18.16). Zones with high stress are called asperities. This model takes into account the previous history of the accumulation of stress in certain zones of the fault (asperities) and the release of stress in other weaker zones. This process is achieved by the production of small earthquakes (foreshocks) that release stress in weak zones and accumulate it in strong ones. The breaking of strong zones or asperities where high stresses have accumulated constitutes the occurrence of the main large earthquake. The complexity of the source in this model is given by the fracture of several asperities in the main event. After the fracture of all asperities, the fault plane remains with a homogeneous distribution of residual frictional stress. This model explains the occurrence of foreshocks, but not that of aftershocks.

In both models, earthquakes are produced by complex fracture processes consisting in the breaking of several asperities or zones between barriers. Since neither model can explain both foreshocks and aftershocks, mixed models in which both asperities and barriers are present must be considered (Kostrov and Das, 1988). Barriers are zones that remain unbroken after the main earthquake and asperities are those that break with high drops in stress. The distribution of stress on the fault plane is heterogeneous before and after an earthquake. Thus we can have both foreshocks that break the weak zones before the main shock and aftershocks that break the barriers that had been left unbroken. The complexity of the source is due both to the heterogeneous distribution of stress that is concentrated on the asperities and to the varying strengths of barriers.

An important factor in the heterogeneity of the fault surface is the distribution of friction. High and low values of friction along the fault plane contribute to the accumulation of stress. Recent dynamic models of fracture accord great importance to the problem of friction (Kostrov and Das, 1988; Cochard and Madariaga, 1994). In the asperity model, motion starts when the applied stress overcomes friction and stops when the friction is larger. For constant friction, motion can only stop arbitrarily at the asperity's border. Authors of models introduce a weakening of the friction with velocity, that is, a decrease in friction with the slip velocity. In this situation the arresting of motion at the fault can be obtained from a healing pulse generated by friction itself (Heaton, 1990). Thus the rupture front advances by stress overcoming friction and is healed by friction behind the front. Energy is radiated, as the rupture front advances, from a narrow strip of the fault while the rest of the fault has already been healed. Thus, the motion does not need to be stopped by a healing pulse from the fault border, as we saw in kinematic models.

The search for more realistic models of earthquake sources leads to the consideration of complexities in the fracturing process, heterogeneities in the distributions of stress, strength, and friction along the fault surface, and the existence of a transition zone at the rupture front. These considerations increase the number of parameters necessary in order to define source models, by introducing the dimensions of asperities, the distances between barriers, a cohesive zone, a critical slip, the distribution of friction, etc. Also we must consider geometric irregularities of the fault surface such as branching, stepping, bending, and junctions that depart from the simple planar model (Andrews, 1989).

18.6.3 Acceleration spectra

We have seen how the dimensions of the source affect the form of the spectra of seismic waves in the far field. Complexities of the source influence the radiation at high frequencies and their effect can be observed in the accelerations in the near field, since they attenuate rapidly with distance. In complex models, rupture is not uniform, but rather has accelerations and decelerations, stopping and restarting with successive breakings of several elementary units (asperities). These irregularities result in the complexities observed in the accelerations of the near field at high frequencies.

The spectrum of acceleration is related to that of displacement by a factor of ω^2. Thus, its form depends on ω^2 for low frequencies, corresponding to the flat part of the displacement spectrum, and is flat for frequencies higher than the corner frequency ω_c. For higher frequencies, we find a maximum value ω_{max} or f_{max}, approximately in the range 8–10 Hz, from which the spectrum decreases rapidly (Hanks, 1982) (Fig. 18.17(a)). This frequency has been related to the attenuation due to propagation in

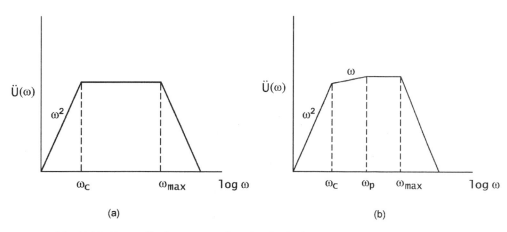

Fig. 18.17. The amplitude spectrum of acceleration in the near field, for (a) a simple source and (b) a complex source.

the upper part of the crust and to processes at the source. In the latter interpretation, this frequency is related to the smallest elementary dimension of asperities approximately about 200 m. Aki (1988), however, related f_{max} to the dimensions of the cohesive zone and the critical value of slip in the form

$$f_{max} \simeq \frac{v}{d} \simeq \frac{4\sigma_c v}{\mu\pi C(v)D} \tag{18.53}$$

The cohesive zone, with dimensions of about 100 m to 1 km, acts as a low-pass filter that is responsible for the attenuation of high frequencies.

Other authors, such as Madariaga (1989), introduce another frequency ω_p with a change in the slope of the envelope of the spectrum which is related to the length b (or its average value $\bar{b}$) of elementary fractures (patches) that form the total fracture ($f_p \simeq v/b$). This frequency has a value between the corner frequency ω_c and the maximum ω_{max}, which is due to attenuation effects (Fig. 18.17(b)). For small earthquakes, with only one elementary fracture, $\omega_p \simeq \omega_c$, whereas for larger events $\omega_p > \omega_c$. The value of ω_c varies with the size of earthquakes, ω_p varies very little, and ω_{max} is practically constant. However, we must realize that the estimation of these frequencies from acceleration spectra is very much affected by attenuation. The study of source complexities has been made possible by digital strong motion records in the near field.

19 METHODS OF DETERMINATION OF SOURCE MECHANISMS

19.1   Parameters and observations

The determination of the source mechanism of earthquakes consists in finding the parameters of the model used in its representation, whose number depends on its complexity. The localization of the source is given by four parameters (ϕ, λ, h, and t_0), the coordinates of the epicenter, depth of focus, and origin time (section 15.2). Generally, in determinations of focal mechanisms, these parameters are assumed to be known, although in some methods some (the focal depth) or all (the centroid) are determined anew. The size given by the magnitude M is independently determined, but, in many methods, the seismic moment M_0 is evaluated as part of the mechanism.

The simplest models of source mechanism are those with point sources. For a point shear dislocation, the orientation of the mechanism is given by the angles ϕ, δ, and λ (section 15.1). For the equivalent double-couple force model, the orientation is given by those of the X and Y axes, that is, by the angles Θ_x, Φ_x, and Φ_y, or by the P and T axes, angles Θ_P, Φ_P, and Φ_T (section 16.2). If the source is not a shear fracture, it can be represented by five components of the moment tensor M_{ij} (for a deviatoric source) or six if there are changes in volume (Chapter 17). Including its size, a shear-fracture point source or DC source is given by four parameters (M_0, ϕ, δ, and λ) plus four for its location, that is a total of eight parameters. If the source is not a DC we need one or two more. If there is a rise time τ in the STF we have to add a new one and the total number is hence nine, ten, or 11.

For an extended source we have to introduce the dimensions of its length and width for rectangular faults or its radius for circular faults. If there is propagation of a fracture we introduce its velocity (Chapter 18). In fracture models, the value of the slip can be obtained from the seismic moment and dimensions. For a rectangular fault with a constant slip, rupture velocity, and rise time, there are five additional parameters (L, D, Δu, v, and τ) that with three for the source's orientation and four for its location sum up to 12 parameters. Thus, even relatively simple models of the source require a considerable number of parameters for its definition. Complex models need new parameters such as the distribution of slip, rise time and rupture velocity on the fault-plane, the number and dimensions of asperities, friction, etc.

The observations used in the determination of focal mechanisms are the displacements of seismic waves recorded at several points on the Earth. Ground displacements are obtained from seismograms in graphical or digital form corrected for the instrument's response. Usually, methods are separated into those that use body and surface waves. Since waves propagate from the focal region to observation points, the elastic properties of the Earth must be known *a priori*. Owing to the heterogeneous nature of the Earth,

359

incomplete knowledge of its structure imposes certain limitations upon the investigation of the source. There is always a latent ambiguity about which characteristics of seismic waves are due to the source and which are effects of the propagating medium. There are ways to isolate these two effects, but there is a trade-off between the details of the source that are to be determined and those of the medium that are supposed to be known. Simple models of the source, with observations at relatively large distances and low frequencies, are, in general, very little affected by propagation effects. However, very detailed models determined by observations from near distances and at high frequencies are more influenced by the heterogeneities of the crustal structure.

An important factor to be considered in observations is that of the development of seismologic instrumentation (Chapter 21). The oldest seismographs (up to 1930) were mechanical ones with smoked paper recording and have very low magnification and not very precise time control. From 1930 onward, electromagnetic seismographs with photographic recordings have increased the amplification and were an important improvement. The installation of the WWSSN global network of seismographs in 1962 provided very good homogeneous observations for studies of mechanisms. Since 1990, modern digital high-dynamic-range broad-band instruments distributed globally have been providing excellent data.

There are many methods of determination of focal mechanisms based on observations of body (P and S) and surface waves (Kasahara, 1981; Udías, 1991). In the following sections, we will present the fundamentals of four of them, which are based on the signs of P waves' first motions, wave-form analysis, moment-tensor inversion, and seismic wave spectra.

19.2   P waves' first motion polarities. Fault-plane solutions

The first method developed to determine the focal mechanism is based on the sign or polarity of the first motion of P waves. The first authors to study data of this type were Omori and Galitzin around 1905, and Shida in 1917 was the first to recognize the alternating distribution of polarities in the four quadrants. Although there were several attempts by European and Japanese seismologists to use these data to study the source, the first operational method was proposed by Byerly (1928). The method is known as the fault-plane solution and also as Byerly's method. The method was simplified by the introduction of the focal sphere proposed by Koning in 1942 and developed by Honda and Ritsema among others in the 1950s. The focal sphere allows a simple graphical resolution of the problem using stereographic projections such as Wulff and Schmidt nets. Between 1950 and 1970, seismologists in the USSR used graphical methods extensively, adding to P polarities those of SV and SH waves. Because of the simplicity of polarity data, fault-plane solutions are still very widely used.

The method is based on the quadrant distribution with alternating signs (compressions and dilations) separated by two orthogonal planes of polarities of the radiation pattern of P waves produced by a shear fracture or double-couple source (section 16.6). The signs of the first motion of P waves are usually read from vertical-component seismograms (upward for compressions, downward for dilations). The method consists in separating the four quadrants of compressions and dilations into two orthogonal

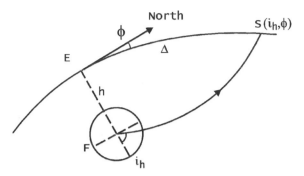

Fig. 19.1. The focal sphere and a ray's trajectory from the focus to a station.

planes, one of which is the fault-plane. Owing to the seismic velocity's dependence on depth, rays are curved and quadrants can not be separated directly by using the distribution of observations on the Earth's surface. To correct for this effect, Byerly proposed extended distances, locating the points on the intersection of straight rays with the surface of the Earth. Today the concept of the focal sphere is more commonly used. The focal sphere is a sphere of homogeneous material and unit radius around the focus. Observation points are located at the surface of the focal sphere corresponding to the rays that arrive at each station, by tracing their trajectories back to the focus (ray back-tracing). If an observing station is located at a distance Δ and azimuth ϕ at the surface of the focal sphere, its location is given by its azimuth ϕ and take-off angle i_h measured from the vertical downward (Fig. 19.1) (section 16.6).

To determine the take-off angle i_h which corresponds to a given distance Δ and focal depth h, it is necessary to know the structure of the medium of propagation. For teleseismic distances ($\Delta > 1000\,\mathrm{km}$), i_h can be calculated from the travel time curves $t(\Delta)$ for a focal depth h, by using Snell's law (section 8.1):

$$\sin i_h = \frac{v_h}{r_h}\frac{\mathrm{d}t}{\mathrm{d}\Delta} \tag{19.1}$$

It is necessary to know the velocity of P waves v_h at the focal depth h ($r_h = R - h$). Errors in this velocity result in errors in i_h and consequently in the mislocation of points on the focal sphere. For local or regional distances ($\Delta < 1000\,\mathrm{km}$), determination of i_h requires knowledge of the velocity distribution in the crust and upper mantle for the region. In general, layered models of constant velocity or with linear velocity distributions are used. Models with velocity gradients are more convenient since they give a continuous relation of i_h to Δ, whereas layers with constant velocities give a discontinuous relation. Once we have determined values of ϕ and i_h for each observation point, they are located on the surface of the focal sphere and the problem is reduced to that of a homogeneous medium. A reliable solution requires a sufficient number of observations distributed over azimuths and take-off angles so that the focal sphere is covered well.

The source model corresponds to a point shear fracture or a double couple (section 16.6). The orientation of the mechanism is given by that of the vectors n (the normal to the fault-plane) and l (the direction of slip), or by that of the axes X and Y of the two couples of forces or also by P and T (the pressure and tension axes). Since all of

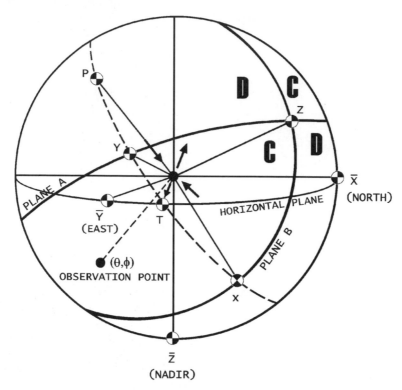

Fig. 19.2. The focal sphere with a vertical axis and a horizontal plane. Planes A and B separate quadrants of compressions and dilations. The locations of the X, Y, Z, P, and T axes are shown.

these pairs are orthogonal unit vectors, the orientation is given by only three angles, $(\Phi_n, \Theta_n, \Phi_l)$, $(\Phi_X, \Theta_X, \Phi_Y)$, and $(\Phi_P, \Theta_P, \Phi_T)$. Also we can define the source in terms of the angles ϕ, δ, and λ for the orientation of the two planes, the fault-plane (A) and auxiliary plane (B) (Fig. 19.2). The method can not distinguish the fault-plane from the auxiliary plane due to the symmetry of the equations for the displacement field with regard to the vectors n and l (section 16.6).

19.2.1   *Graphical methods*

To solve the problem in graphical form, the focal sphere is projected onto a plane by means of a stereographic projection; the most commonly used one is the Schmidt or equal-area projection (Fig. 19.3). Since, in most cases, observations correspond to rays leaving the focus downward, we project onto the lower hemisphere. If there are very near stations or for very deep earthquakes at short distances, stations corresponding to upgoing rays are projected first onto the lower hemisphere along the diameter of the focal sphere. A point s in the lower hemisphere projects as s' on the plane AA$'$ (the horizontal plane through the focus) (Fig. 19.3). The hemisphere projects as a circle of unit radius and a point on the sphere defined by (ϕ, i) projects as (ϕ, r), where $r = \sqrt{2}\sin(i/2)$. Planes through the center project as great circles. In

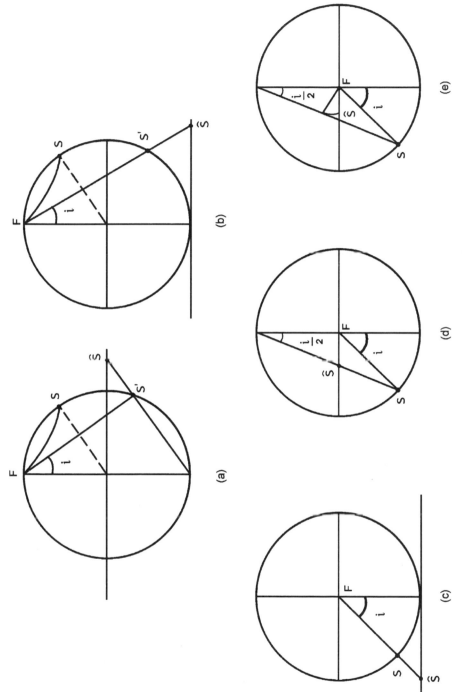

Fig. 19.3. Projections of the Earth ((a) and (b)) and of the focal sphere ((c), (d), and (e)) used in studies of focal mechanisms. (a) Byerly's extended position (s') and projection ($\hat{s}$). (b) Knopoff's (1961) extended position (s') and projection ($\hat{s}$). (c) The central projection. (d) Wulff's nett projection. (e) Schmidt's equal-area nett projection.

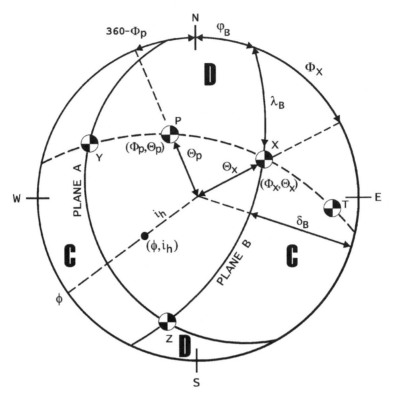

Fig. 19.4. The Schmidt projection of the focal sphere. Planes A and B are the fault and auxiliary planes. The locations of the X, Y, Z, P, and T axes, and an observation point are shown.

practice, the problem is solved by using a Schmidt net or its simulation on a computer screen.

The first step is the location of all observations (ϕ, i_h) on the projection at points (ϕ, r), using different symbols for compressions and dilations (Fig. 19.4). Compressions and dilations are separated in the projection by two orthogonal planes into four quadrants alternating in sign. This is done by drawing first a plane (a great circle) and locating its pole (the normal through the center) and then drawing the second plane such that it must pass through the pole of the first and its pole must be on the trace of the first plane (Fig. 19.4, where X is the pole of plane A and Y is that of plane B). Usually, several attempts are necessary in order to separate all compressions and dilations with a minimum number of inconsistencies. The quality of the solution is given by the score, which is the quotient of the number of correct observations and the total number of observations. Once we have obtained the solution, we measure the orientations of the two planes A and B in terms of the angles ϕ, δ, and λ and those of the axes X, Y, P, and T in terms of the angles θ and ϕ (Fig. 19.4). We can see that all four axes are on the same plane (the plane of forces), which is normal to the A and B planes.

The process we have described can be performed on the screen of a computer by using an interactive program. For example, a program implements the following steps

(Buforn, 1994). Enter observations of the station name, azimuth (ϕ), take-off angle (i_h), and P wave polarity; $+1$ for compression and -1 for dilation. Values of ϕ and i_h have been found previously by another program. The program locates the points on the screen on an equal-area projection. A trial solution is given by specifying the orientation of X and Y, of P and T, or of ϕ, δ, and λ. The program draws the two orthogonal planes and determines the score of the solution. By successive steps, the solution is changed until the one with the maximum score is found.

19.2.2 Numerical methods

Since 1960, with the development of digital computers, several methods for the numerical determination of fault-plane solutions have been proposed. The first formulation is that of Knopoff (1961), reformulated by Kasahara (1963), and applied in a computer program developed by Wickens and Hodgson (1967). This method seeks an orientation of the nodal planes that maximizes the probability of a correct observation at each station. We present briefly a method that uses maximization of a likelihood function and allows for individual and joint solutions for several earthquakes, giving estimations of the variances of the focal parameters (Brillinger *et al.*, 1980; Udías and Buforn, 1988). The method defines the probability that all observations are consistent with a determined orientation of the source as the product of the individual probabilities of each observation and looks for an orientation of the source that maximizes this total probability. The probability of observing a compression at station i is expressed as a function of the theoretical amplitude expected from an orientation of the source as

$$\pi_i = \text{prob}(Y_i = 1) = \gamma + (1 - 2\gamma)\Phi[A_i(\xi_k)] \tag{19.2}$$

where $0 \leq \gamma \leq \frac{1}{2}$ represents reading errors and is given a small value ($\gamma = 0.01$). $\Phi[A_i(\xi_k)]$ is the Gaussian cumulative function of normalized theoretical amplitudes expected from a determined orientation of the source defined by its three parameters (ξ_1, ξ_2, and ξ_3) which correspond to ϕ, δ, and λ or Φ_P, Θ_P, and Φ_T. In this form, the probability is greater for values of A near unity (maximum amplitudes in the radiation pattern).

If we have N observations Y_i for an earthquake, the probability that its mechanism corresponds to a determined orientation of the source given by ξ_k can be written as

$$P = \prod_{i=1}^{N} [\pi_i^{(1+Y_i)/2}(1 - \pi_i)^{(1-Y_i)/2}] \tag{19.3}$$

The method searches for the parameters of the orientation of the model ξ_k that maximize the likelihood function $L = \log P$. Since L is a continuous function of ξ_k, its maximum is found by searching for the values of ξ_k that satisfy $\partial L/\partial \xi_k = 0$. Once these values have been found, the method is used to calculate all the other parameters of the point source which are represented in the focal sphere (Fig. 19.5). The method determines the covariance matrix whose main diagonal gives the standard errors of the parameters and the information matrix. This method is generalized to find also joint solutions for groups of earthquakes with the same mechanism, which is useful in the study of mechanisms of aftershock sequences or swarms of earthquakes.

North Atlantic 16-02-98

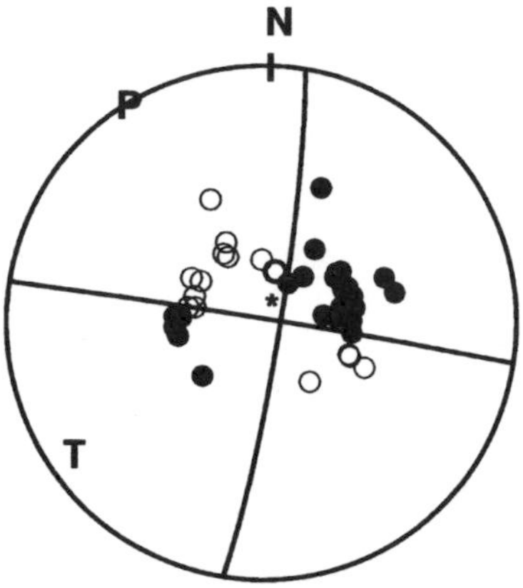

	COPLUNGE	TREND
T	85.66	233.85
P:	87.15	324.06

	STRIKE	DIP	SLIP
A:	9.00	84.92	178.94
B:	278.91	88.95	5.08

N= 49   SCORE= 0.96

Fig. 19.5. The fault-plane solution of the North Atlantic earthquake of 16 February 1998 (courtesy of E. Buforn).

19.3   Wave-form modeling

The method of P wave polarities uses the minimum information contained in seismic waves. Among the methods using more information are those based on the analysis of wave forms (Langston and Helmberger, 1975). The method consists in the comparison of P and S wave forms observed at various azimuths around the focus with those calculated from a point-source model. The comparison is made visually or by a method of minimization of the difference between observed and calculated waves (Nabelek, 1984). Wave-form analysis is frequently applied to earthquakes with observations at teleseismic distances ($\Delta > 30°$). Its application to local or regional earthquakes is more difficult because of the heterogeneity of the crustal structure. A solution for this problem is the use of empirical Green functions as proposed by Hartzell (1978). Empirical Green functions are found from waves of small earthquakes located at the same place and recorded at the same stations as those that record the larger shock whose mechanism is being studied.

Wave-form modeling is based on the determination of theoretical or synthetic seismograms for P and S waves from a given model of the point source defined in terms of its depth, the orientation of the mechanism (DC), the source time function, and the seismic moment. For P waves of long periods, at teleseismic distances (30° to 90°) layering in the focal region and at the recording station does not affect wave forms. Thus the observed P waves are given by the sum of direct P waves and reflected pP and sP waves at the free surface above the focus. Because of the critical influence of the focal depth on these

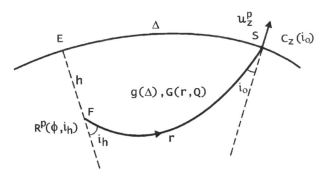

Fig. 19.6. Ray trajectories of P waves from the focus to a station and elements that are used in wave-form modeling.

waves, this is a parameter that is evaluated as part of the analysis (Deschamps *et al.*, 1980; Okal, 1992).

According to (16.79), the displacement of P waves for an infinite medium is given by

$$u_j = \frac{M_0 f(t - r/\alpha)}{4\pi\rho\alpha^3 r} (n_k l_i + n_i l_k)\gamma_i\gamma_k\gamma_j \tag{19.4}$$

For a shear fracture, the components of the vectors l and n can be expressed in terms of the fault angles ϕ, δ, and λ according to (16.99)–(16.104). The STF $f(t)$ represents the time dependence of the slip velocity and γ_i are the direction cosines of the ray that arrives at the station and can be given in terms of its azimuth ϕ_s and take-off angle i_h at the focus (16.105)–(16.107).

If the focus is very deep, reflected pP and sP waves are separated from direct waves and we need consider direct waves only (Fig. 19.6). The vertical component of a P wave at an observation point on the Earth's surface is given by

$$u_z^P = \frac{M_0 f(t - r/\alpha)}{4\pi\rho\alpha^3 r} R^P(\phi, \delta, \lambda, i_h) g(\Delta) G(r, Q) C_z(i_0) \tag{19.5}$$

where r is the distance along the ray from the focus to the station, $R^P(\phi, \delta, \lambda, i_h)$ is the pattern of the radiation for P waves, ϕ is the azimuth at the focus measured from the trace of the fault ($\phi = \phi_{\text{fault}} - \phi_{\text{station}}$), i_h is the take-off angle at the focus, i_0 is the angle of incidence of the ray at the station, $g(\Delta)$ is a factor due to geometric spreading of the wave front (section 8.8), $G(r, Q) = \exp[-\omega r/(2\alpha Q)]$ is the anelastic attenuation (section 14.3), and $C_z(i_0)$ is the effect of the free surface on amplitudes (section 5.5).

In terms of ϕ, δ, and λ, the pattern of the radiation of P waves is given by

$$R^P = A(3\cos^2 i_h - 1) - B\sin(2i_h) - C\sin^2 i_h \tag{19.6}$$

and that of SV waves is given by

$$R^S = -\tfrac{3}{2}A\sin(2i_h) - B\cos(2i_h) - \tfrac{1}{2}C\sin(2i_h) \tag{19.7}$$

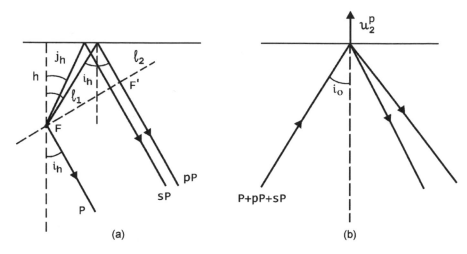

Fig. 19.7. Ray trajectories at the focus (a) and at a station (b) for P, pP, and sP waves.

with

$$A = \tfrac{1}{2}\sin\lambda\sin(2\delta) \tag{19.8}$$

$$B = \sin\lambda\cos(2\delta)\sin\phi + \cos\lambda\cos\delta\cos\phi \tag{19.9}$$

$$C = \sin\delta\cos\lambda\sin(2\phi) - \tfrac{1}{2}\sin\lambda\sin(2\delta)\cos(2\phi) \tag{19.10}$$

The STF $f(t)$ has, for a simple source, a triangular or trapezoidal form, and, for a complex source, a combination of triangular and trapezoidal functions (section 16.7). The duration of the source action is given by the rise time in the first case and by the sum of all rise times in the second.

For a shallow focus or one of intermediate depth, the displacement $u_z^{P}(t)$ at the surface is the sum of P, pP, and sP waves:

$$
\begin{aligned}
u_z^{P}(t) = \frac{M_0}{4\pi\rho_h\alpha_h^3 r} g(\Delta)G(r, Q)C_z(i_0)[&R^{P}(\phi, i_h)f(t - t_{P}) \\
+ &R^{P}(\phi, \pi - i_h)V_{pP}f(t - t_{P} - \Delta t_{pP}) \\
+ &R^{S}(\phi, \pi - j_h)V_{sP}f(t - t_{P} - \Delta t_{sP})]
\end{aligned} \tag{19.11}
$$

The first term corresponds to the direct P wave, the second to the pP wave, and the third to the sP wave (Fig.19.7). R^{P} and R^{S} are the normalized patterns of the radiation for P and SV waves (19.6) and (19.7). For pP and sP waves, rays take-off from the focus in the upward direction and correspond to angles of incidence $\pi - i_h$ and $\pi - j_h$, and, according to Snell's law, $\sin j_h = (\beta_h/\alpha_h)\sin i_h$. If t_{P} is the arrival time of P waves, then the delays with respect to this time of pP and sP waves' arrivals are calculated using the approximation that direct and reflected rays have the same take-off angles at the focus and the surface (Fig. 19.7(a)). Then, from F and F', the traveling times of the two rays are the same and the delay $\Delta t_{pP} = (l_1 + l_2)/\alpha$, where $l_1 + l_2 = 2h\cos i_h$. The delays of sP

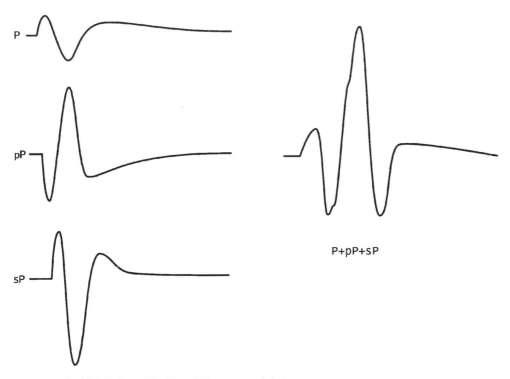

Fig. 19.8. Pulses of P, pP, and sP waves, and their sum.

waves are derived in a similar way:

$$\Delta t_{pP} = t_{pP} - t_P = \frac{2h \cos i_h}{\alpha_h} \tag{19.12}$$

$$\Delta t_{sP} = t_{sP} - t_P = h\left(\frac{\cos j_h}{\beta_h} + \frac{\cos i_h}{\alpha_h}\right) \tag{19.13}$$

V_{PP} and V_{sP} are the coefficients of reflection at the free surface for an incident P and reflected P wave, and an incident SV and reflected P wave. These coefficients are for the vertical component of displacement and can be derived from (5.100) and (5.107) (section 5.4). The observed P wave is the sum of the three waves (Fig. 19.8).

In the process of modeling wave forms, the parameters which are adjusted are the seismic moment M_0, orientation of the source (ϕ, δ, and λ), source time function $f(t)$, and focal depth h. The rest of the parameters are assumed to be known and are kept constant except for those that vary with the focal depth.

The method consists in selecting observations of P waves at different azimuths within the accepted range of distances. At present the best results are obtained by using broad-band digital records. Observed wave forms must be transformed into ground displacements by correcting for the instrument's response. Theoretical wave forms are calculated by assuming some initial values of the variable parameters. A first approximation for the orientation of the source is obtained from the fault-plane solution. Values of focal

AZORES   27 June 1997

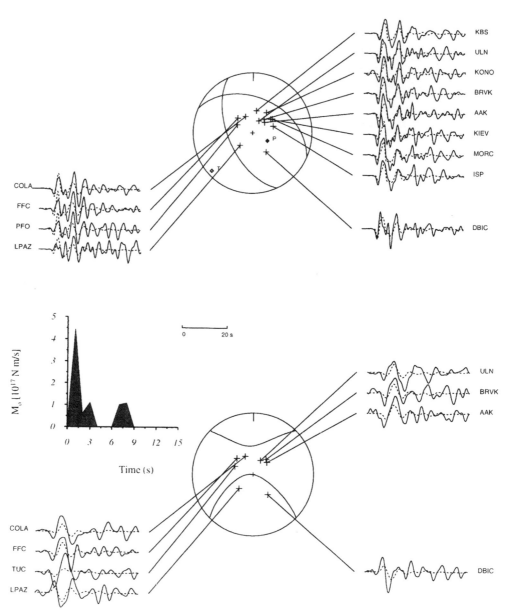

Fig. 19.9. Modeling of P waves for the Azores earthquake of 27 June 1997 (courtesy of E. Buforn).

parameters are changed to improve the fit to observed wave forms. Sometimes the value of Q is also changed. On increasing h delays between arrivals of P, pP, and sP waves increase, and on varying Q the width of the pulses is changed. The fitting is done visually by comparing theoretical and observed wave forms or numerically by an iterative procedure in which the residuals of amplitudes are minimized by a least squares method. A method in which the parameters are adjusted by a least squares procedure is known as wave-form inversion (Nabelek, 1984; MacGraffrey *et al.*, 1991). An example of wave-form modeling is given in Fig. 19.9. The method is also applied to S waves by separating SV and SH components beforehand. Theoretical seismograms are calculated directly in the time domain or are first calculated in the frequency domain and then transformed to the time domain. A problem with these methods, especially if only a few stations with a poor coverage of azimuths are used, is that of how to assign the observed complexity of wave forms to the source that actually belongs to the propagating medium.

19.4 Inversion of the moment tensor

In section 17.7, we saw how the formulation of the source mechanism in terms of the moment tensor allows a linear inversion of its components from the observations of elastic wave displacements. The most common problem concerns the inversion of the first-order time-independent moment tensor that corresponds to a point source. The first method used observations of free oscillations (Gilbert and Dziewonski, 1975). Methods for inversion of surface waves in the time and frequency domains (Mendiguren, 1977; Kanamori and Given, 1981) and of body waves have been proposed (Stump and Johnson, 1977; Strelitz, 1978). A generalized method applicable both to body and to surface waves uses summation of normal modes (Dziewonski *et al.*, 1981). This method, known as the 'centroid moment tensor' (CMT) method because in it one redetermines hypocenter coordinates corresponding to the centroid of the source area in the inversion process, is applied on a routine basis at Harvard University to sufficiently large earthquakes. A method that also uses both body and surface waves is applied in a routine form by the USA's Geological Survey to all earthquakes of magnitude larger than 5 (Sipkin, 1982). Most methods assume the source to be purely deviatoric and afterwards separate the DC part of the solution (section 17.4).

19.4.1 *The inversion of Rayleigh waves*

As an example of the methodology followed in moment-tensor inversions, we present the basic procedure followed in the inversion of Rayleigh waves. Usually vertical-component seismograms recorded at several stations are used. The data must be reduced to ground motion by correcting for the instruments' responses. If all data are reduced to a common distance, in the frequency domain, then the complex spectra are functions only of the azimuth of each station and the frequency:

$$U(\omega, \phi) = A(\omega, \phi)\, e^{i\Phi(\omega,\phi)} = R(\omega, \phi) + iI(\omega, \phi) \tag{19.14}$$

According to (17.51), the spectra can be expressed as the product of the spectrum of the seismic moment and the derivatives of Green's function:

$$U_k(\omega, \phi) = M_{ij}(\omega)G_{ki,j}(\omega, \phi) \tag{19.15}$$

The real and imaginary parts of the spectra are linear combinations of the components of M_{ij} and those of $G_{ki,j}$. The latter are called the excitation functions G_k. For a purely deviatoric source they can be reduced to seven components. These functions are calculated for an appropriate model of a layered medium for $\phi = 0$, and for other values of ϕ are given by their product with the sines and cosines of ϕ. The spectrum of the source time function can be separated into the form $M_{ij}(\omega) = M_{ij}F(\omega)$, assuming that all of the components have the same time dependence. For low frequencies, the problem is practically independent from the frequency, since we are dealing with the flat part of the spectrum. In this case, the real and imaginary parts of the spectrum for the vertical components of displacements and a deviatoric tensor are given by

$$R = G_1[M_{xy}\sin(2\phi) - \tfrac{1}{2}(M_{yy} - M_{xx})\sin(2\phi)] + G_2 M_{zz} - \tfrac{1}{2}G_1 M_{xx} \tag{19.16}$$

$$I = G_3(M_{yz}\sin\phi + M_{zx}\cos\phi) \tag{19.17}$$

where the components of the moment tensor are referred to the geographic axes (section 17.1). The variables are the six components of the moment tensor. For their solution, we need observations at N stations at different azimuths so that the problem is well conditioned. The resulting system of $2N$ equations with six unknowns is solved by a least squares procedure.

The problem can also be solved by using the spectral amplitudes of Rayleigh waves $A(\omega, \phi)$. In this case, their relation to the components of M_{ij} is not linear. If the six components of the moment tensor are M_i, $i = 1, \ldots, 6$, then the spectral amplitudes can be expressed as $A(\omega, \phi, M_i)$. The problem is solved by taking a Taylor expansion about some initial known values M_i^0:

$$A(\omega, \phi, M_i) = A(M_i^0) + \sum_{k=1}^{6} \frac{\partial A}{\partial M_k}\, \delta M_k \tag{19.18}$$

The unknowns are now the six values of the corrections δM_i, and the solution is given by $M_i = M_i^0 + \delta M_i$. Just like in the previous case, we need observations from N stations at different azimuths and the problem is solved by a least squares method. The process is repeated until it converges to a minimum-error solution. The error is defined as the sum of the squares of residuals between observed and calculated spectral amplitudes at each iteration.

In these methods the moment tensor is assumed to be deviatoric and hypocenter coordinates and origination times are supposed known. Since the focal depth affects the excitation functions and its value is often not very precise, it is incorporated into the problem as another variable. In the CMT method, hypocenter coordinates and the origin time are incorporated into the problem as four added unknowns that are recalculated together with the five components of the moment tensor. Coordinates obtained from the minimum-error solution correspond not to the point of initiation of the fracture, as do those obtained from arrival times, but rather to the centroid of the source area. This relocation of the hypocenter results in a better estimation of the moment tensor's components.

In all methods of inversion of the moment tensor, an appropriate model of the Earth is used to calculate the excitation functions. Errors in this model result in errors in the moment tensor obtained. For low frequencies this is not important, but it becomes a problem for small distances and relatively high frequencies. The non-DC part of the solution is in many cases due to these errors rather than to the source itself. As has already been mentioned, moment-tensor inversions give the five components of a deviatoric tensor and must be separated into a DC part and a non-DC part (section 17.4). From the DC part, the parameters of the orientation of the source are obtained according to (17.24). The method gives also the rate of release of the moment as a function of time. In general, solutions are more stable when observations are taken from many different azimuths.

19.5 Amplitude spectra of seismic waves

The amplitude spectra of body and surface waves are used to obtain certain source parameters such as seismic moments and source dimensions (Hanks and Wyss, 1972). According to sections 18.2 and 18.3, the amplitude spectra of seismic waves depend on the source dimensions, practically independently from the model of the source. The seismic moment is determined from the value of the flat part of the spectrum for low frequencies of body and surface waves, corrected for the instrument's response, the pattern of radiation, and all factors that affect wave propagation. For P waves, the seismic moment is given by

$$M_0 = \frac{4U_z^P \pi \rho \alpha^3 r \exp[\omega r/(\alpha Q_\alpha)]}{g(\Delta) C_z(i_0) R_P(\phi, \delta, \lambda, i_h)} \tag{19.19}$$

where U_z^P is the mean spectral amplitude for low frequencies for the vertical component of P waves, r is the distance along the ray from the focus to the station, Q_α is a quality factor for P waves, $g(\Delta)$ is the geometric spreading, $C_z(i_0)$ is the effect of the free surface on amplitudes, and $R_P(\phi, \delta, \lambda, i_h)$ is the pattern of radiation corresponding to the orientation of the source given by ϕ, δ, and λ, where ϕ is the azimuth from the fault trace to the station and i_h is the take-off angle at the focus. The expression for S

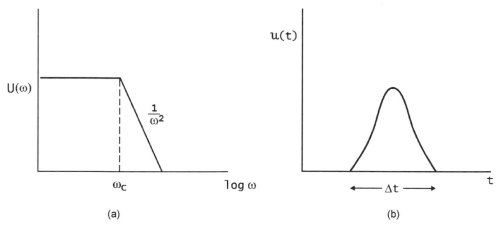

Fig. 19.10. Effects of source dimensions on the spectrum (a) and the pulse width (b) of P waves.

waves is similar. For Rayleigh waves the expression is

$$M_0 = \frac{U_z^R (2\pi R)^{1/2} \exp(\gamma_R R)}{k_R^{1/2} N_z(R, h)(\phi, \delta, \lambda)} \tag{19.20}$$

where $U_z^R(\omega)$ is the mean spectral amplitude for low frequencies of the vertical components of Rayleigh waves, R is the distance from the epicenter, γ_R is an attenuation factor, k_R is the wave number, $N_z(R, h)$ is the excitation function corresponding to the depth h,

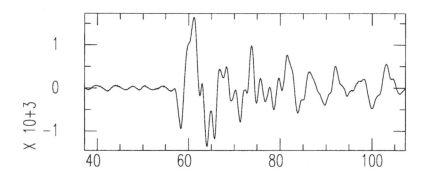

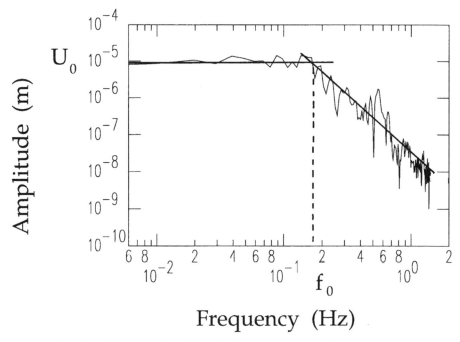

Fig. 19.11. Determination of the seismic moment and the radius of the source from the amplitude spectrum of P waves for the Azores earthquake of 27 June 1997; $M_0 = 3.5 \times 10^{17}$ N m, $r = 10$ km (courtesy of E. Buforn).

and $R_R(\phi, \delta, \lambda)$ is the pattern of radiation. Expressions (19.19) and (19.20) can be simplified by making appropriate approximations for factors.

As we saw in sections 18.2 and 18.3, the source dimensions are related to the corner frequency ω_c (Fig. 19.10). If the spectrum is drawn on a logarithmic scale it can be fitted by two straight lines, one parallel to the frequency axis for low frequencies and another of slope -2 for high frequencies. The intersection of these two lines marks the corner frequency. For a rectangular fault of length L and width W, according to Savage's model,

$$(LW)^{1/2} = \frac{1.7\alpha}{\omega_c^P} = \frac{3.8\beta}{\omega_c^S} \tag{19.21}$$

For a circular fault, according to Brune's model,

$$a = \frac{2.33\beta}{\omega_c^S} \tag{19.22}$$

In this form, both the seismic moment and the source dimension can be estimated from amplitude spectra (Fig. 19.11). Estimations of the seismic moment are very stable, if we use wave lengths larger than the source dimensions. Estimations of dimensions for small earthquakes can be affected by attenuation. For this reason, determinations must be made from data of several stations and a mean value taken. Once the seismic moment and dimensions are known, we can estimate values of the apparent average stress and stress drop (section 15.6).

20 SEISMICITY, SEISMOTECTONICS, AND SEISMIC RISK

20.1   The spatial distribution of earthquakes

The term seismicity was probably used for the first time by Montessus de Ballore in 1906 to describe the distribution of earthquakes and their characteristics within a particular region. The most important aspects of seismicity are given by the geographic distribution of earthquakes' foci, their magnitude, their occurrence over time, their mechanisms, and the damage produced by them. Studies of seismicity are, then, based on seismic catalogs that include parameters such as the dates and times of occurrence, hypocenter coordinates, magnitudes and focal mechanisms of earthquakes and their correlation to regional geological and geophysical characteristics. Historical descriptions of earthquakes can be traced back to written records of old civilizations such as those of China and, for Europe, of Greece and Rome. Among the first universal catalogs of earthquakes were those published by J. Zahn in 1680 and by J. J. Moreira de Mendonça in 1758. Modern catalogs started about 1850 with the work of Perrey, Mallet, and Milne. Today global and regional catalogs are being compiled by various agencies, such as the ISC (International Seismological Centre, Newbury) and NEIC (National Earthquake Information Service, Denver) (Chapter 1). Among the first studies of seismicity were those of Montessus de Ballore between 1850 and 1923, Sieberg between 1923 and 1933, Gutenberg and Richter (1954) and Karnik (1969, 1971). Studies of seismicity are fundamental for understanding the seismotectonic and geodynamic conditions of a region and for the assessment of its seismic risk.

From antiquity it has been known that some regions are more prone to the occurrence of earthquakes than are others. Thus we can separate seismically active regions from those that are more stable. Once the determination of epicenters had become sufficiently accurate, it was observed that earthquakes occur in narrow zones that surround relatively stable regions (Fig. 20.1). These bands or alignments of earthquakes coincide in some cases with the margins of continents, or with oceanic ridges under the oceans. However, not all continental margins correspond to seismic zones, so that they can be divided into active and passive margins, which classification is related to plate tectonics, as we will see later. For example, the western margin of the Americas is active whereas the eastern margin is passive, the northern margin of Eurasia is passive whereas its southern and eastern margins are active, and all margins of Africa are passive.

The depths of earthquakes vary from the surface to about 700 km. According to their depth, earthquakes are classified as shallow (less than 30 km deep), that is, inside the crust, intermediate (30–300 km deep), and deep (more than 300 km deep). Intermediate and deep earthquakes occur in active zones where, according to plate tectonics,

376

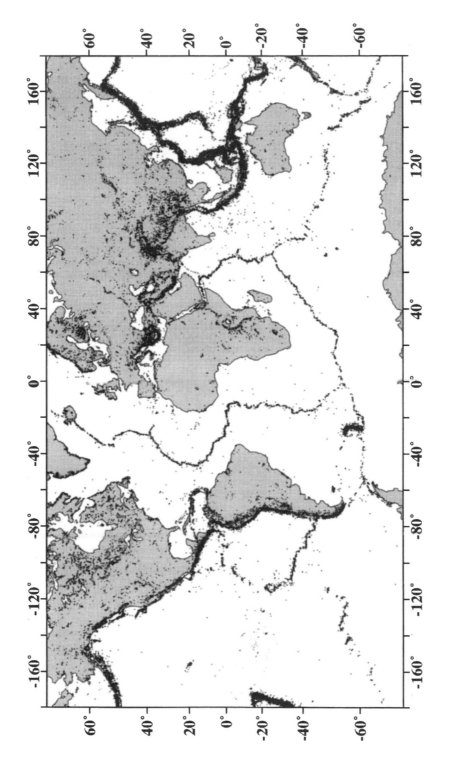

Fig. 20.1. A world map of shallow earthquakes ($h < 60$ km) for the period 1970–90, $M > 4$ (NEIC, USA's Geological Survey).

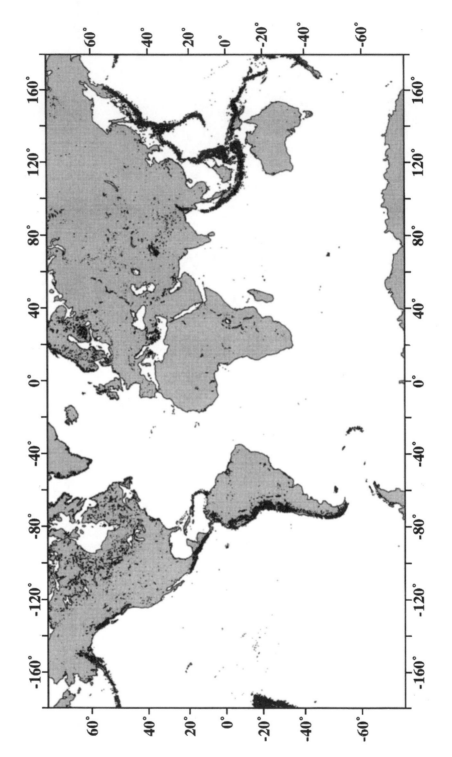

Fig. 20.2. A world map of deep earthquakes ($h > 60$ km) for the period 1970–90, $M > 4$ (NEIC, USA's Geological Survey).

lithospheric plates are introduced into the mantle in zones of subduction. As a general rule, for oceanic ridges there are only shallow earthquakes whereas for island arc regions there are shallow, intermediate, and deep earthquakes (Figs. 20.1 and 20.2). Although shallow earthquakes are classified as those less than 30 km deep, most of them take place at depths less than 20 km, that is, in the upper rigid part of the crust that is called the seismogenic layer (section 15.2). Earthquakes at greater depths correspond to zones where this rigid and relatively cold material of the upper crust is introduced into the upper mantle by lithospheric subduction. However, very deep earthquakes may be produced by processes related to phase changes in focal material instead of brittle fracture.

According to plate tectonics, the global distribution of epicenters is related to boundaries between lithospheric plates (Fig. 20.1). Earthquakes at plate boundaries are called interplate earthquakes. Less commonly, earthquakes also take place in plate interiors and these are called intraplate earthquakes. The most active region in the world corresponds to the margins of the Pacific Ocean. Earthquakes with large magnitudes take place along this zone in the Americas from the Aleutian Islands to southern Chile and from the Kamchatka peninsula in Asia to New Zealand. Besides shallow earthquakes, throughout most of this long region, intermediate and deep shocks take place along the margin of Central and South America and on the other side of the Pacific along the systems of island arcs (Aleutians, the Kuriles, Japan, the Philippines, Fiji, etc.) (Fig. 20.2).

Another large seismically active region is known as the Mediterranean–Alpine–Himalayas region and extends from West to East from the Azores to the eastern coast of Asia. This region is related to the boundary between the plates of Eurasia to the North and Africa, Arabia, and India–Australia to the South (Fig. 20.1). Its seismicity involves shallow, intermediate, and deep earthquakes. A third seismic region is formed by earthquakes located on ocean ridges that form the boundaries of oceanic plates, such as the Mid-Atlantic Ridge, East Pacific Rise, etc. (Fig. 20.1). In these regions earthquakes of shallow depths are concentrated in relatively narrow bands following the trend of the oceanic ridges. On comparing Figs. 20.1 and 20.2, we can see the location of shallow and deep earthquakes and the presence of zones of concentrated seismicity and others with earthquakes spread over wide areas. In general, boundaries between oceanic plates and between oceanic and continental plates have simpler distributions of seismicity than do boundaries between continental plates.

An example of a complex distribution of seismicity in a boundary between two continental plates is the Mediterranean region (Fig. 20.3) (Udías and Buforn, 1994). Inside this region the most active zones in order of activity correspond to Greece, Turkey, Italy and the Alps, the Carpathians and Dynarics, North Africa, southern Spain and the Pyrenees, and the Rhine Graben. Along this region, seismic activity is formed by a continuous occurrence of earthquakes of small and moderate magnitudes ($M < 6$) with the occurrence of large earthquakes ($M > 6$) separated by larger intervals of time. Destructive earthquakes have happened more or less often in all active regions. Most earthquakes are of shallow depths with four zones of intermediate and deep earthquakes, namely the Hellenic arc, the Carpathians, Sicily–Calabria, and southern Spain and Morocco. The deepest earthquakes have taken place in southern Spain (640 km) and Sicily–Calabria (450 km). In the other two regions the maximum depth is about 200 km.

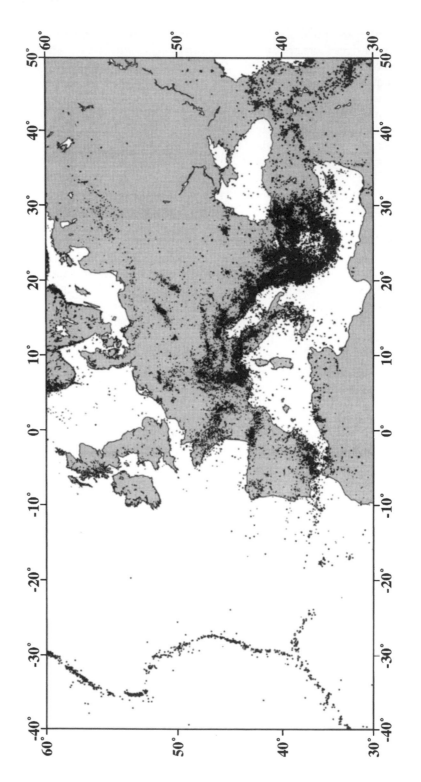

Fig. 20.3. A map of earthquakes in the Azores–Mediterranean region for the period 1970–90, $M > 3$ (NEIC, USA's Geological Survey).

As can be seen from Fig. 20.3, earthquakes are spread over wide areas, indicating the presence of several small plates between the two large plates of Eurasia and Africa.

20.2 The temporal distribution of earthquakes

In studies of seismicity, the temporal distribution of earthquakes is as important as their spatial distribution. In a very general way, it can be said that earthquakes in a region behave as a temporal series of point events resulting from a process of release of stress in the Earth's crust. Statistical studies of temporal series of earthquakes reveal some characteristics of their distribution. The limits of the area being studied and the durations of time intervals are two important factors to consider.

From a statistical point of view, the simplest model of the occurrence of earthquakes with time is a Poisson distribution. This distribution assumes that earthquakes are independent events, that is, their occurrence does not affect that of others. If λ is the rate of occurrence of earthquakes within a time t, the probability that n earthquakes take place within such an interval is

$$p(n) = \frac{\lambda^n e^{-\lambda}}{n} \tag{20.1}$$

If the occurrence of earthquakes follows a Poisson distribution, then the intervals of time δt between consecutive earthquakes have an exponential distribution. The probability that two earthquakes are separated by an interval δt is

$$p(\delta t) = \lambda e^{-\lambda \delta t} \tag{20.2}$$

For a given rate of occurrence, the probability is greater for smaller intervals. Studies of temporal series of earthquakes show that they deviate, to greater or lesser extents, from Poisson's law. The global distribution of large earthquakes follows this distribution quite well, but moderate and small earthquakes and those in a limited region do not. The assumption of independence is not totally correct since earthquakes are grouped in series of various kinds. The interdependence of earthquakes shows up in the occurrence of clusters such as sequences of aftershocks and swarms. Clusters can be better observed in temporal series including small earthquakes in limited areas. Shocks that are near in space and time are necessarily interrelated.

20.2.1 *Foreshocks, aftershocks, and swarms*

The most important clustering of earthquakes is that associated with the occurrence of an earthquake of larger magnitude that is called the main shock. Earthquakes of lesser magnitudes preceding the main shock are called foreshocks and those following immediately thereafter are called aftershocks. The occurrence of aftershocks is a common phenomenon and is related to the release of energy in the fracture zone that is not completed by the main shock. In complex models of earthquake sources, we saw that not all of the accumulated stress is released by the main shock, but rather zones of the source remain unbroken and break afterwards, causing aftershocks. Foreshocks are less frequent, but many earthquakes are preceded by small shocks that break

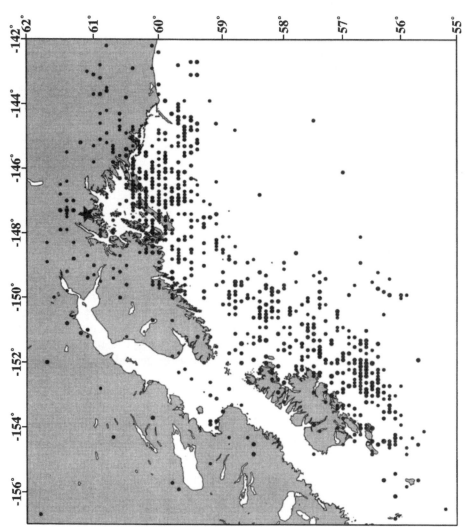

Fig. 20.4. Aftershocks during the year after the Alaska earthquake of 28 March 1964, $M = 8.6$. The star shows the location of the main shock (NEIC, USA's Geological Survey).

weak zones on the fault plane before the main event. When, in a series of earthquakes in a small area, there is no main shock, the series is called a swarm of earthquakes.

In a general way, series of earthquakes can be related to the nature of the conditions of the material of the fracture zone (Mogi, 1963). If the material is very homogeneous and the distribution of stress uniform, there are no foreshocks and the main shock is followed by a series of aftershocks of lesser magnitude. Foreshocks are associated with heterogeneities in the source material that result also in longer sequences of aftershocks. If the material is very heterogeneous and the stress not uniform, then earthquakes happen in swarms without a real main shock.

Both the spatial and the temporal distribution of quakes within series of aftershocks are of interest. Aftershocks are distributed over the fracture area of the main shock and, in some cases, there is a certain migration from one end to another of the fault. Thus, the fault area of large earthquakes can be determined by the area over which the aftershocks are distributed. The number of aftershocks, their magnitudes, and their durations depend on the magnitude of the main shock. An earthquake of magnitude 5 usually has a sequence of aftershocks lasting a few days, whereas for one of magnitude 8, the series may last more than a year. The earthquake in Alaska in 1964 ($M = 8.6$) had a long series of aftershock lasting more than a year and a half over an area of about $360\,000\,\text{km}^2$ (Fig. 20.4). To estimate the total duration of a series of aftershocks is sometimes difficult, especially for very large earthquakes. The last part of the series may last for years. In general, shallow shocks have longer series of aftershocks than do deeper ones. The magnitudes of aftershocks depend on that of the main shock. The largest aftershock has a magnitude about one unit lower than that of the main shock.

The distribution in time of the number of aftershocks follows, generally, an inverse power law of time. In 1894 Omori proposed a dependence of $1/t$ that has been modified to the form (Utsu, 1961)

$$N(t) = \frac{K}{(c + t)^p} \tag{20.3}$$

where K and c vary with each series and p has values in the range 0.7–1.4. Also here, values of p are interpreted in terms of the heterogeneity of the fault material. The number of aftershocks diminishes more rapidly for a very homogeneous material (high values of p) than it does for a heterogeneous material (low values of p).

20.2.2 Seismic cycles

Earthquakes occur when the stress overcomes the strength of the source material at a critical value, producing its fracture. The accumulation of stress in a particular area is produced by tectonic processes that are related ultimately to plate motion. If we assume that stress accumulates at a constant rate, it will reach the critical value and produce earthquakes in a certain fault area at more or less regular intervals. The process of the accumulation of stress and its release by the occurrence of an earthquake on a particular fault constitutes a seismic cycle.

In the simplest case, the rate of accumulation of stress is uniform and reaches the same critical level at which an earthquake is produced with constant drop in stress, reducing the stress to the same level, from which it starts to accumulate again. In this case, all

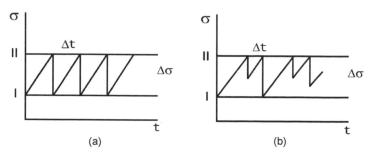

Fig. 20.5. Models of the occurrence of earthquakes. (a) Periodic, equal drop in stress, and equal time interval. (b) Nonperiodic, different drops in stress, and different time intervals (the time-predictable model) (Shimazaki and Nakata, 1980) (with permission from the American Geophysical Union).

earthquakes would have the same magnitude and repeat at constant intervals with periodic seismic cycles. In this ideal case, the magnitudes and times of occurrence of earthquakes could be predicted (Fig. 20.5(a)). A more realistic model was proposed by Shimazaki and Nakata (1970), whereby stress is released when it reaches the same maximum value, but the drops in stress for earthquakes are not always the same; that is, stress is released by earthquakes of different sizes. The intervals between earthquakes are not always the same, but rather depend on the size of the preceding earthquake (Fig. 20.5(b)). This situation, known as the time-predictable model, allows prediction of the time of occurrence of the next earthquake, but not its size. A model with earthquakes happening when the stress reaches various levels and also with different drops in stress will have earthquakes of different sizes occurring at different intervals. In this case, we can not predict the time or magnitude of the next earthquake from knowledge of the preceding seismic cycle.

The concept of a seismic cycle is very important if one is to understand the occurrence of earthquakes on the same fault. Although the accumulation of stress may be approximately uniform, the heterogeneity of the material does not allow the existence of periodic or regular cycles of occurrence of earthquakes of the same magnitude. Owing to the heterogeneous conditions at the fault, stress may be released starting at various levels and producing earthquakes of different sizes, so that the time-predictable model is not valid. Another problem that affects the regularity of seismic cycles is the existence on a fault of aseismic slip caused by creep phenomena. Thus, part of the accumulated stress is released by continuous aseismic slippage of the fault, independently from the occurrence of earthquakes. The aseismic slip depends on the nature of each fault and it is difficult to observe. The result is that the sum of the stresses released by earthquakes does not add up to that accumulated by tectonic processes and so the seismic slippage is only part of the total slippage on the fault.

A concept related to the seismic cycle is that of a characteristic earthquake (Schwartz and Coppersmith, 1984; Aki, 1984). This concept is based on the consideration that a particular fracture, owing to its dimension remaining constant, can only produce earthquakes with a certain maximum size. For each fault, earthquakes with this maximum magnitude are called characteristic earthquakes. These earthquakes have very similar characteristics and repeat themselves on the same fault, at quasi-regular intervals. Implicit in this concept is that each fault is an independent unit that breaks in its totality, producing always an earthquake of the same maximum magnitude. However, it has been observed that large earthquakes may break through several faults, and, thus, the principle of characteristic earthquakes is not always satisfied. Here again, the problem is that the conditions in fault regions are not homogeneous and regularities in the occurrence of earthquakes are difficult to establish.

The main problem with respect to the regularity of seismic cycles and characteristic earthquakes is that observations of the occurrence of large earthquakes are very limited in availability. Instrumental data started to be recorded at the beginning of this century and reliable global data do not extend more than 50 years back. Historical earthquakes dating back several centuries may be known in some regions, but our knowledge of them has many limitations. Even admitting the reliability of data of this type, the record can not be extended back any further than 1000 years ago. In most active regions, large earthquakes ($M > 6.5$) repeat at intervals that vary from tens to hundreds of years. In regions of low seismicity these intervals may be larger, of the order of thousands of years. Moreover, very large earthquakes ($M > 8$) are very rare events that may be separated by even greater intervals. For example, it is difficult to know the return time for the great Lisbon earthquake of 1755, there has been no similar event in historical times. A new source of observations comes from paleoseismicity studies of active faults. Detailed field observations on trenches through faults allow the dating of large earthquakes in the geological past but there are still few of them. In conclusion, with the information we have at present, it is difficult to solve the problem of the periodicity of large earthquakes and to know to within a certain precision the characteristics of seismic cycles.

20.3 The distribution of magnitudes

Observations show that, in seismic regions, during any period of time, the number of small earthquakes is many times that of larger ones. This fact is expressed in the empirical law suggested by Omori in 1889 and proposed by Gutenberg and Richter (1954) in logarithmic form:

$$\log N(M) = a - bM \tag{20.4}$$

where $N(M)$ is the number of earthquakes of magnitude larger than M, a is a constant that represents the number of earthquakes of magnitude larger than zero, and b is the proportion of earthquakes with small and large magnitudes. Although we do not know the exact form in which the elastic energy accumulated in a region is released by earthquakes, the distribution of their magnitudes follows this universal law. The value of b is an important parameter in studies of seismicity. It is deterinined by finding the slope of the straight-line fit to the distribution of the logarithm of the number of earthquakes versus their magnitudes. In general, observations fit this distribution, with some deviations, corresponding to very small and very large earthquakes (Fig. 20.6). For a large area and long interval of time, these deviations are generally due to a lack of completeness in observations at both ends of the magnitude range. For observations from a single fault, deviations may be related to its conditions and the size of the characteristic earthquake.

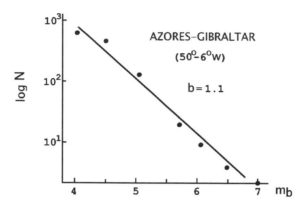

Fig. 20.6. The number of earthquakes together with their magnitudes for the Azores–Gibraltar region.

Values obtained for *b* are very stable, in the range 0.6–1.4, and its most common value is very near to unity. Changes in values of *b* on going from one region to another are related to their mechanical characteristics. High values of *b* indicate a high number of small earthquakes, which is to be expected in regions of low strength and large heterogeneity, whereas low values indicate the opposite, namely high resistance and homogeneity. Changes of *b* with time for the same region are associated with changes in stress conditions and hence *b* has been proposed as a parameter to consider in the problem of the prediction of earthquakes. The degree of significance of these changes has been questioned by some authors.

20.4   Models of the occurrence of earthquakes

Observations of the occurrence of earthquakes in space, time, and magnitude have led to the proposal of models that simulate certain properties of their distribution. These models specify the form in which accumulated elastic stress is released along the fault surface, producing earthquakes of varying magnitude at different times. Models describe the process of stress loading in terms of elastic strain in the focal region and its subsequent release with the occurrence of earthquakes. As we have seen, the process of stress loading and the subsequent release of stress constitutes a seismic cycle. Models must satisfy Gutenberg and Richter's relation and that the universal value of *b* is near unity, and should try to explain this fact.

The first model to simulate the properties of the occurrence of earthquakes consists in a series of blocks connected by springs that slide on a horizontal surface with friction as they are drawn by means of springs that connect them to a block that moves with a constant velocity (Fig. 20.7) (Burridge and Knopoff, 1967). The moving block represents the material on one side of a fault that is drawn by tectonic movements loading springs that connect it to the small blocks. Owing to friction between blocks and the stable surface, motion is not transmitted directly to the blocks, but rather first deforms the springs. Once the springs have accumulated enough energy to overcome the friction of the blocks, these move. When a block moves, it loads elastic energy onto the springs that

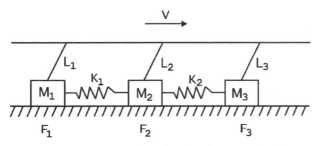

Fig. 20.7. The model of blocks and springs for simulating the occurrence of earthquakes.

link it to adjacent blocks, causing their later displacement. The motion of one or a few adjacent blocks represents small earthquakes and that of many blocks corresponds to a large one. The smallest earthquake is represented by the motion of one block and the maximum size possible is represented by that of all blocks of the system. Owing to friction, the blocks do not move uniformly, but rather by sudden increments, in the form called stick–slip motion.

With models of this type, we can simulate the occurrence of earthquakes with a proportion of small and large earthquakes similar to that of Gutenberg and Richter's law. Another result of these models is that small events take place in a random manner whereas large ones occur in a quasi-periodic way. Originally, the model was only one-dimensional, but it has been generalized to two and three dimensions. More recent models are more complex with viscoelastic elements between blocks, the friction variable with the velocity, and other conditions that allow the simulation of series of aftershocks after large events and other properties of the occurrence of earthquakes (Dieterich, 1972; Mikumo and Miyatake, 1979). The fundamental idea of these models is that each element of a fracture interacts with those surrounding it and thus influences the evolution of the whole system. Although these systems are in themselves deterministic, they behave chaotically as their complexity is increased, due to the increase in their instability. The result is a quasi-chaotic behavior in which very small changes in initial conditions lead to very large effects on the evolution of the system that behaves more and more randomly with time.

Another approach to studying the occurrence of earthquakes is to consider their fractal nature using the fractal theory developed by Mandelbrot (1977) (Andrews, 1980; Aki, 1981; Hirata, 1989). A basic property of all fractal distributions of a variable r is that they obey a power law of the type Ar^{-D}, where the exponent D represents the fractal dimension and is related to the coordinates associated with the elements of the set that represent the phenomenon being studied.

Gutenberg and Richter's law (20.4) can be transformed into a power law for the number of earthquakes as a function of the seismic moment $N(M_0)$ by substituting equation (15.35) into it:

$$N(M_0) = BM_0^{-D} \tag{20.5}$$

where B is constant for a constant drop in stress. Since the seismic moment is proportional to the source dimensions, a similar relation can be written for the fault's length or area using (15.30) and (15.33). According to Aki, $D = 2b$ and, since b has a value near to unity, D, the fractal dimension, is approximately equal to two. This result can be interpreted as representing the fact that earthquakes are distributed over two dimensions (a plane), which agrees with the observation of their occurrence on fault planes.

The fractal nature of the occurrence of earthquakes implies that they have a self-similar stochastic distribution, that is, they behave in a similar form, independently from the range of sizes considered. Although this is the general rule, self-similarity need not be perfectly satisfied. It has been observed that the relation between the fault's length and the seismic moment is different for moderate and small earthquakes ($M < 6.5$) than it is for large ones ($M > 6.5$) (Scholz, 1982). In the first case the seismic moment is proportional to the cube of the source's length whereas in the second it is proportional to its square. This break in the self-similar rule can be explained by invoking the influence of the thickness of the seismogenic layer. As we saw, this layer has an approximate thickness of 20 km and thus affects the geometry of the seismic source, which is nearly circular for small earthquakes and rectangular ($L > W$) for large ones (section 15.1).

Once we have accepted the fractal nature of earthquakes and their self-similarity, we may ask what the dynamic process responsible for this behavior is. Some authors have answered this question in terms of self-organized criticality (SOC) (Bak and Tang, 1989; Sornette and Sornette, 1989; Ito and Matsuzaki, 1990). This term (SOC) is applied to dynamic systems that evolve by themselves until they reach a critical state in which phenomena take place randomly, according to a power law. In the case of earthquakes, the system is the material of the Earth's crust, which evolves under tectonic stresses until it reaches a state of SOC. In this state, earthquakes of all sizes take place with only the limitation of a minimum possible size (a minimum earthquake) and a maximum size imposed by the dimensions of the seismogenic region (the maximum or characteristic earthquake). This behavior is a consequence of the nonlinear dynamic space–time characteristics of the Earth's crustal system in response to stresses generated by lithospheric plate motion.

This type of process can be modeled, in a form similar to that of the models of blocks and springs, by a distribution of cellular automata whose behavior under an applied stress is specified. Each cell receives and transmits stress to its neighbor cells as it moves, changing the stress distribution of the whole system. After a sufficiently long time, the system evolves by itself into a critical state (SOC), in which events of all sizes are produced with a power-law distribution. The size of events is given by the number of adjacent cells that move at a given moment. Models of this type reproduce certain properties of the occurrence of earthquakes, such as the value of b being near unity and the occurrence of aftershocks that decay in number with the inverse of time.

20.5   Seismotectonics

The relation between geological characteristics and the occurrence of earthquakes started to be studied at the end of the 19th century and very early on in the 20th century, by Milne, Mercalli, Sieberg, Montessus de Ballore, Koto, and Omori, among others. The term seismotectonics began to be used around 1910 by Sieberg and Hobbs and is applied to the characteristics of the occurrence of earthquakes in

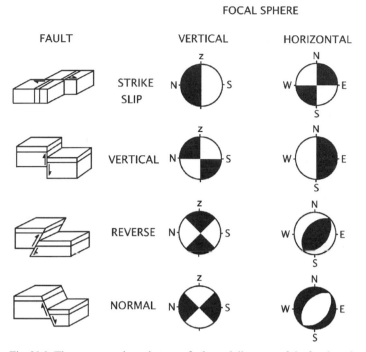

Fig. 20.8. The correspondence between faults and diagrams of the focal mechanism (horizontal and vertical projections of the focal sphere).

relation to regional tectonics and general geodynamic conditions. In seismotectonic studies one tries to integrate earthquake data with other information available from the tectonics, geophysics, and geology of a particular region. Seismologic data used in seismotectonics include the geographic distribution of epicenters, with indications of their magnitude, depth, and focal mechanism. A very common representation of the focal mechanism in seismotectonic studies is shown by means of the projection of the focal sphere with quadrants of compressions and dilations in black and white. The correspondence of diagrams of this type to various types of faults (strike–slip, vertical, normal, and reverse) is shown in Fig. 20.8, together with projections of the focal sphere onto a vertical plane. In strike–slip faults the two planes that separate the quadrants are vertical, whereas in normal and reverse faults, the center of the projection is in a dilation and a compression quadrant, respectively. Another graphical representation of focal mechanisms is by horizontal projection of slip vectors and pressure and tension axes. Slip vectors are related to kinematic aspects of tectonics, that is, the directions of motion of plates or blocks. Pressure and tension axes refer to dynamic aspects, that is, stress orientations is a particular region.

The tectonic interpretations included in seismotectonic studies depend on the accepted theories for the general processes active in the Earth's crust. Historically, the first of these theories, based on vertical movements of crustal blocks related to the contraction of the Earth due to its cooling, were presented by Dana, Hall, and Suess, among others. Authors of early seismotectonic studies used this tectonic framework to interpret seismic

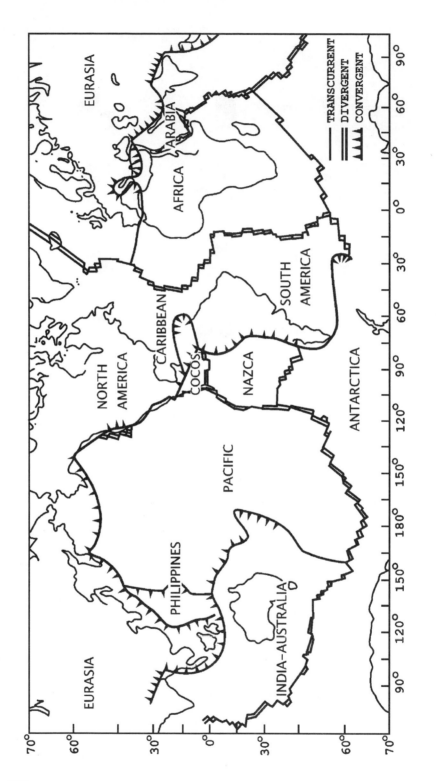

Fig. 20.9. Principal lithospheric plates and types of plate boundaries.

data. In 1912, Wegener proposed his theory of continental drift, whereby tectonic processes are derived from the horizontal motion of continental blocks, but the geophysical objections presented against continental drift (Jeffreys, 1959) were an obstacle to its use in the interpretation of earthquake data. During the 1960s, plate tectonics theory, in whose development seismologic data played an important role, was introduced. Modern seismotectonic studies are based on its principles.

20.5.1 Seismicity and plate tectonics

The basic ideas of plate tectonics can be summarized in a very brief form as follows (Kearey and Vine, 1990). First of all, the basic unit of tectonic motion is the lithosphere–astenosphere system. As was explained in section 9.4, the lithosphere is a layer of about 100 km thickness including the crust and part of the upper mantle that behaves as rigid and cool material in relation to the underlying layer, the astenosphere, a weak warmer layer that behaves with plastic or viscous flow. The viscosity of the astenosphere allows the horizontal motion of the lithosphere with velocities in the range $1-6 \, \mathrm{cm \, yr}^{-1}$. The motion of the lithosphere is due to a number of processes, including thermal convection currents inside the mantle. The lithosphere is divided into a number of plates that, generally, include both continental and oceanic crust (Fig. 20.9). The most important plates are the Pacific, the Americas (sometimes divided into two, North and South), Eurasia, Australia–India (sometimes also divided), Africa, and Antarctica. Other minor plates are Nazca, the Cocos, the Philippines, the Caribbean, Arabia, Somalia, and Juan de Fuca. Other even smaller units are called sub-plates or microplates. Plate boundaries can be established from the distribution of seismic regions, as was mentioned in the discussion of the geographic distribution of seismicity. The first interpretations of seismologic data in terms of plate tectonics revealed the agreement of seismicity and the distribution of focal mechanisms with the expected conditions due to the relative motion at plate boundaries (Isacks et al., 1968; McKenzie, 1972).

The types of plate boundaries can be reduced basically to three, namely, divergent or rift zones, convergent or subduction zones, and transcurrent horizontal slip or transform faults (Fig. 20.10). At divergent boundaries and in rift zones, plates separate themselves from one another, and new oceanic lithosphere is created between them, for example, in ocean ridges. For plate boundaries of this type, earthquakes are of shallow depth and moderate magnitude ($M < 7$), forming a narrow band along boundaries. Their focal mechanisms correspond to normal or strike–slip faulting with the horizontal tension axis normal to the plate's boundary.

At convergent boundaries plates collide and one of them is introduced under the other in the mantle in a process called subduction. Subduction zones are located in zones of deep earthquakes under orogenic belts or island arcs. The subducted lithosphere is always of oceanic nature and its dip varies from case to case. In subduction zones, subducted plates maintain their rigidity and earthquakes take place from the surface to a maximum depth of 700 km. The seismic zone, called the Wadati–Benioff zone, is, generally, limited to the upper part of the subducted plate (Fig. 20.11). Beyond 700 km depth, subducted plates are aseismic and become assimilated into the material of the mantle. If, at a converging boundary, both plates are continental, processes are more complex, with seismicity extended over a wide area, such as for the boundary between

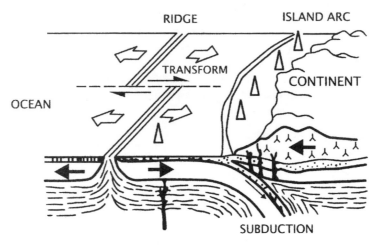

Fig. 20.10. Processes occurring at plate boundaries, namely divergence or extension (at a ridge), convergence or collision (subduction), and transcurrent motion (at a transform fault).

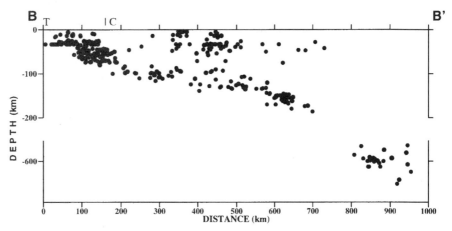

Fig. 20.11. The distribution of earthquakes with depth in the subduction zone of Peru. The line BB′ is perpendicular to the coast. T, trench; and C, coast (courtesy of H. Tavera).

Eurasia and Africa (Fig. 20.3). At convergent boundaries earthquakes reach large magnitudes ($M > 7$), and their mechanisms correspond to thrust or reverse faulting due to horizontal pressure normal to the trend of the boundary. In depth, either the pressure or the tension axis is in the direction of the subducted plate, indicating its condition under compression or tension. Generally, the tension axes are along the plate in the upper part and the pressure axes are in the lower part. Thus, the upper part of the plate is stretched while the lower part is compressed against the mantle. Extension and subduction zones are connected by boundaries of horizontal transcurrent slip or transform faults, formed by strike–slip faults whose motion is transformed into extension or compression at their ends (Wilson, 1965). Along these faults, earthquakes are shallow ones with the strike–slip mechanism and may reach large magnitudes ($M > 7$). Two zones of convergence may be linked by a transform fault. The mechanisms of shallow earthquakes on the convergence front are of reverse or thrust faulting corresponding to the horizontal pressure being normal to the front and of strike–slip motion corresponding to pressure along the transform fault (Fig.20.12(a)). For two zones of extension connected by a transform fault, the mechanisms of earthquakes in the rift zones correspond to normal faulting, whereas those in the transform fault correspond to strike–slip faulting (Fig 20.12(b)).

Figure 20.13 shows an example of a seismotectonic framework for the Mediterranean region based on seismicity and focal mechanism data (Udías and Buforn, 1994). The tectonic situation is due to the converging motion of the plates of Eurasia to the North and Africa and Arabia to the South. The situation is complex due to the continental nature of both plates and the presence of several small plates such as the Italy–Adriatic, Aegean, and Anatolian plates. The seismicity is somewhat diffused over a wide region, indicating that there are some intraplate deformations. The predominant types of focal mechanisms are shown together with the direction of horizontal stresses and slip directions. The regional stresses correspond to horizontal pressure in the NW–SE to N–S direction due to the converging motion of Eurasia, Africa, and Arabia, with two localized regions of horizontal tensions, in the NE–SW direction in central Italy and in the N–S direction in northern Greece. Two regions of right-lateral strike–slip motion are located in the Azores–Gibraltar and north Anatolia faults. Two subduction zones are located in the Hellenic arc and Sicily–Calabria, with another two zones of deep earthquakes in southern Spain–northern Morocco and the Carpathians.

20.6   Seismic hazard and risk

In a general way, the risk associated with the occurrence of earthquakes can be separated into two factors. The first, called the seismic hazard, represents the probability that ground motion of a certain intensity takes place at a certain place due to nearby earthquakes. The second, called the vulnerability, represents the probability that a certain structure suffers an appreciable amount of damage due to the action of earthquakes. Both factors, hazard and vulnerability, are included in the general term of seismic risk. In this form, we separate the purely seismologic aspect of seismic hazard from the engineering considerations regarding the vulnerability of various structures.

The seismic hazard of a site is given in terms of the ground motion (displacement, velocity, and acceleration) produced by nearby earthquakes. Commonly, only accelerations are considered, and, in many instances, ground motion is represented by values of the seismic intensity. The seismic hazard is defined as the probability that ground motion of a certain amplitude takes place at a given site during a certain period of time. The seismic hazard of a site depends on many factors, namely, the distribution of nearby seismic sources, their magnitudes and focal mechanisms, the nature of the propagating medium, especially wave attenuation and geometric spreading, and the characteristics of the shallow layers under the site that increase or decrease the amplitudes of ground motion.

The seismic hazard can be evaluated by deterministic and probabilistic methods. Deterministic methods are based on the assumption that the seismicity near a site in the future will be identical to that in the past. Thus, the seismic hazard is given by the maximum values of the amplitudes of ground motion calculated for past earthquakes.

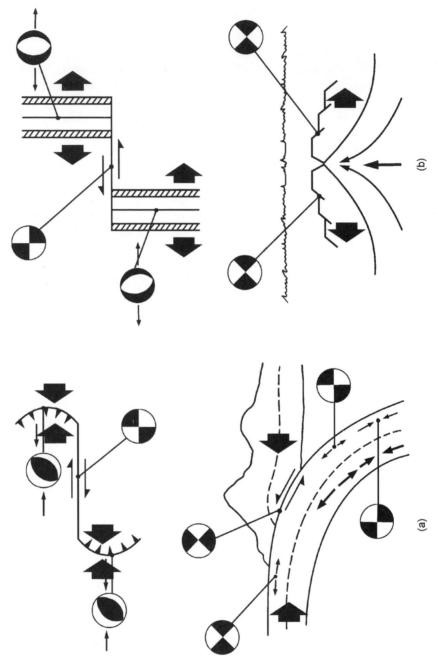

Fig. 20.12. The connection of convergent (subduction) and divergent (ridge) zones by means of transform faults and types of focal mechanisms. (a) Two subduction zones. (b) Two ridges.

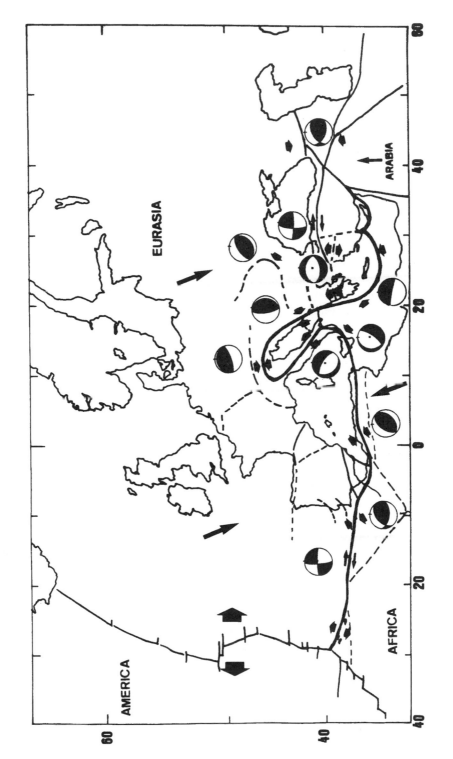

Fig. 20.13. The seismotectonic framework for the Azores–Mediterranean region.

If the seismic history is not well known for a sufficiently long time, maximum values can be extrapolated from an incomplete frequency–intensity relation. These methods depend very strongly on our knowledge of the past seismicity of the region surrounding a given site. The greatest problem is that often we do not know with any accuracy about the occurrence of earthquakes during a very long period of time. This is more crucial for regions of moderate seismicity where large earthquakes are separated by hundreds of years.

Probabilistic methods use statistical distributions to represent the seismic activity of a region on the basis of the occurrence of earthquakes in the past. The result is the determination of the probability of the occurrence of a certain level of ground motion or seismic intensity for a particular time interval. Probabilistic methods have many advantages over deterministic methods. The occurrence of earthquakes in the future is not given directly by that in the past, but rather is estimated from a statistical distribution. In order to calculate the expected ground motion, earthquakes of the past are not considered in terms of their exact location, but rather are assigned to seismogenic zones that group all earthquakes of the same tectonic origin. The determination of these zones (seismic zonation) is an important problem in the determination of seismic hazards. Different zonifications lead to different estimations of hazards from the same set of earthquake data. Probabilistic methods provide estimations of probabilities for various levels of intensities rather than solely maximum values as deterministic methods do.

The seismic hazard is given by the probability that, in a particular region and within a given time interval, ground motion reaches a certain amplitude or the seismic intensity attains a certain degree. This can be formulated also as the probability that, during a certain time interval, the maximum value of the intensity or ground acceleration exceeds a particular value. The inverse of this probability, or the time in which we expect with a certain probability that a certain intensity (or ground acceleration) is exceeded, is called the return period for that intensity. One of the statistical methods used in the assessment of seismic hazard is the method of extreme values. This method consists in using, instead of the complete seismic catalog, only the distribution of the earthquakes with the largest magnitudes for certain time periods. This permits one to extend the analysis further back in time, for which we know about only the largest events. Bayesian methods, in which certain *a priori* probabilities are introduced, can also be used. Each method has its advantages and disadvantages, but ultimately they all depend on the accuracy of our knowledge of the seismic history. From this there follows the importance of instrumental and historical studies of seismicity and of the results from paleoseismicity.

The seismic hazard specifies the probability of ground motion at a particular site and depends on earthquakes at various distances from its neighborhood. We know that ground motion is attenuated with distance due to geometric spreading and anelastic attenuation. Since seismic intensities are used in most hazard studies, we need to know the laws governing the attenuation of intensity with distance (section 15.3, Fig. 15.4). Generally an expression similar to (15.9) is used, whereby the values of the coefficients *a* and *b* are determined from intensity maps for earthquakes in the region. With the availability of dense networks of accelerographs we can determine directly the attenuation of the ground acceleration with distance. From knowledge of the occurrence of earthquakes over time and their maximum intensities, inside each seismogenic zone surrounding a site together with the laws governing the attenuation with distance, we

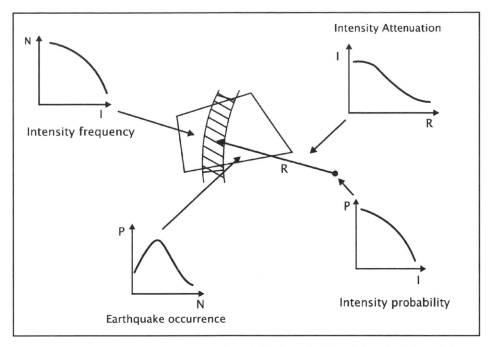

Fig. 20.14. A schematic representation of the determination of the seismic hazard (courtesy of D. Muñoz)

can determine their expected ground motion or intensities at the site (Fig. 20.14). Thus we can calculate the probability of a certain level of intensity (or ground acceleration) being exceeded within a given time interval that defines the seismic hazard, for example, the probability that intensity VIII (or a ground acceleration of 0.4g) is exceeded within 200 years. If we calculate values of the seismic hazard at many sites we can draw maps with intensities or ground accelerations expected with a given probability within a certain interval of time (Fig. 20.15). The ground acceleration is the parameter that is most directly related to engineering design, so modern seismic hazard maps are given in terms of this parameter. However, often ground accelerations are derived from seismic intensities and this relation has a wide margin of variability.

Adequate assessment of seismic hazards is fundamental to prevention of the damage caused by earthquakes, since buildings and other structures can be designed to withstand without collapsing the maximum ground acceleration expected for a given site. As we saw in the intensity scale (Table 15.1), for a given intensity, the damage to buildings depends greatly on the building materials and structural design. For intensities of about VIII, many masonry buildings of adobe or rubble stones suffer very heavy damage and even total destruction, but only a few buildings of reinforced concrete with a minimum of antiseismic design suffer moderate damage. Reinforced concrete and steel-structure buildings with high levels of antiseismic design can escape very heavy structural damage even at maximum levels of intensity. The study of the structural responses of various types of buildings is fundamental for a safe building practice in seismic regions. All countries with an appreciable seismic risk have antiseismic building

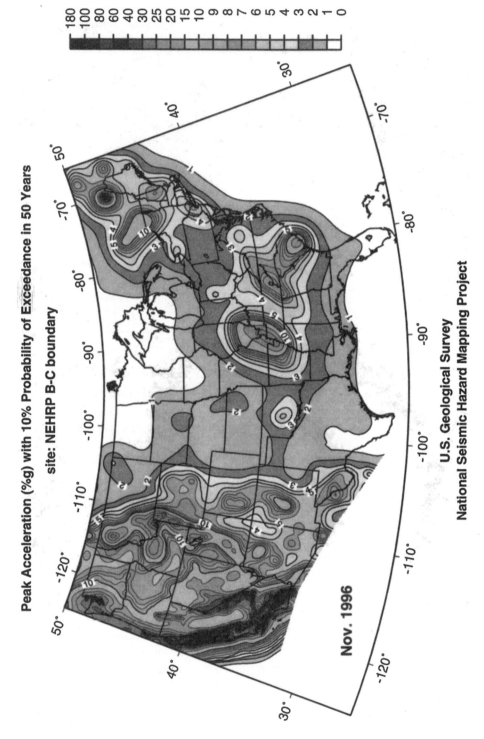

Fig. 20.15. A seismic hazard map for the USA (NEIC, USA's Geological Survey).

codes that are sanctioned by law. Safety norms are stricter for structures whose collapse can produce large catastrophes, such as nuclear plants, dams, public buildings, etc.

20.7   The prediction of earthquakes

The problem of predicting earthquakes has always been on the horizon of seismology (Simpson and Richards, 1981; Rikitake, 1982; Lomnitz, 1994). For many years, this was considered a task beyond the possibilities of true science (Macelwane, 1946). From about 1970 until 1980, attitudes were more optimistic and the solution of the problem was considered to be practically to hand. More recently, a more critical view has returned, recognizing the difficulty of the problem and even hinting at its impossibility (Geller, 1997). However, since forecasting the occurrence of earthquakes with sufficient advance warning is a very efficient way of diminishing the number of casualties and preventing, in part, damage, it remains an unrenounceable task for seismologists.

A true prediction means specifying the location, time, and magnitude of an impending earthquake to within narrow margins and with a high level of probability, with sufficient advance warning that certain measures to reduce its effects can be taken. Prediction can be considered in long, medium, and short terms. Long-term prediction, in practice, can be identified with the assessment of the seismic hazard of a region. In this context, the determination of a high probability for the occurrence of a large earthquake in an area does not constitute a true prediction. Medium- and short-term predictions are really related. For example, a medium-term prediction could be that an earthquake of magnitude about 6 will happen in a certain area within 4–6 months. As this time approaches, prediction requires that the probability increases and the time and place of occurrence are specified more precisely. Finally, enough evidence should make it possible to give a short-term prediction, specifying the size, place, and time of the impending earthquake within relatively narrow margins, so that the authorities can take adequate measures for protection of the population. In the actual state of the problem, there is not yet a methodology that allows medium- and short-term predictions of this type.

The problem of prediction is based on the identification of precursor phenomena that indicate the impending occurrence of an earthquake. Precursors are observables that are related to changes in physical conditions in the focal region. Strictly speaking, some phenomena such as the abnormal behavior of animals can not be considered true seismologic precursors. It is difficult to verify and quantify such phenomena, let alone establish their relation to a future earthquake. Precursors can be divided into seismic and nonseismic ones.

One seismic precursor is the observation of patterns of seismicity in a region with variations in space and time. This observation has shown the existence of seismic gaps, areas in active regions where there is low seismicity during a certain time previous to a large earthquake. In an active region, where the seismicity is expected to be distributed more or less homogeneously, the existence of an area where the level of seismicity is lower than average is called a seismic gap and is considered the most probable place for a large earthquake in the future. The size of the seismic gap and the duration of the lack of activity can be correlated to the magnitude of the future large shock. Correlations between the existence and extent of seismic gaps and the occurrence of large earthquakes have been established *a posteriori* in several instances. However, the use of this type of precursor as a predictor is not very reliable because it is not known when a large earthquake will be produced in an observed seismic gap. A second seismic precursor is the observation of changes in a given active region of the rate of occurrence of small earthquakes, or background seismicity, with intervals of decreasing or increasing activity. In the first case, we have a seismic quiescence that is associated with an impending larger earthquake. Also the opposite effect, an increase in activity, can be associated with a future earthquake. Although it may seem contradictory, both effects can be considered precursors; that is, a change in the rate of occurrence of small shocks may signal the occurrence of a large earthquake. For example, the rate may first diminish and later increase, or vice versa. Just like in the previous case, this type of change in seismicity preceding an earthquake has been observed *a posteriori*, but its use as a true precursor is problematic.

Another seismic precursor that had wide acceptance at one time is the change in the velocity of seismic waves in the neighborhood of the focal region, previous to the occurrence of an earthquake. In some cases, it has been observed that the velocity of body waves that cross the focal region diminishes before a large earthquake. The time interval and amplitude of this decrease are somewhat related to the magnitude of the future earthquake. Sometimes the velocity tends to increase immediately before the occurrence. Although this phenomenon has been observed in some cases, in others changes in velocity were not followed by an earthquake. Thus, although this precursor raised great expectations, more controlled experiments have shown that it has little reliability.

Nonseismic phenomena of diverse types have been investigated, for example, changes in ground elevation or tilting, water levels in wells, strain in rocks, the ground's electric resistivity, electromagnetic fields, and emissions of radon gas. Observations of these physical parameters near an active fault are used to detect consistent patterns that may be correlated to the possible occurrence of an earthquake in the future. Geodetic measurements near faults can detect an accumulation of strain, but do not provide clues about its possible release. The presence of high-frequency electromagnetic signals that is the basis of the proposed VAN method also seems problematic (Varotsos and Lazaridou, 1991).

Some changes in physical parameters that have been proposed as precursors have been justified theoretically in terms of the phenomenon called dilatancy, which is observed in laboratory experiments on the fracture of rocks. According to the dilatancy of a material subject to increasing stress before the occurrence of a fracture, small cracks are produced that are then filled with fluids (water) present in the material and thus its volume increases. Owing to this phenomenon certain physical characteristics, such as the seismic wave velocity, seismicity, electric resistivity, and emission of radon gas, change. If the stress keeps on increasing, water is expelled and variations in some parameters (the seismic velocity and seismicity) change sign. In this state, instabilities that determine its failure occur in the focal region. If this model were universally consistent, precursors would always have the same pattern in all cases, which is not the case.

At present, none of the proposed precursors can be considered a clear indication of the occurrence of a future earthquake and in consequence we can not speak of a proved method (Wyss and Booth, 1997). Some modern programs of prediction have abandoned the idea of a unique precursor and search for a coherent pattern of convergence of several precursors. The rationale behind this practice is that, although no single precursor is a reliable predictor, the convergence of several of them may reveal consistent patterns. This approach was followed in the Parkfield experiments in which many different types of measurements were made in order to detect precursor patterns (Lindh *et al.*, 1979). The present state of prediction research is very provisional and a solution is not foreseen for the near future (Evans, 1997). A question regarding complex nonlinear dynamic systems has recently arisen from the new models of the occurrence of earthquakes. In these systems very small changes of initial or boundary conditions may produce very large effects on their subsequent development. Models of self-organized criticality have these properties and, if the occurrence of earthquakes follows these lines, it is possible that earthquakes may, in practice, not be predictable.

If the prediction of earthquakes remains a distant and problematic question, the prevention of damage is perfectly achievable with the existing state of knowledge. If we can not predict a future large earthquake, we can mitigate or even prevent in great part the damage that it will cause, especially in terms of human casualties, with an adequate building practice. Recent earthquakes in countries with strict antiseismic design practice and in those without such practice produced very different numbers of casualties. The Loma Prieta, California, earthquake of 1989 ($M = 7$) produced only 62 dead whereas the Latur, India, earthquake of 1993 ($M = 6.3$) produced an estimated 11 000 dead. Good assessment of the seismic risk and adequate building practice can avoid the total collapse of buildings and decrease drastically the number of human casualties. We must not forget that it is buildings that produce casualties, not earthquakes *per se*.

21 SEISMOGRAPHS AND SEISMOGRAMS

21.1   The historical evolution of seismographs

The oldest instrument used to detect the occurrence of an earthquake was probably constructed in China during the second century AD and is attributed to the philosopher Chian-hen. This instrument consisted in a bronze figure of a dragon with eight heads in whose mouths there were eight balls. Inside the figure there was some kind of pendular device that pushed the balls and made them fall when it was shaken by an earthquake. The figure was oriented in the geographic directions so that, upon the arrival of seismic waves, the corresponding ball will fall and show the occurrence and orientation of a shock. In Europe, the first instrument was a mercury seismoscope designed by De Haute-Feuille in 1703, consisting in a vessel with mercury connected by eight channels to eight cavities. Earthquakes will make the mercury flow into one or several of the cavities, indicating their orientations and sizes (quantities of mercury spilled). It is not certain that the instrument was actually built, although we have a description of it, but similar instruments were built in 1784 by Cavalli and in 1818 by Cacciatori (Ferrari, 1992). Vertical and horizontal pendulums started to be used around 1750. These instruments have an alarm to indicate the occurrence of an earthquake or a stylus attached to the mass that left a mark on sand or smoked plate of glass in which case they are called seismoscopes.

The first true seismographs with continuous recordings on a rotating drum with smoked paper on which time marks were also recorded were designed at the end of the nineteenth century, mostly in Italy, by Cecchi, Agamennone, Cancani, Grablovitz, and Vicentini. These instruments were horizontal and vertical pendulums with large masses and magnifications under 100. Palmieri, in 1859, was probably the first to use the term seismograph for an electromagnetic instrument that recorded arrival times and durations of motion (Ferrari, 1992). In 1890, Milne introduced the seismograph with an inclined pendulum, which with a very limited length, has a relatively large natural period. Later, in 1915, in collaboration with Shaw, Milne produced the Milne–Shaw seismograph with a mass of 0.5 kg, a period of 8 s, and a magnification of about 200. Similar instruments were designed in Japan by Omori and in Russia by Nikiforov. Toward 1900, Wiechert developed a horizontal seismograph consisting in an inverted pendulum that recorded two components with a single mass and a vertical seismograph with a mass suspended from a spring. Both types of instrument had masses in the range 80–1000 kg, periods of about 12 s, viscous damping and magnifications in the range 100–1000. Another mechanical seismograph that found extended use was developed by Mainka, with a mass in the range 200–500 kg and a magnification of about 300. These were purely mechanical instruments with viscous damping, in which

402

the amplification was produced by a system of levers and recorded on a drum with smoked paper together with time marks from a clock. Wiechert and Mainka seismographs were very popular and many are still in operation (Dewey and Byerly, 1969).

Significant progress in seismologic instrumentation was made due to the electromagnetic seismograph developed in Russia around 1910 by Galitzin. This instrument incorporates a coil attached to the mass of the pendulum that moves in the field of a magnet. The electric current generated in the coil by its motion is passed to a galvanometer whose deflection is recorded on photographic paper by a light beam reflected from a small mirror. Damping of the system is provided by the force opposed to the motion of the coil in the field of the magnet. A further development of this instrument in 1920 resulted in the Galitzin–Wilip seismograph with 12 s period and a magnification of about 1500. These instruments were the first of a series of electromagnetic seismographs based on the combination of a seismometer and a galvanometer. Another instrument of small dimensions, the torsion horizontal seismograph, was developed by Wood and Anderson in 1922. It is formed by a small mass attached to a metallic fiber that is made to oscillate by torsion and records on photographic paper via a light beam that is reflected from a mirror attached to the mass. This instrument with a period of 0.8 s and an amplification of 2800 was used in the first definition of local magnitude by Richter in California.

Toward 1930, Benioff developed the variable-reluctance seismograph, the action of which is based on changes in reluctance due to variations in the distance between two magnets, one fixed to the frame and the other to the mass of a pendulum. Owing to their relative motion, a current is generated in a coil around the magnet of the mass that is passed to a galvanometer and is recorded on photographic paper. This instrument, known as the Benioff seismograph, has a period of 1 s and a magnification in the range 1000–100 000. After around 1945, several models of electromagnetic seismographs with short, intermediate and long periods were developed by Sprengnether with the collaboration of Macelwane. For long periods, Press and Ewing developed in 1953 a seismometer–galvanometer system with seismometer periods in the range 15–30 s and galvanometer periods of 100 s, which operated with magnifications in the range 750–6000. The combination of short-period and long-period seismographs avoided the microseismic noise present for intermediate periods of about 6 s and was used at many seismographic stations (Melton, 1981). Similar systems of electromagnetic seismographs were developed in the USSR by Kirnos, such as those with a seismometer period of 12.5 s and a galvanometer period of 1.2 s, resulting in a nearly flat response in the range 0.2–10 s, and those of seismometer periods of 0.6–1 s and galvanometer periods of 0.2–0.4 s, resulting in a peak response in the range 0.2–1 s (Sawarenski and Kirnos, 1960).

Since 1970, three developments have influenced seismologic instrumentation. The first was the electronic amplification of the electric signal from the seismometer by means of electronic amplifiers that increased the possibilities of the seismometer–galvanometer system. The second was the introduction of digital recordings that could be analyzed directly by computers. Digital recording is carried out by means of an analog–digital converter that transforms the electric output of the seismometer. The first instrument of this type (the Caltech Digital Seismograph) was developed in 1962. The combination of electronic amplification and digital recording is the basis of all modern seismographs. The third was the development of wide-band seismographs, that is, systems with a response curve that is practically flat for a large range of periods, for example 0.1–1000 s. For this purpose a new seismometer that uses a feed-back circuit to extend the response to very low frequencies was developed by Wieland and Streckeisen in 1983. With instruments of this type, the duality of short- and long-period instruments has been overcome. Broad-band seismographs record digitally and, with adequate filters, can simulate a response in any frequency range.

21.2   Seismologic observatories and networks

Early seismologic observations were performed at individual observatories with different kinds of instruments. The exchange of data was by the publication of bulletins and the loan of seismograms. The need to have a large number of observations for hypocenter determinations led to the creation of world and regional centers that received readings of arrival times by mail or telegraph. Seismologic stations kept on updating their instrumentation from early mechanical instruments, such as Wiechert and Mainka seismographs, to later electromagnetic instruments such as Galitzin–Wilip, Benioff, Sprengnether, and Kirnos devices. Stations functioned independently and instrumentation was very heterogeneous, hindering seismologic studies. The first global seismologic network was installed by Milne after his return to England in 1895. It consisted in about 50 stations equipped with his horizontal pendulum seismograph, mainly in territories of the British Empire. The central station was at Shide, on the Isle of Wight. He had the support of the British Association for the Advancement of Science (BAAS). Another early network with homogeneous instrumentation (Wiechert devices, 80 kg) was installed in North America in 16 Jesuit universities in 1911 and reorganized in 1925 under the auspices of the Jesuit Seismological Association (JSA). Its central station was in Saint Louis, Missouri.

In 1962, an important improvement took place with the installation by the USA's government of a world-wide network of 125 seismographic stations with homogeneous instruments (three components, short-period Benioff and long-period Press–Ewing devices) known as the World Wide Standard Stations Network (WWSSN). Data from all stations were sent to a central station in Albuquerque, New Mexico, where they were copied onto microfilm and made available to researchers. This network furthered seismologic research greatly and was in operation from 1962 to 1990. New instrumental developments led to the installation of new types of networks by the USA in various parts of the world, such as the High Gain Long Period stations (HGLP) with digital recording in 1974 and later the Seismic Research Observatories (SRO), the Abbreviated Seismic Research Observatories (ASRO), and, in 1976, the International Deployment of Accelerometers (IDA), with 14 stations consisting in digital force-feedback Lacoste–Romberg gravimeters. Other networks were also installed in Canada, Japan, and the USSR.

With the development of digital broad-band seismographs, new networks replaced the old ones, starting around 1990. Broad-band networks with world-wide coverage are the IRIS (Incorporated Research Institutions for Seismology) installed by the USA, GEOSCOPE installed by France, GEOFON installed by Germany, MEDNET, which was installed in the Mediterranean region by Italy, and POSEIDON installed by Japan (Fig. 21.1). At regional or national level, many countries have installed centralized

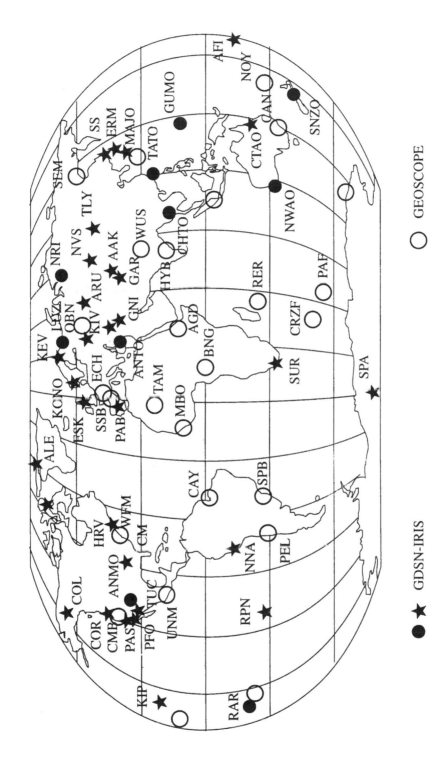

Fig. 21.1. Broad-band seismologic stations of the IRIS and GEOSCOPE networks.

★ GDSN-IRIS

●

○ GEOSCOPE

405

networks with telemetry in order to study the local seismicity. The instrumentation consists, usually, in short-period seismographs with high gain and digital recording that, in some cases, are being replaced by broad-band instruments.

Another type of seismologic observatory is formed by arrays of instruments deployed over a certain area and connected to a central unit where data are processed in digital form. Usually seismometers are distributed in concentric circles of various radii or along lines in specific directions. These arrays can be used as antennas with respect to incoming seismic waves. Examples of seismographic arrays are those of the LASA (Long Aperture Seismic Array) in the USA, the NORSA (Norwegian Seismic Array) in Norway, GERESS in Germany, FINESSA in Finland, and Sonseca in Spain.

21.3   The theory of the seismometer

The physical principles of most types of seismometer are based on the forced motion of a pendulum, be it vertical or horizontal. When the ground moves due to arrivals of seismic waves, it produces a displacement of the frame of the pendulum with respect to the mass due to its inertia. This motion, conveniently amplified, is recorded as a function of time. From this relative displacement, we can deduce the ground motion. In order to understand the basic principle of the theory of a seismometer, we will consider an ideal system consisting in a vertical pendulum with a mass m suspended by a spring of elastic coefficient K and a dashpot with viscous damping c (Fig. 21.2). When the frame of the pendulum is displaced by $x(t)$, the mass moves $y(t)$, and the relative displacement of the mass with respect to the frame is

$$z(t) = y(t) - x(t) \tag{21.1}$$

With respect to a reference system at rest, since the spring and the dashpot are affected only by the relative motion of the frame and the mass $z(t)$, the equation of motion for the mass is given by

$$m\ddot{y} = -Kz - c\dot{z} \tag{21.2}$$

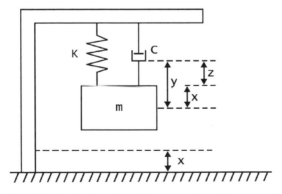

Fig. 21.2. An ideal vertical seismometer formed by a mass hanging from a spring and a dashpot.

If we displace the mass with the frame at rest $y(t) = z(t)$ and the equation of motion becomes, after division by m,

$$\ddot{z} + \frac{c}{m}\dot{z} + \frac{K}{m}z = 0 \tag{21.3}$$

For a system without damping $(c = 0)$, the solution of (21.3) is given by harmonic motion:

$$z(t) = A\sin(\omega_0 t) \tag{21.4}$$

The natural frequency of the undamped system is given by

$$\omega_0 = \left(\frac{K}{m}\right)^{1/2} \tag{21.5}$$

On substituting in ω_0, equation (21.3) can be written as

$$\ddot{z} + 2\beta\omega_0\dot{z} + \omega_0^2 z = 0 \tag{21.6}$$

where β is the damping factor,

$$\beta = \frac{c}{2(Km)^{1/2}} \tag{21.7}$$

If $\beta = 1$, the system is critically damped and $c = 2(Km)^{1/2}$. For a damped system, the solution of (21.6) is

$$z = A e^{-\beta\omega_0 t} \sin[\omega_0(1 - \beta^2)^{1/2}t + \varepsilon] \tag{21.8}$$

where A (the amplitude) and ε (the phase) are constants. Equation (21.8) represents damped harmonic motion of frequency $\omega_0(1 - \beta^2)^{1/2}$.

If the frame is displaced by $x(t)$, then, on making the substitution $y(t) = z(t) + x(t)$ in (21.2), we obtain

$$\ddot{z} + 2\beta\omega_0\dot{z} + \omega_0^2 z = -\ddot{x} \tag{21.9}$$

where $\ddot{x}(t)$ is the acceleration of the ground. This equation is called the seismometer equation, since it relates the relative motion $z(t)$ of the mass and the frame that can be measured to the ground motion $x(t)$. If the motion is very fast, the acceleration term is predominant and we have $\ddot{z} = -\ddot{x}$. Then, $z = -x$; that is, the relative displacement of the mass corresponds to the ground displacement. If the motion is very slow ($\ddot{z}$ and $\dot{z}$ are small), then $z = -\ddot{x}/\omega_0$; the relative displacement of the mass corresponds to the acceleration of the ground.

If we apply to the frame a harmonic motion of frequency ω, $x(t) = X\sin(\omega t)$, equation (21.9) becomes

$$\ddot{z} + 2\beta\omega_0\dot{z} + \omega_0^2 z = X\omega^2 \sin(\omega t) \tag{21.10}$$

The solution of this equation, without considering the transient part, is

$$z = Z\sin(\omega t - \varepsilon) \tag{21.11}$$

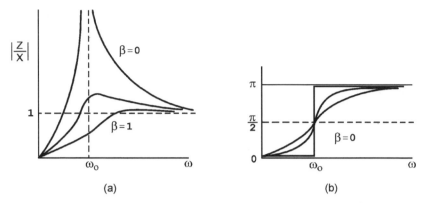

Fig. 21.3. Frequency-response curves of a mechanical seismometer: (a) the amplitude response and (b) the phase response.

where Z and ε, the amplitude and phase of the relative motion between the mass and the frame, are

$$Z = \frac{\omega^2 X}{[(\omega_0^2 - \omega^2)^2 - (2\beta\omega\omega_0)^2]^{1/2}} \tag{21.12}$$

$$\varepsilon = \tan^{-1}\left(\frac{2\beta\omega\omega_0}{\omega_0^2 - \omega^2}\right) \tag{21.13}$$

Thus, the amplitude Z of the relative motion of the mass depends on the amplitude of the ground motion X, its frequency ω, the natural frequency of the pendulum ω_0, and its damping β. The motion $z(t)$ measured by the seismometer is shifted by a phase ε with respect to the ground motion $x(t)$. The ratio Z/X depends on the frequency and damping (Fig. 21.3). When $\omega = \omega_0$, the system is in resonance and, if it is undamped ($\beta = 0$), the ratio becomes infinite and the phase is $\varepsilon = \pi/2$. For critical damping ($\beta = 1$), when $\omega = \omega_0$, the ratio $Z/X = \frac{1}{2}$. For large values of ω, $Z/X = 1$ and $\varepsilon = \pi$. Usually, a seismometer has a damping near its critical value. Equations (21.12) and (21.13) represent the amplitude and phase responses of a seismometer.

Another way to study this problem is to consider the response of the system to an impulsive acceleration, that is, $\ddot{x} = \delta(t)$. By substitution into (21.9), we obtain

$$\ddot{z} + 2\beta\omega_0\dot{z} + \omega_0^2 = -\delta(t) \tag{21.14}$$

If we take the Fourier transform, then, since the transform of $\delta(t)$ is 1 and that of $z(t)$ is $Z(\omega)$,

$$-\omega^2 Z(\omega) + 2i\omega\omega_0\beta Z(\omega) + \omega_0^2 Z(\omega) = -1 \tag{21.15}$$

We obtain for $Z(\omega)$

$$Z(\omega) = \frac{-1}{-\omega^2 + 2i\beta\omega\omega_0 + \omega_0} \tag{21.16}$$

Using $Z(\omega) = |Z(\omega)| \exp[i\varepsilon(\omega)]$, we obtain for the amplitude and phase responses two expressions similar to (21.12) and (21.13) (with $X = 1$):

$$|Z(\omega)| = \frac{1}{[(\omega_0^2 - \omega^2)^2 + (2\beta\omega\omega_0)^2]^{1/2}} \tag{21.17}$$

$$\varepsilon(\omega) = \tan^{-1}\left(\frac{2\beta\omega\omega_0}{\omega_0^2 - \omega^2}\right) \tag{21.18}$$

The response in time $z(t)$ is found by taking the inverse transform of $Z(\omega)$. The response of the seismometer to an impulsive acceleration represents its behavior for all frequencies. The response for an acceleration of arbitrary form $\ddot{x}(t)$ can be obtained by its convolution with the response to the impulsive acceleration or, in the frequency domain, by taking the product of their transforms.

21.4 Recording systems, magnification, and dynamic range

The motion of the mass of the seismometer is recorded in analog or digital form after its amplification. The complete system of the seismometer, amplifier, and recorder is known as a seismograph. The first seismographs were totally mechanical systems in which amplification was carried out by means of systems of levers and recording was by a stylus onto smoked paper fixed onto a revolving drum. This system is today obsolete, but it is useful to study it in order to understand the behavior of a seismograph.

The ratio Z/X in (21.12) is called the dynamic magnification $V_d(\omega)$ of the seismometer for a harmonic input signal of frequency ω:

$$V_d(\omega) = \frac{Z}{X} = \frac{\omega^2}{[(\omega - \omega_0)^2 + 4\omega^2\beta^2]^{1/2}} \tag{21.19}$$

By using a system of levers we can amplify the amplitude of the relative motion of the mass, so that the recording amplitude is $z' = V_s z$, where the factor V_s is called the static magnification. The total magnification of the system is given by the product of both factors $V(\omega) = V_s V_d(\omega)$. The total response of the seismograph is given by the magnification curve in the frequency domain $V(\omega)$ which, according to (21.12) and 21.13), is formed by the amplitude and phase responses. The maximum value of the amplitude magnification V_{max} and its corresponding period are used as characteristics of the response of each seismograph. The ground motion $x(t)$ is obtained from the recorded signal $z'(t)$, dividing its transform by the response of the instrument:

$$X(\omega) = \frac{Z'(\omega)}{V(\omega)} \tag{21.20}$$

To obtain the time function $x(t)$ of the ground motion, we take the inverse transform of $X(\omega)$. An approximation to the amplitude of the ground motion for a given period can be obtained by dividing the recorded amplitude by the corresponding magnification for that period.

Mechanical seismographs are very limited by friction between their parts, and the dimensions of the pendulum and amplification and recording systems. To increase their magnification, their mass was increased in order to overcome friction, reaching several tons in some cases. The Wiechert seismograph of 1000 kg had a maximum magnification of nearly 1000 and the Mainka device of 350 kg had one of about 400. A recording system that avoided friction between the pen and the paper uses a light beam and photographic paper, such as in the Wood–Anderson device that, with a small mass of some grams, attained a magnification of 2800.

In the problem of the amplification of signals, an important concept is the dynamic range, that is, the range between the maximum and minimum amplitudes possible for a particular system. The dynamic range of a system is given by the logarithm of the ratio A/A_0, where A is the maximum amplitude recorded and A_0 is the minimum amplitude or that taken as the zero level. The units used are decibels (dB) such that, for a given ratio, its dynamic range is $20\log(A/A_0)$ dB. For example, if A/A_0 equals 1000, the dynamic range is 60 dB. For a seismograph, A and A_0 are the maximum and minimum recorded amplitudes. In an analog record, A_0 is related to the minimum detectable amplitude and the noise generated by the system itself, and A is related to the dimensions of the record or the saturation level of the recording system. The saturation level is the maximum amplitude possible with a particular recording system. For example, for a photographic analog seismogram, taking into account the thickness of the trace and the size of the record, the minimum appreciable amplitude is about 0.2 mm and the maximum one is 20 cm; thus, $A/A_0 = 1000$ and the dynamic range is 60 dB.

The dynamic range is in itself independent from the magnification. If we increase the magnification for the same dynamic range, the system is saturated for smaller amplitudes of ground motion and we lose information about large amplitudes. For example, if the maximum amplitude of a graphical record is 10 cm and the minimum one is 1 mm, its dynamic range is 40 dB. If the maximum magnification is 10 000, the minimum ground motion recorded is 0.1 μm and the maximum one is 10 μm. If we increase magnification to 100 000 we can detect ground motions of 0.01 μm amplitude, but the system saturates for amplitudes larger than 1 μm. For observations of local seismicity, in order to detect very small earthquakes, short period seismographs with very large magnifications (1 000 000) are used. Since their dynamic range remains limited (60 dB), records are saturated for very small ground motions. These records are useful only for detecting first arrival times; amplitude information is largely lost.

21.5   Electromagnetic seismographs

The foundation of the electromagnetic seismograph consists in adding a coil to the mass of the pendulum that moves in the magnetic field of a magnet. The same effect is produced if the moving part is a magnet inside a fixed coil. In both cases, the relative motion of the coil in the magnetic field generates an electric current in the coil that is proportional to the velocity of the relative movement of the coil and the magnet. This electric current is passed to a galvanometer whose deflection is recorded graphically.

Let us consider the same ideal case of a vertical pendulum as that in section 21.3, with a moving coil and a fixed magnet (Fig 21.4). The force F that acts on the mass of the pendulum due to the motion of the coil in the magnetic field B of the magnet, according

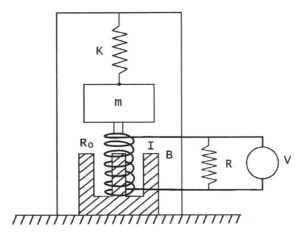

Fig. 21.4. A vertical electromagnetic seismometer.

to the Biot-Savart law, is

$$F = IBl \tag{21.21}$$

where I is the current in the coil, $l = 2\pi r N$ is the coil's length (r is the radius and N is the number of turns), and B is the amplitude of the magnetic induction. If z is the relative displacement of the coil and the magnet, the work done by the force is Fz and the power is $F\dot{z}$. In the electric circuit (Fig. 21.4), power is dissipated by its total resistance $R_0 \mid R$ (R_0 is the coil's resistance and R is an additional resistance in parallel). The power in the circuit is given by IV, where V is the difference in electric potential, thus

$$VI = IlB\dot{z} \tag{21.22}$$

If we make $G = lB$, a constant that depends on the specifications of the instrument, we have that $V = G\dot{z}$ and $F = GI$. Using Ohm's law, $I = V/R$, we obtain

$$I = \frac{G\dot{z}}{R_0 + R} \tag{21.23}$$

$$F = \frac{G^2\dot{z}}{R_0 + R} \tag{21.24}$$

The force that acts on the mass due to the motion of the coil in the magnetic field depends on its velocity and acts in the opposite sense, that is, it is a damping force. The resulting equation of motion has the same form as (21.7), but now the damping coefficient is

$$2\beta = \frac{G^2}{(Km)^{1/2}(R_0 + R)} \tag{21.25}$$

The damping of the system is electromagnetic and its critical value ($\beta = 1$) corresponds to

$$\frac{G^2}{R_0 + R} = 2m\omega_0 \tag{21.26}$$

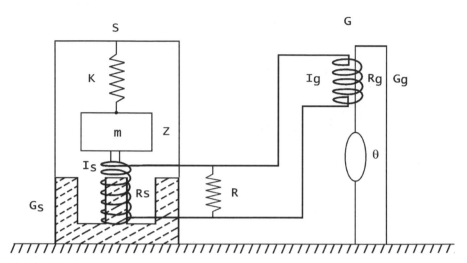

Fig. 21.5. An electromagnetic seismograph formed by a vertical seismometer and a galvanometer.

The damping can be adjusted changing the value of the resistance in parallel R, for each fixed value of the resistance of the coil R_0.

21.5.1 The seismometer–galvanometer system

The recording of the electromagnetic seismograph consists in connecting the current generated in the coil of the seismometer to a galvanometer whose angular deflection is recorded on a drum with photographic paper by using a beam of light that is reflected from a mirror attached to the galvanometer's moving part (Fig. 21.5). Since the seismometer and the galvanometer are connected, the current in the circuit generated by the motion of the mass is affected by the motion of the galvanometer. The motion of the mass of the seismometer is also affected by a force that depends on this current. The motions of the galvanometer and of the mass of the seismometer are coupled through the circuit, so that their equations must be solved together. The equations of motion of the seismometer and galvanometer are

$$\ddot{z} + 2\beta_s\omega_s\dot{z} + \omega_s^2 z = -\ddot{x} - G_s I_s/m \tag{21.27}$$

$$\ddot{\theta} + 2\beta_g\omega_g\dot{\theta} + \omega_g^2\theta = G_g I_g/h \tag{21.28}$$

where z is the relative vertical displacement of the seismometer and θ is the angular deflection of the galvanometer. The subindexes s and g refer to the seismometer and galvanometer, and h is the inertial moment of the moving part of the galvanometer. The electric currents that pass through the coils of the seismometer and galvanometer are interrelated. The recorded motion $\theta(t)$ due to a ground motion $x(t)$ is obtained by the solution of a fourth-order differential equation. In equations (21.27) and (21.28), ω_s and ω_g are the natural frequencies of the seismometer and galvanometer whose values are characteristic to each system.

The seismometer–galvanometer system responds to a ground displacement $x(t)$ with a deflection of the galvanometer $\theta(t)$, which is recorded on photographic paper. Since the

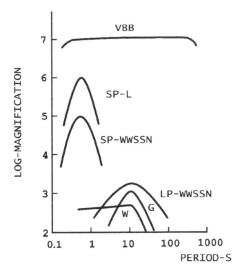

Fig. 21.6. Amplitude-response curves of Weichert (W), Galitzin (G), SP-WWSSN, LP-WWSSN, short period for local seismicity (SP-L), and broad-band (VBB) seismographs.

current generated in the moving coil of the seismometer is proportional to its relative velocity with respect to the magnet fixed to the ground, the instrument records the ground velocity. The total magnification of the system is given by

$$V(\omega) = V_{st} V_s V_g \qquad (21.29)$$

where V_{st} is the static magnification described by the mechanical seismograph, and V_s and V_g are the dynamic magnifications of the seismometer and galvanometer, which are obtained from the solutions to equations (21.27) and (21.28). The result is expressed in terms of the response curves for the amplitude and phase relating the recorded motion to the ground displacement.

The first Galitzin electromagnetic seismographs had natural periods of the seismometer and galvanometer of $T_s = T_g = 12\,s$ and maximum magnifications of the order of 1000. The presence of microseismic noise of meteorologic origin in the period range 4–8 s motivated the avoidance of these periods by separating instruments into two types, of short and long periods. A very commonly used configuration was $T_s = 1\,s$ and $T_g = 0.75\,s$ for short-period instruments, and $T_s = 15\,s$ and $T_g = 100\,s$ for long-period instruments. These instruments have maximum magnifications in the range 10 000–100 000 for short periods and 750–6000 for long periods (Fig 21.6).

A later development of electromagnetic seismographs was the inclusion in the circuit of an electronic amplifier and replacement of the galvanometer by a transducer that transforms the amplified current into the motion of a recording pen. A very common type of recorder involves a heated pen that leaves a mark on thermosensitive paper. Modern seismographs include operational amplifiers of various types, with optional filters. In this way, in theory, the magnification of the instrument can be made as high as desired. The dynamic range, however, remains limited by the noise generated by the system and the dimensions of the graphical record. As was mentioned before, an instrument with graphical recording of the normal type can not exceed a dynamic range of 60 dB. Electromagnetic seismographs of short period have in many instances very high magnifications of the order of a million and, in consequence, are saturated by very small ground motions, providing records with very limited information.

21.6   Digital seismographs

The magnification of old mechanical seismographs was limited by friction between their parts and the size of the mass. Electromagnetic seismographs do not have this limitation and, in theory, using electronic amplifiers, can have magnifications as large as desired. However, as we have seen, because of their analog graphical recording, minimum amplitudes detectable visually and dimensions of recording paper limit the dynamic range to about 60 dB. With the development of electronic digital computers during the 1960s seismologic analysis came to be performed in digital form and digital data were required. The development of digital seismographs took some time; the first instruments were developed around 1965 and it took more than 10 years to solve all the problems associated with their use.

In order to record in digital form, the variable electric current from an electromagnetic seismometer must be amplified and converted into digital form by an analog–digital converter (ADC). The ADC samples the continuous signal at constant intervals that are usually in the range 0.005–10 s, that is, for sampling frequencies of 0.1–200 Hz. The resulting series of digital data are recorded into some kind of magnetic memory. Until recently ADCs were of maximum 12 or 16 bits, so that amplitudes could extend only from zero to 2^{11} or 2^{15} (the first bit is used for the sign), that is, maximum amplitudes without saturation of about 2000 and 70 000, and thus their dynamic ranges were 60 and 96 dB. Modern ADCs of 24 and 32 bits can express maximum amplitudes of 10^7 and 10^9, which correspond to dynamic ranges of 140 and 185 dB. From the point of view of seismic wave amplitudes, with an instrument of 140 dB, saturation of the signal will occur for local earthquakes at less than 10 km distance of magnitude $M > 5$, or teleseisms at a distance of less than 30° of magnitude $M > 9$. Hence the same seismograph can record both near and distant earthquakes.

With a digital seismograph, the amplitudes of seismic waves are given by a series of discrete values $a(t_i)$, at constant intervals of time, of the voltage of the electric current generated in an electromagnetic seismometer, proportional to the relative velocity of the mass and the frame. To obtain the ground displacement, the signal is transformed into the frequency-domain $s(\omega_i)$ by means of discrete Fourier transformation (Appendix 4). The transform $s(\omega_i)$ is divided by the transfer function of the seismograph $T(\omega_i)$ to obtain $g(\omega_i)$, the transform of the ground displacement:

$$g(\omega_i) = \frac{s(\omega_i)}{T(\omega_i)} \tag{21.30}$$

The transfer function is defined in such a way that it relates the voltage output to the ground displacement. In general, this function has the form of a quotient of two polynomials and can be expressed in terms of its poles and zeros (Scherbaum, 1996). For example, in a simple case, corresponding to an electromagnetic seismograph,

$$T(\omega_i) = \frac{1}{2\pi} \frac{G\omega_i^2}{\omega_i^2 + 2h\omega_0\omega_i + \omega_0^2} \qquad (21.31)$$

where G, h, and ω_0 are characteristics of each instrument. As a function of the poles of $T(\omega_i)$, equation (21.29) can be expressed as

$$T(\omega_i) = \frac{1}{2\pi} \frac{G\omega_i^2}{(\omega_i - p_1)(\omega_i - p_2)} \qquad (21.32)$$

where p_1 and p_2 are the roots of the denominator of (21.31). Transfer functions for more complex systems can be expressed in the form

$$T(\omega_i) = G \frac{\prod_{k=1}^{n} (\omega_i - z_k)}{\prod_{k=1}^{m} (\omega_i - p_k)} \qquad (21.33)$$

where p_k are the poles and z_k are the zeros of the transfer function. Ground displacements as a function of time are obtained by taking the inverse Fourier transform of $g(\omega_i)$. Since we are talking about digital systems, the signals are sampled time functions of finite duration and their transforms are also sampled functions that are limited to a finite frequency band (Appendix 4).

An important problem with digital data is their storage. A station with a three-component seismograph with continuous recording sampling at a rate of 20 Hz produces about 100 Mbytes of data per week. These may be copied onto magnetic tapes or optical disks but in some years the amount of storage room required may be fairly great. The stability over time of this type of storage of data may create some problems. Computer systems develop so fast that within a few years they are obsolete and data recorded by them may not be able to be read by more modern systems. Usually the compatibility of systems from one generation to the next is assured, but it may be problematic for distant generations. This may force one to copy all data every so many years, which in time may become an impossible task.

21.7 Broad-band seismographs

As we have already mentioned, the presence of microseismic noise at about a 6 s period resulted in the avoidance of these periods by the separation of short- and long-period instruments with relatively narrow response curves. Modern instrumental development has moved in the direction of an instrument with a response curve that is practically flat for a large range of periods. This type of instrument is called a broad-band seismograph, and using digital recording with a high dynamic range allows the recording of both near and distant earthquakes. The problem of microseismic noise is solved by adequate filtering of data.

The seismometer used in broad-band systems is a force-balanced instrument. This type of seismometer has a feed-back circuit whereby the current generated by motion of the mass is amplified and connected to a device that applies an electromagnetic force to the mass, which moves it to its original equilibrium position, forcing it to follow the ground motion. The voltage generated by this force is proportional to the velocity of the mass and constitutes the output of the instrument. Modern electronic transducers are very sensitive and can detect very small movements of the mass and pass the generated current to the feed-back system. In this way, we can obtain a magnification as large as we like, the only limitation being the noise level. There are several models of instruments of this type that have response curves that are practically flat for the period range 0.05–1000 s (Fig. 21.6).

21.8   Accelerographs

Most seismographs respond to the ground velocity and are designed to record very small ground motions from small near earthquakes or distant large ones. These instruments are saturated by the occurrence of large earthquakes nearby. Near the epicenter of a moderate-to-large earthquake, in what is called the near field, large ground accelerations are produced at high frequencies that may reach values larger than the acceleration due to gravity ($1G$). Seismographs specifically designed to record this type of motion are known as strong motion instruments or accelerographs. In these instruments the relative displacement of the mass corresponds to ground accelerations (section 21.3).

Accelerographs are instruments with very low magnification in order to avoid saturation in response to the very strong motions of accelerations of about $0.1G$ to $2G$ at relatively high frequencies, 1–20 Hz. The seismometer is a force-balanced instrument with transducers of variable capacitance and near-critical electromagnetic damping, resulting in a practically flat response curve. With the old models recording was onto photographic film, whereas for modern instruments it is done in digital form.

Because their purpose is to record only very strong ground motions, accelerographs do not record in a continuous fashion, but rather their recording system is triggered when ground accelerations reach a preset threshold value. Thus, with the old models, the first few seconds of the record were lost. Modern digital instruments have a pre-event memory that allows one to obtain a complete record of the motion. Instruments have an internal clock that, in modern instruments, can adjust itself to real time. Minimum and maximum values of the recorded acceleration are established, depending on the dynamic range of the instrument.

The observation of ground accelerations in the near field is of great interest in earthquake engineering. Antiseismic designs of buildings are based on knowledge of the accelerations expected from near earthquakes. Large arrays of accelerographs allow the observation of ground accelerations at different distances and different sites. Thus the attenuation of acceleration in the near field can be measured directly rather than its having to be determined from intensity data. Accelerographs can be located on the ground (free field) or in buildings at various levels. In this way, responses of buildings can be obtained at various heights. The properties of the near-surface layers (rock, soft or hard soil) are important for high-frequency ground motion and constitute the site effect. This effect is determined by using strong-motion instruments installed on various kinds of soil.

21.9   Other types of seismologic instruments

Besides the instruments we have described, there are other seismologic instruments that will be discussed briefly. Benioff's strain seismometer, developed around 1935, consists in a long (10–100 m) quartz bar with one end fixed to a pillar and the other free and near to a second pillar. By means of a transducer, variations of the distance between the free end of the bar and the pillar fixed to the ground are measured. This distance represents the deformation of the ground between the two pillars. The current from the transducer is passed to a galvanometer and the resulting system has a response similar to that of a long-period seismograph.

A special type of seismograph is portable instruments designed to record small near earthquakes. They are short-period instruments with high gain that can be installed in the field and operated with batteries. The first models, developed during the 1960s, used continuous analog recording (on smoked paper or by pen and ink). Modern instruments employ digital recording and can be used for continuous recording or as event recorders. The latter are triggered when the ground motion reaches a certain threshold value by a triggering algorithm. Some of these instruments form small telemetric arrays sending the signals to a central recording unit by radio. These instruments are very useful for detailed studies of the seismicity of a limited region, studies of aftershocks after a large earthquake, and monitoring volcanic activity.

Seismographs designed to record earthquakes at the bottom of the sea are known as Ocean Bottom Seismographs (OBSs). Owing to the fact that oceans cover 70% of the Earth's surface, these instruments are becoming more and more important as a means to improve the geographic distribution of stations. Also, they are very useful for studying the seismicity of oceanic regions. There are several models of OBS; some are connected to a buoy at the surface of the sea with a radio transmitter whereas others record on an internal system and must be recovered after a certain length of time. Deployment at large depths is difficult and conditions on the high seas may result in the loss of a number of instruments.

Two types of instruments used in studies related to seismology, which are not seismographs, are tiltmeters and dilatometers. The first measure changes in the tilt of the Earth's surface and the second measure changes in the volume of rocks. Both types are used in connection with prediction programs to measure the accumulation of strain in a region. The ground strain and relative displacements of crustal material in a region can be measured directly using GPS (Global Position System) techniques. Repeated measurements of base lines during long periods of time can give us actual displacements resulting from plate motion. The relative motion of tectonic plates can also be measured directly by VLBI (Very Long Base Interferometry) methods with measurements between two stations repeated every several years.

21.10   Seismograms and accelerograms

Graphical recordings of earthquakes by seismographs are called seismograms. Until recently seismographs were analogical on smoked, photographic, thermosensitive, or plain paper. The aspect of these recordings depends on the characteristics of the

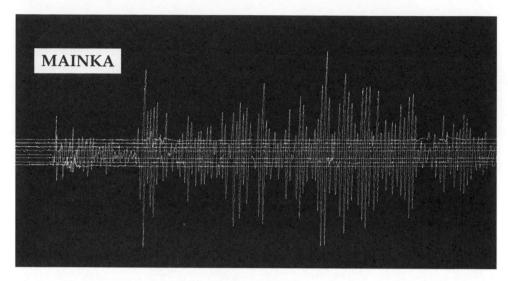

Fig. 21.7. A seismogram from the Mainka seismograph of the EBR station (Observatorio del Ebro, Spain) for the earthquake of 9 February 1948 in Greece.

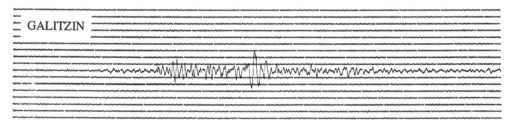

Fig. 21.8. A seismogram of a vertical Galitzin seismograph of the DBN station (Holland) for the earthquake of 19 May 1951 ($M = 5$) in Spain.

instruments' response curves. Old mechanical instruments (Wiechert or Mainka devices) give seismograms on smoked paper that are limited by their very low magnification but have fairly flat response curves over a relatively broad range of periods. Another limitation of these instruments is that the levers which hold the recording stylus produce a curvature on traces with large amplitudes. Their dynamic ranges are also small and good records are limited to sufficiently large earthquakes at intermediate distances (Fig. 21.7). Early electromagnetic seismographs (Galitzin devices) have greater magnifications and their recordings have characteristics between short and long periods (Fig. 21.8). Records from these seismographs are useful for studying old earthquakes (previous to 1960); for Wiechert and Mainka devices, from 1900; and for Galitzin devices, from 1920.

Between 1960 and 1990, seismographs were separated into short- and long-period instruments. Examples of seismograms of instruments of this type are those of the WWSSN stations (Fig. 21.9). Records from these stations are very useful for studying earthquakes that occurred between 1962 and 1990. Short-period instruments used in

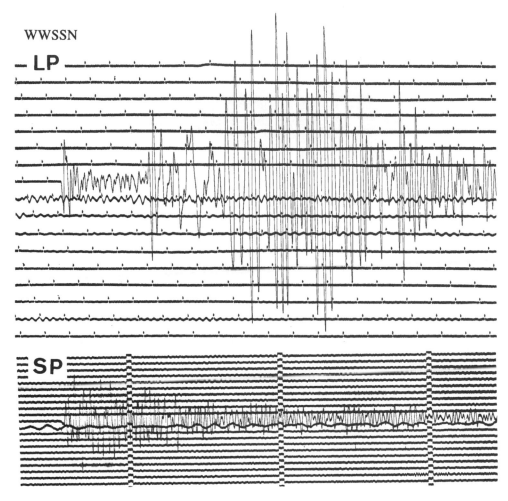

Fig. 21.9. Seismograms of the vertical seismographs of short and long periods of the TOL station (Spain) of the WWSSN network for the earthquake of 17 September 1972 ($M = 6$) in Greece (courtesy of E. Ruiz de la Parte, Observatorio de Toledo).

local seismologic networks with either analog or digital recordings have large magnifications, but their response curves are very narrow, being centered on 1 s periods, producing nearly monochromatic signals that are easily saturated (Fig. 21.10).

Since 1990, digital broad-band seismographs have been installed widely and at present have practically replaced the other types. Owing to their response curves being flat over wide ranges of periods, their large magnifications and their large dynamic ranges, they produce excellent records both of near and of distant earthquakes. The digital form of data allows their direct and easy processing, filtering, etc. (Fig. 21.11).

Besides instrumental characteristics, other aspects of seismograms depend on their size, depth, and distance. The magnitude of an earthquake determines the distance at which it can be recorded by instruments of a certain magnification. Small earthquakes

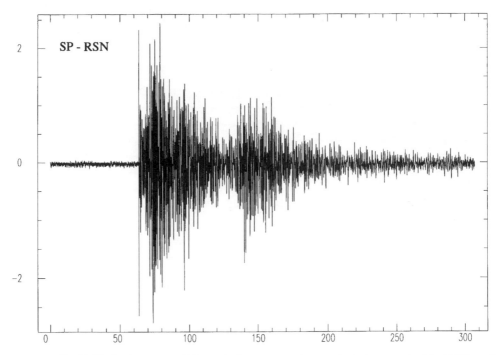

Fig. 21.10. A seismogram of the vertical component short-period seismograph of the ESEL station of the Spanish Seismograph Network of the Instituto Geográfico Nacional (Madrid) for the earthquake of 23 May 1993, in southern Spain.

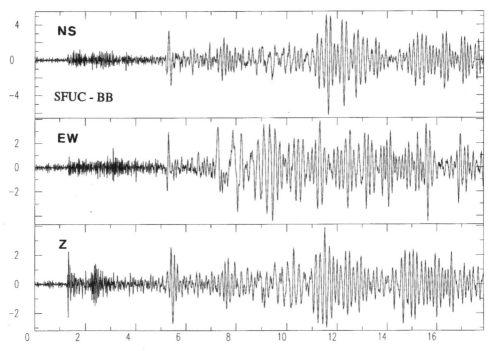

Fig. 21.11. Seismograms of the three-component broad-band seismograph of the SFUC station (Universidad Complutense–Observatorio de la Armada, San Fernando, Spain) for the earthquake of 21 March 1998 in Chile.

420

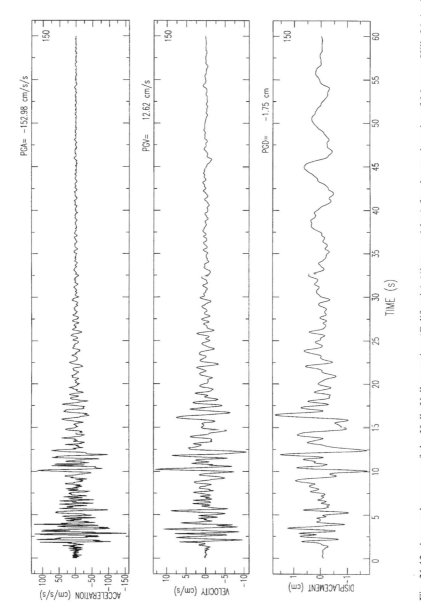

Fig. 21.12. An accelerogram of the Halls Valley station (California) (distance 4 km) for the earthquake of Morgan Hill, 24 April 1984 ($M = 6.2$). Records of acceleration, velocity, and displacement are shown (courtesy of B. Benito).

($M < 4$) are recorded only at short distances by local networks (less than 500 km away), have predominantly high frequencies, and their seismic waves travel mostly through the crust. Larger earthquakes ($M > 5$) are recorded at teleseismic distances by global networks. Their waves travel through the interior of the Earth and along its surface with lower frequencies. The depth of focus determines the generation of surface waves, which are the predominant phases for shallow earthquakes (Figs. 9.19 and 12.11).

Records of strong-motion instruments are called accelerograms. Their appearance is different than that of seismograms in that, because of the short distances involved, they have a high content of high frequencies. From records of acceleration obtained by integration one can obtain ground velocities and displacements (Fig. 21.12). A general aspect of accelerograms is that of a random distribution of impulses of high frequency, in which larger impulses correspond to S waves. The maximum amplitude in each record is called the peak acceleration. This is a parameter that is used in earthquake engineering. For instruments in the free field (not on buildings) the appearance of accelerograms depends mostly on the distance from the epicenter and the type of soil at the site. Instruments on rock have a different response than that of those installed on soft soils of unconsolidated sediments.

APPENDIX 1
VECTORS AND TENSORS

A1.1   Definitions

Quantities representing physical entities can have one, two, three or more components and thus can be defined by scalars, vectors, or tensors. In seismology, certain variables are scalars such as density, vectors such as displacements, and tensors such as stress and strain.

A scalar is a quantity with one component. Its value remains invariant under a change of the coordinate system.

A vector is a quantity with three components that transform under the rotation of the coordinate system as the coordinates of a point. To distinguish them from scalars, we represent them by bold face letters (a), or by a letter with a subindex (a_i) that takes the three values of the components in each coordinate system (a_1, a_2, and a_3). For the Cartesian coordinate system (x, y, z) or (x_1, x_2, x_3), a vector is represented by the three components or by their sum multiplied by the unit vectors in the direction of each axis (i, j, k) (Fig. A1.1):

$$a = a_i = (a_x, a_y, a_z) = (a_1, a_2, a_3) = a_1 i + a_2 j + a_3 k \qquad (A1.1)$$

A vector modulus is a scalar given (in Cartesian coordinates) by

$$a = (a_1^2 + a_2^2 + a_3^2)^{1/2} \qquad (A1.2)$$

If (x_1, x_2, x_3) is a Cartesian coordinate system and (x_1', x_2', x_3') is another with the same origin and rotated by the angles between each pair of axes defined by

$$\cos(x_i, x_j') = \beta_{ij} \qquad (A1.3)$$

then the transformation of the components of a vector a_i referred to system x_i into a_i', corresponding to the system x_i', is given by

$$a_i' = \sum_{j=1}^{3} \beta_{ij} a_j = \beta_{ij} a_j \qquad (A1.4)$$

In the last term, we have used the convention that all repeated subindexes (in this case j) are summed up over their three values ($j = 1, 2, 3$) and it is not necessary to write the summation symbol.

A Cartesian tensor of second rank, represented by a bold face capital letter ($\mathbf{A}$) or one with two subindexes (A_{ij}) can be defined, in a similar way to how we defined a vector, as a quantity of nine components that transform under rotation of the coordinate axes in

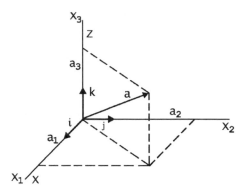

Fig. A1.1. Components of a vector in Cartesian coordinates.

the form

$$B'_{ij} = \beta_{ik}\beta_{jl}B_{kl} \tag{A1.5}$$

An example of a second-rank tensor is the quantity β_{ij} formed by the cosines of the nine angles between the axes of the Cartesian coordinate systems x_i and x'_i. In this textbook, we generally use only vectors and Cartesian second-rank tensors. A more rigorous and general definition of tensors can be found in mathematical textbooks.

A1.2   Operations with vectors and tensors

Addition, subtraction, and multiplication by a scalar of vectors and tensors present no difficulty. In the subindex notation the same subindexes must be used throughout the whole expression:

$$c = a + b - d \tag{A1.6}$$

$$c_i = a_i + b_i - d_i \tag{A1.7}$$

$$d = Ea \tag{A1.8}$$

$$d_i = Ea_i \tag{A1.9}$$

$$C_{ij} = A_{ij} + B_{ij} - D_{ij} \tag{A1.10}$$

The dot product of two vectors is a scalar given by

$$a = b \cdot c = bc \cos(b, c) \tag{A1.11}$$

where b and c are the moduli of the vectors b and c. In the subindex notation, the dot product is

$$a = b_i c_i = b_1 c_1 + b_2 c_2 + b_3 c_3 \tag{A1.12}$$

The dot product can also be expressed using the Kroneker delta tensor δ_{ij}, defined by

$$\delta_{ij} = \begin{cases} 1, & i = j \\ 0, & i \neq j \end{cases} \tag{A1.13}$$

and is given by

$$a = \delta_{ij} b_i c_j \tag{A1.14}$$

The cross product of two vectors is a vector given by

$$a = b \times c = bc \sin(b, c) e \tag{A1.15}$$

where e is the unit vector normal to the plane that contains the vectors b and c, which is positive in the direction of the right-hand rotation of b to c.

The cross product of two vectors in subindex notation can be expressed by using the alternating or permutation tensor:

$$e_{ijk} = \begin{cases} = & 1, & i \neq j \neq k \text{ (even permutations)} \\ = & -1, & i \neq j \neq k \text{ (odd permutations)} \\ = & 0, & \text{(any index repeated)} \end{cases} \tag{A1.16}$$

The relation between the tensors e_{ijk} and δ_{ij} is

$$e_{ijk} e_{ilm} = \delta_{jl} \delta_{km} - \delta_{jm} \delta_{kl} \tag{A1.17}$$

The cross product is given by

$$a_i = e_{ijk} b_j c_k \tag{A1.18}$$

$$a_i = [(b_2 c_3 - b_3 c_2), (b_3 c_1 - b_1 c_3), (b_1 c_2 - b_2 c_1)]$$

We must remember that the cross product of two vectors is noncommutative and that the result is a pseudovector or axial vector. The scalar product of three vectors is

$$d = a \cdot b \times c = e_{ijk} a_i b_j c_k \tag{A1.19}$$

Since all subindexes are repeated, they are summed up and the result is a scalar. The triple cross product of three vectors is a vector, which can be expressed in term of dot products as

$$d = a \times b \times c = (a \cdot c)b - (a \cdot b)c \tag{A1.20}$$

$$d_i = e_{ijk} e_{klm} a_j b_l c_m = a_j c_j b_i - a_j b_j c_i \tag{A1.21}$$

The expression in subindex notation can be derived from equation (A1.17).

Two vectors can be used to construct a second-rank tensor, in what is called the direct product:

$$C = ab \tag{A1.22}$$

$$C_{ij} = a_i b_j \tag{A1.23}$$

Because of this property, second-rank tensors are also called diadics. Vectors and second-rank tensors can be represented by matrices and their operation follows matrix algebra:

$$a_i = (a_1 \quad a_2 \quad a_3) \tag{A1.24}$$

$$A_{ij} = \begin{pmatrix} A_{11} & A_{12} & A_{13} \\ A_{21} & A_{22} & A_{23} \\ A_{31} & A_{32} & A_{33} \end{pmatrix} \tag{A1.25}$$

A1.3   Vector and tensor calculus

Scalars, vectors, and tensors that are functions of spatial coordinates are referred to as fields. Thus, in seismology, we use density, velocity, and stress fields. The most common operations in vector and tensor calculus, involving partial derivatives with respect to spatial coordinates, are gradient, divergence, and curl. In these cases, the symbolic vector operator ∇, called del or nabla, is used. In Cartesian coordinates the nabla operator is defined as

$$\nabla = \left(\frac{\partial}{\partial x_1}, \frac{\partial}{\partial x_2}, \frac{\partial}{\partial x_3} \right) = i\frac{\partial}{\partial x} + j\frac{\partial}{\partial y} + k\frac{\partial}{\partial z} \tag{A1.26}$$

The gradient of a scalar is a vector:

$$a = \nabla f = i\frac{\partial f}{\partial x} + j\frac{\partial f}{\partial y} + k\frac{\partial f}{\partial z} \tag{A1.27}$$

In subindex notation,

$$a_i = \frac{\partial f}{\partial x_i} = \left(\frac{\partial f}{\partial x_1}, \frac{\partial f}{\partial x_2}, \frac{\partial f}{\partial x_3} \right) = f_{,i} \tag{A1.28}$$

In the last term, a comma followed by a subindex is used to indicate the partial derivatives with respect to the three spatial coordinates. Each derivative is the component of the resulting vector.

The gradient of a vector field f or f_i is a second-rank tensor:

$$B = \nabla f \tag{A1.29}$$

$$B_{ij} = \frac{\partial f_i}{\partial x_j} = f_{i,j} \tag{A1.30}$$

The divergence of a vector field is a scalar function:

$$g = \nabla \cdot f = \frac{\partial f_x}{\partial x} + \frac{\partial f_y}{\partial y} + \frac{\partial f_z}{\partial z} = \frac{\partial f_i}{\partial x_i} = f_{i,i} \tag{A1.31}$$

The divergence of a tensor is a vector:

$$a = \nabla \cdot B$$

$$a_i = \frac{\partial B_{ij}}{\partial x_j} = B_{ij,j} \tag{A1.32}$$

The divergence of the gradient of a scalar or vector field is the Laplacian (∇^2), an operator involving the second partial derivatives with respect to the three spatial coordinates. For Cartesian coordinates it is given by

$$\nabla \cdot \nabla f = \nabla^2 f = \frac{\partial^2 f}{\partial x^2} + \frac{\partial^2 f}{\partial y^2} + \frac{\partial^2 f}{\partial z^2} = f_{,ii} \tag{A1.33}$$

The curl of a vector function is another vector function. For Cartesian coordinates it is given by

$$d = \nabla \times f = \left(\frac{\partial f_z}{\partial y} - \frac{\partial f_y}{\partial z}, \frac{\partial f_x}{\partial z} - \frac{\partial f_z}{\partial x}, \frac{\partial f_y}{\partial x} - \frac{\partial f_x}{\partial y} \right) \tag{A1.34}$$

$$d_i = e_{ijk} \frac{\partial f_k}{\partial x_j} = e_{ijk} f_{k,j} \tag{A1.35}$$

The curl of the curl of a vector function is another vector function that can be expressed in terms of the gradient, divergence, and Laplacian:

$$\nabla \times \nabla \times f = \nabla(\nabla \cdot f) - \nabla^2 f \tag{A1.36}$$

This relation can be derived using index notation and equation (A1.17):

$$e_{ijk} e_{klm} f_{m,jl} = f_{j,ji} - f_{i,jj} \tag{A1.37}$$

In equations involving only linear combinations of vectors and tensors, vector notation (bold face letters) is sufficient. In equations with calculus operations of vector and tensor functions, index notation is more convenient.

APPENDIX 2
CYLINDRICAL AND
SPHERICAL COORDINATES

Cylindrical and spherical coordinates are used in many problems in seismology. The most basic equations are presented here in a brief form.

A2.1   Cylindrical coordinates

In problems that have symmetry with respect to an axis, for example, the propagation of surface waves on a flat Earth, cylindrical coordinates are commonly used. The position of a point in space in cylindrical coordinates is given by (ρ, ϕ, z). With respect to the Cartesian coordinates (x, y, z), ρ is measured from the z axis in a plane parallel to (x, y) $(0 \leq \rho \leq \infty)$, with ϕ on the same plane as the x axis $(0 \leq \phi \leq 2\pi)$ and z along the same axis $(-\infty \leq z \leq \infty)$ (Fig. A2.1). The relations between cylindrical and Cartesian coordinates are

$$x = \rho \cos \phi \tag{A2.1}$$

$$y = \rho \sin \phi \tag{A2.2}$$

The inverse transformations are

$$\rho = (x^2 + y^2)^{1/2} \tag{A2.3}$$

$$\phi = \tan^{-1}(x/y) \tag{A2.4}$$

At each point of space of coordinates (ρ, ϕ, z), we can establish a system of Cartesian coordinates in the direction of the unit vectors $(\boldsymbol{e}_\rho, \boldsymbol{e}_\phi,$ and $\boldsymbol{e}_z)$ with their positive senses defined by the positive increment of each coordinate (Fig. A2.1). These vectors change direction for each point of space and, thus, are functions of the coordinates. The relations between the unit vectors $\boldsymbol{e}_\rho, \boldsymbol{e}_\phi,$ and $\boldsymbol{e}_z$ and the unit vectors $\boldsymbol{i}, \boldsymbol{j},$ and $\boldsymbol{k}$ in the directions of the Cartesian coordinates $(x, y,$ and $z)$ are

$$\boldsymbol{e}_\rho = \boldsymbol{i} \cos \phi + \boldsymbol{j} \sin \phi \tag{A2.5}$$

$$\boldsymbol{e}_\phi = -\boldsymbol{i} \sin \phi + \boldsymbol{j} \cos \phi \tag{A2.6}$$

$$\boldsymbol{e}_z = \boldsymbol{k} \tag{A2.7}$$

The elements of distance ds, of the area dS over a cylindrical surface, and of volume dV are

$$\mathrm{d}s^2 = \mathrm{d}\rho^2 + \rho^2 \, \mathrm{d}\phi^2 + \mathrm{d}z^2 \tag{A2.8}$$

$$\mathrm{d}S = \rho \, \mathrm{d}\phi \, \mathrm{d}z \tag{A2.9}$$

$$\mathrm{d}V = \rho \, \mathrm{d}\rho \, \mathrm{d}\phi \, \mathrm{d}z \tag{A2.10}$$

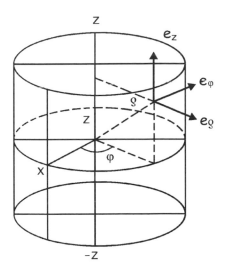

Fig. A2.1. Cylindrical coordinates. Unit vectors in the direction of coordinates.

Expressions for the gradient of a scalar function $f(\rho, \phi, z)$, and for the divergence and curl of a vector function $\boldsymbol{F}(\rho, \phi, z) = (F_\rho, F_\phi, F_z)$ are

$$\nabla f = \boldsymbol{e}_\rho \frac{\partial f}{\partial \rho} + \boldsymbol{e}_\phi \frac{1}{\rho} \frac{\partial f}{\partial \phi} + \boldsymbol{e}_z \frac{\partial f}{\partial z} \qquad (A2.11)$$

$$\nabla \cdot \boldsymbol{F} = \frac{\partial F_\rho}{\partial \rho} + \frac{F_\rho}{\rho} + \frac{1}{\rho} \frac{\partial F_\phi}{\partial \phi} + \frac{\partial F_z}{\partial z} \qquad (A2.12)$$

$$\nabla \times \boldsymbol{F} = \boldsymbol{e}_\rho \left(\frac{1}{\rho} \frac{\partial F_z}{\partial \phi} - \frac{\partial F_\phi}{\partial z} \right) + \boldsymbol{e}_\phi \left(\frac{\partial F_\rho}{\partial z} - \frac{\partial F_z}{\partial \rho} \right)$$

$$+ \boldsymbol{e}_z \left(\frac{\partial F_\phi}{\partial \rho} + \frac{F_\phi}{\rho} + \frac{1}{\rho} \frac{\partial F_\rho}{\partial \phi} \right) \qquad (A2.13)$$

The Laplacian of a scalar function is

$$\nabla^2 f = \frac{\partial^2 f}{\partial \rho^2} + \frac{1}{\rho} \frac{\partial f}{\partial \rho} + \frac{1}{\rho^2} \frac{\partial^2 f}{\partial \phi^2} + \frac{\partial^2 f}{\partial z^2} \qquad (A2.14)$$

Expressions in cylindrical coordinates are greatly simplified when there is symmetry with respect to ϕ. This type of symmetry with the principal axis in the z direction is called cylindrical symmetry.

A2.2 Spherical coordinates

The approximately spherical shape of the Earth makes the use of spherical coordinates very appropriate for problems of seismic wave propagation and normal modes in the Earth. In spherical coordinates, a point in space is given by (r, θ, ϕ). With respect to the Cartesian coordinates (x, y, z), r is the distance from the origin ($0 \le r \le \infty$), θ is the

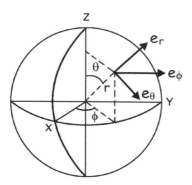

Fig. A2.2. Spherical coordinates. Unit vectors in the direction of coordinates.

angle measured from the z axis $(0 \leq \theta \leq \pi)$, and ϕ is the angle measured from the x axis to the projection of r onto the plane (x, y) $(0 \leq \phi \leq 2\pi)$ (Fig. A2.2). The relations between the spherical and the Cartesian coordinates are

$$x = r \sin \theta \cos \phi \tag{A2.15}$$

$$y = r \sin \theta \sin \phi \tag{A2.16}$$

$$z = r \cos \theta \tag{A2.17}$$

The inverse relations are

$$r = (x^2 + y^2 + z^2)^{1/2} \tag{A2.18}$$

$$\theta = \cos^{-1} \left(\frac{z}{r} \right) \tag{A2.19}$$

$$\phi = \tan^{-1} \left(\frac{y}{x} \right) \tag{A2.20}$$

At each point of space (r, θ, ϕ), we can establish a system of Cartesian coordinates in the directions of the unit vectors e_r, e_θ, and e_ϕ, with their positive senses in the directions of positive increments of the three coordinates (Fig. A2.2). These vectors change direction from point to point and are, thus, functions of the coordinates. The relations between the unit vectors e_r, e_θ, and e_ϕ and the unit vectors i, j, and k in the directions of the Cartesian coordinates (x, y, z) are

$$e_r = i \sin \theta \cos \phi + j \sin \theta \sin \phi + k \cos \theta \tag{A2.21}$$

$$e_\theta = i \cos \theta \cos \phi + j \cos \theta \sin \phi - k \sin \theta \tag{A2.22}$$

$$e_\phi = -i \sin \phi + j \cos \phi \tag{A2.23}$$

The elements of distance ds, of the area dS on a spherical surface of constant radius, and of volume dV are

$$ds^2 = dr^2 + r^2 d\theta^2 + r^2 \sin^2 \theta \, d\phi^2 \tag{A2.24}$$

$$dS = r^2 \sin \theta \, d\theta \, d\phi \tag{A2.25}$$

$$dV = r^2 \sin \theta \, dr \, d\theta \, d\phi \tag{A2.26}$$

The gradient of a scalar function $f(r, \theta, \phi)$, and the divergence and curl of a vector function $\boldsymbol{F}(r, \theta, \phi) = (F_r, F_\theta, F_\phi)$ are

$$\nabla f = \boldsymbol{e}_r \frac{\partial f}{\partial r} + \boldsymbol{e}_\theta \frac{1}{r} \frac{\partial f}{\partial \theta} + \boldsymbol{e}_\phi \frac{1}{r \sin \theta} \frac{\partial f}{\partial \phi} \tag{A2.27}$$

$$\nabla \cdot \boldsymbol{F} = \frac{1}{r} \frac{\partial}{\partial r} (r^2 F_r) + \frac{1}{r \sin \theta} \frac{\partial}{\partial \theta} (\sin \theta \, F_\theta) + \frac{1}{r \sin \theta} \frac{\partial F_\phi}{\partial \phi} \tag{A2.28}$$

$$\nabla \times \boldsymbol{F} = \boldsymbol{e}_r \left(\frac{1}{r \sin \theta} \frac{\partial}{\partial \theta} (\sin \theta \, F_\phi) - \frac{\partial F_\theta}{\partial \phi} \right) + \boldsymbol{e}_\theta \left(\frac{1}{r \sin \theta} \frac{\partial F_r}{\partial \phi} - \frac{\partial}{\partial r} (r \sin \theta \, F_\phi) \right)$$
$$+ \boldsymbol{e}_\phi \left(\frac{1}{r} \frac{\partial}{\partial r} (r F_\theta) - \frac{\partial F_r}{\partial \theta} \right) \tag{A2.29}$$

The Laplacian of a scalar function is

$$\nabla^2 f = \frac{1}{r^2} \frac{\partial}{\partial r} \left(r^2 \frac{\partial f}{\partial r} \right) + \frac{1}{r^2 \sin \theta} \frac{\partial}{\partial \theta} \left(\sin \theta \frac{\partial f}{\partial \theta} \right) + \frac{1}{r^2 \sin^2 \theta} \frac{\partial^2 f}{\partial \phi^2} \tag{A2.30}$$

Expressions in spherical coordinates are simplified when there is some sort of symmetry. Symmetry with respect to ϕ and θ is called spherical symmetry and there is only a dependence on r. In many problems symmetry with respect to ϕ is used and there are only dependences on r and θ.

APPENDIX 3
BESSEL AND LEGENDRE
FUNCTIONS

A3.1 Bessel functions

In many problems with cylindrical symmetry, solutions of the Laplace, wave, and diffusion differential equations are given in terms of Bessel functions. This is the case in seismology with the solutions of wave equations in cylindrical and spherical coordinates. Bessel functions are solutions of differential Bessel equations:

$$\frac{d^2f}{dx^2} + \frac{1}{x}\frac{df}{dx} + \left(1 - \frac{n}{x^2}\right)f = 0 \tag{A3.1}$$

Solutions with finite values for $x = 0$ are Bessel functions of the first class $J_n(x)$, with integer values of n. These functions are expressed as infinite sums of powers of x:

$$J_n(x) = \sum_{m=0}^{\infty} \frac{(-1)^m}{m!\,(m+n)!}\left(\frac{x}{2}\right)^{n+2m} \tag{A3.2}$$

For $n = 0$,

$$J_0(x) = 1 + \frac{x^2}{2^2 1!} + \frac{x^4}{2^4 2!} + \frac{x^6}{2^6 3!} + \cdots \tag{A3.3}$$

For other values of $n > 0$, they can be obtained from those of order zero:

$$J_n(x) = (-x)^n\left(\frac{1}{x}\frac{d}{dx}\right)^n J_0(x) \tag{A3.4}$$

Several recurrence formulas relate Bessel functions of different orders and their derivatives; among them are

$$J_n'(x) = \frac{n}{x}J_n(x) - J_{n+1}(x) \tag{A3.5}$$

$$2J_n'(x) = J_{n-1}(x) - J_{n+1}(x) \tag{A3.6}$$

Solutions of differential Bessel equations with nonfinite values for $x = 0$ are Bessel functions of the second kind $Y_n(x)$. They can be deduced from those of the first kind:

$$Y_n(x) = \frac{\cos(n\pi)\,J_n(x) - J_-(x)}{\sin(n\pi)} \tag{A3.7}$$

From Bessel functions of the first and second kinds, we form Bessel functions of the third kind, or Hankel functions:

$$H_n^1(x) = J_n(x) + iY_n(x) \tag{A3.8}$$

$$H_n^2(x) = J_n(x) - iY_n(x) \tag{A3.9}$$

For large values of x ($|x| \gg 1$ and $|x| \gg n$), we have for Bessel functions of the first and second classes the following asymptotic expressions:

$$J_n(x) \simeq \left(\frac{2}{\pi x}\right)^{1/2} \cos\left(x - \frac{\pi}{4} - \frac{n\pi}{2}\right) \tag{A3.10}$$

$$Y_n(x) \simeq \left(\frac{2}{\pi x}\right)^{1/2} \sin\left(x - \frac{\pi}{4} - \frac{n\pi}{2}\right) \tag{A3.11}$$

A3.2 Spherical Bessel functions

Spherical Bessel functions of the first class are derived from Bessel functions of the first class of noninteger order:

$$j_n(x) = \left(\frac{\pi}{2x}\right)^{1/2} J_{n+1/2}(x) \tag{A3.12}$$

These functions are solutions of the differential equation

$$\frac{d^2 f}{dx^2} + \frac{2}{x}\frac{df}{dx} + \left(1 - \frac{n(n+1)}{x^2}\right)f = 0 \tag{A3.13}$$

This differential equation appears in the resolution of the wave equation in spherical coordinates for the dependence on r. Their name comes from their relation to trigonometric functions:

$$j_n(x) = (-x)^n \left(\frac{1}{x}\frac{d}{dx}\right)^n \frac{\sin x}{x} \tag{A3.14}$$

Complex solutions are given in terms of Hankel spherical functions:

$$h_n(x) = j_n(x) + in_n(x) \tag{A3.15}$$

where

$$n_n(x) = (-1)^{n+1} \left(\frac{\pi}{2x}\right)^{1/2} J_{-n-1/2}(x)$$

Functions $j_n(x)$ and $h_n(x)$ have asymptotic approximations ($x \to \infty$) given by

$$j_n(x) \simeq \frac{1}{x}\cos\left[x - \frac{\pi}{2}(n+1)\right] \tag{A3.16}$$

$$h_n(x) \simeq \frac{1}{x}\exp i\left[x - \frac{\pi}{2}(n+1)\right] \tag{A3.17}$$

A3.3   Legendre functions

Solutions for many problems in spherical coordinates, especially those involving Laplace and wave equations, are expressed in terms of Legendre functions. We start with the differential Legendre equation

$$(1 - x^2)\frac{d^2 f}{dx^2} - 2x\frac{df}{dx} + n(n + 1) = 0 \qquad (A3.18)$$

Solutions for $|x| \leq 1$ can be expressed in terms of Legendre functions of the first and second kind, $P_n(x)$ and $Q_n(x)$. Those of the first kind for integer values of n, called Legendre polynomials, are given by

$$P_n(x) = \sum_{s=0}^{m} \frac{(-1)^s (2n - 2s)!\, x^{n-2s}}{s!\,(n - s)!\,(n - 2s)!\,2^n} \qquad (A3.19)$$

where $m = n/2$ or $(n - 1)/2$, so that it is always an integer. Legendre polynomials can be generated from certain functions, such as

$$(1 - 2xz + z^2)^{-1/2} = \sum_{n=0}^{\infty} z^n P_n(x) \qquad (A3.20)$$

$$P_n(x) = \frac{1}{2^n n!}\frac{d^n}{dx^n}(x^2 - 1)^n \qquad (A3.21)$$

Some recurrence formulas for Legendre polynomials of different order and their derivatives are

$$P'_{n+1}(x) - xP'_n(x) = (n + 1)P_n(x) \qquad (A3.22)$$
$$(2n + 1)xP_n(x) = (n + 1)P_{n+1}(x) + nP_n(x) \qquad (A3.23)$$

Legendre polynomials are orthogonal, so they satisfy

$$\int_{-1}^{+1} P_m(x)P_n(x)\,dx = \delta_{nm}\frac{2}{2n + 1} \qquad (A3.24)$$

The first polynomials are

$$P_0(x) = 1, \qquad P_1(x) = x, \qquad P_2(x) = \tfrac{1}{2}(3x^2 - 1)$$
$$P_3(x) = \tfrac{1}{2}(5x^3 - 3x), \qquad P_4(x) = \tfrac{1}{8}(35x^4 - 30x^2 + 3)$$

A3.4   Associate Legendre functions

A general form of the differential Legendre equation is

$$(1 - x^2)\frac{d^2 f}{dx^2} - 2x\frac{df}{dx} + \left(n(n + 1) - \frac{m^2}{1 - x^2}\right)f = 0 \qquad (A3.25)$$

This equation reduces to (A3.18) for $m = 0$. The solutions of equation (A3.25) are associate Legendre functions of order n and grade m. They can be obtained from

Legendre polynomials via

$$P_n^m(x) = (1 - x^2)^{m/2} \frac{\mathrm{d}^m}{\mathrm{d}x^m} P_n(x) \qquad (A3.26)$$

Some recurrence formulas are

$$(n - m)P_n^m(x) = (2n - 1)xP_{n-1}^m(x) - (n + m - 1)P_{n-2}^m(x) \qquad (A3.27)$$

$$(n - m)P_n^m(x) = nxP_{n-1}^m(x) - (1 - x^2)\frac{\mathrm{d}}{\mathrm{d}x}P_{n-1}^m(x) \qquad (A3.28)$$

The condition of orthogonality is

$$\int_{-1}^{+1} P_n^m(x)P_l^m(x)\,\mathrm{d}x = \delta_{nl}\frac{2(n + m)!}{(2n + 1)(n - m)!} \qquad (A3.29)$$

$$\int_{-1}^{+1} P_n^m(x)P_m^l(x)\,\mathrm{d}x = \delta_{ml}\frac{(n + m)!}{m(n + m)!} \qquad (A3.30)$$

For $m = 0$, associate Legendre functions are equivalent to Legendre polynomials. For $m > 0$, the first associate functions are

$$P_1^1(x) = (1 - x^2)^{1/2}, \qquad P_2^1(x) = 3x(1 - x^2)^{1/2}$$
$$P_2^2(x) = 3(1 - x^2), \qquad P_3^1(x) = \tfrac{1}{2}(1 - x^2)^{1/2}(15x^2 - 3)$$

In spherical coordinates, solutions of Laplace and wave equations using the method of separation of variables result in equation (A3.25) with $x = \cos\theta$, and a harmonic equation for ϕ. Solutions can be given in terms of the products of associate Legendre and trigonometric functions:

$$S_n^m(\theta, \phi) = [A_n^m \cos(m\phi) + B_n^m \sin(m\phi)]P_n^m(\cos\theta) \qquad (A3.31)$$

These functions are called spherical surface harmonics or tesseral harmonics. If the problem has symmetry with respect to ϕ, the resulting equation is (A3.18) with $x = \cos\theta$ and the solution is given in terms of Legendre polynomials, which, in this case, are also called zonal harmonics. If we introduce into (A3.31) a factor that depends on positive or negative powers of r, we obtain the solid harmonics.

APPENDIX 4
FOURIER TRANSFORMS

Seismologic data representing ground displacements, velocities, and accelerations produced by earthquakes and recorded at a point on the Earth's surface are functions of time. A powerful tool for their analysis is the Fourier transform theorem which allows one to change from the time to the frequency domain and vice versa. The fundamental ideas of this important mathematical tool are presented in a very brief form. Although the Fourier theorem can be applied to very general functions of any variable, here we will treat only time functions (Bracewell, 1986).

A4.1   Periodic functions

A periodic function $f(t)$ of period T and frequency $\omega = 2\pi/T$ that satisfies very general conditions of continuity, such as being absolutely integrable, can be expressed, according to Fourier's theorem, as an infinite series of harmonic functions:

$$f(t) = \frac{a_0}{2} + \sum_{n=1}^{\infty} \left[a_n \cos\left(\frac{2\pi n}{T} t\right) + b_n \sin\left(\frac{2\pi n}{T} t\right) \right] \tag{A4.1}$$

The coefficients a_n and b_n are obtained in the forms

$$a_n = \frac{2}{T} \int_{-T/2}^{T/2} f(t) \cos\left(\frac{2\pi n}{T} t\right) dt \tag{A4.2}$$

$$b_n = \frac{2}{T} \int_{-T/2}^{T/2} f(t) \sin\left(\frac{2\pi n}{T} t\right) dt \tag{A4.3}$$

If we make the substitution $\omega_n = 2\pi n/T$, then the arguments of the sine and cosine functions are given in terms of the frequencies $\omega_n = n\omega$, which are integer multiples of the frequency of the function $f(t)$. Equation (A4.1) can be expressed in exponential form as

$$f(t) = \sum_{n=-\infty}^{\infty} F_n e^{i\omega_n t}, \qquad \omega_n = \frac{2\pi n}{T} \tag{A4.4}$$

The coefficients F_n have complex values:

$$F_n = \tfrac{1}{2}(a_n + ib_n) \tag{A4.5}$$

$$F_{-n} = \tfrac{1}{2}(a_n - ib_n) \tag{A4.6}$$

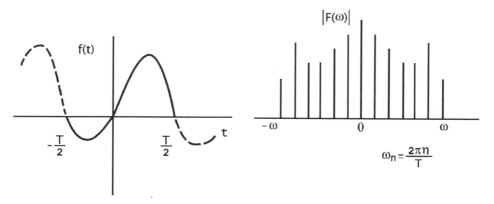

Fig. A4.1. A periodic continuous function and its discrete transform.

These coefficients are found by writing

$$F_n = \frac{1}{T} \int_{-T/2}^{T/2} f(t) \, e^{-\omega_n t} \, dt \tag{A4.7}$$

The complex coefficients F_n can be expressed as

$$F_n = |F_n| \, e^{i\phi_n} = A_n \, e^{i\phi_n} \tag{A4.8}$$

$$A = \tfrac{1}{2} (a_n^2 + b_n^2)^{1/2} \tag{A4.9}$$

$$\phi_n = \tan^{-1}(b_n / a_n) \tag{A4.10}$$

Thus, F_n is the complex Fourier transform of $f(t)$ and, since this is a reciprocal relation, $f(t)$ is the inverse transform of F_n. The values $A_n = |F_n|$ constitute the amplitude spectrum and those of ϕ_n constitute the phase spectrum. The square of the amplitude spectrum is called the power spectrum $P_n = |F_n|^2$. Thus, a periodic continuous function $f(t)$ has a discrete transform with values only for frequencies $\omega_n = n\omega$, which are integer multiples (positive or negative) of its own frequency (Fig. A4.1).

A4.2 Nonperiodic functions

A nonperiodic function $f(t)$ defined for all values of t and that satisfies the already-mentioned continuity conditions can be expressed using the Fourier integral as

$$f(t) = \frac{1}{2\pi} \int_{-\infty}^{\infty} F(\omega) \, e^{i\omega t} \, d\omega \tag{A4.11}$$

where $F(\omega)$, the complex Fourier transform, is given by

$$F(\omega) = \int_{-\infty}^{\infty} f(t) \, e^{-i\omega t} \, dt \tag{A4.12}$$

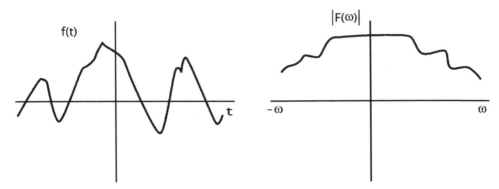

Fig. A4.2. A nonperiodic infinite continuous function and its infinite continuous transform.

Thus, upon substitution of (A4.12) into (A4.11), the Fourier integral theorem is given by

$$f(t) = \frac{1}{2\pi} \int_{-\infty}^{\infty} e^{i\omega t} \int_{-\infty}^{\infty} f(t) e^{-i\omega t} \, dt \, d\omega \qquad (A4.13)$$

The Fourier transform $F(\omega)$ is a continuous function of the frequency ω that is defined over the whole range of positive and negative frequencies. Although $f(t)$ can be a real function, $F(\omega)$ is a complex function with real and imaginary parts, $F(\omega) = R(\omega) + iI(\omega)$, that may be also expressed in the form

$$F(\omega) = A(\omega) \, e^{i\phi(\omega)} \qquad (A4.14)$$

$$A(\omega) = |F(\omega)| = [R(\omega)^2 + I(\omega)^2]^{1/2} \qquad (A4.15)$$

$$\phi(\omega) = \tan^{-1}[I(\omega)/R(\omega)] \qquad (A4.16)$$

where $A(\omega)$ is the amplitude spectrum and $\phi(\omega)$ is the phase spectrum, both of which are continuous functions of the frequency (Fig. A4.2).

A4.3   Convolution and correlation

An important operation in the analysis of temporal functions such as seismologic data is the convolution of two functions, which is given by

$$h(\tau) = \int_{-\infty}^{\infty} f(t)g(\tau - t) \, dt = f(t) * g(t) \qquad (A4.17)$$

The convolution function $h(\tau)$ is a function of the time shift τ. In the shifted function $g(\tau - t)$, the independent variable (t) has changed sign. A property of convolution is that its transform is the product of the transforms of the two functions:

$$H(\omega) = F(\omega)G(\omega) \qquad (A4.18)$$

On passing from the time to the frequency domain, convolution in time changes into a product in frequency. Thus, it may be advantageous instead of doing time convolutions to transform into the frequency domain to take products and then change back into the time domain by means of the inverse transform.

Another operation that can be applied to two functions is their correlation:

$$q(\tau) = \int_{-\infty}^{\infty} f(t)g(t+\tau)\, dt \tag{A4.19}$$

The function $q(\tau)$ is the correlation function of $f(t)$ and $g(t)$. If the functions are different it is called cross correlation, whereas if the operation is performed on the same function the result is the autocorrelation function:

$$p(\tau) = \int_{-\infty}^{\infty} f(t)f(t+\tau)\, dt \tag{A4.20}$$

The transform of the autocorrelation function $p(\tau)$ is the power spectrum of the original function $f(t)$:

$$P(\omega) = |F(\omega)|^2 = A^2(\omega) \tag{A4.21}$$

When we calculate the autocorrelation function, we lose information about the phases of its transform. Autocorrelation functions are always even functions and their transforms have only real parts.

A4.4 Sampled functions of finite duration

The digital form of modern seismologic data and the use of digital computers for their analyses determine that time functions representing them are sampled at certain time increments and have finite durations. Both conditions have their consequences for their Fourier analysis.

If a continuous function $f(t)$ is defined only in the interval $(-t_{\mathrm{m}}, t_{\mathrm{m}})$ and we do not know what values it takes outside of this interval, we can not calculate its transform, since this requires that the function be known for all times. To resolve this difficulty, we must make a hypothesis about the values of the function outside of the known range, either that it is zero or that it repeats as a periodic function of period $2t_{\mathrm{m}}$. The first hypothesis is equivalent to multiplying the function $f(t)$ by a rectangular function of amplitude unity and length $2t_{\mathrm{m}}$ (a rectangular window). Since multiplication in one domain corresponds to a convolution in the other, the spectrum of $f(t)$ is convoluted with the transform of the rectangular window. In the second hypothesis, $f(t)$ becomes a periodic function and, according to (A4.4), can be represented by an infinite series whose coefficients are

$$F_n = \frac{1}{2t_{\mathrm{m}}} \int_{-t_{\mathrm{m}}}^{t_{\mathrm{m}}} f(t) \exp\left(-\frac{i\pi n t}{t_{\mathrm{m}}}\right) dt \tag{A4.22}$$

Although $f(t)$ is a continuous function, since it is considered to be periodic, its transform is discrete with values at increments in sampling frequency of $\Delta\omega = \pi/t_{\mathrm{m}}$:

$$F_n = F\left(\frac{\pi n}{t_{\mathrm{m}}}\right) \tag{A4.23}$$

A sampled function is one that is known only for discrete values of the variable. If the sampling increment Δt is constant, then

$$f_n = f(n\,\Delta t), \qquad n = 0, 1, 2, 3, \ldots \tag{A4.24}$$

According to the relation between a function and its transform, if f_n is a discrete function at increments Δt, then, using (A4.4) and (A4.7), it can be considered as the inverse transform of a periodic function $F(\omega)$ with period $2\pi/\Delta t$ (A4.4). Therefore, the transform function $F(\omega)$ is limited to a band of frequencies between $-\omega_N$ and ω_N, where $\omega_N = \pi/\Delta t$ is called the Nyquist frequency (for $\sigma = \omega/(2\pi)$, $\sigma_N = 1/(2\,\Delta t)$). Frequencies higher than the Nyquist frequency are folded together in the phenomenon known as aliasing. In this way f_n is given in terms of its transform by

$$f_n = \frac{1}{2\omega_N} \int_{-\omega_N}^{\omega_N} F(\omega) \exp\left(\frac{i\pi n\omega}{2\omega_N}\right) d\omega \tag{A4.25}$$

If the number of samples is finite, say equal to $2N$, f_n is a sampled function of finite duration, $2t_m = 2N\,\Delta t$. In practice, all of the functions we use in seismologic analysis are of this type. According to (A4.22), (A4.23) and (A4.25) its transform is discrete and is limited to a band of frequencies:

$$F_n = F\left(\frac{n\pi}{N\,\Delta t}\right), \qquad n = -N, \ldots, -2, -1, 0, 1, 2, \ldots, N \tag{A4.26}$$

Thus, the transform is limited to the frequency band $-\pi/\Delta t \le \omega_n \le \pi/\Delta t$ and the Nyquist frequency is $\omega_N = \pi/\Delta t$ ($\sigma_N = 1/(2\,\Delta t)$). The relation between f_n and F_n is given by the pair of transform equations

$$f_n = \frac{1}{N} \sum_{k=-N}^{N} F_k \exp\left(\frac{i\pi nk}{N}\right), \qquad n = -N, \ldots, N \tag{A4.27}$$

$$F_n = \sum_{k=-N}^{N} f_k \exp\left(\frac{i\pi nk}{N}\right), \qquad n = -N, \ldots, N \tag{A4.28}$$

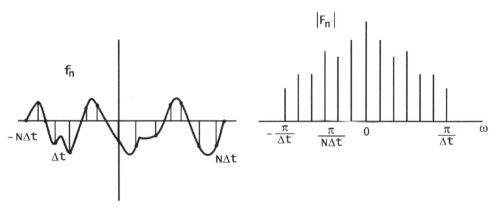

Fig. A4.3. A nonperiodic sampled function of limited length and its discrete limited transform.

These relations are known as the discrete Fourier transforms for sampled functions of limited length. The time function f_n is defined in the interval $-N\,\Delta t \le t \le N\,\Delta t$ and sampled at increments Δt. Its transform F_n is defined in the frequency band $-\pi/\Delta t \le \omega \le \pi/\Delta t$ and sampled at increments $\Delta\omega = \pi/(N\,\Delta t)$. If f_n is real, $F_{-n} = F_n$, and the transform is defined by the N values of positive frequencies (Fig. A4.3). Therefore, if we want to increase the frequency band of the spectrum, we must decrease the value of the increment in sampling time, and if we want to increase its resolution (smaller frequency increments), we must increase the length of the function.

APPENDIX 5
PARAMETERS OF THE EARTH

Equatorial radius	6 378 136 m
Polar radius	6 356 750 m
Radius of the sphere of equal volume	6 370 800 m
Volume	1.083×10^{21} m^3
Mass	5.973×10^{24} kg
Mean density	5.515 g cm^{-3}
Surface area	5.1×10^{14} m^2
Polar moment of inertia	8.0378×10^{37} kg m^2
Equatorial moment of inertia	8.0115×10^{37} kg m^2
Flattening	3.3528×10^{-3}
Angular velocity	7.2921×10^{-5} s^{-1}
Core radius	3486 km
Inner core radius	1271 km
Core mass	1.883×10^{24} kg
Thickness of mantle	2890 km
Mantle mass	4.09×10^{24} kg
Mean distance to the Sun	1.496×10^8 km
Mean distance to the Moon	3.844×10^5 km

APPENDIX 6
THE INTERIOR OF THE EARTH

	Depth (km)	Density (g cm⁻³)	Acceleration due to gravity (m s⁻²)	Pressure (10⁴ MPa)	Bulk (K) (10⁴ MPa)	Shear (μ) (10⁴ MPa)	P wave velocity (α) (km s⁻¹)	S wave velocity (β) (km s⁻¹)	Temperature (K)
Crust	10–30	2.9	9.83	0.06	6.82	4.41	6.57	3.82	280
Upper mantle	40	3.33	9.84	0.12	13.19	5.18	8.12	4.42	650
	220	3.36	9.90	0.71	15.29	7.41	8.06	4.35	1610
	400	3.72	9.97	1.34	18.99	9.06	9.13	5.22	1910
Lower mantle	670	4.38	10.01	2.38	29.26	15.48	10.75	5.95	2375
	1200	4.69	9.94	4.78	38.50	19.96	11.78	6.52	2525
	2885	5.57	10.68	13.58	65.56	29.38	13.72	7.26	2934
Outer core	2890	9.90	10.68	13.58	64.41	0.0	8.06	0.0	3160
	3800	11.11	8.42	22.75	96.33	0.0	9.31	0.0	3650
	5150	12.17	4.40	32.89	130.47	0.0	10.36	0.0	4170
Inner core	5155	12.76	4.40	32.89	134.34	15.67	11.03	3.50	4190
	6371	13.09	0.0	36.39	142.53	17.61	11.26	3.67	4290

APPENDIX 7
IMPORTANT EARTHQUAKES

Date			Region	Casualties	Magnitude
21 07		365	Aegean		
		575	Antioch, Turkey	10 000	
		856	Corinth, Greece	10 000	
27 09		1290	Chilhi, China	100 000	
02 02		1428	Olot, Spain	800	
05 12		1456	Campania, Italy	30 000	
23 01		1556	Shaanxi, China	300 000	
04 02		1729	Quito, Ecuador	40 000	
01 11		1755	Lisbon, Portugal	20 000	
04 02		1783	Calabria, Italy	50 000	
16 08		1868	Peru–Ecuador	100 000	
16 12		1875	Naples, Italy	12 000	
25 12		1884	Andalucia, Spain	749	
31 08		1886	Charleston, USA	60	
28 10		1891	Mino-Owari, Japan	7000	
15 06		1896	Riku-Ugo, Japan	22 000	
12 06		1897	Assam, India	1500	8.7
18 04		1906	San Francisco, USA	700	8.2
28 12		1908	Messina, Italy	120 000	7.5
16 12		1920	Shaanxi, China	30 000	8.5
01 09		1923	Kwanto, Japan	99 300	8.2
15 01		1934	Nepal, India	10 700	8.3
24 01		1939	Chillan, Chile	30 000	7.7
26 12		1939	Erzincan, Turkey	23 000	8.0
15 08		1950	Assam, India	574	8.6
04 11		1952	Kamchatka, USSR	600	8.5
22 05		1960	Valdivia, Chile	5700	8.5
28 03		1964	Anchorage, USA	131	8.5
31 08		1968	Dashy-Bayaz, Iran	11 600	7.4
31 05		1970	Ancash, Peru	68 000	7.8
23 12		1972	Managua, Nicaragua	5000	6.2
04 02		1976	Guatemala	22 000	7.9
27 07		1976	Tangshan, China	650 000	7.6
04 03		1977	Vrancea, Rumania	2000	7.2
16 09		1978	Iran	15 000	7.4
19 09		1985	Michoacan, Mexico	20 000	8.1
07 12		1988	Armenia	25 000	7.0

Date	Region	Casualties	Magnitude
18 10 1989	Loma Prieta, USA	62	7.0
21 06 1990	Iran	50 000	7.7
29 09 1993	Latur, India	11 000	6.3
16 01 1995	Kobe, Japan	5502	7.5

APPENDIX 8
PROBLEMS AND EXERCISES

A8.1 Elasticity

1.1. The coordinate system x_1', x_2', x_3' is obtained by rotation of the system x_1, x_2, x_3 by 90° around x_1 so that $x_2' = x_3$ and $x_3' = -x_2$. If the system x_1, x_2, x_3 corresponds to the principal strain axes, show that: $e_{2'3'} = -e_{32}$, $e_{2'2'} = e_{33}$, and $e_{3'3'} = e_{22}$.

1.2. Given the stress tensor

$$\begin{pmatrix} 2 & -1 & 1 \\ -1 & 0 & 1 \\ 1 & 1 & 1 \end{pmatrix}$$

find the principal stresses, the invariants I_1, I_2, and I_3, the deviatoric stress tensor, and the invariants J_2 and J_3.

1.3. Show that the invariants of the deviatoric stress tensor in terms of its principal values are

$$J_1 = 0; \qquad J_2 = \tfrac{1}{2}(s_1^2 + s_2^2 + s_3^2); \qquad J_3 = \tfrac{1}{3}(s_1^3 + s_2^3 + s_3^3)$$

1.4. Show that the relation between the invariants of the stress tensor and those of its deviatoric tensor are

$$J_2 = 3\sigma_0 + I_2; \qquad J_3 = I_3 - \sigma_0 I_2 + 2\sigma_0^3$$

1.5. Given a stress tensor

$$\begin{pmatrix} 3x_1x_2 & 5x_2^2 & 0 \\ 5x_2^2 & 0 & 2x_3 \\ 0 & 2x_3 & 0 \end{pmatrix}$$

determine the stress vector **T** at point $(2, 1, \sqrt{3})$ through a plane tangential to the cylindrical surface $x_2^2 + x_3^2 = 4$.

1.6. Given the stress tensor

$$\begin{pmatrix} 1/4 & -5/4 & (1/2)^{3/2} \\ -5/4 & 1/4 & -(1/2)^{3/2} \\ (1/2)^{3/2} & -(1/2)^{3/2} & 3/2 \end{pmatrix}$$

(a) determine the principal stresses (eigenvalues); and

447

(b) determine the angles with the coordinate axes of the eigenvector corresponding to the greatest principal stress.

1.7. Show that e_{ij} and ω_{ij} are symmetric and antisymmetric tensors and that $e_{ij}\omega_{ij} = 0$.

1.8. Derive from the stress–strain relation for an isotropic medium that

$$\lambda = \frac{\sigma E}{(1+\sigma)(1-2\sigma)}; \qquad \mu = \frac{E}{2(1+\sigma)}$$

$$E = \frac{\mu(3\lambda + 2\mu)}{\lambda + \mu}; \qquad \sigma = \frac{\lambda}{2(\lambda + \mu)}$$

where σ is Poisson's ratio.

1.9. Derive for the stress–strain relation for an isotropic medium that

$$e_{11} = \frac{(\lambda + \mu)\tau_{11}}{\mu(3\lambda + 2\mu)} - \frac{\lambda(\tau_{22} + \tau_{33})}{2\mu(3\lambda + 2\mu)}$$

1.10. Show that for the case of static equilibrium and no body forces, the components of the elastic displacements in an isotropic medium satisfy the biharmonic equation.

A8.2   Body wave displacements and potentials

2.1. The elastic displacement potentials are

$$\phi = 7\exp[2\mathrm{i}(\tfrac{1}{2}x_1 + \sqrt{3}x_2/2 - 6t)]$$

$$\psi_j = (\sqrt{3}, -1, 6)\exp[3\mathrm{i}(\tfrac{1}{2}x_1 + \sqrt{3}x_2/2 - 4t)]$$

Determine the amplitudes of the components of the P and S waves, and the angles i and α.

2.2. An S wave with only an SV component travels in the $(1/(2\sqrt{2}), 1/(2\sqrt{2}), \sqrt{3}/2)$ direction (x_3 is the vertical) and the modulus of its vector potential ψ is 5. If $k = 1$, find B_1, B_2, and B_3, and u_1^{s}, u_2^{s}, and u_3^{s}.

2.3. The displacement vector potential for S waves is

$$\psi_j = (-8\sqrt{3}, 2, 4)\exp[5\mathrm{i}(\tfrac{1}{4}x_1 + \tfrac{\sqrt{3}}{2}x_2 + \tfrac{\sqrt{3}}{4}x_3 - 4t)]$$

If x_3 is the vertical, determine the SH and SV components, the direction cosines of the SH and SV directions, and the angle of polarization ε.

2.4. The amplitudes of displacements for the P and S waves are

$$u_i^{\mathrm{P}} = (4/(3\sqrt{2}), 4/(3\sqrt{2}), 4/\sqrt{3})$$

$$u_i^{\mathrm{S}} = (-\sqrt{3}, -\sqrt{3}, \sqrt{2})$$

The velocities are $\alpha = 6\,\mathrm{km\,s^{-1}}$ and $\beta = 4\,\mathrm{km\,s^{-1}}$, and the frequency $\omega = 4$. Find the expressions for the potentials ϕ and ψ_i.

2.5. Given the amplitudes of the displacements of P and S waves

$$u_i^P = (4, 4, 8); \qquad u_i^S = [8, 2\sqrt{2}, -(4 + \sqrt{2})]$$

find the components SV and SH and the angles i, α, ε, and γ.

2.6. Given the values $\alpha = 60°$, $\varepsilon = 30°$, and $\gamma = 45°$, if the modulus of the S wave is 5, find the components u_1^S, u_2^S, and u_3^S.

2.7. Given the potentials ϕ with the amplitude $A = 3$, and ψ with amplitudes $B_1 = -2$, $B_2 = 2$, and $B_3 = 0$, and the direction cosines of propagation $(\sqrt{2}/4, \sqrt{2}/4, \sqrt{3}/2)$, $k_\alpha = \frac{2}{3}$, $T = \pi/2$, and $\sigma = \frac{1}{4}$, find the components of the displacements of P, SV, and SH waves.

2.8. At the origin of an infinite medium with $\mu = 3 \times 10^{11}$ dyne cm^{-2}, $\rho = 3$ g cm^{-3} and $\sigma = \frac{1}{4}$, there is a center emitting plane waves. For a point at coordinates (500, 300, 141) km, write the expressions for the cosines of P and S waves of arbitrary amplitudes. Find the arrival times.

2.9. In an elastic medium of $\beta = 4$ km s^{-1} and $\sigma = \frac{1}{4}$ there are P and S waves of frequency 1 Hz propagating in the direction of $(1/3, 1/\sqrt{3}, \sqrt{7}/(3\sqrt{2}))$. If the amplitude of potential ϕ is $A = \sqrt{3}/\pi$ and those of ψ are $B_1 = 3/\pi$ and $B_2 = \sqrt{2}/\pi$, find the amplitudes of the components of the displacements of P and S waves.

2.10. The three components of the record of a P and S wave have the following values (at intervals of 0.5 s):

P wave

    Z: 0, 7, 13, 8, 1, −7, −13, −7, −1.
    NS: 0, 7, 15, 10, 2, −9, −15, −7, −1
    EW: 0, −2, −4, −2, 0, 2, 4, 2, 1

S wave

    Z: 0, −6, −13, −15, −9, 0, 7, 12, 15, 12, 5,
    NS: 1, 6, 15, 20, 15, 2, −7, −15, −20, −10, −2.
    EW: 2, 6, 20, 25, 16, 2, −8, −18, −25, −12, −4.

(a) Draw the forms of the waves.
(b) Draw the particle-motion diagrams for the horizontal and vertical (plane of incidence) planes for P and S waves.
(c) Determine the azimuth and the angles of incidence.
(d) Determine the polarization angles γ and ε.

A8.3   Reflection and refraction

3.1. For two liquid media in contact by the plane $x_3 = 0$, with parameters $\rho_1 = 1$, $K_1 = \frac{1}{2}$, $\rho_2 = \frac{3}{2}$, a P wave of frequency 3 Hz and potential amplitude 6 travels in the direction $(1/(2\sqrt{2}), 1/(2\sqrt{2}), \sqrt{3}/2)$ from medium 1 to medium 2. If the amplitude of the potential of the refracted wave is double that of the reflected one, find the expressions for the incident, refracted, and reflected potentials.

3.2. A wave given by the potential

$$\phi = 4 \exp \left[\frac{1}{4} i \left(\frac{1}{\sqrt{6}} x_1 + \frac{1}{\sqrt{3}} x_2 + \frac{1}{\sqrt{2}} x_3 - 4t \right) \right]$$

travels from medium 1 to medium 2, both liquids, which are separated by the surface $x_3 = 0$. If the velocity of medium 2 is $\alpha_2 = 2 \, \text{km s}^{-1}$, the pressure exerted by the incident wave on the surface is $5 \, \text{dyn cm}^{-2}$ and the refracted energy is four times the reflected energy, find
(a) the energy transmitted to the second medium; and
(b) the expressions for the refracted and reflected potentials.

3.3 In two liquid media in contact, the velocities are $\alpha_1 = 4 \, \text{km s}^{-1}$ and $\alpha_2 = 6 \, \text{km s}^{-1}$, and $\rho_1 = 2 \, \text{g cm}^{-3}$. For normal incidence from medium 1 to medium 2, the reflected potential amplitude is equal to the transmitted one. For a wave of frequency 1 Hz and potential amplitude $A_1 = 2000 \, \text{cm}^2$ with angle of incidence $i_1 = 30°$, the ratio of the transmitted to the reflected potential amplitude is $\frac{1}{3}$. For this wave, find the expression for the transmitted potential.

3.4. Deduce the reflection and transmission coefficients A/A_0, A'/A_0, and B'/A_0 for a P wave incident from a liquid upon an elastic medium with $\sigma = \frac{1}{4}$.

3.5. An incident wave in a semi-infinite elastic medium with Poisson ratio $\frac{1}{4}$ bounded by the surface $x_3 = 0$ is given by the potential

$$\phi = 10 \exp \left[5i \left(\frac{1}{2} x_1 + \frac{1}{2} x_2 + \frac{1}{\sqrt{2}} x_3 - 6t \right) \right]$$

Find the potential of the reflected S wave and the components of the displacement, u_1^S, u_2^S, and u_3^S.

3.6. An incident S wave in a semi-infinite elastic medium of Poisson ratio $\frac{1}{4}$ bounded by the surface $x_3 = 0$ is given by the potential

$$\psi_j = (-10\sqrt{3}, 2, 4) \exp \left[5i \left(\frac{1}{4} x_1 + \frac{\sqrt{3}}{4} x_2 + \frac{\sqrt{3}}{2} x_3 - 4t \right) \right]$$

Find the components of the displacements of the reflected wave (u_1^S, u_2^S, u_3^S).

3.7. An incident wave in a semi-infinite elastic medium with Poisson ratio $\frac{1}{4}$ bounded by the surface $x_3 = 0$ is given by the potential

$$\phi = 15 \exp \left[3i \left(\frac{1}{4} x_1 + \frac{\sqrt{3}}{2} x_2 + \frac{\sqrt{3}}{4} x_3 - 8t \right) \right]$$

Find the direction cosines and the total amplitude of the reflected S wave displacement.

3.8. An incident S wave in a semi-infinite elastic medium of Poisson ratio $\frac{3}{8}$ bounded by the surface $x_3 = 0$ is given by the potential

$$\psi_j = (2, -2, 0) \exp \left[i \left(\frac{1}{2\sqrt{2}} x_1 + \frac{1}{2\sqrt{2}} x_2 + \frac{\sqrt{3}}{2} x_3 - 4t \right) \right]$$

Find the SH and SV components of the reflected wave referred to the same system of coordinate axes as the incident wave.

3.9. An S wave represented by the potential

$$\psi_j = (-10\sqrt{3}, 2, 4) \exp\left[5i\left(\frac{1}{4}x_1 + \frac{\sqrt{3}}{4}x_2 + \frac{\sqrt{3}}{2}x_3 - 4t\right)\right]$$

is incident on the free surface ($x_3 = 0$) of an elastic half-space with $\sigma = \frac{1}{4}$. Find the components of the displacement of the reflected P wave.

3.10. An incident wave in a semi-infinite elastic medium with Poisson ratio $\frac{1}{4}$ bounded by the surface $x_3 = 0$ is given by the potential

$$\phi = 10 \exp\left[5i\left(\frac{1}{2}x_1 + \frac{1}{2}x_2 + \frac{1}{\sqrt{2}}x_3 - 6t\right)\right]$$

Find the components of the total displacement at the surface (U_i) and the apparent angle of incidence $\bar{e}$.

A8.4 Ray theory. Plane media

4.1. For a medium formed by two layers of the same thickness H (1 and 2) and a half-space (3) with velocities $V_2 < V_1 < V_3$, determine the equations for the traveling times of a surface focus for the direct, reflected, and critically refracted waves. Draw the rays' trajectories and travel time curves using units of x/H and H/V_1.

4.2. Consider a medium like that in problem 4.1 with velocities V, $3V$, and $2V$, with a focus at a depth $3H$. Write the parametric equations for x and t as functions of i_0 (the take-off angle at the focus) for the rays that reach the surface without reflection. Give the range of values of i_0 for each ray.

4.3. Given the following first arrival times in a seismic profile x (km), t (s):

x	t	x	t	x	t
9	6	50	30	130	45
20	13	80	36	190	50
30	20	100	40	230	53

determine the model of plane layers with a constant velocity that satisfies the data.

4.4. Write the equations for the travel times of the reflected rays in a medium with an inclined layer with angle θ, in the downward and upward directions. Draw the traveling time curve of t^2 versus x^2.

4.5. Consider a medium with one layer of thickness H over a half-space, of velocity $2V$; the layer is divided vertically into two regions with velocities $V/2$ for $x < H$ and V for $x > H$. Find the distance and arrival time for the ray that crosses the vertical boundary and is reflected on the half-space with the critical angle (write expressions in terms of H and V).

4.6. For a medium with a velocity increasing linearly with depth for a focus at depth Z, write an expression for the take-off angle at the focus i_0, for the ray that arrives at a distance x.

4.7. A medium is formed by a layer of thickness H with velocity distribution $V = V_0 \exp(bz)$, where $b < 1$, over a half-space of velocity $V = 2V_0 \exp(bH)$. Determine the expressions for
(a) the critical angle and distance;
(b) the intersection time of the reflected wave; and
(c) the maximum distance of the direct wave (the wave that travels in the layer).
(d) Draw the travel time curves for the direct, reflected, and refracted waves for $b = 0.1 \,\text{km}^{-1}$, $H = 10 \,\text{km}$, and $V_0 = 1 \,\text{km s}^{-1}$.

4.8. A medium is formed by a layer of thickness H with velocity distribution $V = V_0 + kz$ over a half-space of velocity $V = 2(V_0 + kH)$. Determine the expressions for
(a) the critical angle and distance;
(b) the intersection time of the reflected wave; and
(c) the maximum distance of the direct wave (the wave that travels in the layer).
(d) Draw the traveling time curves for the direct, reflected, and refracted waves for $k = 0.1 \,\text{s}^{-1}$, $H = 10 \,\text{km}$, and $V_0 = 1 \,\text{km s}^{-1}$.

4.9. A medium is formed by a layer of thickness H and velocity V_0 over a half-space of velocity $V = V_0 + k(z - H)$.
(a) Determine the expressions for x and t as functions of i_0, the take-off angle at the focus, for a focus at the surface.
(b) Determine the relation between the angle of incidence at the focus i_0 and the depth h of penetration into the half-space.

4.10. The travel times of P waves in a medium whose velocity is variable with depth are given by the following values; x (km), t (s):

x	t	x	t	x	t
0	0.0	200	28.8	400	41.9
50	9.6	250	33.0	450	44.2
100	17.6	300	36.4	500	46.2
150	23.9	350	39.3	550	48.1

(a) Draw the traveling times curve.
(b) Find the distribution of the velocity with depth.

A8.5   Ray theory. Spherical media

5.1. For a spherical earth of radius 6000 km and constant P wave velocity $V = 10 \,\text{km s}^{-1}$, calculate the travel times and ray parameters for $\Delta = 30°$, $60°$, $90°$, $120°$, $150°$, and $180°$, and draw the curves $t(\Delta)$ and $P(\Delta)$.

5.2. In a spherical Earth of radius R, the velocities of P waves have the form $v = a - br^2$. The surface velocity is v_0 and that at the center is $2v_0$. Determine the angular distance Δ for a ray that penetrates to half of the radius.

5.3. In a spherical medium, the velocity increases with depth according to $v = ar^{-b}$. If $v_0 = 6\,\text{km}\,\text{s}^{-1}$, $r_0 = 6000\,\text{km}$, and $b = 0.2$, calculate the travel times and ray parameters for $\Delta = 30°$, $60°$, $90°$, $120°$, $150°$, and $180°$, and draw the curves $t(\Delta)$ and $P(\Delta)$. Compare your results with the results of problem 5.1.

5.4. In a spherical medium, the velocity increases with depth according to $v = ar^{-b}$. If $v_0 = 6\,\text{km}\,\text{s}^{-1}$, $r_0 = 6000\,\text{km}$, and, at the distance $\Delta = 90°$, the traveling time is $1299\,\text{s}$ and the slope of the travel time curve is $500\,\text{s}$, determine
(a) the value of b; and
(b) the values of r_p and v_p for the ray that arrives at $\Delta = 90°$.

5.5. A spherical Earth of radius R is formed by two regions of constant velocity, the mantle and the core. The radius of the core is $R/2$. The velocity of the mantle is V and that of the core is (A) $2V$ and (B) $V/2$. For each case
(a) draw the trajectories of the direct, reflected, and refracted rays;
(b) calculate the maximum angular distance for the direct ray;
(c) calculate the angular distance for the reflected ray with the critical angle; and
(d) draw the travel time curves for the direct, reflected, and refracted rays using time units of R/V.

5.6. The Earth of radius R and surface velocity v_0 is formed by a mantle of velocity $v = a/\sqrt{r}$ and a core of radius $R/2$ and velocity $4v_0$. Find for a surface focus
(a) the maximum distance Δ_m for the direct ray in the mantle;
(b) the critical angle of the reflected ray at the core and the take-off angle at the focus for the same ray; and
(c) the distance Δ_c for the ray that is critically reflected at the core.
(d) Draw the travel time curve.

5.7. The Earth of radius R is formed by a mantle of constant velocity v_0 and a core of radius $R/2$ and velocity $v = ar^{-b}$ ($b < 1$). The velocity at the surface of the core is $2v_0$. For a surface focus, determine
(a) the take-off angle i_0 and the distance for the reflected ray at the core with the critical angle; and
(b) the distance and travel time for the rays refracted in the core in terms of i_0, v_0, R, and b.

5.8. The Earth of radius R and surface velocity v_0 is formed by a mantle of variable velocity and a core of radius $R/2$ and constant velocity $2v_0$. The velocity at the base of the mantle is $\frac{7}{4}v_0$ and the rays are circular. Find
(a) the critical distance; and
(b) the travel time of the ray that passes through the center.

5.9. The Earth of radius R is formed by a mantle of velocity $v = ar^{-1/2}$ and a core of radius $R/2$ and velocity $v = br^{-1/3}$. The surface velocity is v_0 and that at the top of the core is $2v_0$. Calculate the distance Δ for a ray that leaves the surface focus with the take-off angle $i_0 = 14.5°$.

5.10. For a spherical Earth in which the velocity increases with depth the following travel times have been observed:

Δ (degrees)	t (s)	Δ (degrees)	t (s)
10	130	50	580
20	255	60	661
30	374	70	724
40	483	80	768

The radius is 6370 km and the surface velocity is $8\,\text{km}\,\text{s}^{-1}$.
(a) Using the Herglotz–Wiechert equation, find the distribution of the velocity with the radius.
(b) If the distribution is $v = v_0(r_0/r)^{-\alpha}$, find the value of α.

A8.6   Surface waves

6.1 For a Rayleigh wave in a half-space, if the P wave velocity is $6\,\text{km}\,\text{s}^{-1}$ and Poisson's ratio is $\frac{1}{4}$, determine for a wave of period $T = 20\,\text{s}$
(a) the depth at which $u_1 = 0$; and
(b) the depth at which $u_1 > u_3$.

6.2. Consider a liquid layer of thickness H, density ρ', and velocity α' over a half-space of density ρ and velocity α ($\alpha > \alpha'$). Deduce the dispersion equation for surface waves using the method of constructive interference of supercritical acoustic waves. Draw the dispersion equations for the fundamental mode and the first higher mode.

6.3. Consider a liquid layer of thickness H, density ρ', and velocity α' over an elastic half-space for which $\lambda = \mu$. Write the potentials ϕ and ψ for the layer and half-space, the boundary conditions, and the determinant that gives the equation for the velocity of surface waves.

6.4. Consider an elastic layer of thickness H, coefficients λ and μ, and density ρ, with two free surfaces. Deduce the equation of dispersion $c(\omega)$ for channeled waves of SH and P–SV type in its interior for the fundamental mode and the first higher mode. Draw the dispersion curve for SH wave motion.

6.5. Consider an elastic layer of thickness H, coefficients λ and μ and density ρ over a rigid half-space. Deduce the dispersion equation $c(\omega)$ for guided motion of waves of SH and P–SV type for the fundamental mode and the first higher mode. Draw the dispersion curve for the SH wave motion.

6.6. Consider a layer of thickness H, density ρ, and rigidity μ' over a half-space of density ρ and rigidity μ, with the relations $\mu = 4\mu'$, $\rho = \rho'$, $a = c/\beta$, and $b = H/\lambda$.
(a) Write the dispersion equation of Love waves in terms of a and b.
(b) Find the values of b for $a = 1, \frac{3}{4}$, and $\frac{1}{2}$; and plot a versus b.

6.7. For a layer of thickness H and rigidity $\mu = 0$, over a half-space of $\lambda = 0$, examine whether surface waves exist and the conditions on α, α', and β. Are they dispersed? (Don't solve the determinant.)

6.8. For the third higher mode of Love waves in a layer of thickness H, velocity $\beta/2$, and rigidity $\frac{3}{4}\mu$ over a half-space of velocity β and rigidity μ, find
(a) the maximum wave length;
(b) the depths at which the displacements are null; and
(c) the wave length for which the phase velocity $c = \frac{3}{4}\beta$.
(d) Plot the dispersion curve c/β versus H/λ.

6.9. For the hypothetical case of a layer of thickness H and velocity β' over a half-space of velocity β, if the phase changes on the free surface and at the plane of contact of the layer and the medium are $\pi/4$, and

$$-\tan^{-1}\left(\frac{(c^2/\beta'^2 - 1)^{1/2}}{(1 - c^2/\beta^2)^{1/2}}\right)$$

respectively,
(a) write the dispersion equation using the condition of constructive interference;
(b) find the cut-off frequencies of the fundamental and first higher mode; and
(c) draw the dispersion curves for the fundamental and the first higher mode using c/β and H/λ for $\beta = 2\beta'$.

6.10. On a seismogram of dispersed Rayleigh waves, the times in seconds of the peaks and troughs of the waves (starting with a peak) are 30, 86, 134, 174, 200, 226, 244, 262, 276, 290, 302, 312, 320, 330, 338, 346, and 352. The time for $t = 0$ is 10 h 21 m 55 s. The station is $\Delta = 5000$ km from the epicenter and the origination time of the earthquake is 10 h 10 m 00 s.
(a) Reconstruct approximately the seismogram.
(b) Calculate and draw the dispersion curve of the group velocity U versus the period.

6.11. The same earthquake as that of problem 6.10 has been observed at a second station on a great circle path at a distance of 400 km from the first. The trend of dispersed Rayleigh waves starts for $t = 0$ at 10 h 23 m 35 s. Peaks and troughs are at times in seconds (the first is a peak) of 36, 100, 146, 185, 215, 240, 270, 284, 304, 318, 330, 334, 355, 364, 375, 384, and 394.
(a) Reconstruct approximately the seismogram.
(b) Calculate and draw the dispersion curve of the phase velocity c versus the period.

A8.7 Focal mechanisms

7.1. At the focus of an earthquake situated at depth 100 km a point source consisting of an impulsive force of magnitude 10^{15} N is acting in the direction of the vertical axis. For a point on the surface 100 km distant and at azimuth $45°$ find the components of the displacement of the S wave. Use $\beta = 4$ km s^{-1} and $\rho = 33$ g cm^{-3} and consider the medium infinite.

7.2. In an infinite medium there is a force of magnitude F with a harmonic time dependence acting in the negative direction of the vertical axis (x_3). Find the SH and SV components of the displacement for a point on the plane (x_1, x_3) in terms of the coordinate (r, θ), where θ is measured from x_3.

7.3. Find the far-field amplitude displacements of P waves and S waves at a point at a distance $r = 1000$ km, with $i_h = 45°$ and $\alpha = 45°$, for a shear-fracture point source with $n = (0, 1, 0)$ and $l = (1, 0, 0)$. The medium has $\rho = 3\,\mathrm{g\,cm^{-3}}$, $\alpha = 6\,\mathrm{km\,s^{-1}}$ and $\sigma = \frac{1}{4}$.

7.4. Calculate the components of the moment tensor corresponding to a shear fracture with $\phi = 45°$, $\delta = 60°$, $\lambda = -90°$, and $M_0 = 10^{15}\,\mathrm{N\,m}$. From these values calculate the displacements of P waves at a point at a distance $r = 500$ km, with $i_h = 15°$ and $\alpha = 15°$. The medium has $\alpha = 8\,\mathrm{km\,s^{-1}}$ and $\rho = 3\,\mathrm{g\,cm^{-3}}$.

7.5. The orientations of the T and P axes (Θ and Φ) are T (30°, 60°) and P (70°, 290°). Calculate the components of the moment tensor m_{ij}.

7.6. The components of the seismic moment tensor of an earthquake, with respect to the geographic coordinates (North, East, and down) are

$$m_{11} = 2, \qquad m_{22} = 0, \qquad m_{33} = 2, \qquad m_{12} = -1, \qquad m_{13} = 1, \qquad m_{23} = 1$$

in units of $10^{18}\,\mathrm{N\,m}$.
(a) Find the eigenvalues and eigenvectors.
(b) Separate it into isotropic and deviatoric parts.
(c) Separate the deviatoric part into major and minor DCs.
(d) Separate the deviatoric part into DC and CLVD parts. Find percentages of each.
(e) For the DC part of (d), find the scalar seismic moment M_0 and the orientations of the P and T axes (Φ, Θ). What type of fault is it?

7.7. There is a force per unit surface area acting at the free surface of a half-space given by

$$F = (0, A, 0)\exp[ik(x_1 - ct)]$$

(a) Under what conditions do surface waves of component u_2 and phase velocity c exist.
(b) Find the dispersion equation $c(\omega)$. If D is the amplitude of SH waves in the medium, find the value of c for $\omega = A/(D\mu)$.

7.8. The earthquake of 19 June 1982, $M = 7$, located at 74.2 W, 16.2 N (offshore from El Salvador), has been recorded at teleseismic distances by 14 stations. Their azimuths, take-off angles, and polarities of P waves (compressions and dilations) are

ALQ	327	43	C	GUMO	294	20	D	SLR	112	20	D
DBF	124	35	D	KEV	18	22	C	SNZO	230	20	D
BOCO	118	57	C	LON	327	38	D	TATO	322	20	D
BER	30	24	C	NWAO	227	7	D	TOL	52	25	C
GRFO	40	22	C	SCP	18	43	C				

Using the equal area projection (Schmidt net)
(a) situate the observations on the focal sphere;

(b) find the two nodal planes and give their orientations in terms of ϕ, δ, and λ; and
(c) find the angles Φ and Θ of the axes X, Y, Z, P, and T.

7.9. An earthquake at Granada (Spain) on 20 June 1994, $M = 4.5$, has been recorded at regional distances by 21 stations. Their azimuths, take-off angles, and polarities (compressions and dilations) are

FOR	305	71	C	CAC	261	36	D	ALOJ	135	51	C
TOR	37	42	D	SAL	177	51	C	ASMO	35	80	D
ALM	69	20	D	ELOJ	136	67	C	ACHM	247	70	D
ESC	272	82	D	ERON	277	71	D	RESI	320	78	C
ROM	157	46	C	ECOG	57	77	D	PARA	197	89	C
FRA	299	52	C	APHE	287	88	D	MON	78	78	D
SIL	88	86	D	ATEJ	358	60	C	DOL	28	88	D

Using the equal-area projection (Schmidt net)
(a) situate the observations on the focal sphere;
(b) find the two nodal planes and give their orientations in terms of ϕ, δ, and λ;
(c) find the angles Φ and Θ of the axes X, Y, Z, P, and T.

7.10. The spectrum of P waves recorded at a station at a distance $\Delta = 33°$ and azimuth $\alpha = 55°$ from a surface earthquake of magnitude $M_s = 6$ has a flat part with a ground displacement of $10 \, \mu m$ and a corner frequency of $0.2 \, Hz$. If $i_h = i_o = 30°$, the radiation pattern $R(\phi, \delta, \lambda, i) = 0.453$, $g(\Delta) = 0.48$, and $\alpha = 8 \, km \, s^{-1}$; calculate the seismic moment, the radius of circular source, the apparent average stress, and the drop in stress.

Bibliography

General seismology textbooks

Aki, K. and P. G. Richards (1980). *Quantitative Seismology. Theory and Methods* (2 volumes). W. H. Freeman, San Francisco. 932 pp.

Bath, M. (1973). *Introduction to Seismology*. John Wiley, New York; Birkhäuser, Basel. 395 pp.

Ben Menahem, A. and S. J. Singh (1981). *Seismic Waves and Sources*. Springer, Berlin. 1168 pp.

Bolt, B. A. (1978). *Earthquakes, A Primer*. W. H. Freeman, San Francisco. 241 pp.

Bullen, K. E. (1947). *An Introduction to the Theory of Seismology*. Cambridge University Press, Cambridge. 296 pp.

Bullen, K. E. and B. A. Bolt (1985). *An Introduction to the Theory of Seismology*. Cambridge University Press, Cambridge. 499 pp.

Byerly, P. (1942). *Seismology*. Prentice Hall, New York. 256 pp.

Dahlen, F. A. and J. Tromp (1998). *Theoretical Global Seismology*. Princeton University Press, Princeton. 1025 pp.

Doyle, H. (1995). *Seismology*. John Wiley, New York. 218 pp.

Gershanik, S. (1995). *Sismología*. Universidad Nacional de la Plata, Buenos Aires. 826 pp.

Gubbins, D. (1990). *Seismology and Plate Tectonics*. Cambridge University Press, Cambridge. 339 pp.

Lay, T. and T. C. Wallace (1995). *Modern Global Seismology*. Academic Press, San Diego. 521 pp.

Macelwane, J. B. and F. W. Sohon (1936). *Introduction to Theoretical Seismology. Part I, Geodynamics*, and *Part II, Seismometry*. John Wiley, New York. 366 and 149 pp.

Madariaga, R. and G. Perrier (1991). *Tremblement de terre*. Presses CNRS, Paris. 210 pp.

Pilant, W. L. (1979). *Elastic Waves in the Earth*. Elsevier, Amsterdam. 493 pp.

Richter, C. F. (1958). *Elementary Seismology*. W. H. Freeman, San Francisco. 768 pp.

Sawarensky, E. F. and D. P. Kirnos (1960). *Elemente der Seismologie und Seismometrie*. Akademie Verlag, Berlin. 512 pp.

Udías, A. and J. Mezcua (1996). *Fundamentos de sismología*. UCA Editores, San Salvador. 200 pp.

Special topics in seismology

Babuska, V. and M. Cara (1991). *Seismic Anisotropy in the Earth*. Kluwer Academic, Dordrecht. 217 pp.

Bath, M. (1967). *Mathematical Aspects of Seismology*. Elsevier, Amsterdam.

Bolt, B. A. (1982). *Inside the Earth. Evidence from Earthquakes*. W. H. Freeman, San Francisco. 191 pp.

Boschi, E., G. Ekström and A. Morelli, eds. (1996). *Seismic Modelling of Earth Structure*. Editrice Compositori, Bologna. 572 pp.

Claerbout, J. F. (1985). *Imaging the Earth Interior*. Blackwell Science Press, Oxford. 398 pp.

Coburn, A. and R. Spence (1992). *Earthquake Protection*. John Wiley and Sons, Chichester. 355 pp.

Davison, C. (1927). *The Founders of Seismology*. Cambridge University Press, Cambridge. 240 pp.

Dziewonski, A. M. and E. Boschi, eds. (1980). *Physics of the Earth's Interior*. North-Holland, Amsterdam. 715 pp.

Ewing, W. M., W. S. Jardetzky and F. Press (1957). *Elastic Waves in Layered Media*. McGraw-Hill, New York. 380 pp.

Gibowicz, S. J. and A. Kijko (1994). *An Introduction to Mining Seismology*. Academic Press, San Diego. 399 pp.

Gutenberg, B. and C. F. Richter (1954). *Seismicity of the Earth and Associated Phenomena*. Princeton University Press, Princeton.

James, D. E., ed. (1989) *The Encyclopedia of Solid Earth Geophysics*. Van Nostrand-Reinhold, New York (entries on seismological subjects).

Jeffreys, H. (1959). *The Earth* (5th edn). Cambridge University Press, Cambridge.

Kanamori, H. and E. Boschi, eds. (1983). *Earthquakes: Observation, Theory and Interpretation*. North-Holland, Amsterdam. 608 pp.

Karnik, V. (1969, 1971). *Seismicity of the European Area* (volumes 1 and 2). D. Reidel, Dordrecht.

Kasahara, K. (1981). *Earthquake Mechanics*. Cambridge University Press, Cambridge. 433 pp.

Kearey, P. and F. J. Vine (1990). *Global Tectonics*. Cambridge University Press, Cambridge.

Kennett, B. L. N. (1983). *Seismic Wave Propagation in Stratified Media*. Cambridge University Press, Cambridge. 342 pp.

Kisslinger, C. and Z. Zuzuki (1978). *Earthquakes Precursors*. Center Academic Publisher, Tokyo.

Kostrov, B. V. and S. Das (1988). *Principles of Earthquake Source Mechanics*. Cambridge University Press, Cambridge. 286 pp.

Koyama, J. (1997). *The Complex Faulting Process of Earthquakes*. Kluwer Academic, Dordrecht. 194 pp.

Kulhanek, O. (1990). *Anatomy of Seismograms*. Elsevier, Amsterdam. 178 pp.

Lapwood, E. R. and T. Usami (1981). *Free Oscillations of the Earth*. Cambridge University Press, Cambridge. 243 pp.

Lomnitz, C. (1974). *Global Tectonics and Earthquake Risk*. Elsevier, Amsterdam.

Lomnitz, C. (1994). *Fundamentals of Earthquake Prediction*. John Wiley, New York. 326 pp.

Love, A. E. H. (1911). *Some Problems of Geodynamics*. Cambridge University Press, Cambridge.

Meissner, R. (1986). *The Continental Crust, A Geophysical Approach*. Academic Press, San Diego.

Nolet, G., ed. (1987). *Seismic Tomography with Applications in Global Seismology and Exploration Geophysics*. D. Reidel, Dordrecht. 386 pp.

Officer, C. B. (1958). *Introduction to the Theory of Sound Transmission*. McGraw-Hill, New York. 284 pp.

Payo, G. (1986). *Introducción al análisis de sismogramas*. Instituto Geográfico Nacional, Madrid.

Reiter, C. (1990). *Earthquake Hazard Analysis, Issues and Insights*. Columbia University Press, New York. 254 pp.

Rikitake, T. (1982). *Earthquake Forecasting and Warning*. D. Reidel, Dordrecht. 402 pp.

Sato, H. and M. C. Fehler (1998). *Seismic Wave Propagation and Scattering in the Heterogeneous Earth*. Springer, New York. 308 pp.

Sawarensky, E. (1975). *Seismic Waves*. Mir, Moscow. 350 pp.

Scherbaum, F. (1996). *Of Poles and Zeros. Fundamentals of Digital Seismology*. Kluwer Academic, Dordrecht. 256 pp.

Scholz, C. H. (1990). *The Mechanics of Earthquakes and Faulting*. Cambridge University Press, Cambridge. 433 pp.

Sheriff, R. E. and L. P. Geldart (1982, 1983). *Exploration Seismology* (2 volumes). Cambridge University Press, Cambridge.

Simpson, D. W. and P. G. Richards (eds.) (1981). *Earthquake Prediction: An International Review*. American Geophysical Union, Washington.

Elasticity and wave mechanics

Achenbach, J. D. (1973). *Wave Propagation in Elastic Solids*. North-Holland, Amsterdam. 425 pp.

Brekhovskikh, L. and V. Goncharov (1985). *Mechanics of Continua and Wave Dynamics*. Springer, Berlin. 342 pp.

Lanczos, C. (1986). *The Variational Principles of Mechanics*. Dover Publications, New York. 418 pp.

Landau, L. D. and E. M. Lifshitz (1970). *Theory of Elasticity*. Pergamon Press, Oxford. 134 pp.

Leipholz, H. (1974). *Theory of Elasticity*. Noordhoft, Leiden. 400 pp.

Lindsay, R. B. (1960). *Mechanical Radiation*. McGraw-Hill, New York. 415 pp.

Love, A. E. H. (1945). *The Mathematical Theory of Elasticity* (4th edn). Cambridge University Press, Cambridge.

Malvern, L. E. (1969). *Introduction to the Mechanics of a Continuous Medium*. Prentice Hall, Englewood Cliffs. 713 pp.

Sommerfeld, A. (1964). *Mechanics of Deformable Bodies*. Academic Press, New York. 396 pp.

Timoshenko, I. S. and J. N. Goodier (1951). *Theory of Elasticity*. McGraw-Hill, New York. 506 pp.

Tolstoy, I. (1973). *Wave Propagation*. McGraw-Hill, New York. 458 pp.

References

Please note that citations in the text may refer to books listed in the Bibliography. These are not repeated here.

Aki, K. (1966). Generation and propagation of G waves from Niigata earthquake of June 16, 1964. Estimation of earthquake movement, released energy and stress–strain drop from G wave spectrum. *Bull. Earthq. Res. Inst.* **44**, 23–88.

Aki, K. (1967). Scaling law of seismic spectrum. *J. Geophys. Res.* **73**, 5359–76.

Aki, K. (1969). Analysis of seismic coda of local earthquakes as scattered waves. *J. Geophys. Res.* **74**, 615–31.

Aki, K. (1979). Characterization of barriers on an earthquake fault. *J. Geophys. Res.* **84**, 6140–8.

Aki, K. (1981). A probability synthesis of precursory phenomena. In D. W. Simpson and P. G. Richards (eds.) *Earthquake Prediction: An International Review*. American Geophysical Union, Washington. pp. 566–74.

Aki, K. (1984). Asperities, barriers, characteristic earthquakes and strong motion prediction. *J. Geophys. Res.* **89**, 5867–72.

Aki, K. (1988). Physical theory of earthquakes. In J. Bonnin, M. Cara, A. Cisternas and R. Fantechi (eds.) *Seismic Hazard in Mediterranean Region*. Kluwer Academic, Dordrecht. pp. 566–74.

Aki, K. and B. Chouet (1975). Origin of coda waves: source, attenuation and scattering effects. *J. Geophys. Res.* **80**, 3322–42.

Aki, K., A. Christoffersson and E. S. Husebye (1976). Determination of the three-dimensional seismic structures of the lithosphere. *J. Geophys. Res.* **82**, 277–96.

Alsop, L. E., G. H. Sutton and M. Ewing (1961). Free oscillations of the Earth observed on strain and pendulum seismographs. *J. Geophys. Res.* **66**, 631–41.

Anderson, D. L. and C. B. Archambeau (1964). The anelasticity of the Earth. *J. Geophys. Res.* **69**, 2071–84.

Anderson, D. L. and R. S. Hart (1978). Q of the Earth. *J. Geophys. Res.* **83**, 5869–82.

Andrews, D. J. (1980). A stochastic fault model. I, static case. *J. Geophys. Res.* **85**, 3867–77.

Andrews, D. J. (1989). Mechanics of fault junctions. *J. Geophys. Res.* **94**, 9389–97.

Archuleta, R. J. (1984). A faulting model for the 1970 Imperial Valley earthquake. *J. Geophys. Res.* **89**, 4559–85.

Backus, G. and M. Mulcahy (1976). Moment tensors and other phenomenological descriptions of seismic sources. I Continuous displacements. *Geophys. J. R. Astr. Soc.* **46**, 341–61.

Bak, P. and C. Tang (1989). Earthquakes as a self-organized critical phenomenon. *J. Geophys. Res.* **94**, 15 635–7.

Barenblatt, G. I. (1959). The formation of equilibrium cracks during brittle fracture. General ideas and hypothesis. *J. Appl. Math. Mech.* **23**, 622–36.

Benioff, H., F. Press and S. Smith (1961). Excitation of the free oscillations of the Earth by earthquakes. *J. Geophys. Res.* **66**, 605–19.

Ben Menahem, A. (1961). Radiation of seismic surface waves from finite moving sources. *Bull. Seis. Soc. Am.* **51**, 401–53.

Ben Menahem, A. (1962). Radiation of seismic body waves from finite moving sources in the earth. *J. Geophys. Res.* **67**, 396–474.

Berckhemer, H. (1962). Die Ausdehnung der Bruchfläche im Erdbebenherd und ihr Einfluß auf das seismiche Wellenspektrum. *Gerland Beitr. Geophys.* **71**, 5–26.

Blanco, M. J. and W. Spakman (1993). The P wave velocity structure of the mantle below the Iberian peninsula: evidence for a subducted lithosphere below southern Spain. *Tectonophysics* **221**, 13–34.

461

Bolt, B. A. (1960). Earthquake epicenter, focal depth and origin time using a high speed computer. *Geophys. J. R. Astr. Soc.* **3**, 433–40.

Bolt, B. A. (1982). The constitution of the core: seismological evidence. *Phil. Trans. R. Soc.* A **306**, 11–20.

Bolt, B. A. and R. A. Uhrhammer (1981). The structure, density and homogeneity of the Earth's core. In *Evolution of the Earth*, Geodynamics Series, American Geophysical Union, Washington. Volume 5, pp. 28–37.

Bracewell, R. N. (1986). *The Fourier Transform and its Applications* (2nd revised edn). McGraw-Hill International, New York. 474 pp.

Brillinger, D., A. Udías and B. A. Bolt (1980). A probability model for regional focal mechanism solutions. *Bull. Seis. Soc. Am.* **70**, 149–70.

Brune, J. N. (1970). Tectonic stress and the spectra of shear waves from earthquakes. *J. Geophys. Res.* **75**, 4997–5009.

Brune, J. N., J. E. Nafe and L. Alsop (1961). The polar phase shift of surface waves on a sphere. *Bull. Seis. Soc. Am.* **51**, 247–57.

Buforn, E. (1994). Métodos para la determinación del mecanismo focal de los terremotos. *Física de la tierra* **6**, 113–40.

Buland, R., J. Berger and F. Gilbert (1979). Observations from IDA network of attenuation and splitting during a recent earthquake. *Nature* **277**, 358–62.

Burridge, R. (1969). The numerical solution of certain integral equations with non-integrable kernels arising in the theory of cracks propagation and elastic wave diffraction. *Phil. Trans. R. Soc.* A **265**, 353–81.

Burridge, R. and L. Knopoff (1964). Body force equivalents for seismic dislocations. *Bull. Seis. Soc. Am.* **54**, 1875–88.

Burridge, R. and L. Knopoff (1967). Model and theoretical seismicity. *Bull. Seis. Soc. Am.* **57**, 341–71.

Byerly, P. (1928). The nature of the first motion in the Chilean earthquake of November 11, 1922. *Am. J. Sci.* **16**, 232–6.

Cisternas, A. (1982). *Notas del curso de sismología*. 1a. Escuela Internacional de Riesgo Sísmico, ACIF, Bogotá.

Cochard, A. and R. Madariaga (1994). Dynamic faulting under rate dependent friction. *Pageophysics* **142**, 419–45.

Crampin, S. (1977). A review of the effects of anisotropic layering on the propagation of seismic waves. *Geophys. J. R. Astr. Soc.* **49**, 9–27.

Crampin, S. (1978). Seismic wave propagation through a cracked solid: Polarization as a possible dilatancy diagnostic. *Geophys. J. R. Astr. Soc.* **53**, 467–96.

Dahlen, F. A. (1980). Splitting of the oscillations of the Earth. In A. M. Dziewonski and E. Boschi (eds.), *Physics of the Earth Interior*, North-Holland, Amsterdam. pp. 82–126.

Dainty, A. M. (1981). A scattering model to explain seismic Q observations in the lithosphere between 1 and 30 Hz. *Geophys. Res. Lett.* **8**, 1126–8.

Das, S. and K. Aki (1977). Fault planes with barriers: A versatile earthquake model. *J. Geophys. Res.* **82**, 5648–70.

Derr, J. S. (1969). Free oscillations observations through 1968. *Bull. Seis. Soc. Am.* **59**, 2079–99.

Deschamps, A., H. Lyon-Caen and R. Madariaga (1980). Mise au point sur les méthodes de calcul de séismogrammes synthétiques de long périod. *Ann. Géophys.* **36**, 167–78.

Dewey, J. W. (1972). Seismicity and tectonics of western Venezuela. *Bull. Seis. Soc. Am.* **62**, 1711–51.

Dewey, J. W. and P. Byerly (1969). The early history of seismometry (to 1900). *Bull. Seis. Soc. Am.* **59**, 183–227.

Dieterich, J. H. (1972). Time dependence friction as a possible mechanism for aftershocks. *J. Geophys. Res.* **77**, 3771–81.

Douglas, A. (1967). Joint epicenter determination. *Nature* **215**, 47–8.

Dziewonski, A. M. (1996). Earth's mantle in three dimensions. In Boschi, E., G. Ekström and A. Morelli (eds.). *Seismic Modelling of Earth Structure*. Editrice Compositori, Bologna. pp. 507–72.

Dziewonski, A. M. and D. L. Anderson (1981). Preliminary reference Earth model. *Phys. Earth Planet. Inter.* **25**, 297–356.

Dziewonski, A. M. and F. Gilbert (1972). Observation of normal modes from 84 recordings of the Alaskan earthquake of 1964, March 28. *Geophys. J. R. Astr. Soc.* **27**, 393–446.

Dziewonski, A. M. and A. L. Hales (1972). Numerical analysis of dispersed seismic waves. In B. A. Bolt (ed.) *Methods in Computational Physics, Volume 11, Seismology: Surface Waves and Earth Oscillations.* Academic Press, New York. pp. 39–85.

Dziewonski, A. M. and J. H. Woodhouse (1987). Global images of the Earth's interior. *Science* **236**, 37–48.

Dziewonski, A. M., T. A. Chou and J. H. Woodhouse (1981). Determination of earthquake source parameters from waveform data for studies of global and regional seismicity. *J. Geophys. Res.* **86**, 2825–52.

Evans, R. (1997). Assessment of schemes for earthquake prediction: Editor introduction. *Geophys. J. Int.* **131**, 413–20.

Ferrari, G. (ed.) (1992). *Two Hundred Years of Seismic Instruments in Italy, 1731–1940.* Istituto Nazionale di Geofisica, Rome. 156 pp.

Freund, L. B. (1972). Crack propagation in an elastic solid subjected to general loading. I. Constant rate of extension. *J. Mech. Phys. Solids* **20**, 129–40.

Freund, L. B. (1979). The mechanics of dynamic shear crack propagation. *J. Geophys. Res.* **84**, 2199–209.

Gallart, J., N. Vidal, A. Estévez, J. Pous, F. Sàbat, C. Santisteban, E. Suriñach and ESCI–Valencia Trough Group (1995). The ESCI–Valencia Trough vertical reflection experiment: A seismic image of the crust from NE Iberian Peninsula to the western Mediterranean. *Rev. Soc. Geol. España* **8**, 401–15.

Geiger, L. (1910). Herdbestimung bei Erdbeben aus Ankunftszeiten. *Nachrichten Kön. Gesell. der Wissenschaft Göttingen, Math. Phys. K* 331–49.

Geller, R. J. (1997). Earthquake prediction. A critical review. *Geophys. J. Int.* **131**, 425–50.

Gilbert, F. (1970). Excitation of the normal modes of the Earth by earthquakes sources. *Geophys. J. R. Astr. Soc.* **22**, 223–6.

Gilbert, F. and G. E. Backus (1966). Propagator matrices in elastic wave and vibration problems. *Geophysics* **31**, 326–32.

Gilbert, F. and A. M. Dziewonski (1975). An application of normal mode theory to the retrieval of structural parameters and source mechanisms from seismic spectra. *Phil. Trans. R. Soc. A* **274**, 369–71.

Grünthal, G. (ed.) (1993). European Macroseismic Scale 1992 (up-dated MSK scale). *Cahiers du Centre Européen de Géodynamique et de Séismologie*, Luxembourg. Volume 7, 79 pp.

Gutenberg, B. (1945). Amplitudes of surface waves and magnitudes of shallow earthquakes. *Bull. Seis. Soc. Am.* **35**, 3–12.

Gutenberg, B. and C. F. Richter (1942). Earthquake magnitude, intensity, energy and acceleration. *Bull. Seis. Soc. Am.* **32**, 163–91.

Gutenberg, B. and C. F. Richter (1956). Magnitude and energy of earthquakes. *Annali di geofisica* **9**, 1–15.

Hanks, T. C. (1982). *f*max. *Bull. Seis. Soc. Am.* **72**, 1667–880.

Hanks, T. C. and H. Kanamori (1979). A moment magnitude scale. *J. Geophys. Res.* **84**, 2348–50.

Hanks, T. C. and M. Wyss (1972). The use of body wave spectra in the determination of seismic source parameters. *Bull. Seis. Soc. Am.* **62**, 561–89.

Hartzell, S. (1978). Earthquake aftershocks as Green's functions. *Geophys. Res. Lett.* **5**, 1–4.

Hartzell, S. H. (1989). Comparison of seismic waveforms inversion results for the rupture history of a finite fault. Application to the 1986 North Palm Springs, California earthquake. *J. Geophys. Res.* **94**, 7515–34.

Haskell, N. A. (1953). The dispersion of surface waves on multilayered media. *Bull. Seis. Soc. Am.* **43**, 17–34.

Haskell, N. A. (1964). Total energy and energy spectral density of elastic wave radiation from propagating faults. *Bull. Seis. Soc. Am.* **54**, 1811–41.

Haskell, N. A. (1966). Total energy and energy spectral density of elastic wave radiation from propagating faults. Part II. *Bull. Seis. Soc. Am.* **56**, 125–40.

Heaton, T. H. (1990). Evidence for and implications of self-healing pulses of slip in earthquake rupture. *Phys. Earth Planet. Inter.* **64**, 1–20.

Herraiz, M. and A. F. Espinosa (1987). Coda waves: A review. *Pageophysics* **125**, 499–577.

Herrin, E. (1968). 1968 Seismological tables for P phases. *Bull. Seis. Soc. Am.* **58**, 1193–241.

Hirata, T. (1989). A correlation between the *b* value and the fractal dimension of earthquakes. *J. Geophys. Res.* **94**, 7507–14.

Ida, Y. (1972). Cohesive force across the tip of a longitudinal shear crack and Griffith's specific surface energy. *J. Geophys. Res.* **77**, 3796–805.

Inglada, V. (1928). Die Berechnung der Herdkoordinaten eines Nahbebens aus den Eintrittszeiten der in einigen benachbarten Stationen auf gezeichneten P oder S Wellen. *Beiträge zur Geophysik* **19**, 73–98.

Inoue, H., Y. Fukao, K. Tanabe and Y. Ogata (1990). Whole mantle P wave travel time tomography. *Phys. Earth Planet. Inter.* **59**, 294–328.

Isacks, B. L., J. Oliver and L. R. Sykes (1968). Seismology and new global tectonics. *J. Geophys. Res.* **73**, 5855–99.

Ito, K. and M. Matsuzaki (1990). Earthquakes as self-organized critical phenomena. *J. Geophys. Res.* **95**, 6853–60.

Jackson, D. D. and D. L. Anderson (1970). Physical mechanism of seismic attenuation. *Rev. Geophys. Space Phys.* **8**, 1–63.

Jeanloz, R. (1990). The nature of the Earth's core. *Ann. Rev. Earth Planet. Sci.* **18**, 357–86.

Jeffreys, H. (1931). On the cause of oscillatory movements in seismograms. *Mon. Not. R. Astr. Soc. Geophys.* Suppl. 2, 407–16.

Jeffreys, H. (1957). The damping of S waves. *Nature* **207**, 675.

Jeffreys, H. and K. H. Bullen (1940). *Seismological Tables.* British Association for the Advancement of Science, London. 50 pp.

Jost, M. L. and R. B. Herrmann (1989). A student's guide to and review of moment tensors. *Seism. Res. Lett.* **60**, 37–57.

Kanamori, H. (1977). The energy release in great earthquakes. *J. Geophys. Res.* **82**, 2921–87.

Kanamori, H. and D. L. Anderson (1975). Theoretical basis of some empirical relations in seismology. *Bull. Seis. Soc. Am.* **65**, 1073–95.

Kanamori, H. and J. W. Given (1981). Use of long period surface waves for rapid determination of earthquake source parameters. *Phys. Earth Planet. Inter.* **27**, 8–31.

Kanamori, H. and G. S. Stewart (1978). Seismological aspects of the Guatemala earthquake of February 4, 1976. *J. Geophys. Res.* **83**, 3427–34.

Kasahara, K. (1963). Computer program for fault-plane solutions. *Bull. Seis. Soc. Am.* **53**, 1–13.

Kennett, B. L. N. and E. R. Engdahl (1991). Travel times for global earthquake location and phase identification. *Geophys. J. Int.* **105**, 429–65.

Kennett, B. L. N., E. R. Engdahl and R. Buland (1995). Constraints on seismic velocities in the Earth from travel times. *Geophys. J. Int.* **122**, 108–24.

Keylis-Borok, V. I. (1956). Methods and results of the investigation of earthquake mechanism. *Publ. Bureau Central Sismol. Int. Ser A. Trav. Scient.* **19**, 383–94.

Keylis-Borok, V. I. (1959). On the estimation of the displacement in an earthquake source and of source dimensions. *Ann. Geophys.* **12**, 205–14.

Knopoff, L. (1961). Analytical calculation of the fault plane problem. *Publ. Dom. Obs. (Ottawa)* **24**, 309–15.

Knopoff, L. (1964). A matrix method for elastic wave problems. *Bull. Seis. Soc. Am.* **54**, 431–8.

Knopoff, L. (1967). *The Mathematics of the Seismic Source.* Institut für Geophysik, Universität Karlsruhe (lecture notes).

Knopoff, L. (1972). Observation and inversion of surface wave dispersion. *Tectonophysics* **13**, 497–519.

Knopoff, L. and F. Gilbert (1960). First motions from seismic sources. *Bull. Seis. Soc. Am.* **50**, 117–34.

Knopoff, L. and M. Randall (1970). The compensated vector linear dipole: a possible mechanism for deep earthquakes. *J. Geophys. Res.* **75**, 4957–63.

Kostrov, B. V. (1964). Self-similar problems of propagation of shear cracks. *J. Appl. Math. Mech.* **28**, 1077–87.

Kostrov, B. V. (1966). Unsteady propagation of longitudinal cracks. *J. Appl. Math. Mech.* **30**, 1241–8.

Lamb, H. (1904). On the propagation of tremors over the surface of an elastic solid. *Phil. Trans. Roy. Soc.* A **203**, 1–42.

Lanczos, C. (1957). *Applied Analysis.* Isaac Pitman, London. 539 pp.

Landisman, M., A. Dziewonski and Y. Satô (1969). Recent improvements in the analysis of surface waves observations. *Geophys. J. R. Astr. Soc.* **17**, 369–403.

Langston, C. A. and D. V. Helmberger (1975). A procedure for modelling shallow dislocation sources. *Geophys. J. R. Astr. Soc.* **42**, 117–30.

Lee, W. B. and S. C. Solomon (1978). Simultaneous inversion of surface wave phase velocity and attenuation: Love waves in western North America. *J. Geophys. Res.* **83**, 3389–400.

Lee, W. H. K. and J. C. Lahr (1971). HYPO71: A computer program for determining hypocenter, magnitude and first motion pattern of local earthquakes. USA Geological Survey, Menlo Park. Open-file report. pp. 75–311.

Lee, W. H. K., R. B. Bennet and K. L. Meahger (1972). A method of estimating magnitude of local earthquakes from signal duration. USA Geological Survey, Menlo Park. Open-file report.

Leveque, J. J. and M. Cara (1985). Inversion of multimode surface wave data: Evidence for sublithospheric anisotropy. *Geophys. J. R. Astr. Soc.* **83**, 753–74.

Lindh A., P. Evans, P. Harsh and G. Buhr (1979). The Parkfield experiment. *Earthq. Inf. Bull.* **11**, 205–13.

Lomnitz, C. (1957). Lincar dissipation in solids. *J. Appl. Phys.* **28**, 201–5.

Macelwane, J. B. (1946). Forecasting earthquakes. *Bull. Seis. Soc. Am.* **36**, 1–4.

MacGaffrey, R., G. Abers and P. Zwick (1991). Inversion of teleseismic body waves. In W. H. K. Lee (ed.). *Digital Seismogram Analysis and Wave Form Inversion*. IASPEI Software Library. Volume 3, pp. 81–166.

McKenzie, D. P. (1972). Active tectonics of the Mediterranean region. *Geophys. J. R. Astr. Soc.* **30**, 109–85.

Madariaga, R. (1976). Dynamics of an expanding circular fault. *Bull. Seis. Soc. Am.* **66**, 639–66.

Madariaga, R. (1979). On the relation between seismic moment and stress drop in the presence of stress and strength heterogeneity. *J. Geophys. Res.* **84**, 2243–50.

Madariaga, R. (1983). Earthquake source theory. A review. In H. Kanamori and E. Boschi (eds.). *Earthquakes Observations, Theory and Interpretation*. North-Holland, Amsterdam. pp. 1–44.

Madariaga, R. (1989). Propagación de ondas sísmicas en el campo cercano. *Física de la tierra* **1**, 51–73.

Mandelbrot, B. (1977). *Fractals: Form, Chance and Dimension*. W. H. Freeman, San Francisco.

Marquardt, D. W. (1963). An algorithm for least squares estimation of non linear parameters. *J. Soc. Ind. Appl. Math.* **11**, 431–41.

Melton, B. S. (1981). Earthquake seismograph development: A modern history. *EOS (Trans. Am. Geophys. U.)* **62**, 505–10, 545–8.

Mendiguren, J.A. (1977). Inversion of surface wave data in source mechanism studies. *J. Geophys. Res.* **82**, 889–94.

Mendoza, C. and S. Hartzell (1989). Slip distribution of the 19 September 1985 Michoacan, Mexico, earthquake: Near source and teleseismic constraints. *Bull. Seis. Soc. Am.* **79**, 655–69.

Mikumo, T. and T. Miyatake (1979). Earthquake sequences on a frictional fault model with non-uniform strengths and relaxation times. *Geophys. J. R. Astr. Soc.* **54**, 417–38.

Minster, J. B. (1980). Anelasticity and attenuation. In A. M. Dziewonski and E. Boschi (eds.). *Physics of the Earth Interior*. North-Holland, Amsterdam. pp. 152–212.

Mogi, K. (1963). Some discussions on aftershocks, foreshocks and earthquake swarms, the fracture of a medi-infinite body caused by inner stress origin and its relation to earthquake phenomena. *Bull. Earthq. Res. Inst. Tokyo Univ.* **41**, 615–58.

Molnar, P., B. E. Tucker and J. N. Brune (1973). Corner frequencies of P and S waves and models of earthquake sources. *Bull. Seis. Soc. Am.* **63**, 2091–104.

Morelli, A. and A. M. Dziewonski (1993). Body wave travel times and spherically symmetric P and S wave velocity model. *Geophys. J. Int.* **112**, 178–94.

Nabelek, J. L. (1984). Determination of earthquake source parameters from inversion of body waves. Ph.D. Thesis, Massachusets Institute of Technology, Cambridge, MA.

Nakano, H. (1923). Notes on the nature of forces which give rise to the earthquake motion. *Seis. Bull. Cent. Meteor. Obs. Japan* **1**, 92–120.

Nishimura, C. E. and D. W. Forsyth (1989). The anisotropic structure of the upper mantle in the Pacific. *Geophys. J. Int.* **96**, 203–29.

Nishimura, G. (1937). On the elastic waves due to pressure variations on the inner surface of a spherical cavity in an elastic solid. *Bull. Earthq. Res. Inst. Tokyo* **15**, 614–35.

Nuttli, O. W. (1974). Seismic wave attenuation and magnitude relations for east and north America. *J. Geophys. Res.* **78**, 876–85.

Okal, E. (1992). A student's guide to teleseismic body wave amplitudes. *Seis. Res. Lett.* **63**, 169–80.

Oliver, J. (1962). A summary of observed seismic surface wave dispersion. *Bull. Seis. Soc. Am.* **52**, 81–6.

Orowan, E. (1960). Mechanism of seismic faulting. *Geol. Soc. Am. Mem.* **79**, 323–45.

Palmer, A. C. and J. R. Rice (1973). The growth of slip surfaces in the progressive failure of overconsolidated clays. *Proc. R. Soc.* A **332**, 527–48.

Papageorgiou, A. S. and K. Aki (1983). A specific barrier model for quantitative description of inhomogeneous faulting and the prediction of strong ground motion. I, description of the model. *Bull. Seis. Soc. Am.* **73**, 693–722.

Press, F. (1956). Determination of crustal structure from phase velocity of Rayleigh waves. I Southern California. *Bull. Geol. Soc. Am.* **67**, 1647–58.

Reid, H. F. (1911). The elastic rebound theory of earthquakes. *Bull. Dept Geol. Univ. California* **6**, 412–44.

Richter, C. F. (1935). An instrument earthquake magnitude scale. *Bull. Seis. Soc. Am.* **25**, 1–32.

Ringwood, A. E. (1975). *Composition and Petrology of the Earth's Mantle.* McGraw-Hill, New York.

Satô, Y. (1955). Analysis of dispersed surface waves by means of Fourier transform I. *Bull. Earthq. Res. Inst. Tokyo Univ.* **33**, 33–47.

Savage, J. C. (1966). Radiation from a realistic model of faulting. *Bull. Seis. Soc. Am.* **56**, 577–92.

Savage, J. C. (1972). Relation of corner frequency to fault dimensions. *J. Geophys. Res.* **77**, 3788–95.

Scholte, J. G. J. (1962). The mechanism at the focus of an earthquake. *Bull. Seis. Soc. Am.* **52**, 711–21.

Scholz, C. H. (1982). Scaling laws for large earthquakes: Consequences for physical models. *Bull. Seis. Soc. Am.* **72**, 1–14.

Schwartz, D. and K. Coppersmith (1984). Fault behaviour and characteristic earthquakes: Examples from the Wasacht and San Andreas fault zones. *J. Geophys. Res.* **89**, 5681–98.

Shearer, P. M., K. M. Toy and J. A. Orcutt (1988). Axi-symmetric Earth models and inner-core anisotropy. *Nature* **333**, 228–32.

Shimazaki, K. and T. Nakata (1980). Time predictable recurrence model for large earthquakes. *Geophys. Res. Lett.* **7**, 279–82.

Sipkin, S. A. (1982). Estimation of earthquake source parameters by the inversion of wave form data. Synthetic wave forms. *Phys. Earth Planet. Inter.* **30**, 242–55.

Song, X. and D. V. Helmberger (1992). Velocity structure near the inner core boundary from waveform modeling. *J. Geophys. Res.* **97**, 6573–86.

Sornette, A. and D. Sornette (1989). Self-organized criticality and earthquakes. *Europhys. Lett.* **9**, 197–202.

Sponheuer, W. (1960). *Methoden zur Herdtiefenbestimung in der Makroseismik.* Akademic Verlag, East Berlin.

Steketee, J. A. (1958). Some geophysical applications of the theory of dislocations. *Can. J. Phys.* **36**, 1168–98.

Stoneley, R. (1924). Elastic waves at the surface of separation of two solids. *Proc. R. Soc.* A **106**, 416–28.

Strelitz, R. A. (1978). Moment tensor inversions and source models. *Geophys. J. R. Astr. Soc.* **52**, 359–64.

Strelitz, R. A. (1989). Choosing the best double couple from a moment tensor inversion. *Geophys. J. Int.* **99**, 811–5.

Stump, B. W. and L. R. Johnson (1977). The determination of source properties by the linear inversion of seismograms. *Bull. Seis. Soc. Am.* **67**, 1489–502.

Taylor, S. R. and S. M. McLennan (1985). *The Continental Crust: Its Composition and Evolution.* Blackwell, Boston.

Thomson, W. M. (1950). Transmission of elastic waves through a stratified solid medium. *J. Appl. Phys.* **21**, 89–93.

Tsuboi, C. (1956). Earthquake energy, earthquake volume, aftershock area and strength of the earth's crust. *J. Phys. Earth* **4**, 63–6.

Udías, A. (1991). Source mechanism of earthquakes. *Adv. Geophys.* **33**, 81–140.

Udías, A. and E. Buforn (1988). Single and joint fault-plane solutions from first motion data. In D. Doornbos (ed.). *Seismological Algorithms.* Academic Press, London. pp. 443–53.

Udías, A. and E. Buforn (1994). Seismotectonics of the Mediterranean region. *Adv. Geophys.* **36**, 121–209.

Utsu, T. (1961). Statistical study on the occurrence of aftershocks. *Geophys. Mag.* **30**, 521–605.

Vaněk, J., A. Zápotek, V. Karnik, N. V. Kondorskaya, Y. V. Riznichenko, E. F. Savarensky, S. L. Soloviev and N. V. Shebalin (1962). Standardization of magnitude scales. *Izv. Akad. Nauka SSSR, Geofiz.* **2**, 153–8.

Varotsos, P. and M. Lazaridou (1991). Latest aspects of earthquake prediction in Greece based on seismic electric signals. *Tectonophysics* **188**, 321–47.

Vvedenskaya, A. V. (1956). Determination of displacement fields for earthquakes by means of the dislocation theory. *Izv. Akad. Nauka SSSR, Geofiz.* **3**, 277–84 (in Russian)

Wickens, A. J. and J. H. Hodgson (1967). Computer re-evaluation of earthquake mechanism solutions (1922–1962). *Publ. Dom. Obs. (Ottawa)* **33**, 1–560.

Wilson, J. T. (1965). A new class of faults and their bearing on continental drift. *Nature* **207**, 343.

Woodward, R. L. and G. Master (1991). Global upper mantle structure from long period differential travel times. *J. Geophys. Res.* **96**, 6351-6377.

Wyss, M. and D. Booth (1997). The IASPEI procedure for evaluation of earthquakes precursors. *Geophys. J. Int.* **131**, 423–4.

Young, C. J. and T. Lay (1987). The core–mantle boundary. *Ann. Rev. Earth Planet. Sci.* **15**, 25–46.

Index

468